KINETICS
OF HUMAN MOTION

Vladimir M. Zatsiorsky, PhD
The Pennsylvania State University

Human Kinetics

Library of Congress Cataloging-in-Publication Data

Zatsiorsky, Vladimir M., 1932-
   Kinetics of human motion / Vladimir M. Zatsiorsky.
     p. cm.
   Includes bibliographical references and index.
   ISBN 0-7360-3778-0
   1. Human mechanics.   2. Human locomotion.   I. Title.

   QP303 .Z383 2002
   612.7'6--dc21

                                                     2001051861
ISBN: 0-7360-3778-0

Acquisitions Editor: Loarn D. Robertson, PhD; **Managing Editor:** Amy Stahl; **Assistant Editor:** Derek Campbell; **Copyeditor:** Karen Bojda; **Proofreader:** Erin Cler; **Graphic Designer:** Fred Starbird; **Graphic Artist:** Kathleen Boudreau-Fuoss; **Cover Designer:** Jack W. Davis; **Art Managers:** Craig Newsom, Carl D. Johnson; **Printer:** Sheridan Books

Printed in the United States of America   10 9 8 7 6 5 4 3 2 1

Human Kinetics
Web site: www.humankinetics.com

United States: Human Kinetics, P.O. Box 5076, Champaign, IL 61825-5076
800-747-4457
e-mail: humank@hkusa.com

Canada: Human Kinetics, 475 Devonshire Road Unit 100, Windsor, ON N8Y 2L5
800-465-7301 (in Canada only)
e-mail: orders@hkcanada.com

Europe: Human Kinetics, Units C2/C3 Wira Business Park, West Park Ring Road, Leeds
LS16 6EB, United Kingdom
+44 (0) 113 278 1708
e-mail: hk@hkeurope.com

Australia: Human Kinetics, 57A Price Avenue, Lower Mitcham, South Australia 5062
08 8277 1555
e-mail: liahka@senet.com.au

New Zealand: Human Kinetics, P.O. Box 105-231, Auckland Central
09-523-3462
e-mail: hkp@ihug.co.nz

To Rita

Though human ingenuity may make various inventions
which, by the help of various machines, answer the same end,
it will never devise any invention more beautiful, nor more simple,
nor more to the purpose than nature does.

Leonardo da Vinci (1452–1519)

CONTENTS

Chapter 5
Joint Torques and Forces:
The Inverse Problem of Dynamics

PREFACE

This book is the second volume of a three-volume series on biomechanics of human motion; the first volume is *Kinematics of Human Motion* (1997). The books share a common format; they employ the same notations. It is assumed that these books will be studied in sequence, each book in a one-semester course. The curricula in some universities do not allow for such a comprehensive study of human biomechanics as this book may entail. If biomechanics is taught during only one semester, which does not allow covering all material in the two books, the instructor may have to make choices about what material to skip.

Kinetics of Human Motion is written for graduate students and advanced undergraduates in the human movement sciences: biomechanics, biomedical engineering, motor control, kinesiology, physical and occupational therapy, neurophysiology of movement, orthopedics, and exercise and sport science. Regardless of a student's field of study, he or she is expected to have grounding in Newtonian mechanics and courses in calculus and matrix algebra. Vector analysis is used in the book as a language that allows complex problems to be represented in a compact form. Still, the main focus of the book is not the mathematical methods but understanding the physical concepts.

Biomechanical analysis of human motion can be done at different structural levels, for example, on the level of muscle fibers, individual muscles, or the entire human body. This book is limited to the analysis of motion of the entire body. Throughout the book, the human body is modeled as a kinematic chain with joint actuators. Muscle biomechanics will be covered in the third volume of the series.

Kinetics is the branch of mechanics dealing with forces and their effects on bodies at rest (*statics*) and bodies in motion (*dynamics*). Chapter **1** is devoted to the interaction of the human body with the immediate environment; in particular, it covers external contact forces. The basic concepts of kinetics, such as forces, couples, and equivalent force systems, are also covered in this chapter. Chapters **2** and **3** address the statics of kinematic chains. Chapter **2** deals with transformation analysis, and chapter **3** discusses the equilibrium of kinematic chains and related topics, for example, joint stiffness. Chapter **4** addresses the mass-inertial characteristics of the human body. The chapter is complemented by appendices at the end of the book. The appendices contain quantitative data and technical details on body segment parameters. Chapter **5** is concerned with the dynamics of human motion, specifically with the inverse problem of dynamics. Chapter **6** concentrates on the problems associated with mechanical work and energy in human motion. This is the only chapter in the book that is limited to the discussion of planar motion. Compared with other chapters, the mathematical language in this chapter is simple, and the emphasis is on understanding the physical background of the employed models and methods.

The references in the text are limited to the bare minimum. However, many references pointing to the path to deeper understanding of certain topics follow each chapter. Authors whose publications were inadvertently omitted are asked to be indulgent.

I assure the readers that I placed no deliberate errors in the book and that I did my best to get rid of lapses in the text and equations. However, some errors may still have crept into these pages. Comments and criticism on the current text are most welcome.

ACKNOWLEDGMENTS

This book would not be possible without institutional support from the Pennsylvania State University and its Department of Kinesiology. Gratefully, I wish to acknowledge the university for granting me a sabbatical leave, during which the main parts of this book were written.

Over the period of writing the book, many people were of immense assistance. I would like to extend my sincere appreciation to Drs. M. Latash (Pennsylvania State University), B. Prilutsky (Georgia Tech University), S. Jarich (National Institute for Working Life, Sweden), Z.-M. Li (Walsh University), and M. Duarte (Universidade de São Paulo, Brazil), who sacrificed their time to read the ponderous early drafts and to offer valuable suggestions. Special thanks go to Dr. M. Duarte, who derived the equations for three-link chains included in chapter 5.

I am grateful to the graduate students at the PSU Biomechanical Laboratory, Robert Gregory and Todd Pataky, who helped me to improve the readability of the book.

My debt to my wife, Rita, is immeasurable.

NOTATION AND CONVENTIONS

Sections and subsections of the chapters are numbered, and references to them are printed in boldface. Hence, **4.1.3** means chapter **4**, section **1**, subsection **3** ("Products of Inertia").

A vector is denoted by a letter in boldface type, for example, **F**. The same letter in italic lightface type denotes the magnitude of the vector. Matrices are printed in square brackets, like this: [R]. Newton's notation is used for the successive derivatives with regard to time; thus, when x is a coordinate, $\dot{x}$ means velocity, and $\ddot{x}$ stands for acceleration. A list of the main symbols follows:

C	force couple
F	force
$\mathbf{F}_x, \mathbf{F}_y, \mathbf{F}_z$	vector components of vector **F** along the axes X, Y, and Z
F_x, F_y, F_z	scalar components (magnitudes) of vector **F**
$\overline{\mathbf{F}}$	generalized force, a six-component vector
g	acceleration due to gravity
H	angular momentum
$\dot{\mathbf{H}}$	rate of change of the angular momentum
i, j, k	unit vectors along the orthogonal reference axes
J or [J]	Jacobian matrix
K, Q, E	kinetic, potential, and total mechanical energy, respectively
L	Lagrangian
m	mass
M	moment of force
$\mathbf{M}_C$	moment of a couple **C** (free moment)
$\mathbf{M}_O$	moment of force about a point O
M_X, M_Y, M_Z	scalar components of moment of force $\mathbf{M}_O$
$\mathbf{M}_{OO}$	moment of force about an axis O-O
$\mathbf{M}_w$	wrench moment
O	origin of a Cartesian system of coordinates
P	power

R	resultant force
t	time
$\mathbf{U}, \hat{\mathbf{F}}$	unit vector along the line of force action
v	linear velocity
W	work
w	wrench
$\hat{\mathbf{w}}$	unit wrench
X, Y, Z	axes of a global reference system
x, y, z	axes of a local reference system
θ	angle measured in a global reference system
α	angle measured in a local reference system, usually joint angle
$\dot{\theta}, \dot{\alpha}, \varpi$	angular velocity
$\ddot{\theta}, \ddot{\alpha}$	angular acceleration
λ	in the equilibrium-point hypothesis, the threshold value of the muscle length or joint angle at which the tonic stretch reflex is activated
$[C]$	compliance matrix
$[G]$ or $\mathbf{G}$	muscle Jacobian
$[I]$	tensor of inertia
$[I^c]$	inertia matrix of a kinematic chain
$[K]$	matrix of the (apparent) joint stiffness
$[R]$	rotation matrix
$[S]$	matrix of the (apparent) endpoint stiffness
$[T]$	transformation matrix
$\times$	vector product (cross product) of vectors
CNS	central nervous system
CoM	center of mass
CoP	center of pressure
DoF	degrees of freedom

GRF	ground reaction force
JCC	joint compliant characteristics
MEE	mechanical energy expenditure
PWA	point of wrench application
VC	virtual coefficient

1
CHAPTER

EXTERNAL CONTACT FORCES

Mathematicians are like Frenchmen:
whenever you say something to them,
they translate it into their own language,
and at once it is entirely different.

Johann W. von Goethe (1749–1832)

In biomechanics, forces are classified as *external* (acting between the body and environment) and *internal* (acting between body parts). The external forces can be distant forces (e.g., gravitational force) or contact forces. This chapter deals with external contact forces acting on the human body. According to Newton's third law, for every force acting on the human body, the body exerts an equal, opposite, and collinear reactive force on the environment (action equals reaction). For instance, the push down on the floor by a foot is equal in magnitude and opposite in direction to the push up on the foot by the floor.

Effects produced by external forces on the human body can be roughly classified as (1) mechanical effects that influence the movement of the body (gross mechanical effects) and (2) local biological effects on the contacting tissues. We consider first the gross mechanical effects and then briefly describe the local biological effects.

In section **1.1**, the basic mechanics of contact forces in three dimensions are described. The concepts introduced here are universal. They may also be applied to internal forces acting within and between the body segments and are broadly used in the subsequent chapters. In the discussion, we combine geometric interpretations with vector analysis. Vector notation is a convenient tool for discussing three-dimensional problems. The emphasis, however, remains on explaining the main notions and principles. After defining the basic

••• *MECHANICS REFRESHER* •••

Forces and Moments

Force is a measure of the action of one body on another. Force is a vector quantity. A force can be treated as either a fixed vector or as a sliding vector. When a force is treated as a fixed vector, it is defined by its (a) magnitude, (b) direction, and (c) point of application. When a force is considered a sliding vector, the line of force action rather than the point of application defines the force. Forces are considered sliding vectors when (a) the body of interest is rigid and (b) the resultant external effects, rather than the internal forces and the deformations, are investigated. In this book, if not mentioned otherwise, forces are considered sliding vectors.

The resultant effects of a force do not alter if the force is applied along the line of force action at any point (the *principle of transmissibility*). The principle of transmissibility is valid for a rigid body but not for a system of bodies. For instance, the principle is not valid for kinematic chains: a force, if transmitted along the line of its action from one link to another, may cause different resultant effects. Forces can be added and subtracted according to the *parallelogram rule*. The parallelogram rule has been established experimentally; it cannot be proven or derived mathematically. From the parallelogram rule, it follows that two or several concurrent forces can be reduced to one resultant force.

Moment of force, **M**, is a measure of the turning effect of the force. In two dimensions, the magnitude of the moment of a force **F** about a given center (or "at a center") equals the product of the magnitude F of the force and the perpendicular distance d from the center to the line of force action, $M = Fd$. The distance d is called a *moment arm*. Moments of force in a plane are scalars; moments of force in the space are vectors.

Two systems of *vectors* are said to be *equipollent* if their resultants and resultant moments are equal. Two systems of forces are said to be *equivalent* if they produce the same effect. Note that the term "equipollent" refers to vectors and the term "equivalent" refers to forces. Two equipollent force systems acting on a rigid body are also equivalent. This is not always true if the body is not rigid. For instance, if a string is being pulled with equal and opposite forces at its ends, the effect will be different from that caused by pushing the string with the forces reversed in direction. The two force systems are equipollent but not equivalent. Note that in the mechanics of rigid bodies, the motion and equilibrium of the bodies are analyzed; internal forces and deformations are not considered.

concepts of kinetics, such as forces, moments, and couples, we introduce the notion of screws and twists. With this background, we analyze the external contact forces in human movement, using as two examples interaction with the support and grasping. Section **1.2** addresses friction forces and aerodynamic forces. Some experimental results on the friction forces in human movement are reviewed. Section **1.3** considers local biological effects of the contact forces acting on the human body.

1.1 FORCES AND COUPLES

A *contact* is a collection of adjacent points where two surfaces touch each other over a contiguous area. An individual contact point is formally described by its (a) position, that is, point coordinates; (b) orientation, defined by the *normal* to the surface at the contact point; and (c) curvature, defined by the values of the maximum and minimal radii of curvature at the contact point (the notion of surface curvatures is addressed in *Kinematics of Human Motion*, section **4.1.1.2**). To visualize this definition, think of a pitcher with a baseball in his hand.

Contacts that a human body makes with external objects usually constrain movement in some directions but not in others. For instance, during one-legged standing, the contact with the support ordinarily prevents slipping in antero-posterior and mediolateral directions. It does not, however, prevent rotation around horizontal axes through the contact (rolling).

When two bodies contact each other under pressure, the bodies yield and are in contact over a certain surface. At each point of the surface, forces are acting—the bodies tend to resist compression or adhere to each other. Any set of contact forces is equipollent to a resultant force and a couple (this issue is addressed in more detail in ensuing sections). The resultant force can be resolved into the normal force, which is perpendicular to the contacting surface at the point of contact, and the *tangential* force (shear force) acting along the surface. In human movement, the normal contact forces as a rule act in one direction; for example, a finger or a foot cannot pull on the object. In contrast, grasping allows pulling. In the discussion that follows, if not mentioned otherwise, the absence of sticking is assumed and the normal contact forces are considered unidirectional.

1.1.1 Generalized Contact Force and Its Rectangular Components

Forces and moments acting at a contact area are usually represented, or measured, as three orthogonal force components and three orthogonal moment components. When considered together, these forces and moments are called a *generalized contact force, $\overline{\mathbf{F}}$*.

1.1.1.1 Orthogonal Components of a Force

Consider a force $\mathbf{F}$ applied at the origin of the coordinate system O (figure 1.1). To represent the force in its components, we introduce three vectors of unit magnitude, $\mathbf{i}$, $\mathbf{j}$, and $\mathbf{k}$, directed along the positive X, Y, and Z axes, respectively. The vectors are called *unit vectors*. The orthogonal components of force $\mathbf{F}$ can be written as products

$$\mathbf{F}_X = F_X\mathbf{i}, \ \mathbf{F}_Y = F_Y\mathbf{j}, \ \mathbf{F}_Z = F_Z\mathbf{k} \tag{1.1}$$

where the vectors are printed in bold. The vectors $\mathbf{F}_X$, $\mathbf{F}_Y$, and $\mathbf{F}_Z$ are called the vector components of $\mathbf{F}$; the scalars F_X, F_Y, and F_Z are called the scalar components of $\mathbf{F}$. The scalar components are positive when the vector components have the same sense as the corresponding unit vectors, for example, when $\mathbf{F}_Z$ has the same sense as the unit vector $\mathbf{k}$. Otherwise, the scalar components are negative. When there is no possibility for confusion, the vector and scalar components are called *components* of $\mathbf{F}$. The components of $\mathbf{F}$ can be computed either through the direction angles (the angles between vector $\mathbf{F}$ and the coordinate axes, figure 1.1*a*) or through the projection angles (the angles between vector $\mathbf{F}$ and the coordinate planes, figure 1.1*b*).

Referring to figure 1.1*a*, the following relationships are evident:

$$\mathbf{F} = \mathbf{F}_X + \mathbf{F}_Y + \mathbf{F}_Z = F_X\mathbf{i} + F_Y\mathbf{j} + F_Z\mathbf{k} \tag{1.2a}$$

and

$$F = \sqrt{F_X^2 + F_Y^2 + F_Z^2} \tag{1.2b}$$

where F is the magnitude of force $\mathbf{F}$. Hence, a vector $\mathbf{F}$ equals the vector sum of its rectangular components, $\mathbf{F}_X$, $\mathbf{F}_Y$, and $\mathbf{F}_Z$, and the magnitude F of the vector $\mathbf{F}$ can be obtained by applying the Pythagorean theorem (equation 1.2*b*). Also, because

$$F_X = F\cos\theta_X, \ F_Y = F\cos\theta_Y, \ F_Z = F\cos\theta_Z \tag{1.3}$$

force $\mathbf{F}$ can be expressed as

$$\mathbf{F} = F(\mathbf{i}\cos\theta_X + \mathbf{j}\cos\theta_Y + \mathbf{k}\cos\theta_Z) \tag{1.4}$$

The expression in parentheses in equation 1.4 represents a unit vector $\mathbf{U}$ along the line of action of $\mathbf{F}$. The scalar components of the unit vector are the corresponding direction cosines.

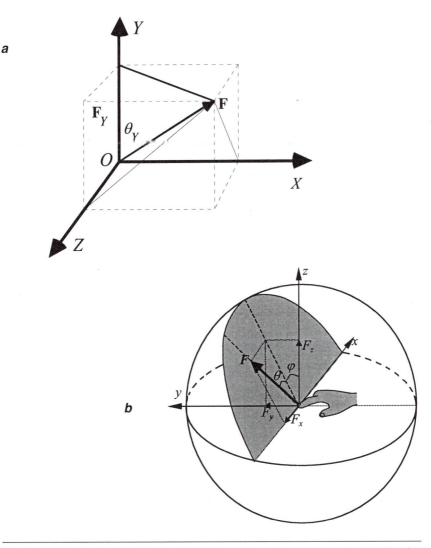

Figure 1.1 Resolution of a force **F** into its rectangular components. (*a*) Resolution through the direction angles, the angles between vector **F** and the coordinate axes. To avoid a messy picture, only the component $\mathbf{F}_Y$ is shown; the angle θ_Y is formed by the vectors **F** and $\mathbf{F}_Y$. The angles θ_X and θ_Z (not shown in the figure) are the angles that vector **F** forms with the *X* and *Y* axes. The unit vectors **i**, **j**, and **k** (also not shown in the figure) are directed along the *X*, *Y*, and *Z* axes, respectively. The cosines of the angles θ_X, θ_Y, and θ_Z are known as the *direction cosines* of the force **F**. (*b*) Resolution through the projection angles, the angles between vector **F** and the coordinate planes. The scalar components of **F** are given by the following equations: $F_X = F\sin\theta$, $F_Y = F\cos\theta\sin\phi$, and $F_Z = F\cos\theta\cos\phi$, where $-\pi/2 \le \theta \le \pi/2$ and $-\pi/2 \le \phi \le \pi/2$.

$$\mathbf{U} = \mathbf{i} \cos\theta_X + \mathbf{j} \cos\theta_Y + \mathbf{k} \cos\theta_Z \qquad (1.5)$$

Any force $\mathbf{F}$ may be represented as a product of its unit vector $\mathbf{U}$ and the force magnitude F, $\mathbf{F} = F\mathbf{U}$. Observe that the direction cosines are not independent.

$$\cos^2\theta_X + \cos^2\theta_Y + \cos^2\theta_Z = 1 \qquad (1.6)$$

For instance, if the angles θ_X and θ_Y equal 60°, the value of θ_Z must be equal to 45° or 135° (since cos 60° = 0.5, $\cos^2\theta_Z = 0.5$).

Two concurrent forces, that is, two forces meeting at the same point, can be added by using the parallelogram rule to obtain a resultant force $\mathbf{R}$ that has the same external effect on the rigid body as the given forces. When several concurrent forces are acting in space, the parallelogram rule is still valid but not practical. It is much simpler to determine the resultant force $\mathbf{R}$ by summing the rectangular components.

$$\mathbf{R} = \sum \mathbf{F} = R_X\mathbf{i} + R_Y\mathbf{j} + R_X\mathbf{k} = \left(\sum F_X\right)\mathbf{i} + \left(\sum F_Y\right)\mathbf{j} + \left(\sum F_Z\right)\mathbf{k} \qquad (1.7)$$

The scalar components of the resultant force $\mathbf{R}$ are equal to the sum of the corresponding scalar components of the given forces, $R_X = \sum F_X, R_Y = \sum F_Y$, and $R_Z = \sum F_Z$.

■ ■ ■ *FROM THE LITERATURE* ■ ■ ■

Push-Off Force in Speed Skating

Source: Koning, J.J. de, R.W. de Boer, G. de Groot, and G.J. van Ingen Schenau. 1987. Push-off force in speed skating. *Int. J. Sport Biomech.* 3: 103–9.

In speed skating, one stroke consists of a gliding phase and a push-off phase during which the support leg is extended (figure 1.2). During the push-off phase, the skate continues to glide. To prevent the backward slipping of the skate during the push-off phase, the push-off force should be directed perpendicularly to the gliding direction of the skate. The velocity increment of the body is the result of the sideward push-off. The authors measured the push-off force $\mathbf{F}$ at different velocities in trained speed skaters. The more horizontally $\mathbf{F}$ is directed, the larger the effective component of force $\mathbf{F}_X$ that is used for propulsion. Hence, the direction angle α should be as small as possible.

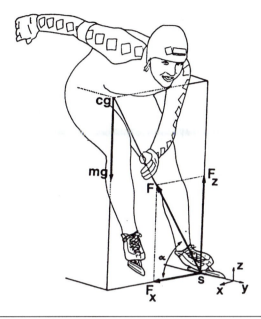

Figure 1.2 The push-off force **F** during skating lies in the plane *s-xz* and is at a right angle to the gliding direction *y*. Force **F** makes a direction angle α with the *x* axis. The center of gravity of the body is indicated by the abbreviation *cg,* and *mg* is body weight.

Reprinted, by permission, from J.J. Koning, R.W. de Boer, G. de Groot, and G.J. van Ingen Schenau, 1987, "Push-off force in speed skating," *International Journal of Sport Biomechanics* 3:103–109.

1.1.1.2 Orthogonal Components of a Moment of Force

A moment of force **M** measures the tendency of a force to impart a rotational motion to the body. This tendency depends not only on the magnitude and direction of the force but also on the location of the line of force action. A force tends to rotate the body about any axis that does not intersect its line of action. The tendency is proportional to the distance of the line of force action from the axis of rotation. Moments of force can be determined about a point or about an axis.

1.1.1.2.1 Moment of Force About a Point

Mathematically, the moment of force M_O about a point *O* is defined as a *cross product* of vectors **r** and **F**, where **r** is the position vector from *O* to the point of force application *P* (figure 1.3).

$$\mathbf{M}_O = \mathbf{r} \times \mathbf{F} \tag{1.8}$$

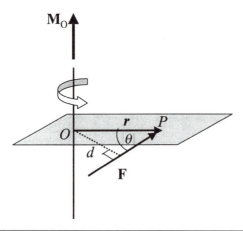

Figure 1.3 Force **F** produces a moment of force M_O at a point O. See explanation in the text.

If the position vector is measured from point O to the force, the moment is defined by the equation $M_O = r \times F$. However, if the position vector is defined from the force to the point, the equation is $M_O = F \times r$. According to the definition of cross product (see *Kinematics of Human Motion,* pp. 9–10), the moment M_O is a vector that possesses the following features:

• The line of action of M_O is perpendicular to the plane containing vectors **r** and **F**. The line represents the axis about which the body tends to rotate at O when subjected to the force **F**.

• The magnitude of moment is $M_O = F(r \sin\theta) = Fd$, where θ is the angle between the vectors **r** and **F** and d is a shortest distance from O to the line of action of **F**, the moment arm. The moment arm is in the plane containing O and **F**. Note that the magnitude of the moment of force does not depend on the actual position of the point of force application along the line of force action. Only the moment arm is important. Hence, **r** is a vector from O to *any* point on the line of action of **F**. Conversely, the moment of force does not specify the exact position of the point of force application along the line of force action.

• The direction of the vector M_O is furnished by the right-hand rule in rotating from **r** to **F**: when the fingers curl in the direction of the induced rotation, the vector is pointing in the direction of the thumb.

We now express the moment of force M_O by its rectangular components. To do that, we consider first the cross products of unit vectors, which were introduced in **1.1.1.1**. To begin, consider the cross product $\mathbf{i} \times \mathbf{j}$. The vectors **i** and **j** are at right angles to each other (for that reason $\sin\theta = 1$), and their magnitude equals 1. Hence, their cross product is also a unit vector. This vector is evidently vector **k**. Because the sense of the cross product of two vectors is reversed when the order of factors is reversed, the product $\mathbf{j} \times \mathbf{i} = -\mathbf{k}$. The vector product of a unit vector by itself, such as $\mathbf{k} \times \mathbf{k}$, is equal to zero because both vectors are along the same line. In general, because the vectors **i**, **j**, and **k** are mutually perpendicular and form a right-handed triad, the following relations between the unit vectors are valid:

$$
\begin{array}{lll}
\mathbf{i} \times \mathbf{i} = 0 & \mathbf{i} \times \mathbf{j} = \mathbf{k} & \mathbf{i} \times \mathbf{k} = -\mathbf{j} \\
\mathbf{j} \times \mathbf{i} = -\mathbf{k} & \mathbf{j} \times \mathbf{j} = 0 & \mathbf{j} \times \mathbf{k} = \mathbf{i} \\
\mathbf{k} \times \mathbf{i} = \mathbf{j} & \mathbf{k} \times \mathbf{j} = -\mathbf{i} & \mathbf{k} \times \mathbf{k} = 0
\end{array}
\tag{1.9}
$$

The vector product $\mathbf{M}_O = \mathbf{r} \times \mathbf{F}$ in terms of the rectangular components of the vectors **r** and **F** is

$$
\mathbf{M}_O = \mathbf{r} \times \mathbf{F} = (X\mathbf{i} + Y\mathbf{j} + Z\mathbf{k}) \times (F_X\mathbf{i} + F_Y\mathbf{j} + F_Z\mathbf{k})
\tag{1.10}
$$

where X, Y, and Z are the coordinates of the position vector **r** (the coordinates of point P in the reference frame located at O). After factoring **i**, **j**, and **k** and recalling the identities in equation 1.9, we obtain

$$
\mathbf{M}_O = (YF_Z - ZF_Y)\mathbf{i} + (ZF_X - XF_Z)\mathbf{j} + (XF_Y - YF_X)\mathbf{k} =
$$
$$
M_X\mathbf{i} + M_Y\mathbf{j} + M_Z\mathbf{k}
\tag{1.11}
$$

where M_X, M_Y, and M_Z are the scalar components of the vector $\mathbf{M}_O$. The components measure the tendency of the force **F** to impart rotation about the axes X, Y, and Z, correspondingly. Hence, the components are

$$
M_X = YF_Z - ZF_Y
$$
$$
M_Y = ZF_X - XF_Z
\tag{1.12}
$$
$$
M_Z = XF_Y - YF_X
$$

The scalar components of $\mathbf{M}_O$ represent the magnitude of the moments of two orthogonal forces acting in the corresponding plane. For instance, the component M_X is due to the forces F_Y and F_Z acting in the plane YZ.

The right-hand part of equation 1.11 corresponds to the expansion of the following *determinant* (the determinants are explained in *Kinematics of Human Motion,* p. 29).

$$
\mathbf{M}_O =
\begin{vmatrix}
\mathbf{i} & \mathbf{j} & \mathbf{k} \\
X & Y & Z \\
F_X & F_Y & F_Z
\end{vmatrix}
\tag{1.13}
$$

Equation 1.13 is the same as equation 1.11. The determinant in equation 1.13 is, however, easier to memorize than equation 1.11, and thus equation

1.13 is used more often. To expand the determinant $\mathbf{M}_O$ (as well as a determinant of any 3×3 matrix), the following three-step procedure, illustrated in equation 1.14, may be used:

1. Repeat the first and second columns.
2. Compute products along each diagonal line.
3. Subtract the sum of the products along the broken lines from the sum of the products along the solid lines.

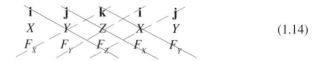

$$(1.14)$$

■ ■ ■ *FROM THE LITERATURE* ■ ■ ■

Moment Components of the Muscle Forces

Source: Carrasco, D.I., J. Lawrence, III, and A.W. English. 1999. Neuromuscular compartments of cat gastrocnemius produce different torques about the ankle joint. *Motor Control* 3: 436–46.

The lateral gastrocnemius muscle (LGm) consists of three compartments: LG1, LG2, and LG3. A primary muscle nerve branch associated with a unique set of *motoneurons* innervates each compartment. Activation of *motor units* in different compartments is task dependent. The goal of the project was to study the mechanical action of individual compartments.

The experiments were performed on deeply anesthetized cats. The limbs were firmly fixed and immobilized. The plantar surface of the foot was secured to a metallic plate that in turn was secured to a six-degrees-of-freedom (6-DOF) force-moment transducer. During the experiment, the primary branches of the lateral gastrocnemius-soleus nerve were isolated and electrically stimulated at frequencies of 40 to 300 Hz.

The origin of the coordinate system was chosen at a point midway between the malleoli, the assumed axis of rotation of the ankle joint (figure 1.4*a*). The moments of the force with respect to the joint center were computed as the cross products of the position vector from the joint center to the center of the transducer and the reaction force vectors (equation 1.8). The pitch, yaw, and roll components of the moments were computed using the formulas presented in equation 1.12. (This was done after extracting an effect of the couple acting at the contact; the procedure is explained later in the text.) Some results are presented in figure 1.4, *b* and *c*.

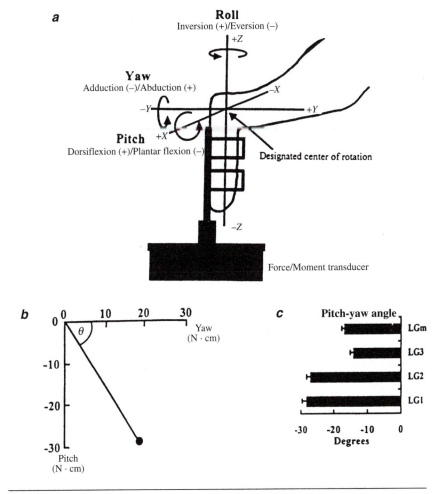

Figure 1.4 Determining the components of the moment of force produced by the individual muscle compartments. (*a*) The measuring system and the system of coordinates. (*b*) Computation of the direction of the moment vector. The direction was characterized by the projection angles that are formed by the projections of the vector on the coordinate planes and the coordinate axes (see section **1.2.5.1.4** in *Kinematics of Human Motion*). In the presented example, components of the total moment about the yaw and pitch axes are plotted. The angle θ is between the moment vector and the abscissa. The angle 0° represents a pure yaw, 90° pure pitch. (*c*) Projection angles in the pitch-yaw plane. Direction of moment vectors produced by the muscle compartments (LG1, LG2, and LG3) and the entire lateral gastrocnemius muscle, LGm.

Adapted, by permission from D.I. Carrasco, J. Lawrence III, and A.W. English, 1999, "Neuromuscular compartments of cat gastrocnemius produce different torques about the ankle joint," *Motor Control* 3:436–446.

Moment of force tends to rotate the body on which the moment is exerted. Any rotation is performed about an axis rather than about a point. The expression "rotation about a point *O*" really means "rotation about an axis passing through *O* in a direction perpendicular to the plane containing *O* and the line of force action." In two dimensions, the line of force action and the axis of rotation are always perpendicular to each other. In three dimensions, they may be at different angles.

••• *Vector Algebra Refresher* •••

Scalar Product of Vectors

The *scalar product* of two vectors, **P** and **Q**, also called *dot product* or *inner product*, is defined as

$$\mathbf{P} \cdot \mathbf{Q} = PQ \cos\theta \qquad (1.15)$$

which is read "**P** dot **Q**." The angle θ is between the tails of the vectors **P** and **Q**. The product is a scalar that equals the magnitude of **P** times the component of **Q** in the direction of **P**, or the magnitude of **Q** multiplied by the component of **P** in the direction of **Q**. Both the *commutative* and *distributive* laws hold for the scalar product. Hence, $\mathbf{P} \cdot \mathbf{Q} = \mathbf{Q} \cdot \mathbf{P}$ and $\mathbf{P} \cdot (\mathbf{Q}_1 + \mathbf{Q}_2) = \mathbf{P} \cdot \mathbf{Q}_1 + \mathbf{P} \cdot \mathbf{Q}_2$. The scalar products of the unit vectors are either 0 or 1:

$$\mathbf{i} \cdot \mathbf{i} = \mathbf{j} \cdot \mathbf{j} = \mathbf{k} \cdot \mathbf{k} = 1; \mathbf{i} \cdot \mathbf{j} = \mathbf{j} \cdot \mathbf{i} = \mathbf{i} \cdot \mathbf{k} = \mathbf{k} \cdot \mathbf{i} = \mathbf{j} \cdot \mathbf{k} = \mathbf{k} \cdot \mathbf{j} = 0 \quad (1.16)$$

The scalar product of vectors can be expressed in their rectangular components:

$$\mathbf{P} \cdot \mathbf{Q} = (P_x\mathbf{i} + P_y\mathbf{j} + P_z\mathbf{k}) \cdot (Q_x\mathbf{i} + Q_y\mathbf{j} + Q_z\mathbf{k}) \qquad (1.17)$$

Making use of the scalar products of the unit vectors (equation 1.16), equation 1.17 can be reduced to

$$\mathbf{P} \cdot \mathbf{Q} = P_xQ_x + P_yQ_y + P_zQ_z \qquad (1.18)$$

A scalar product of a vector by itself is equal to the square of the vector magnitude.

$$\mathbf{P} \cdot \mathbf{P} = P_x^2 + P_y^2 + P_z^2 = P^2 \qquad (1.19)$$

The scalar product can be used to find the component of a vector in a given direction as well as the angle between two vectors. For example, the magnitude of the component of vector **P** along the direction of vector **Q** equals $P \cos\theta$. From equation 1.15, this magnitude is

$$P\cos\theta = \frac{\mathbf{P}\cdot\mathbf{Q}}{Q} = \mathbf{P}\cdot\frac{\mathbf{Q}}{Q} = \mathbf{P}\cdot\mathbf{U}_Q \qquad (1.20)$$

where $\mathbf{U}_Q$ is the unit vector in the direction of the vector $\mathbf{Q}$. To find the angle formed by the two vectors $\mathbf{P}$ and $\mathbf{Q}$, we combine equations 1.15 and 1.18.

$$PQ\cos\theta = P_x Q_x + P_y Q_y + P_z Q_z \qquad (1.21)$$

Hence

$$\cos\theta = \frac{P_x Q_x + P_y Q_y + P_z Q_z}{PQ} = \frac{\mathbf{P}\cdot\mathbf{Q}}{PQ} \qquad (1.22)$$

The scalar product of two vectors $\mathbf{P}\cdot\mathbf{Q}$ defines the same mathematical operation as the scalar product of the row matrix $[P]^T$ and the column matrix $[Q]$, $\mathbf{P}\cdot\mathbf{Q} \equiv [P]^T[Q]$.

••• *VECTOR ALGEBRA REFRESHER* •••

Mixed Triple Product of Three Vectors

The *mixed triple product* $\mathbf{P}\cdot(\mathbf{Q}\times\mathbf{R})$, also called the *triple scalar product*, is a scalar that is computed as a scalar product of a vector $\mathbf{P}$ and a vector $\mathbf{V} = \mathbf{Q}\times\mathbf{R}$. If the vectors $\mathbf{P}$, $\mathbf{Q}$, and $\mathbf{R}$ are coplanar, the mixed triple product is zero. The mixed triple product $\mathbf{P}\cdot(\mathbf{Q}\times\mathbf{R})$ is equal to the mixed triple product $(\mathbf{P}\times\mathbf{Q})\cdot\mathbf{R}$.

1.1.1.2.2 Moment of Force About an Axis

We introduce the moment of force about an axis geometrically and then discuss it in the terms of vector algebra. Consider a force $\mathbf{F}$ acting on a body (figure 1.5). The force does not intersect a fixed axis *O-O* and hence produces a moment $\mathbf{M}_{OO}$ about the axis. To determine the moment, we resolve the force $\mathbf{F}$ into two components, $\mathbf{F}_p$, which is parallel to the axis *O-O*, and $\mathbf{F}_n$, which is normal (perpendicular) to *O-O*. $\mathbf{F}_p$ does not produce any turning effect about *O-O*. The force $\mathbf{F}_n$ generates the moment $\mathbf{M}_{OO}$ about axis *O-O*. The moment $\mathbf{M}_{OO}$ measures the tendency of $\mathbf{F}$ to rotate the rigid body about the axis *O-O*. The magnitude of moment $\mathbf{M}_{OO}$ equals $M_{OO} = F_n d$, where the perpendicular distance d from *O-O* to the line of $\mathbf{F}$ is the moment arm of the force $\mathbf{F}_n$. The magnitude of the normal force, F_n, equals $F\sin\theta$, where θ is the angle between the line of force action $\mathbf{F}$ and the axis of rotation. The equation $M_{OO} = F_n d$ can

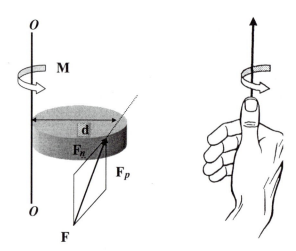

Figure 1.5 A force and its moment in three-dimensional space. Force **F** is applied to a body. The force **F** is resolved into two components, $\mathbf{F}_p$, which is parallel to the axis *O-O*, and $\mathbf{F}_n$, which is normal to *O-O*. $\mathbf{F}_p$ does not produce any turning effect about *O-O*. The force $\mathbf{F}_n$ generates the moment **M** about axis *O-O*. The magnitude of moment **M** is $M = F_n d$, where the shortest distance *d* from *O-O* to the line of force action is the moment arm of the force $\mathbf{F}_n$. The moment **M** can be represented as a sliding vector along *O-O*.

now be written as $M_{OO} = F(d \sin\theta)$, where the product $d \sin\theta$ is the moment arm of the force **F** about the axis *O-O*. In three dimensions, the magnitude of the moment arm of a force about an axis is equal to the perpendicular distance between the axis and the line of force action times the sine of the angle between these two lines. When the angle is zero, the line of force action is parallel to the axis of rotation and the force does not produce any moment about this axis. If the angle is $\pi/2$, the moment arm equals the shortest distance *d* between the two lines. In two dimensions, the line of force action and the axis of rotation are perpendicular to each other, and the moment arm equals *d*.

In the terms of vector algebra, the moment of force **F** about an axis is defined as a component of the moment along this axis. Consider again a force **F** acting on a rigid body (figure 1.6). The force exerts a moment $\mathbf{M}_B = \mathbf{r} \times \mathbf{F}$ about a point *B*, where **r** is a position vector from *A* to *B* (the point of force application). Let *O-O* be an axis through *B* and $\mathbf{U}_{OO}$ be the unit vector along *O-O*. Then the moment of force **F** about the axis *O-O*, $\mathbf{M}_{OO}$, is defined as a component (or projection) of the moment $\mathbf{M}_B$ along this axis. The magnitude of the moment $\mathbf{M}_{OO}$ equals the dot product of the vectors $\mathbf{U}_{OO}$ and $\mathbf{M}_B$.

$$M_{OO} = \mathbf{U}_{OO} \cdot \mathbf{M}_B = \mathbf{U}_{OO} \cdot (\mathbf{r} \times \mathbf{F}) \qquad (1.23)$$

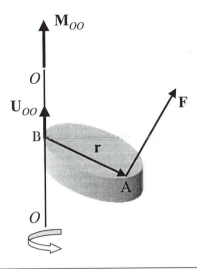

Figure 1.6 Moment of force **F** about an axis *O-O*. See explanation in the text.

According to equation 1.23, the moment of force about an axis is a mixed triple product of three vectors: the unit vector along the axis of rotation, the position vector from an arbitrary point on the axis to any point on the line of force action, and the force vector. The moments of force **F** about the coordinate axes are equal to the components of $\mathbf{M}_O$ about these axes; that is, they are equal to the expressions in equations 1.11 and 1.12.

When dealing with forces all of which act in a single plane, for instance in the plane *YZ*, $X = 0$ and $F_X = 0$, and equation 1.11 is simplified to

$$\mathbf{M}_O = (YF_Z - ZF_Y)\mathbf{i} \tag{1.24}$$

The moment $\mathbf{M}_O$ is normal to the *YZ* plane of figure 1.7. The moment is positive if it tends to rotate the body counterclockwise (in this case, the vector points out of the paper toward the reader). If the vector points into the paper, the force **F** tends to rotate the body clockwise and the moment $\mathbf{M}_O$ is negative.

The right-hand terms of equation 1.24, ZF_Y and YF_Z, have a plain physical meaning. They represent, respectively, the contribution of the horizontal and vertical force components to the turning effect of the force **F**. Consider as an example a "simple" case of creating body rotation to perform a running forward somersault (figure 1.8). After a run, the athlete meets the ground with the balls of the feet in such a way (with the legs "stiff") that a substantial braking force is produced. The horizontal and vertical vector components of the force **F** exerted by the ground on the athlete are $\mathbf{F}_Y$ and $\mathbf{F}_Z$, correspondingly. The coordinates of the *center of mass* (CoM) of the athlete's body are *Y* and *Z*

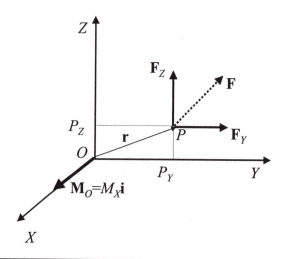

Figure 1.7 Force **F** acting in the plane *YZ* at a point *P* generates a moment of force $\mathbf{M}_o$ about the origin *O*.

(I hope that readers remember what the center of mass is; the concept is explained in detail in chapter **4**). The axis *X* is perpendicular to the plane of the figure (it points out of the paper) and is not shown. The force **F** does not intersect the center of mass of the athlete's body and hence imparts a moment of force $\mathbf{M}_o$ with respect to the center. The scalar component of the moment $\mathbf{M}_o$ about the axis *X* is

$$M_X = ZF_Y - YF_Z = Fd \tag{1.25}$$

where *Z* and *Y* are coordinates of the center of mass, *F* is the magnitude of the force **F**, and *d* is the shortest distance from the CoM to the line of action of the resultant force (*d* is not shown in the figure). Note that the sense of the M_X in equation 1.25 is different from the sense of the scalar components of M_X in equation 1.24 (the expression is $ZF_Y - YF_Z$ versus $YF_Z - ZF_Y$ in the previous equation). The sense is changed because in this example we are analyzing the moment of the force **F** exerted at the origin *O* about the center of mass; in the previous examples, we analyzed the moment of force about the origin *O*.

The moment of force about an axis (equation 1.23) can be written as a determinant:

$$M_{OO} = \begin{vmatrix} \cos_X & \cos_Y & \cos_Z \\ P_X & P_Y & P_Z \\ F_X & F_Y & F_Z \end{vmatrix} \tag{1.26}$$

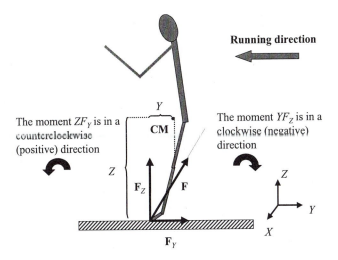

Figure 1.8 Vector components during a takeoff in forward somersaulting. In the position shown, the force applied to the body does not pass through the body's center of mass (CM). The horizontal force $\mathbf{F}_Y$ (the braking force) produces a moment in the counterclockwise direction (in the direction of the forward somersault). The magnitude of the moment is ZF_Y. By convention, a moment that generates rotation in the counterclockwise direction is positive. The vertical force produces a negative moment in the clockwise direction, thus decreasing the body rotation. The magnitude of this moment is YF_Z. The total magnitude of the moment about the axis X is $M_X = ZF_Y - YF_Z$. Because the moment arm Z of the horizontal force $\mathbf{F}_Y$ is much larger than the moment arm Y of the vertical force $\mathbf{F}_Z$, the moment due to the horizontal force dominates and the body rotates counterclockwise.

where $\cos_X$, $\cos_Y$, and $\cos_Z$ are the direction cosines of the axis $O\text{-}O$; P_X, P_Y, and P_Z are the coordinates of the point of force application; and F_X, F_Y, and F_Z are the scalar components of force $\mathbf{F}$. Equations 1.23 and 1.26 yield the scalar magnitude of the moment about the axis. To obtain $\mathbf{M}_{OO}$ as a vector, the magnitude of the vector, M_{OO}, should be multiplied by the unit vector $\mathbf{U}_{OO}$ along the axis:

$$\mathbf{M}_{OO} = M_{OO}\mathbf{U}_{OO} \tag{1.27}$$

There is an evident similarity between the representation of the vectors of force and the vectors of moments in their corresponding rectangular components. The components of a force $\mathbf{F}$ acting on a rigid body measure the tendency of the force to move the body in the X, Y, and Z directions. The components of the moment of force measure the tendency of $\mathbf{F}$ to impart rotation about the X, Y, and Z axes. These components represent the moments acting in the YZ, ZX, and XY planes, respectively.

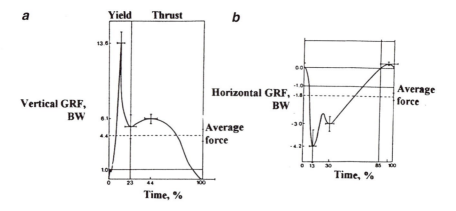

Figure 1.9   (*a*) Vertical and (*b*) horizontal components of the ground reaction force (GRF) exerted on the athlete's body in the running forward somersault, based on average data from eight experienced gymnasts. The force values are normalized in body weights (BW); the time is expressed as a percentage of total support duration.

Reprinted, by permission, from D.I. Miller and M.A. Nissinen, 1987, "Critical examination of ground reaction forces in the triple jump," *International Journal of Sport Biomechanics* 3:189–206.

■ ■ ■ *FROM THE LITERATURE* ■ ■ ■

Takeoff Forces in the Running Forward Somersault

Source: Miller, D.I., and M.A. Nissinen. 1987. Critical examination of ground reaction force in the running forward somersault. *Int. J. Sport Biomech.* 3: 189–206.

The authors studied the ground reaction forces elicited by male gymnasts during a running forward somersault. Very large braking forces in the anteroposterior direction exceeding four body weights (BW) were registered (figure 1.9). In the vertical direction, an initial impact force of 13.6 BW was recorded, followed by a second peak of 6.1 BW. The average duration of the support was 135 ms.

1.1.2 Couples

A *force couple*, or simply a *couple*, consists of two equal, opposite, and parallel forces, **F** and **–F**, that are acting concurrently at a distance *d* apart (figure 1.10). For instance, two equal, parallel, and opposite forces applied by the

■ ■ ■ *From the Literature* ■ ■ ■

Moment Arm of a Muscle Force

Source: Pandy, M.G. 1999. Moment arm of a muscle force. *Exerc. Sport Sci. Rev.* 27: 79–119.

The moment arm of a muscle about an axis O-O was defined as the moment applied by a muscle force of magnitude 1, $\hat{\mathbf{F}}$.

$$\mathbf{r}^M = \frac{\mathbf{M}_{OO}}{F} = \left(\mathbf{U}_{OO} \cdot \mathbf{r} \times \hat{\mathbf{F}}\right)\mathbf{U}_{OO} \qquad (1.28)$$

Equation 1.28 can easily be derived from equations 1.23 and 1.27. The author performed a detailed mechanical analysis of the muscle moment arms in the joints with one and multiple degrees of freedom. The moment arms were determined about the instantaneous screw axes (ISA) for joints with intersecting and nonintersecting joint axes. When movement of the bones is restricted to a single plane and the muscle acts in the plane of bone movement, the magnitude of the moment arm is equal to the perpendicular distance from the instantaneous center of rotation of the joint to the line of action of the muscle force. In all other cases, the magnitude of the moment arm equals the perpendicular distance between the ISA and the line of action of the muscle multiplied by the sine of the angle between these two lines.

hands to a steering wheel during driving or by the thumb and index finger when turning a nut on a bolt form a couple. The plane in which the forces lie is called the plane of the couple. Because the vector sum of the forces constituting a couple is zero in every direction, the couple does not have a tendency to translate the body on which the forces act. The couple makes the body rotate. The measure of this tendency is called the *moment of a couple* or *torque*. A couple C that produces moment of couple $\mathbf{M}_C$ is customarily called the "couple $\mathbf{M}_C$" for brevity. This is similar to calling a person by his or her function, for instance, "teacher" or "plumber."

Consider two equal and opposite forces $\mathbf{F}$ and $-\mathbf{F}$ applied to a rigid body at the corresponding points A and B (figure 1.10). Let $\mathbf{r}_A$ and $\mathbf{r}_B$ be the position vectors of the points A and B, respectively, and $\mathbf{r}$ is the position vector of A with respect to B ($\mathbf{r} = \mathbf{r}_A - \mathbf{r}_B$). The vector $\mathbf{r}$ is in the plane of the couple but need not be perpendicular to the forces $\mathbf{F}$ and $-\mathbf{F}$. The combined moment of the two forces about O is

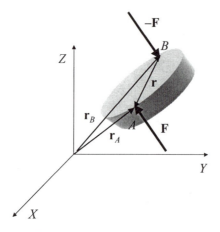

Figure 1.10 Two equal and opposite forces **F** and –**F** a distance *d* apart constitute a couple. The magnitude of their combined moment does not depend on the distance to any point, and hence the couple can be translated to any location in a parallel plane or in the same plane.

$$\mathbf{M}_O = \mathbf{r}_A \times \mathbf{F} + \mathbf{r}_B \times (-\mathbf{F}) = (\mathbf{r}_A - \mathbf{r}_B) \times \mathbf{F} = \mathbf{r} \times \mathbf{F} \qquad (1.29)$$

The product $\mathbf{r} \times \mathbf{F}$ is independent of the vectors $\mathbf{r}_A$ and $\mathbf{r}_B$; that is, it is independent of the choice of the origin *O* of the coordinate reference. Hence, the moment of couple $\mathbf{M}_C = \mathbf{r} \times \mathbf{F}$ does not depend on the position of *O* and has the same magnitude for all moment centers. The magnitude of the moment of couple is $M_C = Fd$, where *d* is the moment arm of the couple—the shortest distance between the lines of action of the two involved parallel forces. The moment arm *d* is perpendicular to the lines of force action. The magnitude of the moment of couple does not depend on the direction of the applied forces. The force magnitude *F* and the moment arm *d* determine the couple.

Because the moment of a couple $\mathbf{M}_C$ is the same for all moment centers and remains unchanged under parallel displacements, it is also called the *free moment*. Two couples having the same moment are equivalent: they produce the same effect on the rigid body on which they act. For instance, two couples $M_{C1} = Fd$ and $M_{C2} = 2F \cdot 0.5d$, if acting in the same or parallel planes, are equivalent, although the acting forces are different.

Couples and moments that they generate can be represented by vectors. A *couple vector* is normal to the plane of the couple. The sense of the couple vector is determined by the right-hand rule. By convention, counterclockwise couples are considered positive, and clockwise couples negative. Couple vectors obey the ordinary rules of vector algebra. For instance, the sum of two couples of moments $\mathbf{M}_1$ and $\mathbf{M}_2$ is a couple of moment $\mathbf{M}_C$ ($\mathbf{M}_C = \mathbf{M}_1 + \mathbf{M}_2$).

Couple vectors can be resolved into the component vectors $\mathbf{M}_x$, $\mathbf{M}_y$, and $\mathbf{M}_z$ along the axes of the coordinates. The component vectors represent the couples acting in the planes that are perpendicular to the corresponding coordinate axes. For instance, the vector $\mathbf{M}_x$ represents a couple acting in the *YZ* plane. Couple vectors are *free vectors;* they can be freely translated in space provided that their orientation remains constant. If so desired, the origin of the reference frame can be selected as the point of application of a couple vector.

In summary, the turning effect of a force depends on the point of force application, while the turning effect of a couple does not depend on the place where the couple is exerted.

1.1.3 Transformation of Forces and Couples

Because forces and couples are vectors, they can be transformed from one Cartesian coordinate system to another in the same manner as the coordinates themselves. In the biomechanics of human motion, a transformation from the global reference system *O-XYZ* (e.g., fixed with a force plate) into a local system *O-xyz* (fixed with a body or a body part) is quite common.

Let $\mathbf{F}_G$ be a column vector whose elements are the three components of force measured in a global system of coordinates *O-XYZ* fixed, for instance, with a force platform. Let $\mathbf{F}_L$ be the same vector expressed in the local reference system *O-xyz* fixed, for instance, with a body part. The global and local systems are related by rotation that is described by an orthogonal *rotation matrix* [R]. The coordinates of point *P* in the two reference systems are related by the equations

$$\mathbf{P}_G = [R]\mathbf{P}_L \qquad (1.30a)$$

and

$$\mathbf{P}_L = [R]^T\mathbf{P}_G \qquad (1.30b)$$

where $\mathbf{P}_G$ and $\mathbf{P}_L$ are the coordinates of point *P* in the global and local systems, respectively. These equations are equations 1.11 and 1.12 from *Kinematics of Human Motion*. The force vectors $\mathbf{F}_G$ and $\mathbf{F}_L$ measured correspondingly in the global and local reference frames are also related by similar equations:

$$\mathbf{F}_G = [R]\mathbf{F}_L \qquad (1.31a)$$

and

$$\mathbf{F}_L = [R]^T\mathbf{F}_G \qquad (1.31b)$$

Hence, to transform the force $\mathbf{F}_G$ from the global system of coordinates into the local coordinates, the force vector should be multiplied by the transpose of the rotation matrix, $[R]^T$.

Transformation of moments and couples is done in a manner similar to the transformation of forces. We discuss this issue on the intuitive level, without a strict mathematical proof. Consider a force $\mathbf{F}_G = (F_x, F_y, F_z)^T$ expressed in the global reference system. If $\mathbf{r}_G$ is a vector from the origin of the global coordinate system to the line of action of the force, then the moment of the force $\mathbf{F}_G$ with respect to the origin is given by the vector product $\mathbf{M}_G = \mathbf{r}_G \times \mathbf{F}_G$ (this is essentially equation 1.8). In the local system of coordinates, the moment is represented by

$$\mathbf{M}_L = \mathbf{r}_L \times \mathbf{F}_L \tag{1.32}$$

The vectors $\mathbf{r}_L$ and $\mathbf{F}_L$ are evidently equal to $\mathbf{r}_L = [R]^T\mathbf{r}_G$ and $\mathbf{F}_L = [R]^T\mathbf{F}_G$. Thus,

$$\mathbf{M}_L = ([R]^T\mathbf{r}_G) \times ([R]^T\mathbf{F}_G) \tag{1.33}$$

Because the rotation matrix $[R]$ is orthogonal, the magnitudes of the vectors $\mathbf{r}_G$ and $\mathbf{F}_G$ do not change as a result of the transformation in equation 1.33. For that reason, the magnitude of their cross product also does not alter, $M_L = M_G$. Only the orientation of the moment of force changes. Consequently, the components of the moment in the global and local system differ from each other. The vector of the moment, however, is still normal to the plane containing $\mathbf{r}_G$ and $\mathbf{F}_G$. Because the orientation of this plane in the two reference systems differs by the rotation $[R]$, the difference in the orientation of the normals to the plane is also defined by $[R]$. Therefore,

$$\mathbf{M}_G = [R]\mathbf{M}_L \tag{1.34a}$$

and

$$\mathbf{M}_L = [R]^T\mathbf{M}_G \tag{1.34b}$$

Equations 1.34a and 1.34b are similar to equations 1.30a and 1.30b. Hence, the moment $\mathbf{M}_G$ is transformed to the local reference system according to the same law of transformation as the coordinates themselves.

1.1.4 Replacement of a Given Force by a Force and a Couple

Any force $\mathbf{F}$ acting on a rigid body produces two effects: it tends to push or pull the body in the direction of the force, and it tends to rotate the body about any axis that does not intersect the line of force action. According to the principle of transmissibility, the force can be moved along its line of action without changing its effect on the rigid body on which it acts. It cannot, however, be moved away from the original line of action without modifying its effect on

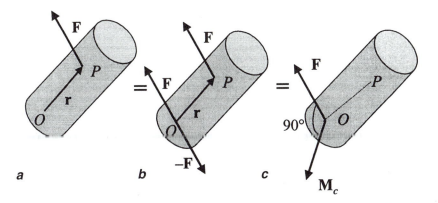

Figure 1.11 A force **F** acting on a rigid body at a point P (a) is replaced by an equal force shifted to a point O and a corresponding couple (c). In (b), two equal and opposite forces **F** and $-$**F** are added at a point O. In (c), the vector of couple $\mathbf{M}_C$ is normal to the plane containing vector **F** and the point P.

the body. Such a parallel displacement changes the moments of force that force **F** generates. The parallel displacement of a force can, however, be done if the change in the moments of the force is compensated by a couple (figure 1.11).

Any force **F** can be replaced by a parallel force of the same magnitude applied at an arbitrary point O and a couple of magnitude $M_C = Fd$, where d is the moment arm from O to the original position of the force. Such a representation is called a force-couple system.

Consider a force **F** acting on a rigid body at a point P (figure 1.11a). To move the force **F** away from its original line of action to a point O, we attach at O two equal and opposite forces **F** and $-$**F**. These forces can be added because they do not change the state of equilibrium or the motion of the body and cancel each other. The equal and parallel forces **F** (applied at P) and $-$**F** (applied at a point O) constitute a couple $\mathbf{M}_C$. The moment of couple $\mathbf{M}_C$ is equal to the moment of force **F** about point O. Hence, the force **F** (applied at the point P) can be replaced by an equal force applied at an arbitrary point O and a couple $\mathbf{M}_C = \mathbf{r} \times \mathbf{F}$, where **r** is a position vector of P with respect to O. The couple is added to compensate for the change in the moment of force. The plane of the couple coincides with the plane containing the vectors **r** and **F**. Therefore, the vector of the couple $\mathbf{M}_C$ is perpendicular to this plane. We can conclude that any force **F** acting on a rigid body at a point P can be replaced by the same force acting at another point O and the corresponding couple represented by a vector $\mathbf{M}_C$ perpendicular to **F**. Conversely, if a force **F** and a couple $\mathbf{M}_C$ are mutually perpendicular, a single equipollent force (and a couple of zero magnitude) can replace them.

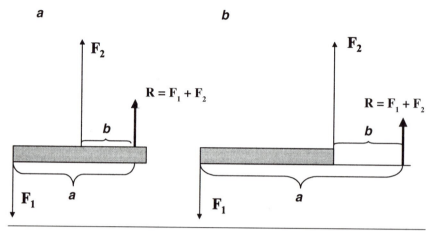

Figure 1.12 Two parallel, opposite, and unequal forces, F_1 and F_2, are acting on a rigid body. The magnitude of the resultant **R** equals the algebraic sum of F_1 and F_2. The distances from the action line of **R** to the action lines of F_1 and F_2 are inversely proportional to the magnitudes of the forces, $F_1 a = F_2 b$. (The reader is encouraged to prove this statement.) The resultant force **R** produces the same effect as the forces F_1 and F_2 combined. (*a*) This resultant force can actually be applied to the body (instead of pulling two ropes in opposite directions, for example, just one rope can be pulled with the same effect). (*b*) This resultant force **R** can be computed but cannot be actually applied to the body because the line of its action is outside the body. In general, when two parallel forces have similar senses, their resultant lies between them; when they have different senses, the resultant lies outside the space between them. When the forces are opposite and equal, they form a couple, and their resultant force is zero.

The preceding statement does not necessarily mean that the equipollent force can actually be exerted on the body. If the magnitude of the original couple is large and the size of the rigid body is small, the point of application of the resultant force that corresponds to a zero couple lies outside the body (figure 1.12*b*). Nevertheless, the equipollent force, which in this case is a purely theoretical construct, can be computed.

1.1.5 Replacement of a Given Force and Couple by Another Force and Couple: Invariants in Statics

Consider a force **F** applied at a point P and a couple **C** that exerts a moment M_C. The moment of couple is a free vector and can be applied anywhere. For convenience, we draw it from the point P (figure 1.13). As we just discussed, when the point of force application is changed from P to another point O, the

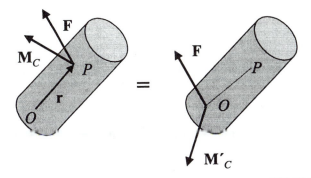

Figure 1.13 Representation of a force-couple system about points P and O. The system is initially given by a force $\mathbf{F}$ and a couple of moment $\mathbf{M}_C$ at P and then moved to a point O.

force $\mathbf{F}$ at P should be replaced by a similar force $\mathbf{F}$ at O and a corresponding couple $\mathbf{C}'$. The moment of couple $\mathbf{C}'$ equals the moment of $\mathbf{F}$ about O. Hence, with the force $\mathbf{F}$ applied at a new location O, two couples, $\mathbf{C}$ and $\mathbf{C}'$ are exerted on the body. Their combined effect is equal to the moment of a single couple $\mathbf{M}'_C$. The moment $\mathbf{M}'_C$ can be found by adding the moment of $\mathbf{F}$ about O to $\mathbf{M}_C$:

$$\mathbf{M}'_C = \mathbf{M}_C + \mathbf{r} \times \mathbf{F} \qquad (1.35)$$

where $\mathbf{r}$ is the position vector from O to P. Thus, any force $\mathbf{F}$ and couple $\mathbf{M}_C$ exerted at a point P on a rigid body can be replaced by an equal and parallel force applied at an arbitrary point O and a couple $\mathbf{M}'_C$, provided that equation 1.35 is satisfied.

The magnitude of force F is *invariant* (does not change) under a parallel translation to a new position. The dot product $\mathbf{F} \cdot \mathbf{M}_C$ is also invariant; it does not alter when the point of force application changes, $\mathbf{F} \cdot \mathbf{M}_C = \mathbf{F}' \cdot \mathbf{M}'_C$. Because the vectors $\mathbf{F}$, $\mathbf{r}$, and $\mathbf{F}'$ are in the same plane, the mixed triple product $\mathbf{F}' \cdot (\mathbf{r} \times \mathbf{F})$ equals zero, and therefore

$$\mathbf{F}' \cdot \mathbf{M}'_C = \mathbf{F}' \cdot (\mathbf{M}_C + \mathbf{r} \times \mathbf{F}) = \mathbf{F}' \cdot \mathbf{M}_C + \mathbf{F}' \cdot (\mathbf{r} \times \mathbf{F}) = \mathbf{F} \cdot \mathbf{M}_C \quad (1.36)$$

Hence, the scalars F and $\mathbf{F} \cdot \mathbf{M}_C$ are invariant with respect to the choice of point of force application. If $\mathbf{F} \cdot \mathbf{M}_C = 0$, the system of forces can be reduced to a single force; if $F = 0$, the system is equivalent to a single couple.

1.1.6 Equivalent Force-Couple Systems: Varignon's Theorem

Any system of forces acting on a rigid body can be reduced to a single result-ant force and a single resultant couple acting at a given point O. (It can also be reduced to a *space cross,* two nonintersecting forces, in infinite number of ways. The space cross is illustrated later in figure 1.16.) Such a system is the simplest equivalent force combination that produces the same result as the real forces acting on a rigid body. If original forces $\mathbf{F}_i$ are concurrent at O, they can be directly summed up to a single force and, hence, generate a zero moment, $\mathbf{M}_O = 0$. The single resultant force produces the same effect as the forces it replaces.

In general, the reduction can be seen as a three-stage procedure: (1) Each original force $\mathbf{F}_i$ acting on the body is replaced by a similar force at an arbitrary point O and the corresponding couple vector perpendicular to $\mathbf{F}_i$. (2) The concurrent forces at O are added according to the parallelogram rule and reduced to one resultant force. (3) The couple vectors are also added according to the parallelogram law and thus reduced to one resultant couple vector (we can do this because the couple vectors are free vectors and can be moved to one point). The magnitude and direction of the resultant force are the same regardless of the selected point of force application. In contrast, the magnitude and direction of the resultant couple depend on the particular point selected.

Usually, the resultant force vector and the resultant couple vector are not mutually perpendicular. If they are, the resultant force-couple system can be reduced to a single equivalent force or, if the resultant force is zero, to a single couple. The resultant force vector and resultant couple vector are perpendicular to each other when all the original forces are either coplanar (act in the same plane) or parallel.

By the definition of equivalent force systems (see Mechanics Refresher, p. 2), the equivalent sets of forces acting on a rigid body produce the same effect. Therefore, they exert the same moment of force about any arbitrarily chosen point in space. In particular, *if a set of forces acting on a rigid body is reduced to one resultant force, the moment of the resultant force about any point O is equal to the moment of the original force system about O.* This statement is credited to the French scientist Pierre Varignon (1654–1722) and is known as *Varignon's theorem,* or the *theorem of moments.* For easy memorization, the theorem can be simplified: *the sum of moments equals the moment of the resultant.* Varignon's theorem is valid for both coplanar and parallel force systems. It is also valid for concurrent forces (figure 1.14). The orthogonal force components of a force $\mathbf{F}$ applied at a point P are the concurrent forces. Therefore, the moment of a force $\mathbf{F}$ about O is equal to the sum of the components of that force, F_x, F_y, and F_z, about O.

Varignon's theorem is widely used to analyze parallel force systems. For parallel forces, the moment of a resultant force about any point is equal to the sum of the moments of the original forces about the same point. Because the moment of the resultant force about its line of action equals zero, the sum of the moments of all the contributing forces about this line is also zero. This

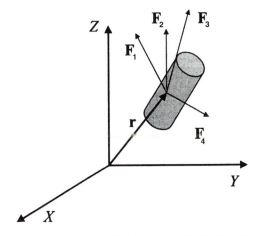

Figure 1.14 Proof of Varignon's theorem for concurrent forces. Consider several forces $\mathbf{F}_i (i = 1, 2, \ldots, n)$ exerted on a rigid body at a point P. Denoting the position vector of P as $\mathbf{r}$ and applying the distributive property of vector products, we obtain $\mathbf{r} \times \mathbf{F}_1 + \mathbf{r} \times \mathbf{F}_2 + \ldots + \mathbf{r} \times \mathbf{F}_n = \mathbf{r} \times (\mathbf{F}_1 + \mathbf{F}_2 + \ldots + \mathbf{F}_n)$, which proves the theorem.

property of the resultant force of parallel force systems is used to find the location of the resultant force. In the biomechanics of human motion, the most common parallel forces are the gravity forces that act on different parts of the body. Varignon's theorem provides a tool for determining the *center of gravity,* a point at which the resultant gravity force acting on the entire body is exerted. When external contact forces are parallel, Varignon's theorem can be used to determine the point of application of the resultant.

A special case of a parallel force system called a *lever* deserves particular mention. A lever is a rigid body revolving about an axis, the *fulcrum.* Levers involve a system of three forces: a resistance force, an effort force, and the force that is exerted on the fulcrum. In human movement, the effort force is the force generated by a subject or a muscle. Levers are classified by the position of the fulcrum with respect to the resistance and effort forces. In *first-class levers,* the resistance force and the effort force are on opposite sides of the fulcrum, as in a pair of scissors or in a seesaw. In *second-class levers,* the resistance force is applied between the fulcrum and the effort force, as in a nutcracker or wheelbarrow. In *third-class levers,* the effort force is applied between the fulcrum and resistance. Such an arrangement is typical for muscles in the human body; the point of muscle insertion is between the joint (fulcrum) and the point of application of the resistance force, for example, the external force acting on the *end effector.* Levers are just systems with three parallel forces; hence, Varignon's theorem can be applied. In levers, the moments are

■ ■ ■ *FROM THE LITERATURE* ■ ■ ■

Rising on the Toes: Does the Foot–Ankle System Work As a Lever of the First or Second Class?
or
A Century-Long Discussion About Nothing

Source: Fenn, W.O. 1957. The mechanics of standing on the toes. *Am. J. Phys. Med.* 36: 153–56.

In earlier research, the lifting of the body by the gastrocnemius–soleus muscle group was explained as either a first- or a second-class lever system (figure 1.15). When the system is considered a first-class lever, the ankle joint is regarded as a fulcrum, the ground reaction force is treated as the resistance force, and the force exerted by the *Achilles tendon* on the calcaneus is the effort force. When the system is analyzed as a second-class lever, the fulcrum is at the ball of the foot, and the resistance (body weight) acts at the talus, that is between the fulcrum and the point of exertion of the Achilles tendon force.

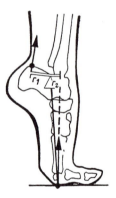

Figure 1.15 Rising on the toes can be modeled as a lever of the first or second class. Either approach is correct.

The discussion of this movement has a long history, starting with a paper by E. Weber published in 1846. Weber's problem was calculating the force exerted by the triceps surae when rising on the toes. For this analysis, he weighted the body using a lever attached to the belt of the subject and determined the load on the lever when the subject rose on his toes. In calculating the results, Weber considered the foot as a lever of the second class, with the fulcrum at the ball of the foot. He obtained a surprisingly low value for the absolute muscle force (force per unit of cross-sectional area of the muscle), about 1 kg of force per square centimeter. Later, other authors suggested that the foot–ankle system should be regarded as a lever of the first class. Using this interpretation, an absolute muscle force of about 4 kg per square centimeter was obtained.

According to Varignon's theorem, either interpretation is correct, provided that all the forces involved are included. The real error made by Weber (and some other authors) was in forgetting that when the calf muscles pull up on the calcaneus, they simultaneously pull down with an equal force on the tibia and femur. While it is somewhat simpler to regard the ankle as a lever of the first class, it is by no means wrong to regard it as a lever of the second class. In static equilibrium, any point can be considered a fulcrum, and around any such point the sum of the clockwise moments equals the sum of the moments acting in the counterclockwise direction. The discussion of whether rising on the toes should be modeled as a first- or second-class lever system is beside the point.

usually computed with respect to the fulcrum, although this is not necessary for an equilibrium equation. The theorem is valid for moments computed about any point.

As mentioned previously, a force system acting on a rigid body can be reduced to a single resultant force if, and only if, the vector of the resultant couple $\mathbf{M}_C$ is perpendicular to the vector of the resultant force $\mathbf{F}$. In the three-dimensional case, this usually does not happen, and the original force system commonly cannot be reduced to a single resultant force or a single couple.

1.1.7 Wrenches

Representative paper: VanSickle et al. (1998)

In three dimensions, an arbitrary set of forces can be reduced to a resultant force and a corresponding couple. Innumerable equivalent force-couple representations are possible. These resultant forces would have the same magnitude and direction but different points of application, which can be selected arbitrarily. When the point of force application is changed, the corresponding couple also changes. Consider, as an example, figure 1.16*a*, where two forces, a vertical force $\mathbf{F}_v$ and a horizontal force $\mathbf{F}_h$, are exerted on a rigid body at points *A* and *B*, correspondingly (a space cross). These forces can then be reduced to a force $\mathbf{R}$ at *A* and a couple $\mathbf{M}_C$ acting in the horizontal plane (figure 1.16*b*) or to force $\mathbf{R}$ at *B* and a couple $\mathbf{M}_C$ in the vertical plane (figure 1.16*c*). All three representations are equipollent.

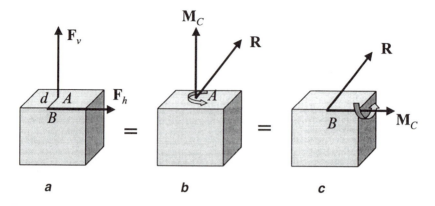

Figure 1.16   Reduction of forces $\mathbf{F}_v$ and $\mathbf{F}_h$ to a resultant force $\mathbf{R}$ and couple $\mathbf{M}_C$ in two different ways. In (*b*), the horizontal force $\mathbf{F}_h$ is moved to *A* and the corresponding couple $\mathbf{M}_C$ acting in the horizontal plane is added. In (*c*), the vertical force $\mathbf{F}_v$ is moved to *B* and the couple $\mathbf{M}_C$ in the vertical plane is added. In both cases, the forces $\mathbf{F}_v$ and $\mathbf{F}_h$ are resolved into one resultant force $\mathbf{R}$ applied at either *A* or *B*. The forces $\mathbf{F}_v$ and $\mathbf{F}_h$ represent the space cross that was mentioned in **1.1.6**.

There exists, however, a simplest equipollent form of the force-couple system into which a generalized contact force can be uniquely reduced. The representation is called the *wrench*. In a wrench, the vectors of the force and the couple are along the same line. Hence, in the wrench, the total effect of the force system is expressed as a "push" or "pull" along a certain line and a "turning" about this line. Figures 1.17*a* and 1.17*b* illustrate a wrench representation of a general force system, while parts *c* through *e* provide a specific case of a space cross: the two nonintercepting forces, $\mathbf{F}_v$ and $\mathbf{F}_h$, act at right angles to each other (compare with figure 1.16).

For a given force system, there is only one equipollent wrench representation. The moment (torque) produced by the couple is called a wrench moment. The wrench moment is a free moment.

Any particular wrench applied to a rigid body has an associated (1) singular line, the wrench axis; (2) *pitch*; and (3) magnitude, or intensity. The pitch and the wrench intensity are scalars. The pitch, p_w, is the ratio of the magnitude of the moment about the wrench axis (wrench moment) to the magnitude of the force along this axis. If $p_w = 0$, the wrench degenerates into a single force. When the pitch is infinite, the wrench is a pure couple. Because the pitch of a wrench is $p_w = M/F$, the wrench can be written as $\mathbf{w} = (\mathbf{F}, p_w \mathbf{F})$. Any wrench $\mathbf{w}_i$ can also be represented as a product of a unit wrench $\hat{\mathbf{w}}_i$ and the wrench intensity F_i, $\mathbf{w}_i = \hat{\mathbf{w}}_i F_i$.

In a wrench, the relation between force $\mathbf{F}$ and wrench moment $\mathbf{M}_w$ is expressed by

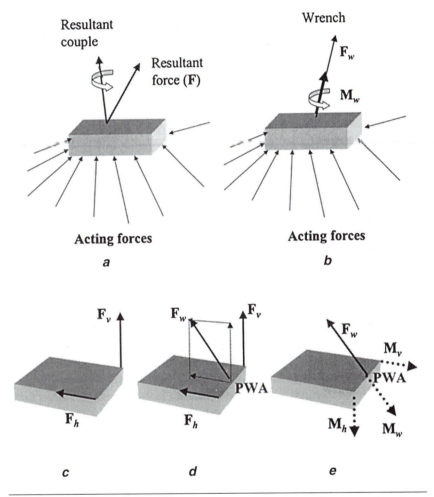

Figure 1.17 Wrench representation of a force system: (*a–b*) a general case and (*c–e*) a space-cross case. The force system acting on the body is reduced to (*a*) a resultant force and a couple and (*b*) a wrench, the combination of the force **F** and the collinear couple. (*a*) The resultant force and couple are applied at a point that is chosen arbitrarily. The resultant couple can be resolved into a component parallel to the resultant force and a component perpendicular to it. The latter component of the couple and the resultant force can be replaced by the force **F**. (*b*) The resultant force **F** and the component of the couple that is parallel to **F** constitute a wrench. In this example, the wrench is positive. (*c*) Two nonintercepting forces, F_v and F_h, are exerted on a rigid body. (*d*) The components of the wrench force F_w are equal to F_v and F_h. PWA is the point of wrench application. (*e*) The moments about the PWA that are due to the vertical and horizontal forces, M_v and M_h, respectively, can be replaced by a moment of force about the wrench axis, M_w. The details of the computation are explained in the text. In this example, the wrench is negative.

$$\mathbf{M}_w = p_w \mathbf{F} \tag{1.37}$$

where p_w is a scalar having a dimension of length. Hence, the scalar product of vectors $\mathbf{F}$ and $\mathbf{M}_w$ is

$$\mathbf{F} \cdot \mathbf{M}_w = \mathbf{F} \cdot p_w \mathbf{F} = p_w F^2 \tag{1.38}$$

and

$$p_w = \frac{\mathbf{F} \cdot \mathbf{M}_w}{F^2} \tag{1.39}$$

According to equation 1.36, the expressions just obtained are independent of the *point of wrench application* (PWA).

A wrench is said to be positive if the vector of resultant couple is in the same direction as the resultant force vector; if the vectors point in opposite directions, the wrench is negative. When the pitch of a wrench is zero (the wrench is a pure force), $\mathbf{F} \cdot \mathbf{M}_w = 0$. When a wrench degenerates into a couple, p is infinite and $F = 0$. In each of these cases, a force system is equipollent to a plane system of forces or to a system of parallel forces.

For a given force system acting on a rigid body, the moment about the wrench axis has a minimum magnitude. For any point of application of the resultant force, the projection of the resultant moment onto the wrench axis is constant. Any wrench can be written as a six-dimensional vector, with the first three components having units of force and the last three components units of moment: $\mathbf{w} = (F_x, F_y, F_z, M_x, M_y, M_z)^T$, where F_x, F_y, and F_z are the components of the contact force and M_x, M_y, and M_z are the components of the free moment acting at the contact. If a contact restricts movement in certain directions, some elements of the wrench are zero.

In human movement, understanding wrench moment is important because this moment is exerted by the end effector of the kinematic chain, usually the hand or foot, and consequently represents the moment produced or transmitted by the most distal joint, the wrist or ankle.

1.1.8 Wrenches and Helical Axes

An evident similarity exists between wrenches (force systems) and the helical representation used to describe body motion (see *Kinematics of Human Motion,* section **1.2.5.3**). Helical representation of motion is also called twist representation, or simply *twist*. Twist is a combination of the angular velocity and the linear velocity parallel to the axis of rotation. The resemblance between a description of body motion in space (the twist representation) and a system of forces resisted at a contact (the wrench representation) is an example of the *duality of statics and kinematics* that is addressed in chapter **2**. The duality is

Wrenches Exerted on the Hand During Wheelchair Propulsion

Source: VanSickle, D.P., R.A. Cooper, M.L. Boninger, R.N. Robertson, and S.D. Shimada. 1998. A unified method for calculating the center of pressure during wheelchair propulsion. *Ann. Biomed. Eng.* 26 (2): 328–36.

Wheelchair users suffer frequently from carpal tunnel syndrome and other hand pathologies. Knowledge of the forces and moments acting at the hand during wheelchair propulsion is important in preventing these maladies. Because the direction of the forces and moments exerted on the hand changes during the stroke and also because the rim can be grasped and then either pushed or pulled, the task of determining the force system acting on the hand is not trivial.

With specially designed dynamometric equipment, the authors registered the force **F** and moment of force **M** acting on the hub of the push rim. Then they expressed the force and moment in the hand-fixed coordinate system as a wrench. To do that, the total moment at the hub **M** was presented as a sum

$$\mathbf{M} = \mathbf{M}_f + \mathbf{M}_w$$

where $\mathbf{M}_f$ is the moment at the hub due to the hand force **F** exerted on the rim ($\mathbf{M}_f = \mathbf{r} \times \mathbf{F}$) and $\mathbf{M}_w$ is the wrench moment applied to the rim. To find $\mathbf{M}_f$, the authors needed to know the moment arm of the force **F** with respect to the hub. By definition, the moment arm is perpendicular to both **F** and $\mathbf{M}_f$. To find the direction of a vector along the moment arm, the authors computed the cross product **F** × **M** (the direction of vector **F** × **M** is the same as the direction of vector **F** × $\mathbf{M}_f$). The magnitude of the moment arm was then determined. The line of the wrench was resolved as the line where the vector of wrench moment was parallel to the force vector. The point of application of the wrench to the hand was determined as a point of intersection of the wrench line with the three planes of the hand-fixed reference system. Figure 1.18 presents the time-series plots of the moments $\mathbf{M}_f$ and $\mathbf{M}_w$. The coordinate system was centered at the hub with the *x* axis forward, the *y* axis up, and the *z* axis to the wheelchair rider's right.

The magnitude of the $\mathbf{M}_f$ was much larger than the magnitude of $\mathbf{M}_w$, which was less than 1 N · m. Consequently, the moment at the hub, **M**, was mainly due to the applied force and not the wrench moment. The wrench line lay approximately perpendicular to the transverse plane of the hand, and hence the force exerted by the hand did not produce a

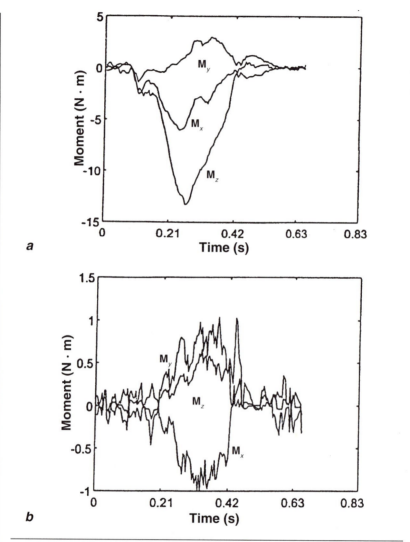

Figure 1.18 Wrench representation of the forces and moments exerted on the hand during wheelchair propulsion. *(a)* Moments about the hub generated by a force exerted by the hand on the rim, $\mathbf{M}_f$. These moments are equal to the cross product $\mathbf{r} \times \mathbf{F}$, where $\mathbf{r}$ is the position vector from the hub to the line of force action. The wheelchair velocity was 1.34 m/s. *(b)* Moments about the hub that are due to the wrench moment. Numerically, these moments are computed as a difference between the total moment at the hub $\mathbf{M}$ and the $\mathbf{M}_f$. The data represent the moments that must have been generated by the wrist.

> substantial moment with respect to the wrist center. Although the amount
> of wrench moment (< 1 N · m) was too small to propel the rim, it may still
> be important as a cause of wrist injury.

not a complete coincidence. For instance, *pitch* in kinematics is the magnitude
of translation per unit of rotation, and pitch in statics is the magnitude of the
moment per unit of force. In both cases, however, the units of measurement are
the same: pitch is expressed in units of length. An analogy exists between
linear velocity in kinematics and moment of force in statics. Consider, for simplicity, a planar case. If a body has an angular velocity ω, the magnitude of the
linear velocity at a point is $v = \omega r$, where r is the distance to the point from the
axis of rotation. If, on the other hand, a body has a force F imposed upon it, the
magnitude of the moment of force with respect to the axis of rotation is $M =
Fd$, where d is the shortest distance from the axis of rotation to the line of force
action. In both cases ($v = \omega r$ and $M = Fd$), the velocity v and the moment M are
the effects that occur at points in the bodies when either angular velocity or
force are applied to the whole body.

Similar to a wrench, a helical motion (twist) can be represented as $\mathbf{h} = (\omega,
p_h \omega)$. In the vectors $\mathbf{w}$ and $\mathbf{h}$, the first three elements have the same units of
measurement, which are different from the last three. Such vectors are called
dual vectors, or simply *duals*. In the literature, wrenches are sometimes called
action duals, and twists are called *motion duals*. Both wrenches and twists are
examples of *screws*. By definition, a screw is a line with a pitch.

A wrench representation of the forces acting at a contact, when combined
with a helical representation of motion, allows a joint analysis of forces
and movement in one procedure. This is especially convenient when the task
is to determine the power, the rate of doing work, transmitted at the contact.
Consider a moving body with a force and a couple exerted on it, shown in
figure 1.19.

As an example, think about a baseball or another implement in an athlete's
hand. Two screws are imposed by the athlete on the implement, a twist $\mathbf{h} = (\omega,
p_h \omega)$ and a wrench $\mathbf{w} = (\mathbf{F}, p_w \mathbf{F})$. We are interested in the power transmitted to
the ball. Considering the products *force times linear velocity along the twist
axis* and *torque times angular velocity about the twist axis,* we can write

$$P = \underbrace{(F\cos\theta)p_h\omega}_{\text{Force times linear velocity}} + \underbrace{(p_w F\cos\theta - Fd\sin\theta)\omega}_{\text{Torque times angular velocity}}$$

$$= F\omega[(p_w + p_h)\cos\theta - d\sin\theta] \tag{1.40}$$

where P is power, d is the shortest distance between the screws, θ is the angle
between the screws measured according to the right-hand rule, and p_h and p_w

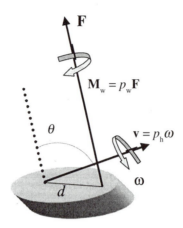

Figure 1.19 A wrench of magnitude F and pitch p_w is acting on the body that is moving at an angular velocity ω and pitch p_h. Two screws are presented, a twist $\mathbf{h}$ = (ω, $p_h\omega$) and a wrench $\mathbf{w}$ = ($\mathbf{F}$, $p_w\mathbf{F}$). Each screw acts along its own line. The common perpendicular distance between the screws is d, and the angle between the screws taken according to the right-hand rule is θ.

are the pitches of the helical motion and the wrench, correspondingly. Hence, the power transmitted at the contact depends on the product $F\omega$, that is, the magnitude of force times the magnitude of angular velocity, and the expression in the square brackets, which has a dimension of length. The expression is called the *virtual coefficient* (VC):

$$VC = (p_w + p_h)\cos\theta - d\sin\theta \qquad (1.41)$$

The magnitude of the VC equals the power transmitted by the duals with unit force and unit angular velocity (1 N of force acting on the body rotating at 1 rad/s). When VC = 0, the power is zero and the wrench is doing zero work. The screws with VC = 0 are said to be reciprocal, or orthogonal. Note that the virtual coefficient does not depend on the intensities of the wrench and helical motion. If VC = 0, then disregarding of the magnitude of F and ω, the wrench does not produce any work.

The product $d\sin\theta$ is simply a moment arm of the force $\mathbf{F}$ with respect to the helical axis. When the wrench and helical axes are parallel, $\sin\theta = 0$ and the force does not produce any moment about the helical axis. When this is the case, the force $\mathbf{F}$ does not transmit any power to change the angular velocity of the body. The expression $(p_w + p_h)\cos\theta$ represents the effect of the couple acting at the area of contact; when the wrench axis is collinear with the twist axis, the amount of transmitted power is maximal. Note that when only the translation is considered and the couple is zero (a force without a couple is acting on a body that moves in translation), p_h is infinite and p_w is zero. In such a situation, the virtual coefficient as defined by equation 1.41 cannot be computed. The same holds true when a body rotates without translation and only a couple is applied (p_h is zero and p_w is infinite).

Some authors define the virtual coefficient slightly differently from equation 1.41. The virtual coefficient can be seen as a kind of dot product of the dual vectors, in one of which the first three and the last three elements are interchanged. This operation is called the *reciprocal product* between screws. For two screws, $S_1 = (a_1, a_2, a_3, a_4, a_5, a_6)$ and $S_2 = (b_1, b_2, b_3, b_4, b_5, b_6)$, the virtual coefficient is defined as

$$VC = S_1 \langle \cdot \rangle S_2^T = a_1 b_4 + a_2 b_5 + a_3 b_6 + a_4 b_1 + a_5 b_2 + a_6 b_3 \qquad (1.42)$$

where the symbol $\langle \cdot \rangle$ designates the reciprocal product. Following this definition, if a wrench is $\mathbf{w} = (F_x, F_y, F_z, M_x, M_y, M_z)^T = (\mathbf{F}, \mathbf{M})^T$, where $\mathbf{F}$ and $\mathbf{M}$ are respectively the vectors of the contact force and couple, and a relative helical motion of the contacting bodies is represented as $\mathbf{h} = (\omega_x, \omega_y, \omega_z, v_x, v_y, v_z)^T = (\boldsymbol{\omega}, \mathbf{v})^T$, where $\boldsymbol{\omega}$ and $\mathbf{v}$ are the vectors of the angular and linear velocity, then the virtual coefficient is

$$VC = \left(F_x \ F_y \ F_z \right) \cdot \begin{pmatrix} v_x \\ v_y \\ v_z \end{pmatrix} + \left(M_x \ M_y \ M_z \right) \cdot \begin{pmatrix} \omega_x \\ \omega_y \\ \omega_z \end{pmatrix} \qquad (1.43)$$

Equation 1.43 is analogous to equation 1.41. For instance, the product of force and linear velocity is

$$\mathbf{F}^T \cdot \mathbf{v} = (F \cos\theta) p_h \omega = (F\omega) p_h \cos\theta \qquad (1.44)$$

When the virtual coefficient is defined according to equation 1.43, it equals the power, the rate of work done by the wrench. When $\mathbf{w}$ and $\mathbf{h}$ are reciprocal, $VC = 0$ and no work is done by the wrench. Hence, the virtual coefficients presented in equations 1.41 and 1.43 differ from each other by a multiplier $F\omega$.

In the following sections, we consider contact forces using two typical examples, the interaction with the support and grasping force exerted by individual digits.

1.1.9 Interaction With the Support

In the general case during a contact with a support, both forces and moments are exerted. The contact forces are commonly registered with force plates.

■ ■ ■ **FROM THE LITERATURE** ■ ■ ■

Forces Between the Foot (Boots) and the Binding in Alpine Skiing

Source: Yee, A.G., and C.D. Mote, Jr. 1997. Forces and moments at the knee and boot top: Models for an alpine skiing population. *J. Appl. Biomech.* 13: 373–84.

In the United States, approximately 10 million people participate in alpine skiing annually. Of these 10 million skiers, between 200,000 and 500,000 are injured. Current bindings designed to reduce the injury risk are of limited effectiveness because they do not limit the moments at the boot top and knee, where injuries most commonly occur. Instead, they attempt to limit forces at the boot toe and heel. The goal of this study was to predict moments at the boot top and knee from the measured force and moment components at the bindings. Two six-component force and moment sensors were used. One measured the force and moment components at the toe binding, and the other measured the force and moment components at the heel binding (figure 1.20). The angle at the ankle joint was measured with an electrogoniometer. A mathematical model was developed to determine the moments at the boot tops and knee from the registered parameters of a sample of skiers. The authors concluded that binding designs and standards for injury prevention must account for forces and moments at the sites of potential injury: at the knee and boot top.

1.1.9.1 Basic Mechanics of Force Platforms

Representative papers: Bobbert and Schamhardt (1990); Begg and Rahman (2000)

Force plates are customarily built as rectangular constructions supported at four corners. In each corner, a three-component force sensor is mounted (figure 1.21). The sensors are able to register the force in the lateromedial (X), anteroposterior (Y), and vertical (Z) directions. In total, there are 12 signals F_{ik}, where $i = X, Y, Z$ and $k = 1, 2, 3, 4$.

The components of the force **F** can be found by algebraic summation of the force values registered by the individual sensors.

$$F_i = \sum_{k=1}^{4} F_{ik} \ (i = X, Y, Z) \tag{1.45}$$

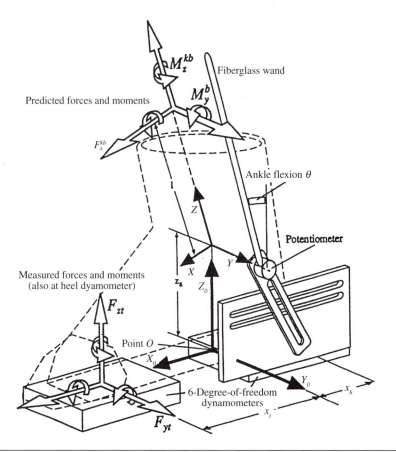

Figure 1.20 Kinetic and kinematic variables used in alpine skiing tests.

Reprinted, by permission, from A.G. Yee and C.D. Mote, Jr., 1997, "Forces and moments at the knee and boot top: Models for an alpine skiing population," *Journal of Applied Biomechanics* 13:373–384.

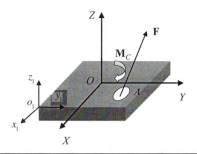

Figure 1.21 A schematic of a force plate. A reference system *O-XYZ* is fixed at the center of the supporting surface. The local reference systems o_k-$x_k y_k z_k$ (k = 1, 2, 3, 4) are fixed with the force sensors (only system o_1-$x_1 y_1 z_1$ is shown). Force **F** = $(F_X, F_Y, F_Z)^T$ is applied in contact area *A*. A couple **C** that produces a free moment $\mathbf{M}_C$ is also applied in this area.

Force **F** and moment of couple $\mathbf{M}_C$ produce a moment (rotational effect) with respect to the origin of the system of coordinates O. The moment equals the sum of the moments of force from the four sensors. Writing equations in projections on the reference axes, we have

$$M_i = \sum_{k=1}^{4} F_{ik} a_{ik} \ (i = X, Y, Z) \tag{1.46}$$

where M_i is the magnitude of the moment about the axis i (X, Y, or Z) and a_{ik} is the moment arm (the shortest distance) from a sensor k to the axis i. Both the force components and moment arms can be either positive or negative numbers.

Moments $\mathbf{M}_i$ are due to the interaction of the foot or feet with the support, in particular due to the force **F** and moment of the couple $\mathbf{M}_C$. A moment of the couple requires a combination of pulling and pushing forces. In practice, the free moments $\mathbf{M}_X$ and $\mathbf{M}_Y$ are customarily neglected (no sticking is assumed between the feet and the ground). However, in some athletic activities, such as when athletes perform fast body rotations while wearing shoes with spikes, these components of the free moment are large and cannot be ignored.

If the point of force application P (discussed later in the text) is known, the moment about the axis Z can be represented as

$$M_Z = \underbrace{-F_X P_Y + F_Y P_X}_{\text{Moment of force } \mathbf{F} \text{ about the platform center}} + \underbrace{M_{CZ}}_{\text{Free moment about axis } Z} \tag{1.47}$$

From equation 1.47, the component of the free moment about the vertical axis M_{CZ} can be determined.

Commercially available software packages for analysis of the *ground reaction force* (GRF) use either action-oriented coordinate systems or reaction-oriented systems. In the first case, the force exerted on the plate by the performer is analyzed, while in the second the reaction force acting on the body is presented. When the aim of the research is to determine the effect of the external forces on the human body, for instance, whether the body accelerates or decelerates, the reaction-oriented approach is more convenient.

Note that the real mechanics of force plates are much more complex than was outlined in the preceding paragraphs. In particular, the real center of the force platform usually does not coincide with the center of the supporting surface. Therefore, the equations become more complicated. Each force plate has its own unique calibration matrix. When four three-component transducers are used, the equation is

$$\overline{\mathbf{F}} = [B]\,\mathbf{S} \tag{1.48}$$

where $\overline{\mathbf{F}}$ is a 6-component vector of the generalized contact force, $\mathbf{S}$ is a 12-component vector of the signals, and $[B]$ is the 6×12 calibration matrix.

■ ■ ■ *FROM THE LITERATURE* ■ ■ ■

Free Moment in Running

Source: Holden, J.P., and P.R. Cavanagh. 1991. The free moment of ground reaction in distance running and its changes with pronation. *J. Biomech.* 24 (10): 887–97.

Excessive twisting moments exerted on the support leg in running may cause injuries. The authors registered the ground reaction during the support period in running at 4.5 m/s and determined the free moment about the vertical axis acting on the foot using equation 1.47. Because the authors were interested in the mechanical loading of the foot, they computed the total (net) ground reaction moment about an axis AA' fixed in the shoe and passing approximately through the ankle joint (figure 1.22). The net moment was calculated by adding the free moment to the moment of the resultant shear force about the axis AA'. Again, equation 1.47 was used. In this case, however, the horizontal coordinates of the point of force application were computed about the axis AA' (and not about

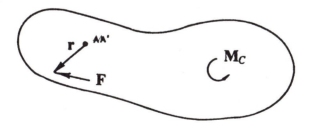

Figure 1.22   A net ground reaction moment M_z about the vertical axis AA' (shown in end view). The moment is equal to a sum $M_z = M_C + \mathbf{r} \times \mathbf{F}$, where M_C is a free moment (shown in arbitrary location), $\mathbf{F}$ is the horizontal force, and $\mathbf{r}$ is any vector from the point of intersection of the axis AA' with the supporting surface and the line of force action. The fixed axis AA' was chosen to pass approximately through the ankle joint; it was on the midline of the shoe at 25% of shoe length from the heel.

Adapted from *Journal of Biomechanics*, 24 (10), J.P. Holden and P.R. Cavanagh, The free moment of ground reaction in distance running and its changes with pronation, 887–897, 1991, with permission from Elsevier Science.

the platform center). The resultant shear force was supposed to act through the center of pressure (the center of pressure is explained later, in **1.1.9.3**).

On average, the free moment acted to resist foot abduction during the first 70% of support and in the opposite direction during the late support (figure 1.23).

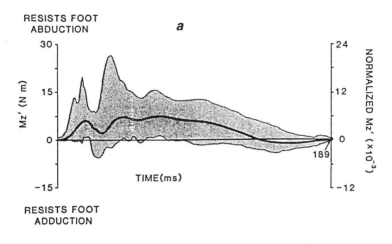

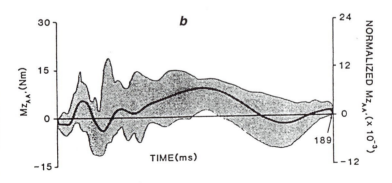

Figure 1.23 *(a)* The average pattern of the free moment (Mz$'$) and *(b)* the net moment (Mz$_{AA'}$) for 10 subjects. The moment values were normalized by dividing by the product of subject body weight and height. Absolute and normalized moment values are shown on the two vertical axes. The shaded areas represent the range of the normalized patterns for 20 individual feet. The moment about AA$'$ contributed by the resultant shear force equals the difference between the two curves shown.

Adapted from *Journal of Biomechanics*, 24 (10), J.P. Holden and P.R. Cavanagh, The free moment of ground reaction in distance running and its changes with pronation, 887–897, 1991, with permission from Elsevier Science.

1.1.9.2 Wrench Representation of the Ground Reaction Force

Representative paper: Shimba (1996)

We discuss the wrench representation of a force-couple system using as an example the interaction of the foot or feet with the ground. For the sake of illustration, think about a takeoff performed with either one or two legs or a turning maneuver in alpine skiing. When an individual presses down on the support (a force plate), the plate presses up on the individual with an equal and opposite force (the law of action and reaction). Hence, by measuring the force exerted on the plate, we know the force exerted on the individual, the ground reaction force (GRF). Compared with other external force systems, the GRF registered in such activities as walking and running is relatively uncompli- cated. In the absence of sticking, the legs only push (and do not pull) on the support, and the vertical and horizontal directions provide a natural frame for the analysis. These advantages are lost when more complex cases, for example, manual forces exerted on tools during manual work or the forces in downhill skiing, are investigated.

In the general case, the resultant force has no definite point of application and can be considered to act through an arbitrary point (with the addition of the corresponding couple). In a wrench, the location of a resultant force is defined uniquely. The point of wrench application (PWA) is the point where the wrench vector intersects the area of contact between the body and the sup- port surface. To determine the wrench, in particular to find the PWA location, the following procedure can be employed.

Force platforms usually have a system of coordinates with the origin at the geometric center of the platform surface. For such a platform, measurements are customarily expressed as the resultant force $\mathbf{F} = (F_x, F_y, F_z)^T$ and resultant moment $\mathbf{M}_O = (M_x, M_y, M_z)^T$ about the origin O (figure 1.24). The moment about O depends on $\mathbf{F}$, the PWA, and the couple acting at the contact area that produces the free moment $\mathbf{M}_C$. The usual goal is to find the couple (free mo- ment) and the PWA.

To find the couple $\mathbf{M}_w$, we resolve the moment $\mathbf{M}_O$ into two components: $\mathbf{M}_{par}$, which is parallel to force $\mathbf{F}_O$, and $\mathbf{M}_{nor}$, which is normal to $\mathbf{F}_O$ in the plane defined by $\mathbf{F}_O$ and $\mathbf{M}_O$ (figure 1.25). The component $\mathbf{M}_{par}$ immediately deter- mines the couple $\mathbf{M}_w$ acting in the plane perpendicular to $\mathbf{F}_w$. The vector of the couple is parallel to the resultant force, which is a requirement for the wrench. Hence, $\mathbf{M}_{par} = \mathbf{M}_w$.

To find the location of the wrench force $\mathbf{F}_w$, we recall that this force pro- duces the same mechanical effect as the force $\mathbf{F}_O$ applied at the origin O and moment $\mathbf{M}_{nor}$ combined. Force $\mathbf{F}_w$ is equal and parallel to force $\mathbf{F}_O$ and exerts the moment $\mathbf{M}_{nor}$ with respect to the origin O:

$$\mathbf{M}_{nor} = \mathbf{d} \times \mathbf{F}_w \qquad (1.49)$$

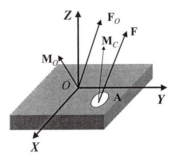

Figure 1.24   A contact force **F** and a couple of moment $\mathbf{M}_C$ are exerted somewhere in area *A* of the force platform. For simplicity, it is assumed that the contact takes place over a single area (a single foot contact). The combined effect of the contact force and the couple is registered as the force $\mathbf{F}_O$ at the origin and the moment $\mathbf{M}_O$ about the origin *O*. The moment $\mathbf{M}_O$ depends on the force $\mathbf{F} = (F_x, F_y, F_z)^T$, the PWA, and the couple $\mathbf{M}_C$. The task is to resolve $\mathbf{F}_O$ and $\mathbf{M}_O$ into the wrench exerted at the contact, that is, to find the PWA and the free moment $\mathbf{M}_w$ along the line of force action (subscript *w* stands for wrench). The magnitude and direction of $\mathbf{F}_w$ are already known; they do not change due to the parallel displacement of the force vector.

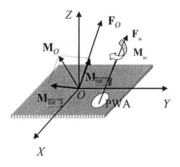

Figure 1.25   Decomposition of the vector $\mathbf{M}_O$ into two components: $\mathbf{M}_{par}$, which is parallel to the force $\mathbf{F}_O$, and $\mathbf{M}_{nor}$, which is normal to $\mathbf{F}_O$. The vector $\mathbf{M}_{par}$ represents the moment around an axis that is parallel to the resultant force $\mathbf{F}_O$. This moment is evidently equal to $\mathbf{M}_w$, the component of the moment of the couple acting at the area of contact along the line of force action. The curved arrow in the figure and the vector $\mathbf{M}_{par}$ represent the turning effect of the couple $\mathbf{M}_w$.

where **d** is a position vector perpendicular to both $\mathbf{F}_w$ and $\mathbf{M}_{nor}$. Vector **d** specifies the unique line of action of the resultant force $\mathbf{F}_w$. The resultant force $\mathbf{F}_w$ and the moment of couple $\mathbf{M}_w$ (or the parallel moment $\mathbf{M}_{par}$) constitute a wrench. When the shoe or foot is not sticking to the ground, the components of $\mathbf{M}_w$ around the horizontal axes *X* and *Y* are zero.

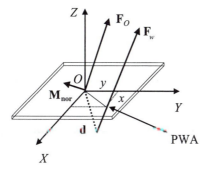

Figure 1.26 Locating the PWA. The vector $\mathbf{M}_{nor}$ is perpendicular to the plane that contains the two force vectors $\mathbf{F}_O$ and $\mathbf{F}_w$. Therefore, it defines the moment in this plane. The magnitude of the moment is $M_{nor} = Fd$, and consequently the distance is $d = M_{nor}/F$. Note that vector $\mathbf{d}$ is not in the horizontal plane. To find the PWA, the projection of $\mathbf{d}$ onto the support surface should be determined. The PWA is defined by the x and y coordinates where $\mathbf{F}_w$ intersects the support plane.

Generally, the vectors $\mathbf{F}_w$ and $\mathbf{M}_w$ are not directed in a strictly vertical direction; that is, the force may have horizontal components, and the torque may be produced around an oblique axis. Because vector $\mathbf{d}$ is perpendicular to $\mathbf{F}_w$ and $\mathbf{M}_w$, it is not obliged to lie in the horizontal plane. Hence, vector $\mathbf{d}$ does not immediately specify the PWA that is a piercing point, that is, the point at which the resultant force vector intersects the support surface. To find the PWA, the projection of vector $\mathbf{d}$ onto the support surface should be determined (figure 1.26).

In general, the PWA coordinates can be computed from the following equations (Shimba 1996):

$$X_{PWA} = \frac{F_X M_Z - F_Z M_Y}{F^2} - \frac{F_X^2 M_Y - F_X F_Y M_X}{F^2 F_Z}$$
$$X_{PWA} = \frac{F_Z M_X - F_X M_Z}{F^2} - \frac{F_X F_Y M_Y - F_Y^2 M_X}{F^2 F_Z}$$

(1.50)

where F is the magnitude of vector $\mathbf{F}$. If the horizontal components F_X and F_Y of the resultant force are small and can be neglected, equation 1.50 can be simplified.

$$X_{PWA} = \frac{M_Y}{F_Z}$$
$$Y_{PWA} = \frac{M_X}{F_Z}$$

(1.51)

Fortunately, in the majority of human movements, including walking and running, this simplification is justified. However, in some athletic activities where large horizontal forces are exerted, for instance, during the first steps in sprinting or in the tug of war, equation 1.51 may be not valid. It may also be not applicable in the beginning and ending periods of the ground contact in walking and running, when the values of the F_z are close to zero. Because in equation 1.51 the force appears in the denominator, the application of the equation may result in large inaccuracies and is not recommended for these periods of movement.

1.1.9.3 Center of Pressure and Contact Pressure

Representative papers: Cavanagh, Rodgers, and Iiboshi (1987); Shimba (1996); Fuller (1999)

In biomechanics research, the point of application of the vertical component of the GRF is often computed. This point is usually called the *center of pressure* (CoP) and is defined as the point where the resultant of the vertical force components intersects the support surface. Because the vertical force components are parallel by definition, the CoP can be computed by using Varignon's theorem (see **1.1.6**). The moment of the vertical forces about the CoP is zero. In three dimensions, the CoP and PWA are generally different points. The simple example presented in figure 1.27 illustrates the difference between the CoP and PWA. In figure 1.27*a*, a resultant vertical force $\mathbf{F}_z$ is applied at the CoP. In figure 1.27*b*, besides the force $\mathbf{F}_z$ acting at the CoP, a horizontal force $\mathbf{F}_{hor}$ is exerted on the body. The addition of the horizontal force does not change the point of application of $\mathbf{F}_z$. In figure 1.27*c*, the two acting forces are replaced by a resultant wrench. The PWA, the point of application of the wrench force $\mathbf{F}_w$, is evidently different from the CoP, the point of application of $\mathbf{F}_z$.

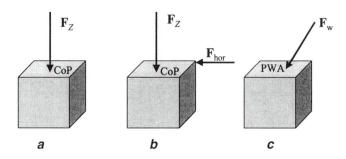

Figure 1.27   The difference between the CoP and PWA. See explanation in the text.

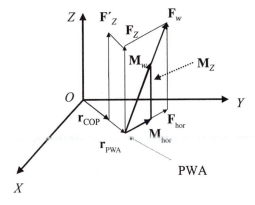

Figure 1.28 The difference between the PWA and CoP. The PWA is the point where the wrench force $\mathbf{F}_w$ intersects the support surface, and the CoP is the point where the resultant of the vertical force components intersects the support surface. The location of the PWA is specified by the position vector $\mathbf{r}_{PWA} = (x, y)^T$, where x and y are the coordinates of the PWA in the XY plane. Vector $\mathbf{r}_{CoP}$ defines the position of the CoP.

The relationship between the PWA and CoP is explained in figure 1.28. In this figure, a wrench is represented by the force $\mathbf{F}_w$ and the couple $\mathbf{M}_w$. The force and the couple act along the same line. The force $\mathbf{F}_w$ is resolved into the vertical and horizontal components $\mathbf{F}_z$ and $\mathbf{F}_{hor}$, and the couple is resolved into $\mathbf{M}_z$ and $\mathbf{M}_{hor}$. At the CoP, the moment $\mathbf{M}_{hor}$ (i.e., the moment of the vertical force components) should be equal to zero. To find the CoP, the vertical force component $\mathbf{F}_z$ of the wrench force $\mathbf{F}_w$ and the horizontal moment component $\mathbf{M}_{hor}$ are reduced to one force, $\mathbf{F}'_z$. The CoP is the point where $\mathbf{F}'_z$ intersects the support surface.

The component $\mathbf{M}_{hor}$ defines the moment of the vertical component of the resultant force $\mathbf{F}_z$ in the plane containing the position vector $\mathbf{r}_{PWA}$. The moment is about the PWA and is due to the difference in the locations of the PWA and the CoP.

$$\mathbf{M}_{hor} = (\mathbf{r}_{CoP} - \mathbf{r}_{PWA}) \times \mathbf{F}_z \tag{1.52}$$

When the system of forces acting on a force plate is expressed with respect to the CoP rather than the PWA, the system is reduced to the vertical force $\mathbf{F}_z$, the horizontal force $\mathbf{F}_{hor}$, and the moment about the CoP acting in the horizontal plane $\mathbf{M}'_z$. This moment is equal to

$$\mathbf{M}'_z = \mathbf{M}_z + (\mathbf{r}_{CoP} - \mathbf{r}_{PWA}) \times \mathbf{F}_{hor} \tag{1.53}$$

Readers are encouraged to compare this equation with equation 1.47. The equations are similar; equation 1.47 is written with respect to the center of the force plate O, and equation 1.53 with respect to the CoP. The moment $\mathbf{M'}_z$ is usually not equal to zero. If it is zero, the original resultant force and moment are mutually perpendicular to each other and hence can be reduced to one resultant force. In this case, the CoP and the PWA are the same point, and the term "point of force application" is appropriate.

When two feet are on the ground, separate measurement of the force exerted by each foot is advantageous. The location of the CoP either in the anteroposterior or mediolateral directions can be computed in this case by using Varignon's theorem:

$$\text{CoP}_{\text{total}} = \text{CoP}_{\text{right}} \frac{F_{\text{Zright}}}{F_{\text{Zright}} + F_{\text{Zleft}}} + \text{CoP}_{\text{left}} \frac{F_{\text{Zleft}}}{F_{\text{Zright}} + F_{\text{Zleft}}} \qquad (1.54)$$

where F_{Zright} and F_{Zleft} are the vertical components of the GRF under the right and left feet, correspondingly. For a symmetric or near symmetric stance with wide foot placement, the total CoP lies between the feet and outside the areas of foot contact with the support.

For parallel forces acting over a certain contact area, the amount of force per unit of area, or contact *pressure,* can be determined. In the International System (SI) of metric units, the unit of pressure is the pascal (Pa; after the French scientist Blaise Pascal, 1623–1662), which is equal to 1 N/m^2. The pressure corresponding to 1 Pa is very small. For instance, pressures underneath the foot are typically hundreds of thousands of pascals, customarily expressed in kilopascals (1 kPa = 1,000 Pa). When local pressure over an area of contact is known, the CoP location can be computed.

The values of contact pressure reported in the literature may depend on the area of the sensitive elements used for the measurements. For instance, if a force is applied over a very small area (e.g., a marble in the shoe) but reported as an average force over a larger area, the reported values are distorted. The measurement of contact pressure is important because it characterizes the local load imposed on individual areas of the contacting surface (figure 1.29). The local biological effects of the contact pressure are considered later in this chapter.

1.1.9.4 Gross Mechanical Effects of the Ground Reaction Force: Interplay of the Internal and External Forces

Representative paper: Blickhan (1989)

If biomechanics research were limited to only one method, the registering of GRF would be the technique of choice. In many cases, for example, in walking, running, and jumping, there are only three external forces that act on the human body: (1) air resistance, which is usually small and can be neglected;

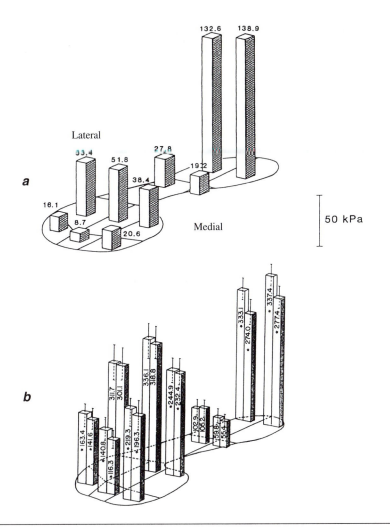

Figure 1.29 Peak plantar pressure during (*a*) standing and (*b*) walking. (*a*) Mean regional peak pressures during standing (107 adult subjects, 30.1 ± 9.9 years, body weight 701 ± 146.7 N, and body height 174 ± 10 cm). The ratio of peak rear-foot pressure to peak forefoot pressure was approximately 2.6:1. (*b*) Regional pressure means and standard deviations during first-step (shaded columns) and midgait walking in a group of men, ages 40 to 81. Regions where the difference between the two conditions was statistically significant are marked with an asterisk.

Figure 1.29*a* Reprinted, by permission, from P. Cavanagh et al., 1987, "Pressure under symptom-free feet during barefoot standing," *Foot & Ankle* 7 (5):262–276.

Figure 1.29*b* Reprinted, by permission, from Rodgers, M.M. (1985). *Plantar pressure distribution measurement during barefoot walking: normal values and predictive equations.* Unpublished doctoral dissertation, The Pennsylvania State University, University Park.

••• *MECHANICS REFRESHER* •••

Linear and Angular Momenta

Linear momentum of a body (a system of material particles) is a vector:

$$\mathbf{L} = m\mathbf{v}_{CoM} \tag{1.55}$$

where m is the body mass and $\mathbf{v}_{CoM}$ is the velocity of the CoM. The direction of the momentum is defined by $\mathbf{v}_{CoM}$. If the mass m of the system is constant, the rate of change in the linear momentum equals the *acceleration* of the CoM, $\mathbf{a}_{CoM}$.

Angular momentum of a rigid body pivoted about a fixed axis is a vector $\mathbf{H}$ that has a magnitude $H = I\omega$, where I is the moment of inertia of the body about the axis of rotation and ω is the magnitude of the angular velocity. The direction of the angular momentum is defined by the angular velocity vector $\boldsymbol{\omega}$. The angular momentum of a system about a line is the sum of angular momenta of the particles composing it. The rate of change in the angular momentum of a system about a fixed line is equal to the total moment of the external forces about this line.

The angular momentum can be computed relative to the CoM. The following principle is valid: the rate of change of angular momentum relative to the center of mass is equal to the sum of the moment of the external forces about the center of mass and the torque of the external couple (free moment). The angular momentum during airborne movements is constant.

Hence, only external forces acting on the body determine changes in the linear and angular momenta.

---

(2) gravity force; and (3) the GRF. Since the force of gravity is constant, when whole-body movement is the object of interest, both body translation and rotation are highly influenced by the GRF. Track and field coaches know this from practical experience. They formulated the following sensible rule: athletic performance is determined by an interaction with the support. For the biomechanical analysis of sport technique, the following rule of thumb is useful: analyze the GRF first, and try to understand the factors that influence the propelling and braking forces. The *propelling forces,* also called *driving forces* or *thrust,* are defined as external forces that act in the direction of progression (figure 1.30); the *braking forces* are external forces acting opposite to the direction of progression.

The term "propelling force" should not be taken literally. With the exception of a takeoff performed from an elastic support (presented in figure 1.30),

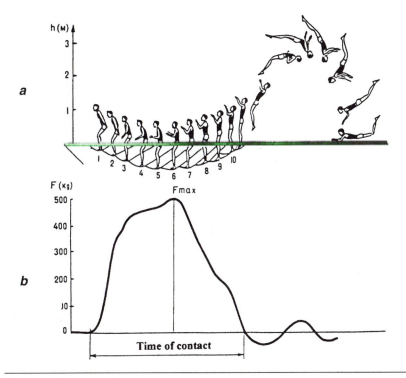

Figure 1.30 Propelling forces in human movement: the force during the takeoff on a trampoline. During contact, the athlete's body initially is moving down with decreasing speed (during this time, the net is deflecting) and then moves up; therefore, the acceleration is directed upward the whole time. Note the strict correspondence between (*a*) the net deflection and (*b*) the force. The force is measured in kilograms. One kilogram of force equals 9.8 N.

Reprinted from D. Donskoi and V. Zatsiorsky, 1979, *Biomechanics* (Moscow: FiS Publishers).

the GRF—being a reaction force—cannot by itself propel a performer. Schematically, during a takeoff, the chain of events (shown in figure 1.31) is as follows:

1. Muscle forces are applied to the upper part of the body, thereby inducing its acceleration.

2. Equal and opposite forces are applied to the bottom part of the body (the foot) that is in contact with the support and cannot move down.

3. The foot transmits the force to the ground.

4. An equal and opposite force (the ground reaction force) acts on the foot.

Muscle forces are internal forces. They can displace one body part with respect to another, but without the GRF (an external constraint), they cannot

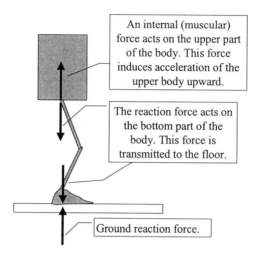

An internal (muscular) force acts on the upper part of the body. This force induces acceleration of the upper body upward.

The reaction force acts on the bottom part of the body. This force is transmitted to the floor.

Ground reaction force.

Figure 1.31 A schematic representation of the interplay of the internal and external forces during a takeoff. As a result of muscle activity, the upper part of the body moves upward. In the air, the lower body would move in the opposite direction (since linear momentum is conserved). However, the performer's feet are in contact with the floor, and the floor exerts a force on the performer. This force resists the downward movement.

change the movement of the body's CoM. The GRF is evidently not a cause of movement (the muscle force is), but without the GRF whole-body movement would not be possible.

The GRF is a convenient indicator of the total muscular effort; it also can be used to ascertain acceleration or deceleration of the entire body in a given direction. In gross body movement, the GRF represents whole-body acceleration, both linear and angular.

1.1.9.4.1 Linear Acceleration

The linear acceleration of the body's CoM is determined by the magnitude and direction of the GRF taken in the global system of coordinates. When equations are written in projections on the horizontal and vertical axes

$$a_{hor}^{CoM} = \frac{F_{hor}}{m} \tag{1.56a}$$

and

$$a_{ver}^{CoM} = \frac{F_{ver}}{m} - g \tag{1.56b}$$

••• *MECHANICS REFRESHER* •••

Linear Impulse

Linear impulse (**I**) is the definite integral of force over time:

$$\mathbf{I} = \int_{t_1}^{t_2} \mathbf{F}dt \qquad (1.57)$$

where t_1 and t_2 are the times at which force application begins and ends (the limits of integration). The impulse is a vector quantity that measures the effect of a force during the time that the force acts. The linear impulse equals the increment in linear momentum (the *principle of linear impulse and momentum*):

$$\mathbf{I} = m\mathbf{v}_2 - m\mathbf{v}_1 \qquad (1.58)$$

If the mass of a body is constant, the linear impulse is proportional to the increment of the linear velocity of the body's CoM.

---

where $a_{\text{hor}}^{\text{CoM}}$ and $a_{\text{ver}}^{\text{CoM}}$ are the horizontal and vertical components of the acceleration of the center of mass, respectively, m is the body mass, and g is acceleration due to gravity.

In the literature, the GRF is presented in various forms, for example, as a sequence of vectors in space (figure 1.32), as vector dynamograms (figure 1.33), or as time-history curves with three components—*X*, *Y*, and *Z*—called dynamograms (figure 1.34), which are most customarily used.

The area under the force versus time curve of a dynamogram is a linear impulse. According to the principle of linear impulse and momentum, the impulse divided by the mass of the body represents the change in the velocity of the CoM in a given direction during the time the GRF acts. We consider first the horizontal GRF and then the vertical GRF.

During contact in level walking and running, the foot pushes first in the anterior direction and then in the posterior direction. Consequently, during the first half of contact, the GRF is in the posterior direction, and it is in the anterior direction in the second half. Hence, the horizontal anteroposterior components of the GRF are negative (opposite to the movement direction) at the beginning of the support period and positive at the end. When people walk or run at a constant velocity, the impulse of the propelling force equals the impulse of the braking force. When the body accelerates, for instance, during the first steps in sprinting, the positive impulse is larger than the negative impulse.

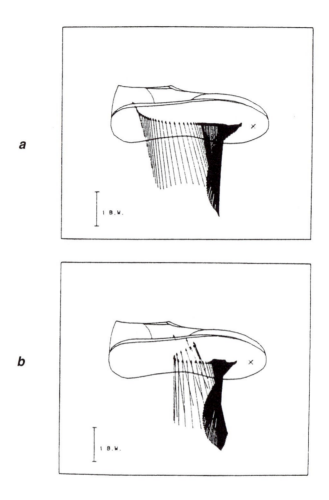

Figure 1.32 Vector representation of the GRF in running for two subjects. (*a*) A rear-foot runner: a subject whose first contact with the ground was made in the rear third of the shoe. (*b*) A midfoot runner: a subject whose first contact was in the middle third of the shoe. The view is of the right shoe from the lateral aspect with the shoe tilted 30° upward. The length of the force vectors are proportional to the resultant of the vertical and horizontal anteroposterior components of the GRF. The lateromedial force component was neglected due to its small magnitude. The point of force application is at the center of pressure corresponding in time to the force value. A scale of 1 body weight (B.W.) is indicated.

Reprinted from *Journal of Biomechanics*, 13, P.R. Cavanagh and M.A. Lafortune, Ground reaction forces in distance running, 397–406, 1980, with permission from Elsevier Science.

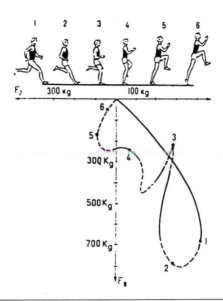

Figure 1.33 Vector dynamogram of a takeoff in long jumping, projected onto the sagittal plane. The numbers correspond to the consecutive positions of the jumper. The force exerted by the athlete on the force plate is presented. The GRF is opposite to this force. Note that the horizontal GRF is mainly negative (braking). As a result, the horizontal velocity of the athlete's body during the takeoff decreases. The force is measured in kilograms. To convert the force to *newtons,* the values should be multiplied by 9.8. Note that with this representation, the displacement of the PWA is neglected.

Reprinted from Donskoy, D., and V. Zatsiorsky. 1979. *Biomechanics.* Moscow: FiS.

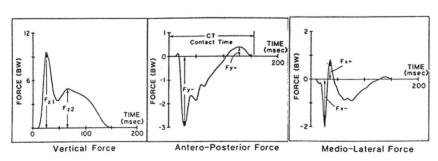

Figure 1.34 A typical example of ground reaction forces in triple jumping. The subjects were collegiate triple jumpers. The vertical force records showed two peaks with magnitudes of 7 to 12 times body weight (BW) and 3.3 to 5.0 BW, respectively. The horizontal braking force is close to 3 BW. Note the difference between the magnitudes of the braking and propelling anteroposterior force.

Reprinted, by permission, from M.R. Ramey and K.R. Williams, 1985, "Ground reaction forces in the triple jump," *International Journal of Sport Biomechanics* 1:233–239.

During the takeoff in long and triple jumping, the impulse is mainly negative: the athlete loses horizontal velocity (see figures 1.33 and 1.34).

When an individual is standing still, the vertical component of the GRF equals the body weight. When the vertical GRF exceeds the body weight, the CoM of the body accelerates upward. If the vertical component of the GRF is smaller than the gravity force acting on the body, the body accelerates downward. A standing person can achieve downward acceleration by fast leg bending; this maneuver is seen in the beginning of countermovement vertical jumps (figure 1.35). A similar "unloading" is customarily used in alpine skiing to decrease the load on the skis during the turns.

In walking, the vertical component of the GRF usually has two peaks (figure 1.36). The first peak is associated with the downward deceleration of the CoM (immediately before the heel strike, the CoM velocity is directed downward). The second peak is associated with the upward acceleration of the CoM. When a person walks on a sand beach, the impressions left on the wet sand represent these two peaks of the vertical force (figure 1.37).

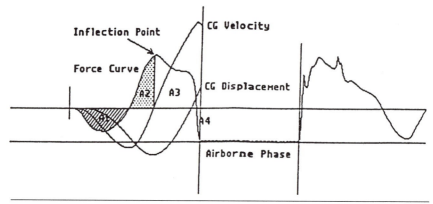

Figure 1.35   Vertical GRF, velocity, and displacement of the CoM of the body during a typical countermovement jump. The different scales for GRF, velocity, and displacement values are omitted. In the force–time curve, area A1 (below the body weight) represents the impulse in the downward direction. During this time, the downward vertical velocity increases. Area A2, above the body weight, represents the impulse that is spent to decelerate the downward movement of the body mass. At the end of the A2 period, the vertical velocity is zero, the GRF is maximal, and the CoM is in its lowest position. Areas A1 and A2 are equal. Area A3 corresponds to the impulse that is used to accelerate the body mass upward from zero velocity at the bottom position to the maximum velocity. A small area A4 corresponds to the end of takeoff when the GRF is less then body weight and the vertical velocity of the CoM decreases.

Adapted, by permission, from A. Kibele, 1998, "Possibilities and limitations in the biomechanical analysis of countermovement jumps: A methodological study," *Journal of Applied Biomechanics* 14:105–125.

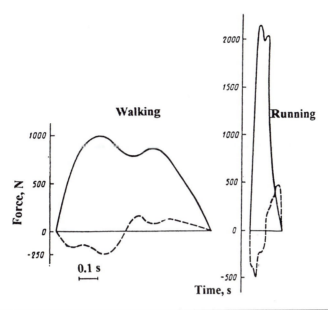

Figure 1.36 Ground reaction force in walking and running. The solid line indicates the vertical component of the GRF; the dotted line indicates the horizontal anteroposterior component of the GRF.

Figure 1.37 Impressions left on the wet sand that represent two peaks of the vertical GRF in walking.

Reprinted from Demeny, G. 1909. *Le bases scientifiques de l'éducation physique*. Paris: F. Alcan.

In running (except in toe running at slow speed) and in running jumps, the vertical force component usually has two peaks, but their origins are different. The first peak is called the impact peak; it represents the impact forces that occur during the 25 to 50 ms after the first contact of the foot with the ground. Similar impact peaks may occur in the horizontal force–time curves. Impact forces are discussed later in the text, see **1.1.9.4.3**. The second peak is called the active peak; it represents the acceleration of the CoM and is entirely

---

▪ ▪ ▪ FROM THE LITERATURE ▪ ▪ ▪

Ground Reaction Forces in Obese Patients

Source: Messier, S.P., W.H. Ettinger, Jr., T.E. Doyle, T. Morgan, M.K. James, M.L. O'Toole, and R. Burns. 1996. Obesity: Effects on gait in an osteoarthritic population. *J. Appl. Biomech.* 12: 161–72.

The association between obesity and GRFs was examined in 101 older adults with knee osteoarthritis. Increased body mass index was significantly correlated with increases in the braking and propelling forces, vertical force minimum and maximum, vertical impulse, and maximal medially and laterally directed forces. There was no correlation between obesity and hip or ankle kinematic variables.

---

controlled by muscle activity (Nigg 1988). Both the vertical and the horizontal peaks in the GRF depend on gait velocity. For instance, when the running velocity changed from 3.25 m/s to individual maximum (8.31 ± 0.75 m/s), the peak vertical GRF in the braking period increased from 1,665 to 2,088 N, and the peak horizontal GRF increased from 237 to 675 N (Kyröläinen, Belli, and Komi 1999).

If the initial velocity and location of the CoM are known, for instance, if the takeoff is performed from an initial resting position, the liftoff velocity (v_z) and the position of the CoM (P_z) can be found by computing the integrals

$$v_z = v_{z0} + \int_{t_0}^{t} \left(\frac{F_z(t)}{m} - g \right) dt \qquad (1.59)$$

and

$$P_z = P_{z0} + \int_{t_0}^{t} v_z(t)\, dt \qquad (1.60)$$

where v_{z0} and P_{z0} are the initial velocity and initial position, correspondingly.

Because in terrestrial conditions two forces are acting on the body, the GRF and gravity, the resultant force that determines the acceleration of the body's CoM is a vector sum of these two forces (figure 1.38). Therefore, to propel a body or an implement at a desired inclination angle to the horizon, the voluntary force should be exerted at a higher inclination angle to compensate for the action of gravity.

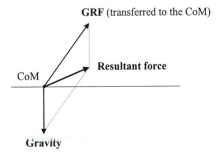

Figure 1.38 The effect of gravity on the direction of the resultant force.

1.1.9.4.2 Angular Acceleration

Representative paper: Hof (1992)

The rate of change of the angular momentum of the body about its CoM is determined by the moment of the external force and the external couple (free moment). Suppose that both the GRF wrench, $\mathbf{w} = (\mathbf{F}_w, \mathbf{M}_w) = (\mathbf{F}, p_w\mathbf{F})$, and the location of the center of mass $\mathbf{r}_{CoM}$ are known. The following equation (refer to figure 1.39) represents the rate of change of the angular momentum, $\dot{\mathbf{H}}$.

$$\dot{\mathbf{H}} = (\mathbf{r}_{PWA} - \mathbf{r}_{CoM}) \times \mathbf{F}_w + \mathbf{M}_w \qquad (1.61)$$

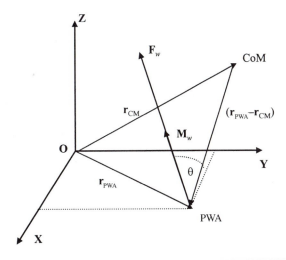

Figure 1.39 Generating a body rotation. A wrench, that is, GRF $\mathbf{F}_w$ and couple $\mathbf{M}_w$ (free moment), is acting at the PWA. The line of force action does not pass through the body's CoM.

If $\mathbf{M}_w = 0$, that is, if an original force system can be reduced to a single resultant force, the rate of change of angular momentum of the body depends only on the magnitude of the GRF and on the location of the line of force action with respect to the CoM of the body, that is, on the difference $\mathbf{r}_{PWA} - \mathbf{r}_{CoM}$. Figure 1.40 illustrates this. A similar case was discussed previously: generating body rotation in the running forward somersault (see figure 1.8).

When a somersault is performed from a standing position, a moment of force with respect to the body's CoM is necessary to produce the whole-body rotation. It is provided by the GRF (figure 1.41). Somersaulting is initiated by

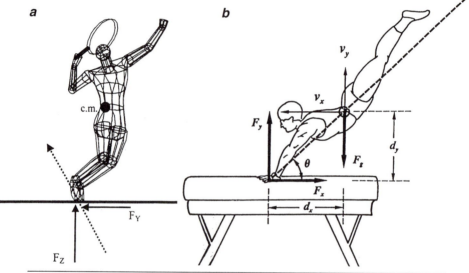

Figure 1.40 The GRF and the development of angular momentum during (*a*) the tennis serve and (*b*) horse vaulting. (*a*) Relationship between the GRF and the development of angular momentum about the *X* axis of rotation. Location of the CoM is represented by a filled circle. (*b*) Influence of the horse contact on the angular momentum and linear velocity of the gymnast's body, showing forces acting on the gymnast, moment arms (d_x and d_y), and body angle θ during the horse-support phase of the vault. Since the CoM is posterior and superior to the point of contact on the horse, vertical GRF $\mathbf{F}_y$ exerted by the horse acts to increase the gymnast's vertical velocity $\mathbf{v}_y$ and decrease the angular momentum during the course of blocking. Similarly, the horizontal reaction force $\mathbf{F}_x$ acts to decrease the horizontal velocity $\mathbf{v}_x$ and increase the angular momentum.

Figure 1.40*a* reprinted, by permission, from R.E. Bahamonde, 2000, "Changes in angular momentum during the tennis serve," *Journal of Sports Sciences* 18:579–592. By permission of Taylor & Francis Ltd, **http://www.tandf.co.uk/journals**

Figure 1.40*b* reprinted, by permission, from Y. Takei, 1992, "Blocking and postflight techniques of male gymnasts performing the compulsory vault at the 1988 Olympics," *International Journal of Sport Biomechanics* 8:87–110.

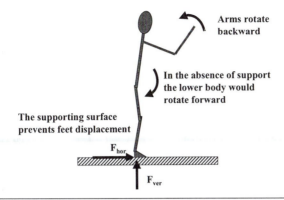

Figure 1.41 A schematic representation of the interplay of the internal and external forces during backward somersaulting. As a result of "throwing" the arms backward, the arms and upper body rotate. In the air, the lower body would rotate in the opposite direction (since angular momentum is conserved). However, with the feet on the support, the "throw" creates a ground reaction force. The force produces a moment around the body's CoM that induces the whole-body rotation. Compare this figure with figure 1.31. The mechanisms for creating linear and angular acceleration of the body are quite similar.

■ ■ ■ *From the Literature* ■ ■ ■

Ground Reaction Forces During Multiple Back-Somersaulting Dives

Source: Miller, D.I., E. Henning, M.A. Pizzimenti, I.C. Jones, and R.C. Nelson. 1989. Kinetic and kinematic characteristics of 10-m platform performances of elite divers: Back takeoffs. *Int. J. Sport Biomech.* 5: 60–88.

During the Fifth World Diving Championship, the authors registered the GRF in elite divers performing multiple back-somersaulting dives. A force plate was mounted flush on the 10-m platform. The GRF was related to the patterns of motion (figure 1.42). During the final period of the takeoff, the line of action of the vertical reaction force was in front of the CoM, and the moment arm of the force continued to increase. As a result, the vertical component of the GRF promoted backward-somersaulting angular momentum. Contrarily, the angular impulse produced by the horizontal component of the GRF retarded the development of backward-somersaulting angular momentum. Because the angular impulse of the vertical GRF exceeded that produced by the horizontal component, the resultant angular impulse promoted backward-somersaulting angular momentum.

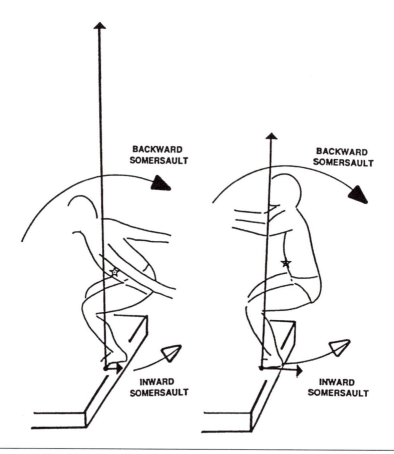

Figure 1.42 Moments of vertical and horizontal components of platform reaction force with respect to the CoM (star) at two instants during the final portion of the takeoff. The data are for the Olympic champion G. Louganis, who was performing a back three-and-a-half-tuck somersault.

Reprinted, by permission, from D.I. Miller et al., 1989, "Kinetic and kinematic characteristics of 10-m platform performances of elite divers: Back takeoffs," *International Journal of Sport Biomechanics* 5:60–88.

a "throw" of the arms in the desired direction, which involves a rotation of the arms. The supporting surface prevents the opposite rotation of the bottom part of the body. As a result, the whole body rotates in the direction of the throw. The vertical force makes the athlete airborne. It also may contribute to the change of the angular momentum of the body.

The foregoing analysis of somersaulting is limited to a planar motion. When movement is considered only in the *sagittal plane* and the couple acting at the contact (free moment) is neglected, the moment at the CoM that may induce somersaulting or stumbling is determined by the GRF, in particular, by its line of action. To avoid unwanted body rotation, the GRF should not produce a moment of force about the body's CoM. In particular, during walking and running, the GRF direction and point of application change in a coordinated manner that ensures that the line of GRF action is near the CoM. Because of that, the point of force application in walking and running is sometimes called a "zero-moment point." The term does not mean that the moment at the CoM is zero in three dimensions. For example, during the double-support period in walking, the horizontal components of the GRFs produce a couple acting in the horizontal plane (figure 1.43). The couple induces a body rotation (change in angular momentum) around the vertical axis.

1.1.9.4.3 Impact Effects in Landing

In every step of walking and running and in landing after jumps and dismounts, the stance phase starts with a sharp drop of the velocity of the free foot to zero. This drop occurs during a short interval of time during which the foot and the supporting surface impinge on each other. Such an interaction is an example of an *impact*. Other examples of impacts are striking movements in racket sports and collision with an obstacle during a falling accident. Only a brief, intuitive description of impact is given here.

Impact takes place when two bodies collide with each other during a very short period of time. During the impact, the contact force is of very large magnitude. It is easy to hammer a nail into a wooden wall, whereas to produce the same effect with static loading requires thousands of newtons of force on the hammer head. During impact, Newton's laws are correct, as always, and the GRF represents the acceleration of the CoM of the body as usual. However, acceleration of the CoM should not be confused with the trunk acceleration (a common mistake among students). The influence of landing on the trunk velocity is not as profound as it is on the CoM velocity or foot velocity. The reason is that during the landing, the body parts deform (e.g., the heel pad is compressed, muscles dislocate with respect to the skeleton, etc.) and displace with respect to each other.

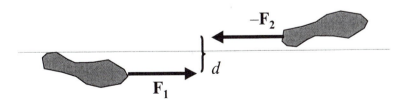

Figure 1.43 In walking, the GRFs produce a couple in the horizontal plane during the double-support period.

It is known from classical mechanics that the effect of the collision of a moving body with an unmovable obstacle (e.g., with a force plate firmly fixed to a massive support) depends on the body's linear momentum, that is, on the product of body mass and velocity. During a collision, the velocity of the body drops to zero. The impact forces in human movement depend on the momenta of both the *impactor,* that is, the body part that is in immediate contact with the environment (e.g., the foot during landing, the hand during boxing blows), and the rest of the body. During the collision, the velocity of the impactor drops to zero very fast. However, the impactor (e.g., foot, fist) usually has small mass. The rest of the body has large mass, but its deceleration during the collision is usually much smaller. The relative contribution of the two momenta into the impact force depends on the body position and the relative displacement of the body parts during the collision.

In walking and running, the velocity of the foot during landing determines some features of the GRF, in particular, the peaks observed during the first 30 to 50 ms of ground contact. For instance, during walking, an initial positive spike in the horizontal GRF is often seen (figure 1.44). The spike is in the opposite direction of the braking force, which dominates the early stance phase. The spike is due to the backward motion of the rear edge of the shoe, which creates a forward-directed friction force. Hence, although the whole body is moving in the direction of progression, immediately before the heel strike the foot can move in the backward direction. The rearward foot motion induces the impact force.

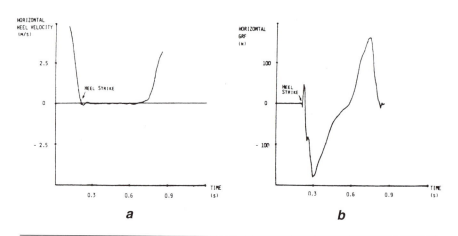

Figure 1.44 (*a*) Horizontal velocity of the heel rear end and (*b*) the horizontal fore-aft GRF.

Reprinted, by permission, from H. Lanshammar and L. Strandberg, 1983, Horizontal floor reaction forces and heel movement during the initial stance phase. In *Biomechanics VIII-B,* edited by H. Matsui and K. Kobayashi (Champaign, IL: Human Kinetics), 1125.

> ■ ■ ■ *FROM THE LITERATURE* ■ ■ ■
>
> ## Vertical Impact Force in Heel-Toe Running
>
> Source: Bobbert, M.F., M.R. Yeadon, and B.M. Nigg. 1992. Mechanical analysis of the landing phase in heel-toe running. *J. Biomech.* 25 (3): 223–34.
>
> The landing phase in trained heel-toe runners running at their preferred speed and style was studied (figure 1.45*a*). The vertical component of the GRF vector reaches a first peak (impact peak) approximately 25 ms after the touchdown (figure 1.45*b*). Thereafter, the vertical force decreases and then increases again. The origin of the impact peak is in the upward acceleration of the support leg segments; the support leg segments collide with the ground. The segmental contribution of the support leg to the vertical GRF is maximal at approximately 25 ms after the heel strike; it almost disappears at 50 ms (figure 1.45*c*). Rotation of the support leg reduces the vertical acceleration of the upper part of the body and thereby decreases the contribution of the upper part of the body to the impact force and its magnitude (figure 1.45*d*). Note that until 0.055 s, the acceleration vectors pertaining to the head and sternum markers were so small that they do not even appear in the figure. The magnitude of the impact peak correlates significantly with the downward velocity of the ankle joint (figure 1.45*e*).

Padding decreases the forces between the colliding bodies by increasing the time of collision. Heel-contact-first running on turf (stiffness below 100 kN/m) induces much smaller impact forces than running on concrete (stiffness around 4,500 kN/m). Consequently, joggers are less prone to injuries when they run on grassy fields than on city pavements.

1.1.10 Forces in Grasping

Representative papers: Kinoshita et al. (1995); Li et al. (1998)

When a human holds and manipulates an external object, such as a tool or a glass of liquid, contact forces are exerted by the palm of the hand and the individual fingers. Each finger produces a six-component wrench, a generalized contact force. All the wrenches combined produce a resultant wrench that should correspond to the requirements of the task. For instance, to hold a glass filled with liquid without spilling, rotation of the glass should be prevented. To do that, the torque of the resultant wrench must be equal to zero. Hence, when the glass does not move, the forces and moments from the different fingers

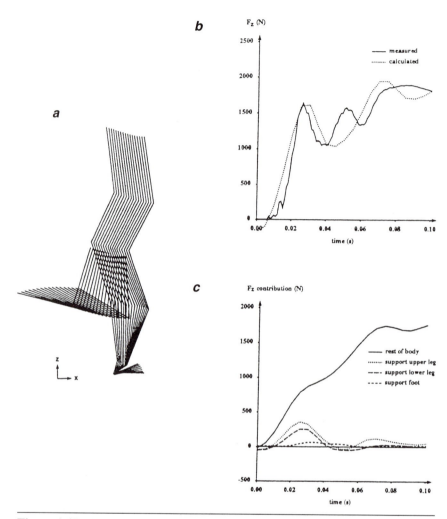

Figure 1.45 Impact forces in heel-toe running. (*a*) Consecutive positions of the body during the first part of the support period in heel-toe running. The leftmost figures show the segment orientation on the last frame before touchdown. The time between two subsequent stick figures is 5 ms. (*b*) Vertical component of the GRF. (*c*) Segmental contribution to the vertical GRF. (*d*) Acceleration vectors. The vectors are shown for the markers at the support-leg ankle, support-leg knee, support-leg hip, sternum, and head. For clarity, the stick figures have been separated, and the swing leg has been deleted. (*e*) Contributions of the support lower leg to the first peak in the vertical GRF as a function of the maximal downward velocity of the ankle joint.

Reprinted from *Journal of Biomechanics*, 25 (3), M.F. Bobbert, M.R. Yeadon, and B.M. Nigg, Mechanical analysis of the landing phase in heel-toe running, 223-234, 1992, with permission from Elsevier Science.

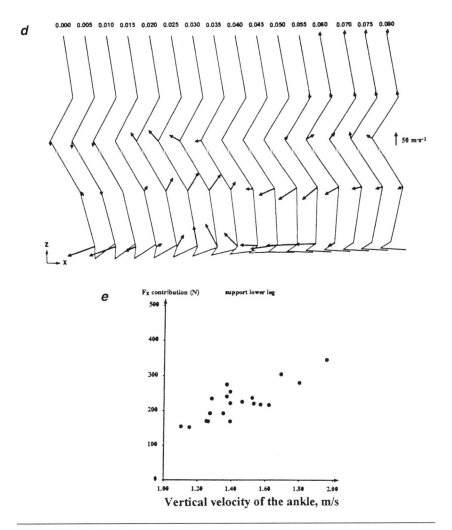

Figure 1.45 *(continued)*

cancel each other. Note that the person's individual fingers are free to exert different forces and moments to produce the correct resultant wrench. When grasping an object, the fingers work in parallel; they represent a *parallel manipulator* and form closed kinematic chains. Other examples of parallel manipulators in the human body are the two arms when manipulating an object, the two legs when performing a takeoff, and several muscles that serve the same joint.

Any grasp constrains the movement of the handheld object. For instance, an object that is pinched by the thumb and the index finger can rotate around the axis through the points of the contacts, provided that the friction is low. A

How Do People Hold an Object in the Hand?

Source: Li, Z.M., M.L. Latash, K.M. Newell, and V.M. Zatsiorsky. 1998. Motor redundancy during maximal voluntary contraction in four-finger tasks. *Exp. Brain Res.* 122 (1): 71–77.

Experimental subjects produced maximal voluntary contractions (MVCs) while holding an instrumented handle (figure 1.46) with the thumb at seven different locations—from opposite the index finger, L0, to opposite the little finger, L6. They also produced MVCs while holding the handle with the thumb at a self-selected comfortable location and while pressing the four fingers against a secured handle. The maximal force in single-finger tasks was also measured.

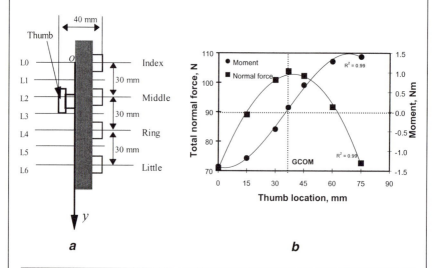

Figure 1.46   Forces exerted on the handle. *(a)* The instrumented handle. The dimensions of the handle and the seven locations (L0 to L6) for the thumb. The coordinate axis *y* originates from the level of the index finger. Force acting along the axis *x* was called the normal force, and a force along the axis *y* was called the shear force. *(b)* Total normal force and moment due to the normal force. The six data sets at abscissa of 0, 15, 30, 45, 60 and 75 mm are for the six fixed thumb locations. The vertical dotted line indicates the comfortable thumb location.

It was found that to stabilize an object in space at various thumb positions, the *central nervous system (CNS)* changes the force sharing among the fingers as a percentage of the total force and the contribution of the normal and shear forces to the total force. There exists a functional neutral line of the hand with respect to which the moment of force in a four-finger press task is zero. In the gripping tasks, when the thumb was along this line, (1) the moment of force and the shear force were zero, (2) the total normal force and the total resultant force were maximal, (3) the relative peak force—in percentage of the maximal force exerted by the digit in the single-finger test—was alike in all the fingers, and (4) this position was preferred by the subjects as the most comfortable. The data were in agreement with the *secondary moment hypothesis* (figure 1.47).

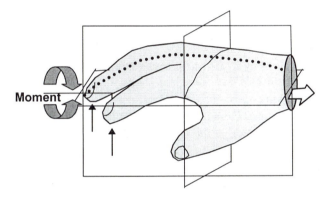

Figure 1.47 Three cardinal planes in the hand and the secondary moment hypothesis. Consider the moments acting on the hand in the segmental transversal plane, and imagine a functional, neutral axis of the hand perpendicular to the transverse plane. Any finger force that does not pass through this axis generates a moment with respect to this axis. When several fingers are acting, they produce a certain resultant moment, the *secondary moment*, which is unnecessary for the task. To maintain the equilibrium, if a secondary moment exists, it must be counterbalanced by additional muscle activity. The hypothesis assumes that the CNS is trying to minimize the secondary moment or bring it to zero in order to avoid needless muscle activity.

It seems that the CNS tends to select a pattern of muscle activation that reduces unnecessary, additional moments that act with respect to degrees of freedom that are not immediately involved in the motor task.

grasp that immobilizes the object completely is called a closure grasp, or simply *closure*. A closure imposes six constraints on the object; it is a zero-freedom contact. Closures are classified as either *form closures* or *force closures*. In a form closure, the movement of an object is completely restricted by the fingers; the object cannot move without changing the finger configuration. An example is grasping a small object, such as a ball for table tennis, in the hand. Force closures are grasps that can be broken under external forces without changing the finger position. In the discussion that follows, we deal with force closures. An example is an object touched by two contact points with friction. In the absence of friction and sticking, a minimum of seven contact points is necessary to completely restrain an object in three-dimensional space. A minimum of four contact points is needed to constrain an object in a plane without friction and sticking.

1.1.10.1 Two-Point Contacts

Representative paper: Wells and Carnahan (1998)

If a rigid object is held by a two-finger contact (pinch), as in figure 1.48*a*, and the weight of the object is neglected, the action of one contact wrench $\mathbf{w}_1$ is counterbalanced by an equal and opposite wrench $\mathbf{w}_2$. The magnitude of two wrenches can be changed by the same amount without disturbing the object's equilibrium. The magnitude of the wrenches affects the internal stress of the handheld object, but if the wrenches cancel each other, no motion occurs. The counterbalancing effect holds even if other forces are acting on the object. When the object's weight is taken into account, there are three forces that should be considered: two finger forces and the gravity force. It is known from me-

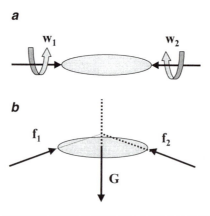

Figure 1.48 Two fingers exert the wrenches/forces on a body. See explanation in the text.

chanics that for a body to be at equilibrium under the action of three nonparallel forces, the forces must be concurrent. Furthermore, to cancel each other, the three forces must lie in a plane. If the forces are not meeting at the same point, then one of the forces would exert a moment about the point of interception of the other two forces, and the equilibrium would not be maintained. Hence, the three forces intercept at one point (figure 1.48*b*).

When several fingers exert wrenches on an object, the situation becomes complex. For instance, if an object is manipulated by five digits, each of the digits exerts a six-component wrench on the object. In total, 30 parameters are represented in five reference frames. When an object is manipulated by two hands with all ten fingers and external forces act on the object, the situation becomes even more complicated. To analyze such complex cases, Plücker's coordinates and grip matrices are used. The main ideas of this approach are presented next.

1.1.10.2 Plücker's Coordinates

Plücker's coordinates, also called *line coordinates,* are convenient for designating vectors and lines. The coordinates are named after Julius Plücker (1801–1868), a German scientist. We introduce them for sliding vectors (directed line segments). Recall that when a force is acting on a rigid body, the force can be considered a sliding vector (the principle of transmissibility). First the coordinates for a planar case are introduced and then those for three-dimensional space.

In a plane, Plücker's coordinates are an ordered triple of real numbers $\{L, M; R\}$. L and M have the dimension of length, while R has the dimension of length squared (area). Because the dimensions are not the same, a semicolon is used to separate R from L and M. The geometric meaning of Plücker's line coordinates is as follows (figure 1.49): L and M are projections of vector $\mathbf{F}$ onto the X and Y axes, respectively ($L = X_2 - X_1$ and $M = Y_2 - Y_1$). When $F = 1$, $L = \cos\theta$ and $M = \sin\theta$. R equals the magnitude of the vector product:

$$R = |\mathbf{r} \times \mathbf{F}| \tag{1.62}$$

where $\mathbf{r}$ is any vector from O to any point on the straight line passing through points 1 and 2 and the vector $\mathbf{F}$ can be located anywhere along the line. If $\mathbf{F}$ is force, R is the magnitude of the moment of force $\mathbf{F}$ about origin O.

The coordinates $\{L, M\}$ determine the magnitude of the vector $\mathbf{F}$, $F = \sqrt{L^2 + M^2}$. The triple $\{L, M; R\}$ defines the line 1-2 and the direction of this line. Translation of vector $\mathbf{F}$ along the line does not change the triple. Note that the Plücker line coordinates $\{L, M; R\}$ do not define the position of vector $\mathbf{F}$ along the line (if $\mathbf{F}$ is force, Plücker's coordinates do not define exactly the point of force application). Plücker's coordinates are convenient because they

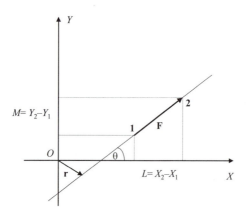

Figure 1.49 The Plücker line coordinates. Vector **F** is from point 1 to point 2.

are homogeneous: multiplication of a triple $\{L, M; R\}$ by a nonzero scalar λ yields the same line. The triples $\{L, M; R\}$ and $\{\lambda L, \lambda M; \lambda R\}$ define the vectors of different magnitude along the same line. Such vectors are called line-bound vectors.

The line-bound vector $\{L, M; R\}$ can be represented by three 2×2 determinants of the matrix

$$\begin{bmatrix} 1 & X_1 & Y_1 \\ 1 & X_2 & Y_2 \end{bmatrix} \tag{1.63}$$

obtained by deleting each of the three columns in turn. Hence,

$$L = \begin{vmatrix} 1 & X_1 \\ 1 & X_2 \end{vmatrix}, \qquad M = \begin{vmatrix} 1 & Y_1 \\ 1 & Y_2 \end{vmatrix}, \text{ and } \qquad R = \begin{vmatrix} X_1 & Y_1 \\ X_2 & Y_2 \end{vmatrix} \tag{1.64}$$

In three-dimensional space, Plücker's coordinates are an ordered sequence of six real numbers $\{L, M, N; P, Q, R\}$. For a line segment of a unit length, the first three Plücker's coordinates, L, M, and N, are simply direction cosines of the line. For a line segment of a length λ, these coordinates equal the product $\lambda \cos\theta_i$, where θ_i is a direction angle, that is, the angle that the line makes with each of the coordinate axes ($i = 1, 2, 3$). The coordinates P, Q, and R are proportional to the cross product of a vector from the origin to any point on the line and a unit vector directed along the line (see equation 1.62). They are projections of the moment of the line about the origin of the reference system onto the reference axes (figure 1.50).

1.1.10.3 Wrenches in Plücker's Coordinates

Plücker's coordinates are convenient for working with screws (lines with pitches) and in particular for working with wrenches. A wrench can be characterized

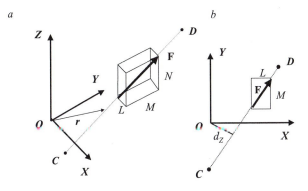

Figure 1.50 (*a*) Plücker's coordinates of the line segment/vector **F** in the three-dimensional space and (*b*) their projections onto the *O-XY* plane. Coordinates *L*, *M*, and *N* are projections of the line segment onto the reference axes. When the line segment has a unit length, the coordinates are simply the direction cosines of the corresponding angles. Coordinates *P*, *Q*, and *R* (not shown in the figure) are projections of the cross product of vectors **r** and **F** (equation 1.62) onto the reference axes. They are equal to the moments made by the vector **F** at the origin of the reference frame. For instance, a moment around the *Z* axis computed in the *O-XY* plane is $R = d_z F$, where d_z is the shortest distance (normal) from *O* to the line *CD* in the *O-XY* plane. The moment represents the sixth Plücker's coordinate, *R*.

by three components of the contact force, $\mathbf{F}_x$, $\mathbf{F}_y$, and $\mathbf{F}_z$, and three components of the free moment acting at the contact, $\mathbf{M}_x$, $\mathbf{M}_y$, and $\mathbf{M}_z$ (see **1.1.7**). Note that the free moment components can be represented as $p_w \mathbf{F}_x$, $p_w \mathbf{F}_y$, and $p_w \mathbf{F}_z$, where p_w is the pitch of the wrench. However, these components cannot be used as coordinates of the wrench because they do not provide necessary information about the location of the wrench with respect to the origin of the reference system. Any screw, such as a wrench or a twist, can be represented by a coordinate system known as the *screw coordinates*. We introduce the screw coordinates for a wrench. In the screw coordinates, the unit wrench $\hat{\mathbf{w}}$ is defined by a pair of three-component vectors (or by a six-component vector):

$$\hat{\mathbf{w}} = \begin{bmatrix} \hat{\mathbf{F}} \\ \mathbf{r} \times \hat{\mathbf{F}} + p_w \hat{\mathbf{F}} \end{bmatrix} = \begin{bmatrix} \hat{F}_1 \\ \hat{F}_2 \\ \hat{F}_3 \\ \hat{T}_1 \\ \hat{T}_2 \\ \hat{T}_3 \end{bmatrix} \qquad (1.65)$$

where $\hat{\mathbf{F}}$ is a unit force (a unit vector pointing in the direction of the screw axis), $\mathbf{r}$ is the position vector of any point on the screw axis, and p_w is the pitch. The vector $\mathbf{r} \times \hat{\mathbf{F}}$ defines the moment of the force $\hat{\mathbf{F}}$ about the origin of the reference system with the coordinate axes 1, 2, and 3. The product $p_w \hat{\mathbf{F}}$ defines the free moment (effect of the couple) at the area of the contact. Hence the vector

$$\mathbf{T} = \mathbf{r} \times \hat{\mathbf{F}} + p_w \hat{\mathbf{F}} \qquad (1.66)$$

represents the moment at the origin of the reference system that is due to the wrench force $\hat{\mathbf{F}}$ and the wrench moment $p_w \hat{\mathbf{F}}$. (To avoid confusion with one of the Plücker coordinates, we designate this moment by a letter $\mathbf{T}$ instead of the usual $\mathbf{M}$.) The relationship between the screw coordinates and the Plücker coordinates of a unit wrench is

$$
\begin{aligned}
L &= \hat{F}_1 \\
M &= \hat{F}_2 \\
N &= \hat{F}_3 \\
P &= T_1 - p\hat{F}_1 \\
Q &= T_2 - p\hat{F}_2 \\
R &= T_3 - p\hat{F}_3
\end{aligned}
\qquad (1.67)
$$

From equation 1.67, it follows that the coordinates P, Q, and R represent the moment of force $\hat{\mathbf{F}}$ about the origin of the reference system. Because vectors of a force and its moment are always perpendicular to each other, the following equation is valid:

$$\mathbf{F} \cdot (\mathbf{r} \times \mathbf{F}) = LP + MQ + NR = 0 \qquad (1.68)$$

Hence, only five of the six Plücker's coordinates are independent. Following are advantages of representing a wrench in Plücker's coordinates:

• The coordinates are homogeneous: multiplication of all six of the coordinates by any nonzero scalar does not change the orientation of the screw line. Therefore, proportional increase or decrease of the force does not change the wrench itself; only the wrench intensity changes.

• When several wrenches are acting on an object, the Plücker coordinates of the resultant wrench equal the sum of coordinates of individual wrenches, that is,

$$\Sigma w_i = (\Sigma L_i, \Sigma M_i, \Sigma N_i; \Sigma P_i, \Sigma Q_i, \Sigma R_i) \qquad (1.69a)$$

Equation 1.69a does not account for the moments of couples $p_i \hat{F}_i$ acting at the individual contact areas. To account for them, the equation should be expressed in screw coordinates:

$$\Sigma \mathbf{w}_i = [\Sigma L_i, \Sigma M_i, \Sigma N_i; \Sigma(P_i + p_i \hat{F}_{i1}),$$
$$\Sigma(Q_i + p_i \hat{F}_{i2}), \Sigma(R_i + p_i \hat{F}_{i3})] \qquad (1.69b)$$
$$= \Sigma L_i, \Sigma M_i, \Sigma N_i; \Sigma T_{i1}, \Sigma T_{i2}, \Sigma T_{i3})$$

where the subscript i refers to an individual wrench and the subscripts 1, 2, and 3 refer to coordinate axes. The remarkable simplicity of the Plücker and screw coordinates can be used to analyze the grip forces exerted by several fingers.

1.1.10.4 Multifinger Contacts: Grip Matrices

In multifinger grasps, each finger exerts a wrench $\mathbf{w}_i$ on the grasped object. The corresponding unit wrenches are $\hat{\mathbf{w}}_i$. Each wrench has a given intensity F_i. Such a system of unit wrenches and wrench intensities is called the *grip, G*. The grip G generates a resultant wrench $\mathbf{W}$ that acts on the handheld object:

$$\mathbf{W} = \sum_{i=1}^{n} F_i \hat{\mathbf{w}}_i \qquad (1.70)$$

In matrix notation, equation 1.70 is written as

$$\mathbf{W} = \left[\hat{W}\right] \mathbf{F}_i \qquad (1.71)$$

where $[\hat{W}]$ is a $6 \times n$ *grip matrix* whose columns are the unit wrenches $\hat{\mathbf{w}}_i$ of the grip G and $\mathbf{F}_i$ is the vector of the wrench intensities. In multifinger manipulation, matrix $[\hat{W}]$ plays a role similar to the role played by the *Jacobian* matrix $\mathbf{J}$ for multilink open chains in kinematics. In kinematics, the Jacobian relates the vector of joint angular velocities to the velocity of the end effector. In a similar way, the grip matrix relates the finger-contact wrenches with the resultant wrench exerted on the object.

1.2 FRICTION FORCES AND AIR RESISTANCE

As already mentioned, force acting at a point of contact can be resolved into a normal component, $\mathbf{F}_n$, and a tangential component, $\mathbf{F}_t$. The component of the contact force between two bodies that lies in the tangent plane of the contact point is called the *friction force*. Assuming that contacting bodies are rigid and do not interpenetrate, whether and how contacting surfaces move with respect to each other depends on the friction between the contacting bodies. Friction variously acts as friend and enemy in human movement. Without sufficiently large friction between the soles and the ground, we would not be able to walk or run. A slipping accident after stepping on an icy patch is an example of what may happen if the friction is too low for ambulating or is underestimated. The tangential force of friction is always opposite the direction of motion. Frictional force is not conservative; energy lost due to friction dissipates as heat. In society, the amount of energy lost by friction in various machines is tremendous; it is about one-third of the present energy production in the world (Moor 1975).

1.2.1 Friction Angle and Friction Cone

During sliding, the tangential friction force F_t is a product of the *coefficient of dynamic friction*, μ_d, and the normal force F_n:

$$F_t = \mu_d F_n \tag{1.72}$$

Equation 1.72 is credited to French scientists G. Amonton (1663–1705) and C.A. de Coulomb (1736–1806). It is called *Amonton's law* or *Coulomb's law.* According to Amonton's law, the force of friction is directly proportional to the load and does not depend on the contact area or the skidding velocity. Amonton formulated the law in 1699. Coulomb explained friction in terms of cobblestones in a rough road: the bigger the stones, the higher the friction. According to Coulomb's theory, which was later confirmed, the friction is determined by the area of real contact. On the microscopic scale, even the smoothest surface is not perfectly smooth; it has peaks and hollows. The contact occurs at the tips of peaks only. The area of real contact is a very small proportion of the apparent area of the contact between the surfaces. As the load increases, the peaks crush down until the area of contact is sufficient to support the load. The coefficient of friction depends on the contacting materials. For many materials, the area of real contact is proportional to the load. Therefore, the friction is also proportional to the load. Coulomb was the first to distinguish *dynamic* and *static friction.*

If the contacting surfaces do not move with respect to each other, the *coefficient of static friction*, μ_s, is used. When there is no motion, the tangential force, F_t, is smaller than the product of the normal force, F_n, and the coefficient of static friction: $F_t < F_n \mu_s$. The coefficient of friction, dynamic or static, can be interpreted geometrically as the tangent of an angle formed by the normal force as an adjacent side and the tangential force as an opposite side, $\mu = F_t/F_n$ (figure 1.51a). The angle $\theta = \tan^{-1} (F_t/F_n)$ at which $F_t = F_n \mu_s$ is called the *friction angle.* If $F_t \leq F_n \mu_s$, the contacting surfaces glide with respect to each other. As long as the applied force is within the friction angle, there will be no slippage. In three-dimensional space, the friction angles form a *friction cone* at a point on a surface (figure 1.51b). To avoid skidding, the acting force must be within the cone of friction.

The friction that must be overcome to set a body in motion (static friction) is larger than the friction between the same bodies when they slide with respect

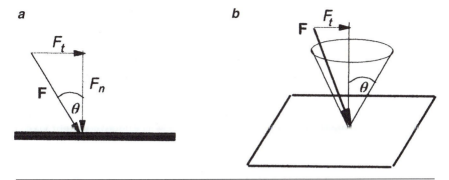

Figure 1.51   (*a*) The angle and (*b*) the cone of friction.

to each other (dynamic friction). Note that the ratio of shear force to normal force measured during movement of two surfaces defines the dynamic friction. When this ratio is measured in static conditions, it does not define the static friction and cannot be called the coefficient of static friction. In static conditions, the ratio is always smaller than or equal to the coefficient of friction. As very accurate experiments demonstrate, the coefficient of friction decreases gradually from its static to its dynamic value when the velocity slowly increases. However, in a majority of practical applications, this continuous transition is disregarded, and discontinuity of friction is admitted.

■ ■ ■ *FROM THE LITERATURE* ■ ■ ■

Optimal Angle of Cervical Spine Traction

Source: Pio, A., M. Rendina, F. Benazzo, C. Castelli, and F. Paparella. 1994. The statics of cervical traction. *J. Spinal Disord.* 7 (4): 337–42.

Cervical traction is broadly used in clinical practice. An immediate goal of the procedure is to separate the cervical vertebrae. The separation of the intervertebral foramens reduces pain and aligns the vertebral bodies. The maneuver is performed in both supine and sitting postures. When a patient is in a supine position and a pulling force is applied to the skull, resistance is provided not only by the cervical spine but also by a friction force between the head and the bed. The frictional resistance should be overcome to induce the displacement of the head. The frictional resistance depends on the coefficient of friction and the "effective weight" of the head, that is, the normal force. The effective weight depends on the angle of pull. If the pulling force is applied in a strictly horizontal direction, the effective weight is large, and a large force should consequently be applied to the head. If the pulling force is applied at a certain angle, the vertical component of the pulling force decreases the effective weight of the head. The goal of the study was to determine the optimal angle of the pull to achieve head displacement with a minimal pulling force.

The authors measured the effective weight of the head in 12 patients. They also measured the minimal force that caused motion of the head on the bed surface during chin traction. The static friction coefficient was determined from the ratio of the two quantities. The effective weight was equal to 4.69% ± 0.84% of the body weight. (According to the data in the literature the real weight of the head and neck is 6.94% ± 0.70% on average; see chapter **4**. However, in this study, the effective weight was measured in people lying supine, and the head was supported in part by the neck structure.) The coefficient of friction was 0.62 on average, and the friction angle was about 32°. The authors concluded that the optimal angle of cervical traction corresponds approximately to the angle of static friction.

In technical devices, the sequence of events "from rest to gliding," and consequently from static to dynamic friction, is quite common. For instance, when a car engine starts working, the friction between rotating parts is initially large and then decreases. A similar transition from static to dynamic friction occurs in human joints. However, in human movement, slipping should usually be avoided (except in winter sports such as skiing and skating). Therefore, the events typically occur in the opposite sequence, with a transition from dynamic to static friction; for example, the relative movement of the foot with respect to the supporting surface comes to a stop during gait.

The concepts of static and dynamic friction are only valid as first approximations. They do not completely represent the complexity of the interaction between contacting bodies. For instance, static friction is measured when the bodies are in contact for a long period of time and might adhere to each other. If the time of contact is very short, the coefficient of friction may be different. In *tribology* (the field of mechanics that studies friction), the *kinetic coefficient of friction* is used. The kinetic coefficient depends on the sliding velocity, normal force, and the time of force application.

Similar to the contact force, a couple acting at the contact can also be resolved into two components: (1) the *couple of the twisting friction* about the normal common to the two contacting surfaces and (2) the *couple of the rolling friction* about an axis lying in the tangent plane. The couple of the twisting friction produces tangential torque about the axis normal to the surface of contact. For instance, when a figure skater performs multiple rotations on one leg, the angular momentum of the body decreases due to the twisting friction. If the contacting bodies are not sticky, the couple of the rolling friction in human movement is usually neglected. The moment of the couple of the twisting friction depends on the coefficient of friction and the area of contact. When the pressure between two contacting surfaces is uniform and the contact takes place over a full circular area of radius r, the magnitude of the friction moment M_{fr} is

$$M_{fr} = \tfrac{2}{3}\mu_d F_n r \tag{1.73}$$

where F_n is the normal force that is acting perpendicularly to the surface of the contact and μ_d is the coefficient of dynamic friction.

To characterize resistance of rolling wheels, as on bicycles, to a propelling force, the *coefficient of rolling resistance* is used. Formally, the coefficient is similar to the coefficient of friction; it designates the ratio of the horizontal force necessary to induce the wheel rotation to the weight (vertical force) that impinges on the wheel. To explain the effect under discussion, first consider a

---

■ ■ ■ *From the Literature* ■ ■ ■

Tangential Torque Effects on the Control of Grip Forces

Source: Kinoshita, H., L. Bäckström, J.R. Flanagan, and R.S. Johansson. 1997. Tangential torque effects on the control of grip forces when holding objects with a precision grip. *J. Neurophysiol.* 78: 1619–30.

When people manipulate small objects, the fingertips are usually subjected to tangential torques about the axis normal to the grasp surface. To prevent slipping, the tangential torques must be counterbalanced by twisting friction, which, according to equation 1.73, depends on the normal force, the friction coefficient, and the area of contact. The authors investigated the effects of tangential torques (as well as tangential forces) on the normal forces employed by subjects to hold an object in a stationary position with the tips of the index finger and thumb. By changing the location of the object's center of gravity in relation to the grasp surface, various levels of tangential torque (0–50 N · mm) were created while the subject counteracted object rotation. Changing the weight of the object varied tangential force (0–3.4 N). The flat grasp surfaces were covered with rayon, suede, or sandpaper, providing differences in friction in relation to the skin. Tangential torque strongly influenced the normal force required for grasp stability (figure 1.52). The employed normal force increased in proportion to tangential torque with a slope that reflected the current frictional condition.

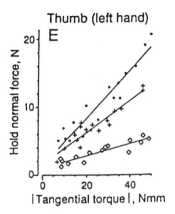

Figure 1.52 The relationship between the tangential torque and the normal force when holding a flat object with the index finger and thumb. Dots represent a rayon surface, plus signs represent suede, and circles represent sandpaper.

Reprinted, by permission, from H. Kinoshita et al., 1997, "Tangential torque effects on the control of grip forces when holding objects with a precision grip," *Journal of Neurophysiology* 78:1619–1630.

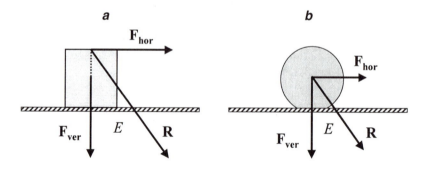

Figure 1.53 Mechanics of rolling resistance. (*a*) A block under the influence of weight and a horizontal force. The block will tip about the edge *E* if the resultant force intersects the ground outside the edge. (*b*) A wheel deforms under pressure and behaves like a block.

block with the forces acting on it (figure 1.53*a*). The forces are the weight of the block $\mathbf{F}_{ver}$, the horizontal pulling force $\mathbf{F}_{hor}$, and their resultant force $\mathbf{R}$.

It is immediately seen from figure 1.53*a* that (1) if the magnitude of the force F_{hor} is smaller than the product of the vertical force F_{ver} and the coefficient of static friction, $F_{hor} < F_{ver}\mu_s$, the object will stay at rest; (2) if $F_{hor} > F_{ver}\mu_s$ and the resultant force $\mathbf{R}$ intersects the ground within the area of support, the block will slide; and (3) if **R** intersects the supporting surface outside the area of support, the block will tip. The tipping depends on the magnitudes F_{ver} and F_{hor}, the height of the point of application of the horizontal force, and the horizontal dimension of the block. When a wheel (as of a bicycle) contacts the ground, the wheel flattens locally, and the ground gives slightly (figure 1.53b). The wheel contacts the ground not at a point, as it would if the wheel and the road were made from infinitely hard materials, but over a certain area. As a result, while rolling, the wheel "tips" over the leading edge of its ground contact. The wheel's tipping depends on the radius of the wheel and the dimension of the contact area, which in turn depends on the hardness of the ground and the tires (or air pressure in the tires). The larger the radius of the wheel and the higher the pressure in the tires, the smaller the rolling resistance. To obtain the coefficient of rolling resistance, the magnitude of the resistance should be divided by the radius of the wheel. The coefficient is measured in units of length. As we can see, the coefficient of rolling resistance has a little to do with friction.

1.2.2 Friction Forces in Human Movement

Friction forces in human movement do not exactly obey Amonton's law (equation 1.72). Often, the friction depends on movement velocity and contact area. In some motor tasks, for instance, walking, the applicability of Amonton's law is debatable.

1.2.2.1 Walking and Running

Representative paper: Lanshammar and Strandberg (1983)

During walking and running, slipping should be prevented. The supporting foot does not usually displace relative to the supporting surface. During the support period, the supporting foot exerts force on the ground at different angles (figure 1.54a). The force can be represented by its horizontal and vertical components, $\mathbf{F}_{hor}$ and $\mathbf{F}_{ver}$. To prevent slipping, the ratio F_{hor}/F_{ver} should be smaller than the coefficient of static friction, μ_s. The risk of skidding (RS) can be estimated from the equation

$$RS = \frac{F_{hor}}{F_{ver}} - \mu_s \qquad (1.74)$$

During the support phase, neither F_{hor}/F_{ver} nor μ_s is constant. An example of the alteration in F_{hor}/F_{ver} during gait is presented in figure 1.54b.

From figure 1.54, it is evident that two periods are the most unsafe: the beginning and the end of the support. Slipping at the beginning of the support period may result in falling down on the back and a serious injury.

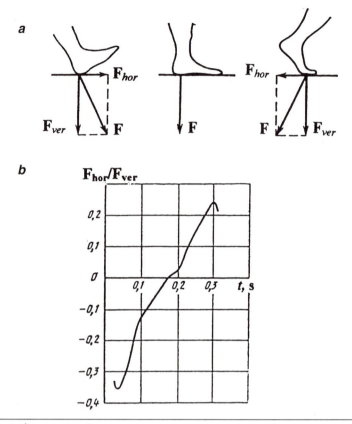

Figure 1.54 (*a*) The direction of the GRF and (*b*) the ratio F_{hor}/F_{ver} during a single support period in walking. The subject was a young, healthy man.

Reprinted from A.S. Aruin and V.M. Zatsiorsky, 1989, *Occupational biomechanics* (Moscow: Mashinostroenie Publishing House [In Russian]).

All the friction estimates presented so far are for the static friction. The numbers are based on the postulate that during the support phase, the foot does not displace with respect to the ground. However, in walking and running, the foot approaches the surface with a certain velocity. Also, most slipping accidents occur on "lubricated" surfaces, for instance, on a concrete floor covered with a film of water. It takes some time for the foot to travel downward from the first fluid contact to the hard surface. This time is very short, but it is exactly during this period that slipping can start. Hence, not static friction but so-called *kinetic friction,* which depends on the movement velocity, is acting.

Measurement of kinetic friction is a challenging task. The measurement method should duplicate human gait as closely as possible: the velocity, force, shape, and area of the contacting bodies should mimic natural conditions as in

■ ■ ■ *From the Literature* ■ ■ ■

Preventing Slipping Accidents

It is generally accepted that, to prevent slipping accidents, the coefficient of friction between the heel of the shoe, especially at its rear edge, and the floor should be not less than 0.3 for slow walking and 0.4 for fast gait (Carlsöö 1975). For amputees who use crutches, the coefficient of friction should be above 0.65 (Buczek et al. 1990). Unfortunately, in some conditions, especially in industrial settings, the requirement for high friction is not easy to satisfy (table 1.1).

Table 1.1   Coefficients of Friction Between Rubber Heels and a Floor Soiled by Different Substances

Floor	Coefficient of friction	Friction angle[a]	Risk of slipping
Dry	0.65	33.8°	No risk
With layer of water	0.50	26.6°	No risk
Smeared with brawn jelly[b]	0.25	14.0°	Risk
Moistened with glycerin	0.12	7.0°	Major risk
Smeared with minced fat	0.12	7.0°	Major risk

[a]The angle is measured from the vertical. At a slightly larger angle, slipping would occur.
[b]Brawn is cooked boar's flesh.
The data are from Carlsöö (1975, p. 178).

In athletics, friction between the shoes and supporting surface should satisfy two opposing requirements. It should be large enough to prevent skidding, and it should be small enough to prevent the overloading of the leg tissues, particularly the ankle and the knee joint, during fast stops. In sports in which the athletes stop suddenly (e.g., basketball, football), excessive friction increases the risk of injury (Torg, Quedenfeld, and Landau 1974; Bonstingl et al. 1975; Rheinstein, Morehouse, and Niebel 1978). The following friction values are recommended for athletics (track and field events): more than 0.5 for wet surfaces and more than 0.8 for dry surfaces, for sport games more than 0.5 for wet surfaces and not less than 1.1 for dry surfaces (Deutsche Normen 1978). In some sports (e.g., soccer), in addition to linear foot-on-surface motion, foot rotation occurs. To measure friction resistance during rotation, a torsional traction test is used (Canaway and Bell 1986). The mean rotational friction torque during pivoting (rotation on one leg) on a number of playing surfaces is about 30 N · m (reviewed by Bell, Baker, and Canaway 1985).

■ ■ ■ *FROM THE LITERATURE* ■ ■ ■

How Do Slipping Accidents Occur?

Source: Strandberg, L., and H. Lanshammar. 1983. On the biomechanics of slipping accidents. In *Biomechanics VIII-A,* ed. H. Matsui and K. Kobayashi, 397–402. Champaign, IL: Human Kinetics.

According to official statistics, more people are killed in falls than in motor vehicle accidents in Sweden. In the experiments, a level track was covered with stainless steel. The force plate in the center of the track was either lubricated with soap solution or cleaned between the trials without the subject's knowledge. A rail-suspended safety harness was used to prevent falls and injuries. Loss of balance occurred only when the foot slid forward in the beginning of the stance phase. Skidding occurred in 39 trials of 124 (figure 1.55). Sliding started on average 50 ms after the heel strike. At this instant in time, the vertical GRF was on average 60% of the body weight, the shoe's angle of contact was 6°, and the coefficient of friction was only 0.09 (Strandberg and Lanshammar 1981, 1983). The authors concluded that kinetic friction (not static friction) acts during the heel strike in walking.

walking. In a comparison of 27 different methods of measuring walking slipperiness (the methods selected were the best of 70 different skid tests reported in the literature), very low correlations were found between many of them (Lanshammar and Strandberg 1985). The authors concluded that the static coefficient of friction seems to be an irrelevant quantity for predicting slipping accidents. They suggested an indirect measure of friction during human gait, the time-based estimate of the friction utilization (TFU):

$$TFU = k/t^2 \qquad (1.75)$$

where t is the time of traveling around a closed triangular path as fast as possible and k is a constant. The larger the TFU (the smaller the lap time), the higher the friction (Lanshammar and Strandberg 1985).

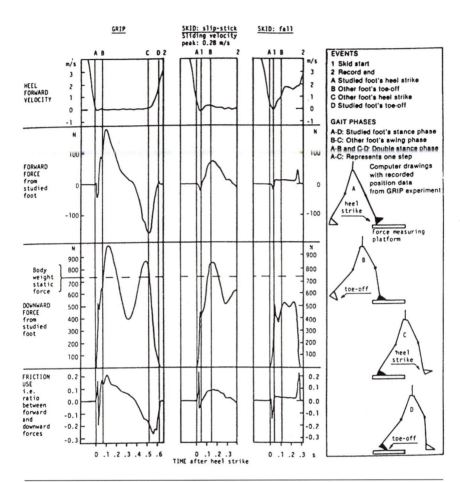

Figure 1.55 Skidding at the beginning of the support phase in walking. The results of three trials by the same subject with different outcomes are presented. The trials were classified as skids if the foot-sliding velocity exceeded 0.5 m/s and the sliding distance was more than 4 mm. Among the 39 skid trials, 23 resulted in a "fall" (subject was restrained by a safety harness). In the remaining 16 trials, called "slip-stick," the subjects regained their balance and continued walking. The remaining 85 trials were called "grips."

Adapted, by permission, from L. Strandberg and H. Lanshammar, 1983, On the biomechanics of slipping accidents. In *Biomechanics VII-A*, edited by H. Matsui and K. Kobayashi (Champaign, IL: Human Kinetics), 399.

Sufficiently large friction between the shoe and ground is essential for walking and running. However, excessive friction between the foot and footwear can result in blisters. Resistance inside the shoe should be provided mainly by the shoe upper.

1.2.2.2 Winter Sports: Skating, Skiing, Bobsledding

Representative papers: Smith (2000); Koning and van Ingen Schenau (2000)

In winter sports based on gliding (e.g., skating, skiing, bobsledding, luge), friction resistance to the athlete's advancement is low. In general, low friction is observed when the two skidding surfaces are separated by a lubricant, a low-viscosity fluid. The most popular theory for the low values of the retarding force is that the blade or ski movement causes frictional heating that is adequate to melt some of the ice or snow. The skate or ski is thus moving over a thin film of melted water. The water serves as a lubricant. The rate of heat production is equal to the *mechanical energy* loss due to friction:

$$\text{Heat production (in } watts\text{)} = \mu_d F_{\text{nor}} v \qquad (1.76)$$

where μ_d is the coefficient of friction, F_{nor} is the force normal to the surface of gliding, and v is the gliding velocity. The amount of melted ice or snow depends on the temperature and on the distribution of the produced heat between the blade or ski and the ice or snow. The more heat that goes to the surface substance, the better. The coefficient of friction heavily depends on weather conditions, in particular, on the ice and snow temperature. Therefore, during outdoor competitions, if the temperature is not constant, some athletes may compete in better conditions than others. Heating the blades sharply decreases the friction and improves performance. In bobsledding and luge, blade heating is forbidden, and blade temperature is measured immediately before the start.

▪▪▪ *From the Literature* ▪▪▪

Friction in Skating

Source: Koning, J.J. de, G. de Groot, and G.J. van Ingen Schenau. 1992. Ice friction during speed skating. *J. Biomech.* 25: 565–71.

In skating, the coefficient of friction μ_d, when measured on well-prepared ice and with good skate blades, was in the range 0.003 to 0.007 during the straights. This resistance corresponds to friction force of 3 to 6 N. On the curves, the coefficient of friction was on average 35% higher than on the straights (Jobse et al. 1990). Even smallest changes in μ_d result in a visible difference in performance. For instance, an increase in μ_d from 0.004 to 0.006 increases the performance time in a 500-m sprint by 0.8 s and in a 10,000-m race by 23.5 s. These differences are larger than the difference between the sixth-place performance at the Olympic Games and the gold-medal performance.

The coefficient of friction between the blade and ice is not constant; it depends on the ice temperature (figure 1.56*a*) and increases with skating velocity (figure 1.56*b*). Hence, in speed skating, Amonton's law does not hold.

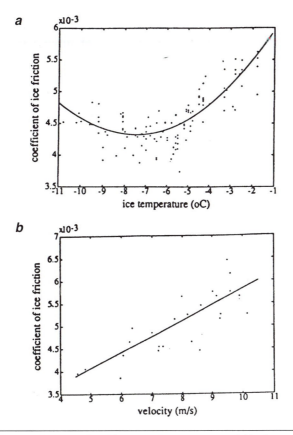

Figure 1.56 (*a*) Relationship between the coefficient of friction and ice temperature during skating the straight parts of the track with a velocity of 8 m/s. (*b*) Relationship between the coefficient of friction and skating velocity during skating the straight parts of the track at an ice temperature of –4.6°C.

Reprinted from *Journal of Biomechanics*, 25 (8), J.J. de Koning, G. de Groot, and G.J. van Ingen Schenau, Ice friction during speed skating, 565–571, 1992, with permission from Elsevier Science.

In skiing, three friction values are particularly important: the friction of (1) the boot over the ski, (2) the ski over the snow during ski advancement, and (3) the ski over the snow during propulsion. Boot–ski friction should be low so as not to prevent the release of the boot by the binding during risky situations. To prevent injury, the boot must be released at safe loads. The friction in downhill skiing is approximately 10 times larger than in skating and depends on both the quality of the ski surface and the snow conditions. To decrease friction, the ski surface is usually made of low-friction materials, mainly Teflon. The most difficult part of ski construction is to find a compromise between the low friction required for ski advancement and the high friction needed to avoid ski backsliding during push-offs. The higher the ski friction over the snow during the propulsion phase, the better the skier can propel himself or herself and the steeper the incline that can be traveled without putting the skis at an angle (the herringbone technique) when moving uphill. Several creative engineering solutions have been suggested to solve these problems. As a result, contemporary skis, for both alpine and cross-country skiing, are much better now than they were a quarter of a century ago.

1.2.3 Air Resistance

Representative papers: Pugh (1971); Davies (1980a, b)

Due to interaction with the air, human bodies experience considerable forces while moving at high speeds or in windy weather. A brief account on the role of air resistance in human movement is given here. The resistive force exerted on a body along the airflow is called *aerodynamic drag,* or simply *drag.* Drag works to slow the forward motion of a body. Two main sources contribute to the total drag: friction drag and form drag. Friction drag is caused by the movement of the air over the body surface. The smoother the surface, the smaller the friction drag. Form drag depends on the shape of the body. A streamlined shape generates less form drag than a flat body. The drag force F_{drag} equals

$$F_{\text{drag}} = \frac{1}{2}\rho V^2 A C_d \qquad (1.77)$$

where F_{drag} is drag force; ρ is the air density in kilograms per cubic meter, which depends on the ambient temperature and the barometric pressure, that itself depends on the altitude; V is the body's velocity with respect to the air in meters per second; A is the area of the body projected on a plane perpendicular to the airflow in square meters; and C_d is a dimensionless drag coefficient that represents both the friction drag and the form drag. For the human body, C_d depends on the clothes worn and body posture, and it varies from 0.35 to 1.36. These values correspond approximately to the drag coefficients of vehicles with bad streamlining. For instance, for trucks, C_d is usually between 0.3 and 0.6. If the body posture and the drag coefficient do not change, drag is proportional to the velocity squared. For a given velocity and air pressure, drag depends on (1) body size, (2) body posture, and (3) smoothness of the body surface. Due to the variety of factors that influence drag, equation 1.77 is sel-

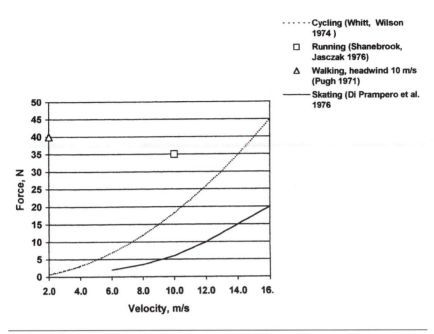

Figure 1.57 Dependence of the total drag on the movement velocity in cycling, running, walking, and skating: representative examples from the literature.

Adapted from Zatsiorsky, V.M., S.Y. Aleshinsky, and N.A. Yakunin. 1982. Biomechanical basis of endurance. Moscow: FiS.

dom used. Customarily, air resistance to human movement is determined experimentally, usually in a wind tunnel. Data on the air resistance experienced by the human body in various activities are presented in figure 1.57.

Streets around tall buildings erected in city centers are subjected to strong winds near ground level. To determine safe and comfortable conditions for pedestrians, the force of the wind on people dressed in normal outdoor clothes has been measured. The results are presented in table 1.2. The results heavily depend on the clothes worn. For instance, for females wearing trousers, C_d was larger than for skirt-wearers. The drag coefficients for females in trousers were identical to those for males.

Athletes moving at a high speed expend energy to overcome air resistance. The cost of overcoming air resistance on a calm day outdoors is about 7.8% of the total energy cost for sprinting (10 m/s), 4% for middle-distance running (6 m/s), and 2% for marathon running (5 m/s; Davies 1980b). When subjects exercise on a stationary bicycle ergometer or run on a treadmill, they do not need to overcome aerodynamic resistance, and thus they spend less energy (figure 1.58).

In sport, athletes are usually interested in decreasing drag. They do this by assuming streamlined body postures (as in downhill skiing) and wearing smooth,

Table 1.2 Wind Drag on Standing People

Posture	Projected area, m²	Wind speed, m/s	Force, N	Drag coefficient, C_d
Facing the wind	0.55	8.3	25.6	1.18
Facing the wind	0.55	5.0	10.6	1.36
Sideways to the wind	0.38	8.3	16.6	1.11
Sideways to the wind	0.38	5.0	5.6	1.05

Note. Average data from 239 males and 92 females measured in a wind tunnel.

The data are from Penwarden, A.D., P.F. Grigg, and R. Rayment. 1978. Measurement of wind drag on people standing in a wind tunnel. *Building Environ.* 13: 75-84.

low-friction clothes. Wind-tunnel tests of clothing materials, hair, and shoes show that it is possible to lower the wind resistance of a runner by about 0.5% to over 6% by improving aerodynamics (Kyle and Caiozzo 1986). In cycling, *drafting* (i.e., riding behind another athlete) is a common technique. Drafting greatly reduces aerodynamic drag and gives the rider in back an advantage through its shielding effect. The athletes riding in the front of a pack spend 30% to 40% more energy than the riders in the middle of the pack. Drafting may also be useful in middle-distance running. Running 1 m behind another runner virtually eliminates air resistance at middle-distance speeds (figure 1.59). When a competition is performed at altitude (e.g., at 2,000 m above sea level), air resistance decreases because of the decreased air density. This may result in improved performance in sports in which aerodynamic resistance plays a substantial role.

1.3 LOCAL BIOLOGICAL EFFECTS OF CONTACT FORCES

We consider first the biological effects of the compression forces and then the effects induced by the shear forces.

1.3.1 Compression Forces

Contact forces exerted on body parts induce compression of the subcutaneous soft tissues. The deformation has several biological effects. It increases the

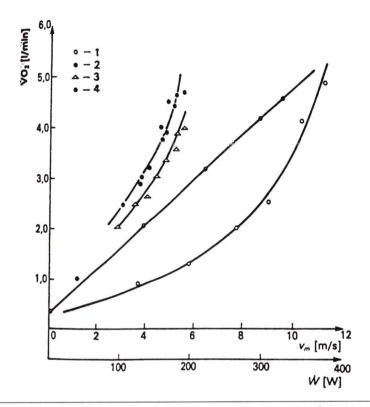

Figure 1.58 Influence of air resistance on oxygen consumption. (1) Cycling outdoors, (2) cycling on a bicycle ergometer, (3) running on a treadmill, (4) running outdoors.

Reprinted, by permission, from L. Pugh, 1971, "The influence of wind resistance in running and walking and the mechanical efficiency of work against horizontal or vertical forces," *Journal of Physiology* 213 (2):255–276.

acuity of tactile perception, such as at the fingertips. It is also important for reducing impact forces. The deformation of the heel pad during the heel strike in walking and running decreases the impact load on the bones and joints. The deformation depends on the mechanical qualities of the underlying tissues, their geometry, and other factors. The relationship between the applied force and the amount of deformation is highly nonlinear. Because of the nonlinearity, the stiffness values are not constant; they are usually reported at a given magnitude of the applied force. For instance, the stiffness values of normal heel pads are about 900 kN/m at body weight (Aerts et al. 1995). The biomechanics of deformation is not covered in this book.

An immediate consequence of high contact pressure is occlusion of the blood vessels. The routine medical procedure of blood pressure measurement is based

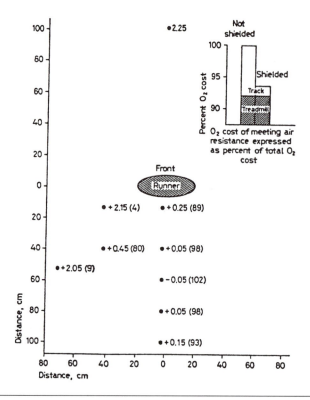

Figure 1.59 Air pressure at various distances from the runner, in kilograms of force per square meter. Observations were made at a height of 126 cm and wind speed of 6 m/s. Percentage reduction in air pressure is shown in parentheses.

Reprinted, by permission, from L. Pugh, 1971, "The influence of wind resistance in running and walking and the mechanical efficiency of work against horizontal or vertical forces," *Journal of Physiology* 213 (2):255–276.

on this fact. When the pressure in the cuff wrapped around the upper arm exceeds the systolic pressure in the brachial artery, the artery is pinched off and the blood flow is arrested. When the cuff pressure is greater than the diastolic blood pressure but smaller than the systolic pressure, the artery is constricted. The blood flow becomes turbulent, and the artery vibrates and produces "tapping" sounds known as the sounds of Korotkoff. The sounds occur because the blood flow stops during diastole and resumes during systole. In adults, the normal value of systolic blood pressure is 120 to 130 mmHg (approximately 16.0–17.3 kPa; 1 mmHg = 133.322 Pa), and the value of diastolic pressure is 60 to 80 mmHg (8.0–10.7 kPa). Therefore, any contact pressure that exceeds 60 to 80 mmHg may occlude blood flow. In reality, the magnitude of this pressure is even less because the blood pressure in capillaries is much smaller than

■ ■ ■ *From the Literature* ■ ■ ■

Choosing the Right Mattress

Source: Garfin, S.R., S.A. Pye, A.R. Hargens, and W.H. Akenson. 1980. Surface pressure distribution of the human body in the recumbent position. *Arch. Phys. Med. Rehabil.* 61: 409–13.

When lying on a supporting surface, the majority of the body's weight is borne by five bony prominences: the sacrum, heels, scapulae, occiput, and trochanters. These sites are at high risk of pressures sores in people who are bedridden. Good mattresses should provide as even a pressure distribution as possible to avoid a concentration of pressure. The authors studied the surface pressure distribution of various mattresses. The pressure was most evenly distributed on an orthopedic 720-coil bed (not shown) and a waterbed (figure 1.60).

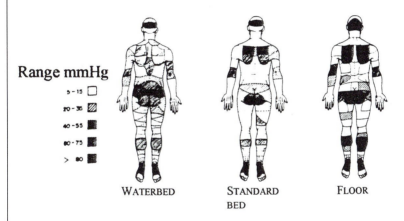

Range mmHg

Range	
5 – 15	□
20 – 35	▨
40 – 55	■
60 – 75	■
> 80	■

WATERBED   STANDARD BED   FLOOR

Figure 1.60 Pressure distribution patterns in the supine position on three different surfaces: waterbed, standard 500-coil bed, and floor. Sixty-five pressure transducers were employed to measure the pressure distribution. The data are averaged for the group (*n* = 5). The standard bed and the floor supported the body essentially on only five points (occiput, shoulders, buttocks, calves, and heels). A more uniform pressure distribution over the entire body was observed on the waterbed. The skin pressures measured over the bony prominences were significantly greater than the critical values for ischemia (30 mmHg), thereby enhancing the development of decubiti.

Adapted, by permission from S.R. Garfin et al., 1980, "Surface pressure distribution of the human body in the recumbent position," *Archive of Physical Medicine & Rehabilitation* 61:409–413.

in arteries. External pressure as small as 30 mmHg (4.0 kPa) can cause tissue *ischemia* (Webster 1991, p. 86). Commercially available mattresses yield pressure in excess of 30 mmHg (Garfin et al. 1980).

The first sign of occlusion, specifically the occlusion of the superficial vessels in the skin, is *ischemic pallor* (blanching). When sufficiently large pressure is exerted on the skin through a transparent surface, a white spot is seen at the pressure area. The whiteness is due to the squeezing of the blood out of superficial vessels in the skin and a blockage in the blood flow. Transient occlusion of the arteries and capillaries similar to that which occurs during walking usually does not have undesirable consequences. However, when a high contact pressure is maintained for a long period, it may induce pressure sores. Shear forces, if present, aggravate the risk of pressure sores. Human skin can resist stresses up to 100 kg/cm^2 (about 10 MPa). The force of a pinprick, for example, is concentrated on the small area at the tip and thus produces a very large pressure that the skin cannot resist.

In bedridden patients and wheelchair users, pressure sores (*decubitus ulcers*) are a serious medical problem. To prevent pressure sores, the person's position or the sites of pressure application should be altered periodically. From a biomechanical standpoint, the task is to distribute the body weight over a large area and prevent load concentration at sensitive regions. This goal is achieved by using custom, contoured cushions for wheelchair users or a supporting surface with different components that have different physical properties in different regions of contact. Custom, contoured cushions decreased the average pressure on the buttocks to 47 mmHg, compared with 57 mmHg for flat foam (Webster 1991). Able-bodied people do not suffer from pressure sores because they periodically move during sitting and sleeping. During an 8-h sleep on a soft bed, healthy people perform one postural change every 11-12 min (Keane 1978–1979). The most stable positions are the supine position and the lateral position with the legs flexed.

1.3.2 Shear Forces and Skin Friction

The skin is the first line of defense in protecting the body from changes in the environment. The damaging effects of friction and shear forces on skin and underlying tissues are classified into two groups: those without slide and those with slide relative to the skin (rubbing).

1.3.2.1 Shear Forces in Sitting

Large shear forces without slide are exerted during sitting. During relaxed sitting, the vertical line passing through the center of gravity of the upper body does not pass through the point of application of the resultant force to the seat. A force couple exerts a shear force on the buttocks. For a person with an aver-

■ ■ ■ *FROM THE LITERATURE* ■ ■ ■

Contact Pressure in Patients

Measurement of contact pressure has many applications, such as identifying areas of concentrated plantar pressure that may lead to pressure ulcers in diabetic patients with insensitive feet (Cavanagh et al. 1985; Cavanagh, Sims, and Sanders 1991; figure 1.61) and rheumatic patients (Grieve and Rashdi 1984), improving the handles of hand tools to prevent compression of sensitive areas of the palm (Tichauer 1978), determining seat pressure patterns that induce lipoatrophia semicircularis (the local atrophy of the subcutaneous fat tissue; Haex et al. 1998), determining the effects of seat angle and lumbar support on seated buttock pressure to prevent sores in people with spinal cord injuries (Shields and Cook 1988), and others. Several computer-based models of pressure distribution in soft tissues have been developed; for a review, see Webster (1991).

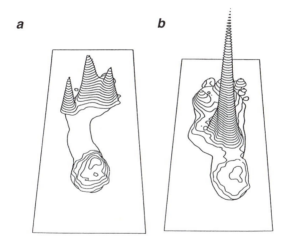

Figure 1.61　Peak plantar pressure diagrams from two diabetic patients, showing the largest pressure at each point during the first step of walking, regardless of the time of occurrence. Contour interval is 50 kPa. (*a*) Three local areas of pressure in medial forefoot. (*b*) An extremely high concentration of pressure underneath midfoot prominence (Charcot foot). Patient was 65 years old (body mass, 96.2 kg; peak pressure, 1,138 kPa; duration of non-insulin-dependent diabetes, 20 years).

Adapted, by permission, from P. Cavanagh, D.S. Sims, and L.J. Sanders, 1991, "Body mass is a poor predictor of peak plantar pressure in diabetic men," *Diabetes Care* 14 (8):750-755.

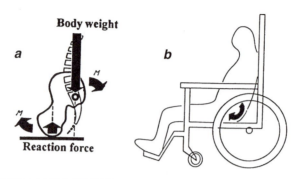

Figure 1.62 (*a*) A force couple acting on the body during relaxed sitting and (*b*) a reclined posture frequently assumed by wheelchair users.

age body weight of 73.8 kg sitting in a wheelchair with a soft cushion, shear forces up to 17.69 kg have been reported (Gilsdorf et al. 1990). Shear stress during sitting in a wheelchair was reported at 4.8 kPa in the anteroposterior direction and 8.5 kPa in the transverse direction (Goossens et al. 1997). If not supported by the backrest and the armrest, the body tends to slide and, if permitted to slide, assumes a reclined posture in the chair (figure 1.62).

Such a reclined posture, with a forward sliding of the buttocks on the seat and the low settling of the back on the backrest, is highly undesirable for wheelchair users. The pressure and the shear forces on the ischial tuberosities increase. Skin shear forces promote blood occlusion. When high skin shear forces are acting, the normal force needed to occlude blood flow in the underlying tissues is only half the force needed for occlusion when no shear is present (Bennett et al. 1979). The deeper soft tissue slides forward with the bone, while the superficial sacral fascia remains attached to the skin. This shear deformation increases the risk of the pressure sores. The tilted posture is also associated with elevated intradisk pressure and decreased respiratory capacity. To prevent wheelchair users from sliding on the seat, several measures are recommended; among them are strapping the chest to the backrest, reclining the seat 10° and the backrest 15°, and using high-friction materials for the seat. High-friction materials are also recommended for work chairs where sliding is undesirable.

1.3.2.2 Skin Rubbing

Repetitive rubbing and repetitive application of tangential or friction forces to human skin result in blisters; if rubbing is repeated over years, it can induce calluses and clavi (corns). Blisters result from frictional forces that mechanically separate epidermal cells at the external surface of the skin. The probability of blister development depends on the magnitude of the friction forces and the number and duration of exposures. Foot blisters can disable soldiers in

Table 1.3   Dynamic Coefficients of Friction of Human Skin on Teflon

Region	Friction coefficient
Forehead	0.34 ± 0.02
Upper arm	0.23 ± 0.01
Volar forearm	0.26 ± 0.01
Dorsal forearm	0.23 ± 0.01
Palm	0.21 ± 0.01
Abdomen	0.12 ± 0.01
Upper back	0.25 ± 0.02
Thigh	0.15 ± 0.01
Ankle	0.21 ± 0.01

Notes. The friction meter in this study consisted of a Teflon wheel rotating at constant speed and under constant load so that the measurements were performed at a fixed surface pressure and rubbing velocity. The friction coefficient values should be regarded as approximate; they are subject to considerable variation, depending on the measurement conditions. The values can be used, however, as an estimate of the magnitude of the friction effect.

$n = 29$ (Cua, Wilhelm, and Maibach 1990).

minutes and thus have been studied seriously. Blisters also occur frequently in athletes.

Skin friction has been studied primarily to prevent friction injuries. Skin friction measurements have also been used in dermatology and cosmetics to evaluate skin quality. The values of the coefficient of dynamic friction for skin on different parts of the body are presented in table 1.3.

Skin obeys Amonton's law over a limited range of loads only. In a broader range, a logarithmic relationship exists between the friction force F and the normal force W, $\log F = \log \mu + n \log W$, where the coefficient $n < 1.0$ (Comaish and Bottoms 1971; figure 1.63).

Because of the extensive variety of materials that interact with human skin and the dependence of the skin–material friction on many factors (age, humidity, wetness, sweating, etc.), comprehensive data on skin friction and its mechanisms cannot be given here. The interested reader is directed to review articles (Cua, Wilhelm, and Maibach 1990; Krouskop and van Rijswijk 1995; Knapik et al. 1995) and other examples from the scientific literature.

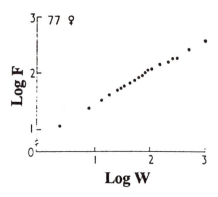

Figure 1.63 A logarithmic relationship between the static friction force *F* and the normal force *W*. The data are from postmortem measurements on 77 females.

Reprinted, by permission, from S. Comaish, E. Bottoms, 1971, "The skin and friction: Deviations from Amonton's laws and the effects of hydration and lubrication," *British Journal of Dermatology* 84: 37–43.

■ ■ ■ *From the Literature* ■ ■ ■

Avoid Wearing Wet Diapers

Source: Zimmerer, R.E., K.D. Lawson, and C.J. Calvert. 1986. The effects of wearing diapers on skin. *Pediatr. Dermatol.* 3 (2): 95–101.

Two types of diapers are currently used: reusable cloth and disposable. Disposable diapers are usually covered with an impenetrable material that contains excretory waste but obstructs skin ventilation and increases hydration of the skin. The goal of the study was to compare the effects of wearing wet diapers and dry diapers on skin friction and susceptibility to abrasion and to quantify performance of different diaper materials. Skin wetness was found to be proportional to diaper wetness. As skin wetness increased, the coefficients of friction, abrasion damage, skin permeability, and microbial growth increased. In infants who wore diapers, the coefficient of skin friction was 0.30 ± 0.14 for dry diapers and 0.96 ± 0.30 for wet diapers. During the experiments, the diapers were loaded with a solution of chemicals in water instead of natural urine. At equivalent volume loading, cloth diaper material produced wetter skin than did disposable diaper material.

1.3.2.3 Friction Forces in Grasping

In everyday activities and in industrial tasks, people exert forces on hand tools and workpieces. A high requirement for pinching or grasping in a job is a major risk factor for cumulative trauma disorders. The minimal pinch or grasp force that is needed to resist slippage of the tool when effort is transmitted from the tool to the workpiece defines the force requirement of a task. The force is influenced by the friction between the palmar skin and the contact-surface material of the tool. People usually exert larger normal forces than are necessary to prevent the slippage on a tool or a handheld object. The difference between the produced force and the minimally necessary force is called a *safety margin*.

■ ■ ■ **FROM THE LITERATURE** ■ ■ ■

Safety Margin in Precision Grip

Source: Westling, G., and R.S. Johansson. 1984. Factors influencing the force control during precision grip. *Exp. Brain Res.* 53: 277–84.

Subjects gripped a small object between the tips of the index finger and the thumb and held it stationary (figure 1.64). The friction conditions be-

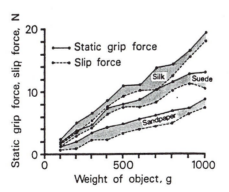

Figure 1.64 Grip adjustment to the friction condition. Relation between the weight of object and grip forces (solid curves) and slip forces (segmented curves) for lifting series with sandpaper, suede, and silk, respectively. Each series consisted of 10 consecutive trials with the weight increased in 100-g steps for each succeeding trial. Dashed areas illustrate safety margins.

tween the object and the skin varied in different trials. The grip force and the vertical lifting force acting on the object were recorded. The minimal grip force necessary to prevent slipping (*slip force*) was also measured. The safety margin, the difference between the minimal force and the employed grip force, was computed. It was observed that the applied grip force depended on friction and was adjusted to satisfy two conflicting demands: to prevent slipping and not to produce unduly high forces.

Experiments with local anesthesia indicated that the nervous system relies on skin receptors to measure the friction condition between the surface structure of the object and the fingers.

The gripping forces of working with tools are also influenced by such factors as the size of the handheld object and the area of the palm and fingers in contact with the handle.

The skin friction depends not only on the surface material but also on whether the skin is dry or moist (friction sharply decreases when the fingers are soaped; Taylor and Lederman 1975) and the person's age. Friction in elderly individuals is nearly twice the average for young people (measured against a sandpaper

▪ ▪ ▪ *From the Literature* ▪ ▪ ▪

Hand and Handle Interaction in Gripping and Turning

Source: Pheasant, S., and D. O'Neill. 1975. Performance in gripping and turning: A study in hand/handle effectiveness. *Appl. Ergon.* 6: 205–8.

The goal of the study was to generate data for the optimal design of screwdrivers and other devices that are gripped and used forcefully against resistance. In the experiment, handle diameters from 1 to 7 cm were used. Ten subjects exerted maximal torques about and maximal thrusts along the axes of the handles. Thrust shear force was measured directly, and the twist shear force was calculated from this equation:

$$\text{Torque} = G \times \mu_s \times d \qquad (1.78)$$

where G is the grip force (defined as the sum of the normal components of compressive force exerted on the handle, figure 1.65*a*), μ_s is the coefficient of static friction, and d is the handle diameter. Note that when the shear force was computed, the leverage effect of the handle size was eliminated.

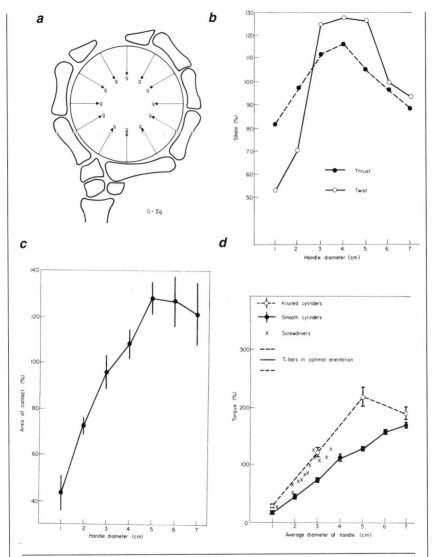

Figure 1.65 Hand/handle effectiveness. *(a)* Definition of the grip strength. *(b)* The relationship between twist and thrust shear force and handle diameter *d*. *(c)* The dependence of the area of hand/handle contact on the handle diameter. *(d)* The torques that were exerted on smooth cylinders, knurled cylinders, and a range of commercially available screwdrivers. The torque exerted on a T-bar handle is also shown for comparison. The values are normalized against the smooth cylinder average. The vertical lines represent ±1 standard deviation.

Reprinted from *Applied Ergonomics*, 6, S. Pheasant and D. O'Neill, Performance in gripping and turning: A study in hand/handle effectiveness, 205-208, 1975, with permission from Excerpta Medica Inc.

The force that the hand can bring to bear was optimal in the middle range of handle size, $d \approx 4$ cm. It was concluded that the difference was due to different grip strengths at various handle sizes. The handle diameter influenced both the area of contact (figure 1.65c) and the force exerted at individual points of contact (because of changes in the joint angles). When handles of similar diameter but different smoothness were used, the torque was larger for rougher surfaces. The torque exerted with smooth cylinders (low friction) and with knurled cylinders (high friction) was smaller than that developed when a T-bar handle was used (figure 1.65d). Hence, several factors—such as friction, area of contact, and finger configuration—determine the maximal torque and thrust.

surface; K.J. Cole 1991). Due to the diversity of experimental conditions, it is not possible to provide exact data on the coefficients of friction between the palmar skin and various surface materials. A good ballpark estimate is 1.21 for sandpaper (#320 grade), 0.68 for suede, and 0.35 for silk (Westling and Johansson 1984). In some sports (e.g., gymnastics, Olympic weightlifting), the athletes are interested in increasing the friction. To do that, various powdered materials that prevent sweating (mainly magnesia, MgO) are used.

In psychophysiology, studies of friction between the fingers or palm and handheld objects are an important part of research on the mechanisms of tactile perception. It seems that tactile detection of slip is realized separately from the perception of other qualities of the contacting surfaces, for instance, their texture and roughness.

1.4 SUMMARY

This chapter deals with the contact forces in human movement, specifically with the external contact forces. The chapter first introduced the concepts of force and moments of force and then couples and wrenches. Vector analysis is used throughout the presentation. Force **F** is a vector quantity that can be represented in its orthogonal components. A moment of force $\mathbf{M}_O$ about a point O is defined as a cross product of the position vector from O to the point of force application and vector **F**. The moment of force **F** about an axis is a component of the moment $\mathbf{M}_O$ along this axis. It is a mixed triple product of the unit vector along the axis of rotation, the position vector from an arbitrary point on the axis to any point on the line of force action and the force vector. A force couple, or simply a couple, consists of two equal, opposite, and parallel forces that act concurrently at a distance d apart. The entire effect of the couple is the tendency to make the body rotate. The measure of this tendency is called the moment of a couple. The moment of a couple is a free moment; it remains unchanged under parallel displacements. Any force **F** acting on a rigid body at a point P can be replaced by the same force acting at another point O and the corresponding couple represented by a vector $\mathbf{M}_C$ perpendicular to **F**. Such a representation is called a force-couple system. Conversely, if a force **F** and a couple $\mathbf{M}_C$ are mutually perpendicular, a single equipollent force can replace them. If a set of forces acting on a rigid body can be reduced to one resultant force, the moment of the resultant force about any point O is equal to the moment of the original force system about O (Varignon's theorem). Any system of forces acting on a rigid body can be reduced to a single resultant force and a single resultant couple acting at an arbitrarily selected point O.

A wrench is a force-couple system in which the vectors of the force and couple are parallel. For a given force system, there is only one equivalent wrench representation. A wrench is characterized by the wrench axis, pitch, and intensity. The pitch is the ratio of the magnitude of the wrench moment to the magnitude of the wrench force. A wrench representation of the forces acting at a contact, when combined with a helical representation of motion, allows analysis of forces and movement in one procedure.

Interaction with the support and grasping force exerted by individual digits were considered as typical examples of the contact forces in human movement. The wrench representation of the ground reaction forces (GRFs) allows the determination of not only the magnitude and the direction of the resultant force but also the point of wrench application (PWA) and the wrench moment, a component along the wrench axis of the free moment acting at the support. The PWA is different from the center of pressure (CoP), which is a point of application of the resultant of the vertical forces. Linear and angular momenta of the body can be changed only under the influence of the external forces. Gross mechanical effects of the GRF on the linear acceleration of the center of mass (CoM) of the body and on the rate of change of its angular momentum were discussed. The interaction between the internal and external forces was also addressed.

When grasping an object, the fingers work in parallel. If an object is manipulated by five digits, each of the digits exerts a six-component wrench on the object. In total, 30 parameters are represented in five reference frames. To analyze such complex cases, Plücker's coordinates and grip matrices are used.

Section **1.2** deals with the friction forces and air resistance. The friction forces in human movement usually do not follow Amonton's law. Contact forces between the human body and external objects result in various biological effects, including skin injuries, compression of the blood vessels, and changes in tactile perception. These issues were discussed in section **1.3**.

1.5 QUESTIONS FOR REVIEW

1. Is the principle of transmissibility valid for kinematic chains? Can a ground reaction force be transmitted to the center of mass of the body? If the answer is yes, discuss the physical meaning of this operation.

2. Determine the X, Y, and Z components of a 10-N force that makes the angles $\theta_X = \theta_Y = 60°$ with the coordinate axes X and Y.

3. Determine the magnitude and direction of the force $\mathbf{F} = 30\mathbf{i} - 36\mathbf{j} + 52\mathbf{k}$ (N). Answer: -70 N, $64.6°$, $120.9°$, $42.0°$.

4. The position vector from an origin of the coordinate system O to a point P is $\mathbf{r} = -\mathbf{i} + 3\mathbf{j} + 5\mathbf{k}$. Determine the moment at O of the force $\mathbf{F} = 3\mathbf{i} + 9\mathbf{j} + 15\mathbf{k}$ that acts at P. Answer: 0.

5. A force $\mathbf{F}$ is directed from a point P_1 with the coordinates x_1 and y_1 to a point P_2 with the coordinates x_2 and y_2. Determine the moment of the force about the origin.

   Answer: $\mathbf{M}_o = \dfrac{F(x_1 y_2 - x_2 y_1)\mathbf{k}}{\sqrt{(x_2 - x_1)^2 + (y_2 - y_1)^2}}$

6. The expression $M_X = ZF_Y - YF_Z$ appears frequently in various applications. Explain the meaning of individual terms, ZF_Y and YF_Z, in the right-hand part of this equation.

7. During a takeoff, the vertical component of the ground reaction force equals 1.5 times body weight (1.5 BW) and the horizontal component is 0.5 BW. The force is applied at the origin of the global reference system. The angle between the position vector of the body's center of mass and the vertical is 30°. Determine the magnitude of the force component that passes through the center of mass. What is the magnitude of acceleration of the center of mass? Hint: Use the rotation matrix.

8. Why can a couple be represented as a free vector?

9. Explain how a force $\mathbf{F}$ can be replaced by a force-couple system.

10. Discuss Varignon's theorem and its applicability to different force systems.

11. A wrench intensity is F and its pitch is p_w. What is the magnitude of the wrench moment?

12. What is a virtual coefficient?

13. Why cannot a unique point of application of the resultant force be determined for an arbitrary force system in three dimensions?

14. Explain the difference between the point of wrench application and the center of pressure.

15. What is the SI unit for measuring pressure?

16. During a takeoff, the vertical force component of the ground reaction force is twice the body weight. Determine the upward acceleration of the center of mass of the body.

17. In long jumping, the horizontal takeoff velocity is regularly smaller than the horizontal velocity prior to the takeoff by 1 to 2 m/s. Compare the impulses of the braking and propelling forces. Which one is larger?

18. According to classical mechanics, internal forces acting between the parts of a system of bodies *cannot* induce acceleration of the center of mass of the system; an external force is necessary to do that. For the human body, the muscular forces are evidently internal forces. What then is the role of muscular forces in propelling (accelerating) the human body?

19. Explain the mechanism of creating body rotation during somersaulting.

20. Explain the Plücker coordinates.

21. Write an equation for the grip matrix.

22. What is the cone of friction?

23. Discuss friction forces between the foot and the support in walking.

24. Name the two sources that contribute to air resistance.

25. What are the main local biological effects of large contact compressive forces?

26. What are the main local biological effects of large contact friction forces acting between the skin and the environment?

1.6 BIBLIOGRAPHY

Aerts, P., R.F. Ker, D. De Clercq, D.W. Isley, and R.M. Alexander. 1995. The mechanical properties of the human heel pad: A paradox resolved. *J. Biomech.* 28 (11): 1299-1308.

Aissaoui, R., S. Heydar, J. Dansereau, and M. Lacoste. 2000. Biomechanical analysis of legrest support of occupied wheelchairs: Comparison between a conventional and a compensatory legrest. *IEEE Trans. Rehabil. Eng.* 8 (1): 140–48.

Akers, W.A. 1985. Measurement of friction injuries in man. *Am. J. Ind. Med.* 8 (4–5): 473–81.

Aruin, A.S., and V.M. Zatsiorsky. 1989. *Occupational biomechanics* (in Russian). Moscow: Mashinostroenie.

Bahamonde, R.E. 2000. Changes in angular momentum during the tennis serve. *J. Sports Sci.* 18 (8): 579–92.

Begg, R.K., and S.M. Rahman. 2000. A method for the reconstruction of ground reaction force-time characteristics during gait from force platform recordings of simultaneous foot falls. *IEEE Trans. Biomed. Eng.* 47 (4): 547–51.

Bell, M.J., S.W. Baker, and P.M. Canaway. 1985. Playing quality of sports surfaces: A review. *J. Sports Turf Res. Inst.* 61:26–45.

Bennett, L., D. Kavner, B.K. Lee, and F.A. Trainor. 1979. Shear versus pressure as causative factors in skin blood occlusion. *Arch. Phys. Med. Rehabil.* 60 (7): 309–14.

Besser, M.P., D.L. Kowalk, and C.L. Vaughan. 1993. Mounting and calibration of stairs on piezoelectric force platforms. *Gait Posture* 1:231–35.

Betts, R.P., C.I. Franks, and T. Duckworth. 1980. Analysis of pressures and loads under the foot: 2. Quantification of the dynamic distribution. *J. Clin. Phys. Physiol. Meas.* 1 (2): 113–24.

Bhatia, M.M., and K.M. Patil. 1999. New on-line parameters for analysis of dynamic foot pressures in neuropathic feet of Hansen's disease subjects. *J. Rehabil. Res. Dev.* 36 (3): 264–72.

Blickhan, R. 1989. The spring-mass model for running and hopping. *J. Biomech.* 22:1217–27.

Bobbert, M.F., and H. Schamhardt. 1990. Accuracy of determining the point of force application with piezoelectric force plates. *J. Biomech.* 23:705–10.

Bobbert, M.F., M.R. Yeadon, and B.M. Nigg. 1992. Mechanical analysis of the landing phase in heel-toe running. *J. Biomech.* 25 (3): 223–34.

Bobjer, O., S. Johansson, and S. Piguet. 1993. Friction between hand and handle: Effects of oil and lard on textured and non-textured surfaces; perception of discomfort. *Appl. Ergon.* 24 (3): 190–202.

Bonstingl, R.W., A. Chaungey, C.A. Morehouse, and B.W. Niebel. 1975. Torques developed by different types of shoes on various playing surfaces. *Med. Sci. Sports Exerc.* 7 (2): 127–31.

Borzelli, G., A. Cappizzo, and E. Papa. 1999. Inter- and intra-individual variability of ground reaction forces during sit-to-stand with principal component analysis. *Med. Eng. Phys.* 21 (4): 235–40.

Bosco, C., P. Luhtanen, and P.V. Komi. 1976. Kinetics and kinematics of the take-off in the long jump. In *Biomechanics V-B,* ed. P.V. Komi, 174–80. Baltimore: University Park Press.

Buchholz, B., L.J. Frederick, and T.J. Armstrong. 1988. An investigation of human palmar skin friction and the effects of materials, pinch force and moisture. *Ergonomics* 31 (3): 317–25.

Buczek, F.L., P.R. Cavanagh, B.T. Kulakowski, and P. Pradhan. 1990. Slip resistance needs of the mobility disabled during level and grade walking. In *Slips, stumbles, and falls: Pedestrian footwear and surfaces,* ASTM STP 1103, ed. B.E. Gray, 39–54. Philadelphia: American Society for Testing and Materials.

Cadoret, G., and A.M. Smith. 1996. Friction, not texture, dictates grip forces used during object manipulation. *J. Neurophysiol.* 75:1963–69.

Cairns, M.A., R.G. Burdett, J.C. Pisciotta, and S.R. Simon. 1986. A biomechanical analysis of racewalking gait. *Med. Sci. Sports Exerc.* 18 (4): 446–53.

Canaway, P.M., and M.J. Bell. 1986. An apparatus for measuring traction and friction on natural and artificial playing surfaces. *J. Sports Turf Res. Inst.* 64:211–14.

Carlsöö, S. 1975. *How man moves.* London: Heinemann.

Carrasco, D.I., J. Lawrence, III, and A.W. English. 1999. Neuromuscular compartments of cat gastrocnemius produce different torques about the ankle joint. *Motor Control* 3:436–46.

Cavagna, G.A. 1975. Force platforms as ergometers. *J. Appl. Physiol.* 39:174–91.

Cavanagh, P.R. 1980. A technique for averaging center of pressure paths from a force platform. *J. Biomech.* 11:397–406.

Cavanagh, P.R., and E.M. Hennig. 1983. Pressure distribution measurement: A review and some new observations on the effect of shoe foam materials during running. In *Biomechanical aspects of sports shoes and playing surfaces,* ed. B.M. Nigg and B.A. Kerr, 187–90. Calgary, AB: University of Calgary Press.

Cavanagh, P.R., E.M. Hennig, M.M. Rodgers, and D.J. Sanderson. 1985. The measurement of pressure distribution on the plantar surface of diabetic feet. In *Biomechanical measurements in orthopaedic practice,* ed. M. Whittle and D. Harris, 159–66. Oxford, UK: Clarendon Press.

Cavanagh, P.R., and M.A. Lafortune. 1980. Ground reaction forces in distance running. *J. Biomech.* 13:397–406.

Cavanagh, P.R., M.M. Rodgers, and A. Iiboshi. 1987. Pressure distribution under symptom-free feet during barefoot standing. *Foot Ankle* 7:262–80.

Cavanagh, P.R., D.S. Sims, and L.J. Sanders. 1991. Body mass is a poor predictor of peak plantar pressure in diabetic men. *Diabetes Care* 14 (8): 750–55.

Cavanagh, P.R., J.S. Ulbrecht, and G.M. Caputo. 1999. Elevated plantar pressure and ulceration in diabetic patients after panmetatarsal head resection: Two case reports. *Foot Ankle Int.* 20 (8): 521–26.

Chung, K.C. 1987. Tissue contour and interface pressure on wheelchair cushion. Doctoral diss., University of Virginia.

Clayton, H.M., H.C. Schamhardt, M.A. Willemen, J.L. Lanovaz, and G.R. Colborne. 2000. Kinematics and ground reaction forces in horses with superficial digital flexor tendinitis. *Am. J. Vet. Res.* 61 (2): 191–96.

Cole, G.K., B.M. Nigg, G.H. Fick, and M.M. Morlock. 1995. Internal loading of the foot and ankle during impact in running. *J. Appl. Biomech.* 11:25–46.

Cole, K.J. 1991. Grasp force control in older adults. *J. Motor Behav.* 23 (4): 251–58.

Cole, K.J., and R.S. Johansson. 1993. Friction at the digit–object interface scales the sensorimotor transformation for grip responses to pulling loads. *Exp. Brain Res.* 95 (3): 523–32.

Collins, J.J., and M.W. Whittle. 1989. Impulsive forces during walking and their clinical implications. *Clin. Biomech.* 4:179–87.

Comaish, S., and S. Bottoms. 1971. The skin and friction: Deviations from Amonton's laws, and the effects of hydration and lubrication. *Br. J. Dermatol.* 84:37–43.

Cooper, R.A., R.N. Robertson, D.P. VanSickle, M.L. Boninger, and S.D. Shimada. 1996. Projection of the point of force application onto a palmar plane of the hand during wheelchair propulsion. *IEEE Trans. Rehabil. Eng.* 4 (3): 133–42.

Crossley, K., K.L. Bennell, T. Wrigley, and B.W. Oakes. 1999. Ground reaction forces, bone characteristics, and tibial stress fracture in male runners. *Med. Sci. Sports Exerc.* 31 (8): 1088–93.

Cua, A.B., K.P. Wilhelm, and H.I. Maibach. 1990. Frictional properties of human skin: Relation to age, sex and anatomical region, stratum corneum hydration and transepidermal water loss. *Br. J. Dermatol.* 123 (4): 473–79.

Davies, C.T. 1980a. Effect of air resistance on the metabolic cost and performance of cycling. *Eur. J. Appl. Physiol.* 45 (2–3): 245–54.

Davies, C.T. 1980b. Effects of wind assistance and resistance on the forward motion of a runner. *J. Appl. Physiol.* 48 (4): 702–9.

Davis, B.L., J.E. Perry, D.C. Neth, and K.C. Waters. 1998. A device for simultaneous measurement of pressure and shear force distribution on the plantar surface of the foot. *J. Appl. Biomechanics.* 14 (1): 93–104.

De Lorenzo, D.S., and M.L. Hull. 1999. A hub dynamometer for measurement of wheel forces in off-road bicycling. *J. Biomech. Eng.* 121 (1): 132–37.

Demeny, G. 1909. *Le bases scientifiques de l'éducation physique.* Paris: F. Alcan.

Demes, B., J.G. Fleagle, and W.L. Jungers. 1999. Takeoff and landing forces of leaping strepsirhine primates. *J. Hum. Evol.* 37 (2): 279–92.

Deutsche Normen. 1978. *DIN 18035. Sportplätze. Kunstoff-Flächen. Anforderungen, Prüfung, Pflege.* Berlin: Deutches Institut für Normung.

Devita, P., and W.A. Skelly. 1992. Effect of landing stiffness on joint kinetics and energetics in the lower extremity. *Med. Sci. Sports Exerc.* 24 (1): 108–15.

Di Prampero, P.E., G. Cortili, P. Mognoni, and F. Saibene. 1976. Energy cost of speed skating and efficiency of work against air resistance. *J. Appl. Physiol.* 40 (4): 584-91.

Dixon, S.J., M.E. Batt, and A.C. Collop. 1999. Artificial playing surfaces research: A review of medical, engineering and biomechanical aspects. *Int. J. Sports Med.* 20 (4): 209–18.

Donskoy, D., and V. Zatsiorsky. 1979. *Biomechanics.* Moscow: FiS.

Elftman, H. 1934. A cinematic study of the distribution of the center of pressure in the human foot. *Anat. Rec.* 59:481–91.

El-Shimi, A.F. 1977. In-vivo skin friction measurements. *J. Soc. Cosmet. Chem.* 28:37–51.

Fenn, W.O. 1957. The mechanics of standing on the toes. *Am. J. Phys. Med.* 36:153–56.

Frederick, E.C. 1992. Mechanical constraints on Nordic ski performance. *Med. Sci. Sports Exerc.* 24 (9): 1010–14.

Fuller, E.A. 1999. Center of pressure and its theoretical relationship to foot pathology. *J. Am. Podiatr. Med. Assoc.* 89 (6): 278–91.

Garfin, S.R., S.A. Pye, A.R. Hargens, and W.H. Akenson. 1980. Surface pressure distribution of the human body in the recumbent position. *Arch. Phys. Med. Rehabil.* 61:409–13.

Gilsdorf, P., R. Patterson, S. Fisher, and N. Appel. 1990. Sitting forces and wheelchair mechanics. *J. Rehabil. Res. Dev.* 27 (3): 239–46.

Gonzalez, L.J., S.V. Sreenivasan, and J.L. Jensen. 1999. A procedure to determine equilibrium postural configurations for arbitrary locations of the feet. *J. Biomech. Eng.* 121 (6): 644–49.

Goossens, R.H., C.J. Snijders, T.G. Holscher, W.C. Heerens, and A.E. Holman. 1997. Shear stress measured on beds and wheelchairs. *Scand. J. Rehabil. Med.* 29 (3): 131–36.

Grabiner, M.D., J.W. Feuerbach, T.M. Lundin, and B.L. Davis. 1995. Visual guidance to force plates does not influence ground reaction force variability. *J. Biomech.* 28 (9): 1115–17.

Grieve, D.W., and T. Rashdi. 1984. Pressures under normal feet in standing and walking as measured by foil pedobarography. *Ann. Rheum. Dis.* 43:816–18.

Groot, G. de, A. Sargeant, and J. Geysel. 1995. Air friction and rolling resistance during cycling. *Med. Sci. Sports Exerc.* 27 (7): 1090–95.

Haas, B.M., and A.M. Burden. 2000. Validity of weight distribution and sway measurements of the Balance Performance Monitor. *Physiother. Res. Int.* 5 (1): 19–32.

Haex, B., G. Van der Perre, V. Hermans, V. Hautekiet, and A. Spaepen. 1998. Lipoatrophia semicircularis induced by posture related seat pressure. *J. Biomech.* 31 (Suppl. 1): 177.

Hall, M.G., H.E. Fleming, M.J. Dolan, S.F.D. Millbank, and J.P. Paul. 1996. Static in situ calibration of force plates. *J. Biomech.* 29 (5): 659–65.

Han, T.R., N.J. Paik, and M.S. Im. 1999. Quantification of the path of center of pressure (COP) using an F-scan in-shoe transducer. *Gait Posture* 10 (3): 248–54.

Hanke, T.A., and M.W. Rogers. 1992. Reliability of ground reaction force measurements during dynamic transitions from bipedal to single-limb stance in healthy adults. *Phys. Ther.* 72 (11): 810–16.

Harris, N.J., J. Chell, and T.W. Smith. 1999. Patterns of weight distribution under the metatarsal heads [letter; comment]. *J. Bone Joint Surg. Br.* 81 (4): 744.

Hattori, T. 1998. Body up-down acceleration in kinematic gait analysis in comparison with the vertical ground reaction force. *Biomed. Mater. Eng.* 8 (3–4): 145–54.

Heise, G.D., P.E. Martin, and P.S. Carroll. 1996. Ground reaction force characteristics and running economy. In *Twentieth Annual Meeting of the American Society of Biomechanics, Conference Proceedings,* 103–4. Georgia Tech, Atlanta.

Hicks, J.H. 1955. The foot as a support. *Acta Anat.* 25:34–41.

Hill, A.V. 1927. The air resistance to a runner. *Proc. Roy. Soc.* B102:380–85.

Hof, A.L. 1992. An explicit expression for the moment in multibody terms. *J. Biomech.* 25 (10): 1209-11.

Hoffman, J.R., D. Liebermann, and A. Gusis. 1997. Relationship of leg strength and power to ground reaction forces in both experienced and novice jump trained personnel. *Aviat. Space Environ. Med.* 68 (8): 710–14.

Holden, J.P., and P.R. Cavanagh. 1991. The free moment of ground reaction in distance running and its changes with pronation. *J. Biomech.* 24 (10): 887–97.

Holt, R.R., D. Simpson, J.R. Jenner, S.G. Kirker, and A.M. Wing. 2000. Ground reaction force after a sideways push as a measure of balance in recovery from stroke. *Clin. Rehabil.* 14 (1): 88–95.

Hosein, R., and M. Lord. 2000. A study of in-shoe plantar shear in normals. *Clin. Biomech.* 15 (1): 46–53.

Houdijk, H., J.J. de Koning, G. de Groot, M.F. Bobbert, and G.J. van Ingen Schenau. 2000. Push-off mechanics in speed skating with conventional skates and klapskates. *Med. Sci. Sports Exerc.* 32 (3): 635–41.

Hsi, W.L., J.S. Lai, and P.Y. Yang. 1999. In-shoe pressure measurements with a viscoelastic heel orthosis. *Arch. Phys. Med. Rehabil.* 80 (7): 805–10.

Inlow, S., T.P. Kalla, and J. Rahman. 1999. Downloading plantar foot pressures in the diabetic patient. *Ostomy Wound Manage.* 45 (10): 28–34, 36, 38, 39–40.

Jansen, M.O., A.J. van den Bogert, D.J. Riemersma, and H.C. Schamhardt. 1993. In vivo tendon forces in the forelimb of ponies at the walk, validated by ground reaction force measurements. *Acta Anat.* 146 (2–3): 162–67.

Jobse, H., R. Schuurhof, F. Cseser, A.W. Schreurs, and J.J. de Koning. 1990. Measurement of push-off force and ice friction during speed skating. *Int. J. Sport Biomech.* 6:92–100.

Karlsson, A., and G. Frykberg. 2000. Correlations between force plate measures for assessment of balance. *Clin. Biomech.* 15 (5): 365–69.

Keane, F.X. 1978–1979. The minimum physiological mobility requirement for man supported on a soft surface. *Paraplegia* 16:383–89.

Kibele, A. 1998. Possibilities and limitations in the biomechanical analysis of countermovement jumps: A methodological study. *J. Appl. Biomech.* 14 (1): 105–25.

King, D.L. 2000. Jumping in figure skating. In *Biomechanics in sport,* ed. V.M. Zatsiorsky, 312-25. Oxford, UK: Blackwell Science.

Kinoshita, H., L. Bäckström, J.R. Flanagan, and R.S. Johansson. 1997. Tangential torque effects on the control of grip forces when holding objects with a precision grip. *J. Neurophysiol.* 78:1619–30.

Kinoshita, H., S. Kawai, and K. Ikuta. 1995. Contributions and co-ordination of individual fingers in multiple finger prehension. *Ergonomics* 38 (6): 1212-30.

Kirpensteijn, J., R. van den Bos, W.E. van den Brom, and H.A. Hazewinkel. 2000. Ground reaction force analysis of large breed dogs when walking after the amputation of a limb. *Vet. Rec.* 146 (6): 155–59.

Knapik, J.J., K.L. Reynolds, K.L Duplantis, and B.H. Jones. 1995. Friction blisters. Pathophysiology, prevention and treatment. *Sports Medicine* 20 (3): 136-47.

Komi, P.V., and M. Virmavirta. 2000. Determinants of successful ski-jumping performance. In *Biomechanics in sport,* ed. V.M. Zatsiorsky, 349–62. Oxford, UK: Blackwell Science.

Koning, J.J. de, R.W. de Boer, G. de Groot, and G.J. van Ingen Schenau. 1987. Push-off force in speed skating. *Int. J. Sport Biomech.* 3:103–9.

Koning, J.J. de, G. de Groot, and G.J. van Ingen Schenau. 1992. Ice friction during speed skating. *J. Biomech.* 25:565–71.

Koning, J.J. de, B.M. Nigg, and K.G. Gerritsen. 1997. Assessment of the mechanical properties of area-elastic sport surfaces with video analysis. *Med. Sci. Sports Exerc.* 29 (12): 1664–68.

Koning, J.J. de, and G.J. van Ingen Schenau. 2000. Performance-determining factors in speed skating. In *Biomechanics in sport,* ed. V.M. Zatsiorsky, 232–46. Oxford, UK: Blackwell Science.

Kromodihardio, S., and A. Mital. 1986. Kinetic analysis of manual lifting activities: Part 1. Development of a three-dimensional computer model. *Int. J. Ind. Ergon.* 1:77–90.

Kromodihardio, S., and A. Mital. 1987. Biomechanical analysis of manual lifting tasks. *J. Biomech. Eng.* 109 (2): 132–38.

Krouskop, T., and L. van Rijswijk. 1995. Standardizing performance-based criteria for support surfaces. *Ostomy Wound Manage* 41 (1): 34-36.

Kyle, C.R., and V.J. Caiozzo. 1986. The effect of athletic clothing aerodynamics upon running speed. *Med. Sci. Sports Exerc.* 18 (5): 509–15.

Kyröläinen, H., A. Belli, and P.V. Komi. 1999. Lower limb mechanics with increasing running speed. In *International Society of Biomechanics, XVIIth congress, Book of abstracts,* ed. W. Herzog and A. Jinha, 262. Calgary, AB: University of Calgary.

Lafortune, M.A., M.J. Lake, and E. Hennig. 1995. Transfer function between tibial acceleration and ground reaction force. *J. Biomech.* 28 (1): 113–18.

Lafortune, M.A., M.J. Lake, and E.M. Hennig. 1996. Differential shock transmission response of the human body to impact severity and lower limb posture. *J. Biomech.* 29 (12): 1531–37.

Lanka, J. 2000. Shot putting. In *Biomechanics in sport,* ed. V.M. Zatsiorsky, 435–57. Oxford, UK: Blackwell Science.

Lanshammar, H., and L. Strandberg. 1983. Horizontal floor reaction forces and heel movements during the initial stance phase. In *Biomechanics VIII-B,* ed. H. Matsui and K. Kobayashi, 1123–28. Champaign, IL: Human Kinetics.

Lanshammar, H., and L. Strandberg. 1985. Assessment of friction by speed measurement during walking in a closed path. In *Biomechanics IX-B,* ed. D.A. Winter, R.W. Norman, R.P. Wells, K.C. Hayes, and A.E. Patla, 72–75. Champaign, IL: Human Kinetics.

Levin, O., J. Mizrahi, and M. Shoham. 1998. Standing sway: Iterative estimation of the kinematics and dynamics of the lower extremities from force-platform measurements. *Biol. Cybern.* 78:319–27.

Li, Y., R.H. Crompton, R.M. Alexander, M.M. Gunther, and W.J. Wang. 1996. Characteristics of ground reaction forces in normal and chimpanzee-like bipedal walking by humans. *Folia Primatol.* (Basel) 66 (1–4): 137–59.

Li, Z.M., M.L. Latash, K.M. Newell, and V.M. Zatsiorsky. 1998. Motor redundancy during maximal voluntary contraction in four-finger tasks. *Exp. Brain Res.* 122 (1): 71–77.

Li, Z.M., M.L. Latash, and V.M. Zatsiorsky. 1998. Force sharing among fingers as a model of the redundancy problem. *Exp. Brain Res.* 119 (3): 276–86.

Liddle, D., K. Rome, and T. Howe. 2000. Vertical ground reaction forces in patients with unilateral plantar heel pain: A pilot study. *Gait Posture* 11 (1): 62–66.

Lord, M., and R. Hosein. 2000. A study of in-shoe plantar shear in patients with diabetic neuropathy. *Clin. Biomech.* 15 (4): 278–83.

Lyu, S.R., K. Ogata, and I. Hoshiko. 2000. Effects of a cane on floor reaction force and center of force during gait. *Clin. Orthop.* 375:313–19.

MacKenzie, C.L., and T. Iberall. 1994. *The grasping hand.* Amsterdam: North Holland.

McCaw, S.T., and P. Devita. 1995. Errors in alignment of center of pressure and foot coordinates affect predicted lower extremity torques. *J. Biomech.* 28 (8): 985–88.

McNair, P.J., and H. Prapavessis. 1999. Normative data of vertical ground reaction forces during landing from a jump. *J. Sci. Med. Sport* 2 (1): 86–88.

McPoil, T.G., M.W. Cornwall, L. Dupuis, and M. Cornwell. 1999. Variability of plantar pressure data: A comparison of the two-step and midgait methods. *J. Am. Podiatr. Med. Assoc.* 89 (10): 495–501.

Menard, M.R., M.E. McBride, D.J. Sanderson, and D.D. Murray. 1992. Comparative biomechanical analysis of energy-storing prosthetic feet. *Arch. Phys. Med. Rehabil.* 73 (5): 451–58.

Messier, S.P., W.H. Ettinger, Jr., T.E. Doyle, T. Morgan, M.K. James, M.L. O'Toole, and R. Burns. 1996. Obesity: Effects on gait in an osteoarthritic population. *J. Appl. Biomech.* 12:161–72.

Miller, D.I. 2000. Springboard and platform diving. In *Biomechanics in sport,* ed. V.M. Zatsiorsky, 326–48. Oxford, UK: Blackwell Science.

Miller, D.I., and M.A. Nissinen. 1987. Critical examination of ground reaction force in the triple jump. *Int. J. Sport Biomech.* 3:189–206.

Miller, D.I., E. Henning, M.A. Pizzimenti, I.C. Jones, and R.C. Nelson. 1989. Kinetic and kinematic characteristics of 10-m platform performances of elite divers: Back takeoffs. *Int. J. Sport Biomech.* 5:60–88.

Mita, K., K. Akataki, K. Itoh, H. Nogami, R. Katoh, S. Ninomi, M. Watakabe, and N. Suzuki. 1993. An investigation of the accuracy in measuring the body center of pressure in a standing posture with a force plate. *Front. Med. Biol. Eng.* 5 (3): 201–13.

Mital, A., and R. Vinayagamoorhty. 1984. Three-dimensional dynamic strength measuring device: A prototype. *Am. Ind. Hyg. Assoc. J.* 45 (12): B9–10, B12.

Moor, D.F. 1975. *Principles and applications of tribology.* Oxford, UK: Pergamon Press.

Mueller, M.J. 1999. Application of plantar pressure assessment in footwear and insert design. *J. Orthop. Sports Phys. Ther.* 29 (12): 747–55.

Murray, M.P., A. Seireg, and R.C. Scholz. 1967. Center of gravity, center of pressure, and supportive forces during human activities. *J. Appl. Physiol.* 23:831–38.

Nagano, A., Y. Ishige, and S. Fukashiro. 1998. Comparison of new approaches to estimate mechanical output of individual joints in vertical jumps. *J. Biomech.* 31 (10): 951–55.

Nieuwboer, A., W. De Weerdt, R. Dom, L. Peeraer, E. Lesaffre, F. Hilde, and B. Baunach. 1999. Plantar force distribution in Parkinsonian gait: A comparison between patients and age-matched control subjects. *Scand. J. Rehabil. Med.* 31 (3): 185–92.

Nigg, B.M. 1988. The assessment of loads acting on the locomotor system in running and other sport activities. *Semin. Orthop.* 3 (4): 196–206.

Nigg, B.M. 1994. Force. In *Biomechanics of the musculo-skeletal system,* ed. B.M. Nigg and W. Herzog, 200–24. Chichester, UK: Wiley.

Oggero, E., G. Pagnacco, D.R. Morr, S.Z. Barnes, and N. Berme. 1997. The mechanics of drop landing on a flat surface: A preliminary study. *Biomed. Sci. Instrum.* 33:53–58.

Oggero, E., G. Pagnacco, D.R. Morr, S.R. Simon, and N. Berme. 1997. Probability of valid gait data acquisition using currently available force plates. *Biomed. Sci. Instrum.* 34:392–97.

Orlin, M.N., and T.G. McPoil. 2000. Plantar pressure assessment. *Phys. Ther.* 80 (4): 399–409.

Outwater, J.O. 1970. On the friction of skis. *Med. Sci. Sports* 2 (4): 231–34.

Pagnacco, G., A. Silva, E. Oggero, and N. Berme. 2000. Inertially compensated force plate: A means for quantifying subject's ground reaction forces in non-inertial conditions. *Biomed. Sci. Instrum.* 36:397–402.

Pandy, M.G. 1999. Moment arm of a muscle force. *Exerc. Sport Sci. Rev.* 27:79–119.

Penwarden, A.D., P.F. Grigg, and R. Rayment. 1978. Measurement of wind drag on people standing in a wind tunnel. *Building Environ.* 13:75–84.

Pheasant, S., and D. O'Neill. 1975. Performance in gripping and turning: A study in hand/handle effectiveness. *Appl. Ergon.* 6:205–8.

Pio, A., M. Rendina, F. Benazzo, C. Castelli, and F. Paparella. 1994. The statics of cervical traction. *J. Spinal Disord.* 7 (4): 337–42.

Pitei, D.L., M. Lord, A. Foster, S. Wilson, P.J. Watkins, and M.E. Edmonds. 1999. Plantar pressures are elevated in the neuroischemic and the neuropathic diabetic foot. *Diabetes Care* 22 (12): 1966–70.

Pollard, J.P., L.P. Lequesne, and J.W. Tappin. 1983. Forces under the foot. *J. Biomed. Eng.* 5:37–40.

Prapavessis, H., and P.J. McNair. 1999. Effects of instruction in jumping technique and experience jumping on ground reaction forces. *J. Orthop. Sports Phys. Ther.* 29 (6): 352–56.

Pugh, L.G. 1971. The influence of wind resistance in running and walking and the mechanical efficiency of work against horizontal or vertical forces. *J. Physiol.* (London) 213 (2): 255–76.

Ramey, M.R. 1970. Force relationships of the running long jump. *Med. Sci. Sports Exerc.* 2:146–51.

Ramey, M.R., and K.R. Williams. 1985. Ground reaction forces in the triple jump. *Int. J. Sport Biomech.* 1 (3): 233–39.

Randolph, A.L., M. Nelson, M.P. de Araujo, R. Perez-Millan, and T.T. Wynn. 1999. Use of computerized insole sensor system to evaluate the efficacy of a modified ankle-foot orthosis for redistributing heel pressures. *Arch. Phys. Med. Rehabil.* 80 (7): 801–4.

Rheinstein, D.J., C.A. Morehouse, and B.W. Niebel. 1978. Effect on traction of outsole composition and hardnesses of basketball shoes and three types of playing surfaces. *Med. Sci. Sports* 10 (4): 282–88.

Riener, R., M. Rabuffetti, C. Frigo, J. Quintern, and G. Schmidt. 1999. Instrumented staircase for ground reaction measurement. *Med. Biol. Eng. Comput.* 37 (4): 526–29.

Rodgers, M. 1993. Biomechanics of the foot during locomotion. In *Current issues in biomechanics,* ed. M.D. Grabiner, 33–52. Champaign, IL: Human Kinetics.

Rodgers, M.M., and P.R. Cavanagh. 1989. Pressure distribution in Morton's foot structure. *Med. Sci. Sport Exerc.* 21 (1): 23–28.

Rudy, T.W., M. Sakane, R.E. Debsk, and S.L. Woo. 2000. The effect of the point of application of anterior tibial loads on human knee kinematics. *J. Biomech.* 33 (9): 1147–52.

Sacco, I.C., and A.C. Amadio. 2000. A study of biomechanical parameters in gait analysis and sensitive cronaxie of diabetic neuropathic patients. *Clin. Biomech.* 15 (3): 196–202.

Sanderson, D.J., E.M. Hennig, and A.H. Black. 2000. The influence of cadence and power output on force application and in-shoe pressure distribution during cycling by competitive and recreational cyclists. *J. Sports Sci.* 18 (3): 173–81.

Schmiedmayer, H.B., and J. Kastner. 1999. Parameters influencing the accuracy of the point of force application determined with piezoelectric force plates. *J. Biomech.* 32 (11): 1237–42.

Seyfarth, A., A. Friedrichs, V. Wank, and R. Blickhan. 1999. Dynamics of the long jump. *J. Biomech.* 32 (12): 1259–67.

Shanebrook, J.R., and R.D. Jaszczak. 1976. Aerodynamic drag analysis of runners. *Med. Sci. Sports* 8 (1): 43-45.

Shields, R.K., and T.M. Cook. 1988. Effect of seat angle and lumbar support on seated buttock pressure. *Phys. Ther.* 68:1682–86.

Shimba, T. 1983. Ground reaction forces during human standing. *Eng. Med.* 12 (4): 177–82.

Shimba, T. 1996. Consequences of force platform studies. *Nat. Rehab. Res. Bull. Jpn.* 17:17–23.

Simon, S.R., I.L. Paul, J. Mansour, M. Munro, P.J. Abernethy, and E.L. Radin. 1981. Peak dynamic force in human gait. *J. Biomech.* 14 (12): 817–22.

Simpson, K.J., and P. Jiang. 1999. Foot landing position during gait influences ground reaction forces. *Clin. Biomech.* 14 (6): 396–402.

Sloboda, W., and V.M. Zatsiorsky. 1996. Time-frequency (wavelet) analysis of ground reaction data in walking and running. In *Twentieth Annual Meeting of the American Society of Biomechanics, Conference Proceedings,* 257–58, Atlanta, Georgia Tech.

Smith, G.A. 2000. Cross-country skiing: Technique, equipment and environmental factors affecting performance. In *Biomechanics in sport,* ed. V.M. Zatsiorsky, 247–70. Oxford, UK: Blackwell Science.

Soames, R.W. 1985. Foot pressure patterns during gait. *J. Biomech. Eng.* 7:120–26.

Soutas-Little, R.W. 1987. Center of pressure plots for clinical uses. In *Biomechanics of normal and prosthetic gait,* ed. J.L. Stein, 69-76. New York: ASME.

Stijnen, V.V., A.J. Spaepen, and E.J. Willems. 1981. Models and methods for the determination of the center of gravity of the human body from film. In *Biomechanics VII-A,* ed. A. Morecki, K. Fidelus, K. Kedzior, and A. Wit, 558–64. Baltimore: University Park Press.

Strandberg, L., and H. Lanshammar. 1981. The dynamics of slipping accidents. *J. Occup. Accid.* 3:153–62.

Strandberg, L., and H. Lanshammar. 1983. On the biomechanics of slipping accidents. In *Biomechanics VIII-A,* ed. H. Matsui and K. Kobayashi, 397–402. Champaign, IL: Human Kinetics.

Strandberg, L., and H. Lanshammar. 1985. Walking slipperiness compared to data from friction meters. In *Biomechanics IX-B,* ed. D.A. Winter, R.W. Norman, R.P. Wells, K.C. Hayes, and A.E. Patla, 76–81. Champaign, IL: Human Kinetics.

Suzuki, N., T. Shinohara, M. Kimizuka, K. Yamaguchi, and K. Mita. 2000. Energy expenditure of diplegic ambulation using flexible plastic ankle foot orthoses. *Bull. Hosp. Jt. Dis.* 59 (2): 76.

Takei, Y. 1992. Blocking and postflight techniques of male gymnasts performing the compulsory vault at the 1988 Olympics. *Int. J. Sport Biomech.* 8:87–110.

Tanaka, T., H. Takeda, T. Izumi, S. Ino, and T. Ifukube. 1999. Effects on the location of the centre of gravity and the foot pressure contribution to standing balance associated with ageing. *Ergonomics* 42 (7): 997–1010.

Taylor, M.M., and S.J. Lederman. 1975. Tactile roughness of grooved surfaces: A model and the effect of friction. *Percept. Psychophys.* 17 (1): 23–36.

Tichauer, E. 1978. *The biomechanical basis of ergonomics.* New York: Wiley.

Torg, J.S., T.G. Quedenfeld, and S. Landau. 1974. The shoe-surface interface and its relationship to football injures. *J. Sports Med.* 2:261–69.

VanSickle, D.P., R.A. Cooper, M.L. Boninger, R.N. Robertson, and S.D. Shimada. 1998. A unified method for calculating the center of pressure during wheelchair propulsion. *Ann. Biomed. Eng.* 26 (2): 328–36.

Viitasalo, J.T., A. Salo, and J. Lahtinen. 1998. Neuromuscular functioning of athletes and non-athletes in the drop jump. *Eur. J. Appl. Physiol.* 78 (5): 432–40.

Webster, J.G., ed. 1991. *Prevention of pressure sores.* Bristol, UK: Adam Hilger.

Wells, C., and H. Carnahan. 1998. A mathematical model of grasp. In: *Proceedings of NACOB '98: The third North American congress on biomechanics, August 14–18, 1988* (pp. 561-62). Waterloo, Ontario: University of Waterloo.

Westling, G., and R.S. Johansson. 1984. Factors influencing the force control during precision grip. *Exp. Brain Res.* 53:277–84.

White, R., I. Agouris, R.D. Selbie, and M. Kirkpatrick. 1999. The variability of force platform data in normal and cerebral palsy gait. *Clin. Biomech.* 14 (3): 185–92.

White, S.C., H.J. Yack, C.A. Tucker, and H.Y. Lin. 1998. Comparison of vertical ground reaction forces during overground and treadmill walking. *Med. Sci. Sports Exerc.* 30 (10): 1537–42.

Whitt, F.R., and D.G. Wilson. 1974. *Bicycling science.* Cambridge, MA: MIT Press.

Williams, G.E., B.W. Silverman, A.M. Wilson, and A.E. Goodship. 1999. Disease-specific changes in equine ground reaction force data documented by use of principal component analysis. *Am. J. Vet. Res.* 60 (5): 549–55.

Willson, J.D., and T.W. Kernozek. 1999. Plantar loading and cadence alterations with fatigue. *Med. Sci. Sports Exerc.* 31 (12): 1828–33.

Woude, L.H.V. van der, H.E.J. Veeger, and A.J. Dallmeijer. 2000. Manual wheelchair propulsion. In *Biomechanics in sport,* ed. V.M. Zatsiorsky, 609–36. Oxford, UK: Blackwell Science.

Xu, H., M. Akai, S. Kakurai, K. Yokota, and H. Kaneko. 1999. Effect of shoe modifications on center of pressure and in-shoe plantar pressures. *Am. J. Phys. Med. Rehabil.* 78 (6): 516–24.

Yee, A.G., and C.D. Mote, Jr. 1997. Forces and moments at the knee and boot top: Models for an alpine skiing population. *J. Appl. Biomech.* 13:373–84.

Yu, B., E.S. Growney, F.M. Schultz, and K.-N. An. 1996. Calibration of measured center of pressure of a new stairway design for kinetic analysis of stair climbing. *J. Biomech.* 29 (12): 1625–28.

Zatsiorsky, V.M., S.Y. Aleshinsky, and N.A. Yakunin. 1982. *Biomechanical basis of endurance.* Moscow: FiS.

Zatsiorsky, V.M., and B.I. Prilutsky. 1987. Soft and stiff landing. In *Biomechanics X-B,* ed. B. Johnson, 739–44. Champaign, IL: Human Kinetics.

Zhang, M., A.R. Turner-Smith, V.C. Roberts, and A. Tanner. 1996. Frictional action at lower limb/prosthetic socket interface. *Med. Eng. Phys.* 18 (3): 207–14.

Zimmerer, R.E., K.D. Lawson, and C.J. Calvert. 1986. The effects of wearing diapers on skin. *Pediatr. Dermatol.* 3 (2): 95–101.

2
CHAPTER

STATICS OF MULTILINK SERIAL CHAINS: TRANSFORMATION ANALYSIS

Make the model as simple as possible but not simpler.

Albert Einstein (1879–1955)

The biomechanics of the human body is complex. Various models are used to study the body in motion and at rest. One of the most useful models of the human body is a *kinematic chain* consisting of rigid links connected by *revolute joints* (figure 2.1). This and subsequent chapters are chiefly concerned with such a model. Throughout the ensuing chapters, complex human anatomy, in particular muscle anatomy, is neglected, and analysis is limited to the following model:

- The body consists of rigid links.
- The links are connected by ideal revolute joints. (In this context, the adjective "ideal" means that there is no energy dissipation due to friction or deformation in the joints.)
- The lines of action of joint reaction forces pass through the joint centers.
- The masses and moments of inertia of body parts do not change during movement.
- Each joint is driven by a *torque actuator*, an abstract generator that produces a moment of force with respect to the joint axis. By assumption, moments of force equal in magnitude and opposite in direction act on the two adjacent body links that form the joint. The term *joint torque* (or *joint moment*) refers to these two equal in magnitude and opposite moments of force about the joint rotation axis.

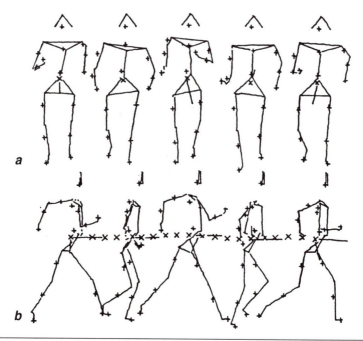

Figure 2.1 Human body during fast walking, represented as a branched kinematic chain (a stick figure): (*a*) front view and (*b*) side view. The chain consists of rigid links connected by revolute joints. The crosses (+) mark the centers of mass of the individual body segments; the oblique cross (×) designates the general center of mass of the entire body.

Adapted from D.D. Donskoi and V.M. Zatsiorsky, 1979, *Biomechanics* (Moscow: FiS Publishers).

Such a model is called a *basic model* of the human body. The model can be visualized as a *stick-figure model*. A more sophisticated version of the basic model is the basic model with muscles. In such a model, the muscles rather than the joint torque actuators are viewed as the actuators. The muscles are, however, highly idealized. They are represented by threadlike cords possessing only one feature: they produce a force at the points of origin and insertion.

None of the aforementioned simplifying assumptions is absolutely correct. Body segments, especially the trunk and the feet, are not rigid; they often are deformed. The body tissues, such as at the fingertips, deform at the area of contact with an external object. There is a certain passive resistance to motion and thus energy dissipation in the joints—for example, passive resistance in the finger joints can be rather high. Movement at the joints can be more complex than pure rotation about fixed joint axes. For instance, the shoulder joint allows sizable translation. The mass-inertial characteristics of the body parts during movement may vary due to the displacement of the internal organs with respect to the spine in the trunk and the muscles with respect to the bones in the body limbs. The replacement of the muscles, especially multijoint muscles, by torque actuators may also be inadequate. Nevertheless, when a displacement of the body is large compared with tissue deformation, the basic model is justified. The basic model has been used both as an independent research tool and as a first step in a more detailed biomechanical analysis. In what follows, bodies are considered rigid and joints frictionless. This is always an approximation to the real state of affairs, but a very convenient and useful one.

This chapter and chapter 3 deal with the static analysis of biokinematic chains. Statics is the branch of mechanics that studies bodies at equilibrium subjected to the action of forces. Section **2.1** addresses the notion of joint torques. The difference between actual and equivalent torques is stressed. The actual joint torques are produced by one-joint muscles; *two-joint muscles* generate equivalent torques. The transformation of joint torques into the endpoint force and the inverse transformation of the endpoint force into the joint torques are considered in section **2.2**. Section **2.3** deals with producing an endpoint force of a desired magnitude and in a desired direction. Finally, section **2.4** establishes an analogy between statics and kinematics and introduces the concept of null space.

• • • *MECHANICS REFRESHER* • • •

Static Equilibrium

Bodies in *equilibrium* either are at rest or move along a straight line with constant velocity. In this chapter, we consider bodies at rest.

The conditions for equilibrium can be worded in different ways. In particular, a body is at equilibrium if and only if

1. the resultant force vector and the resultant couple are both zero and
2. the sums of the X, Y, and Z components of all forces acting on the body are zero and the sums of the moments of all the forces with respect to the X, Y, and Z axes are zero. (In total, there are six equations: three for the forces and three for the moments.)

Static equilibrium can be analyzed in three-dimensional space or in a plane. For instance, a freely gliding figure skater can maintain a perfect static equilibrium in the *frontal plane* that is perpendicular to the movement direction.

2.1 TORQUE AND FORCE ACTUATORS AT REVOLUTE JOINTS

By definition, *actuators* are devices that generate force or convert one form of energy into another. Each joint in a kinematic chain is driven or powered by one or more actuators. Examples of actuators in technical systems are electric motors and hydraulic or pneumatic cylinders. In humans, muscles are the actuators.

Joint torques that are generated by actuators are called *active joint torques*. In biomechanics, the active torques are customarily called *muscle-tendon torques*. Joint torques can also be due to the passive resistance provided by the bones and ligaments; those torques are not produced by torque actuators. This situation occurs when the limit of the joint range of motion is achieved. (In the model under consideration, the joints are frictionless, and hence passive resistance to motion in the working range of the joint is neglected.) For instance, if the elbow joint is completely extended, the bones and ligaments provide resistance to further extension. Another example is joints with one degree of freedom (1-DoF joints), when joint torque is not about the joint axis. Such torques are called *passive joint torques. Net joint torque* is the total torque produced at a joint. The net torque is the sum of the active and passive torques. This section deals with active (muscle-tendon) torques, that is, with the torques generated by the actuators.

Actuators are classified as torque actuators (*rotary actuators*) or *force actuators* (*linear actuators*). This is not a literal classification. Strictly speaking, torque actuators produce force. The force, however, is always normal to the long axis of the link. The sole effect of this force is a moment of force at the

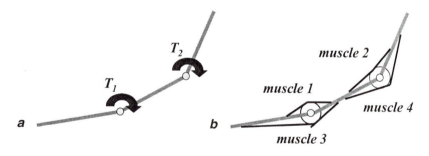

Figure 2.2 Models of kinematic chains with (*a*) two torque actuators, T_1 and T_2, and (*b*) four force actuators, muscles 1, 2, 3, and 4. Muscles 1 and 2 are single-joint flexors and muscles 3 and 4 are single-joint extensors. Forces applied by the muscles on the links produce moments of force about the joint axes. The forces exerted by the muscles on the links may be resolved into components acting along the links and components acting perpendicularly to them.

joint. Force generated by a torque actuator does not have a component acting along the link; it neither increases nor decreases the force acting at the joint bearing. An example is an electric motor in which the axis of rotation coincides with the joint rotation axis. Torque actuators are broadly used in various technical systems. They do not increase the force acting on the axle and do not waste energy for unnecessary displacement of rotating parts toward or away from the rotation axis (figure 2.2).

Muscles, however, are not torque actuators. They are force actuators. A muscle force can always be resolved into two components: (1) normal to the long axis of the bone (this force component contributes to the moment of force at the joint) and (2) along the bone (this force component increases or decreases the force exerted by one of the adjacent links at the joint). Hence, the tension in the muscles determines both the moments and the forces at the joints.

2.1.1 Joint Torques at Joints Served by One-Joint Muscles

Representative papers: Zajac and Gordon (1989); Zajac (1993)

Consider a one-joint muscle crossing a frictionless revolute joint (figure 2.3). A planar case is considered. The muscle develops collinear forces at the points of its effective origin and insertion. The forces can be resolved into two components: along the link and normal to the link. The forces acting along the links pass through the joint center and, in *ideal joints* (frictionless joints with nondeformable joint surfaces), do not affect the motion of the segments about this joint. The normal forces produce moments of force about the joint center. The forces, $\mathbf{F}^{mus}$ and $-\mathbf{F}^{mus}$, are equal in magnitude but point in opposite directions and act on the different links. Their moment arms, d, about the joint center are the same. Hence, the forces $\mathbf{F}^{mus}$ and $-\mathbf{F}^{mus}$ produce moments of force about the joint center of the same magnitude and opposite in direction; a clockwise moment is applied to one link and a counterclockwise moment to another.

Each force, $\mathbf{F}^{mus}$ or $-\mathbf{F}^{mus}$, can be represented by a force applied at the joint center and a force couple acting on the segment (figure 2.3c). The moment of each force couple is equal to the moment of the muscle force about the joint center. These two moments, when considered together, are called the joint moment or joint torque. In figure 2.3d they are represented by one curved arrow. The sign of a joint torque is selected arbitrarily; for example, a moment acting in flexion can be designated positive and a moment acting in extension, negative.

The model presented in figures 2.3a–c illustrates the idea of joint torque. Representation of a muscle as a straight line from the origin to insertion (like a bow string) is an obvious oversimplification. The lines of action of many muscles, such as the gluteus, are curved. The muscles are customarily curved

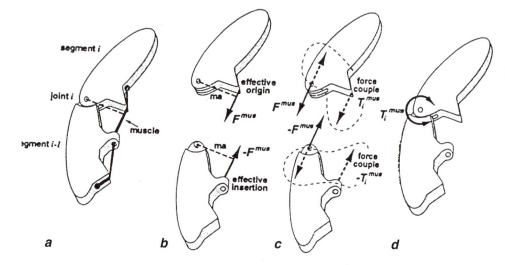

Figure 2.3 A two-link system with one single-joint muscle. (*a*) The path of the muscle is the heavy, solid line. The shortest distance from the muscle to the joint center is the moment arm of the muscle, the dashed line. (*b*) The forces at the effective origin of the muscle ($\mathbf{F}^{mus}$) and its effective insertion ($-\mathbf{F}^{mus}$) produce the moments of force about the joint center. The moments are equal and opposite. They act on the different segments. (*c*) Each of the two muscle forces in (*b*) can be equivalently represented by a force acting at the joint center and a force couple acting on one of the segments. The torques of these two couples are equal to the moments of the two forces in (*b*) about the joint center. (*d*) The two segmental torques in (*c*) are collectively referred to as the joint torque produced by the force in the muscle.

Reprinted from *Journal of Biomechanics*, 26 Suppl. 1, F.E. Zajac, Muscle coordination of movement: A perspective, 109–124, 1993, with permission from Elsevier Science.

when they act through an angle larger than π. Also, the synovial sheaths or retinacula restrain tendons of many muscles, for instance, the finger flexors, so that they stay close to the skeleton. As a result, in the majority of joints, the two forces that the muscle develops at its effective origin and insertion points are of equal magnitude but are not collinear. Still, there exists a minimal distance from the center of rotation to the (curved) line of force action along the curved muscle or the curved tendon. Therefore, as in the previously described straight-line case, two equal and opposite moments of force about a common axis of joint rotation are applied to the two adjacent segments; by and large, they can be referred to as the joint moment.

The knee joint is, however, unique (figure 2.4). During knee extension, the patella acts as a lever. Force produced by the quadriceps—the only extensor at the joint—is applied to the top of the patella. This force is not equal to the

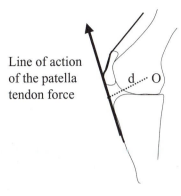

Line of action
of the patella
tendon force

Figure 2.4 Forces and moments at the knee joint (a schematic). The patellar tendon is presumed to be nonextensible. The patella is in equilibrium under the action of three forces: from the quadriceps tendon, from the patellar tendon, and from the patellofemoral contact. The moment arm of the patella tendon force d is the shortest distance from the axis of rotation O to the tendon line of action.

force acting on the tibia through the patellar tendon. The moment arms of the two forces are also of different magnitude. However, knee flexion–extension takes place about an axis located in the femur; hence, only the moment arm of the patella tendon is important. The moments exerted by the patella tendon force on the tibia and the "patella-plus-femur" system are of equal magnitude and opposite direction. Hence, the concept of a joint torque can be applied.

2.1.2 Joint Torques at Joints Served by Two-Joint Muscles

Representative paper: Zatsiorsky and Latash (1993)

In joints served by only uniarticular muscles, the existence of two equal-in-magnitude and opposite-in-direction moments of force acting on the adjacent segments is a straightforward consequence of Newton's third law. When a joint is spanned by a two-joint or multijoint muscle, the situation becomes more confusing. The notion of joint torque cannot be immediately applied. Consider a two-joint muscle spanning hinge joints J_1 and J_2 (figure 2.5).

The muscle, shown by a bold line, attaches to segments S_1 and S_3 and spans segment S_2. The dashed lines show the moment arms, d_1, d_2, and d_3, the shortest distances from the joint centers to the line of force action. The forces acting on S_1 and S_3 (figure 2.5*b*) can be equivalently represented by a resultant force acting at the joint contact point and a force couple acting on the segment (figure 2.5*c*). The forces acting on S_2 do so through joint centers J_1 and J_2 and therefore do not generate moments about the joint through which they pass. Thus, biarticular muscles do not immediately create moments of force about

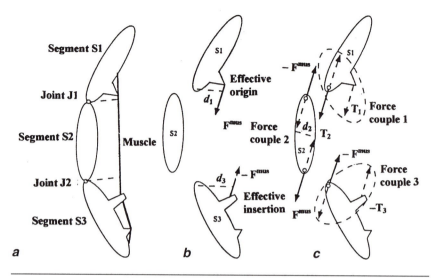

Figure 2.5   Forces and couples acting in a three-link chain served by one two-joint muscle. The muscle does not exert a moment of force on an intermediate segment. The segment rotation is due to force couple 2, two equal and opposite forces acting at the joint centers J_1 and J_2. (*a*) A three-segment system (S_1, S_2, and S_3) with one *biarticular muscle* crossing two frictionless revolute joints, J_1 and J_2. (*b*) Muscle force acting on segments S_1 and S_3 and the different moment arms. (*c*) An equivalent representation of forces produced by the biarticular muscle. Forces acting on S_1 and S_2 are equivalently represented by a force acting at the joint center and a force couple. Two equal, opposite, and parallel forces act on segment S_2 at the joint centers. They may cause the segment to rotate.

Adapted, by permission, from V.M. Zatsiorsky and M.L. Latash, 1993, "What is a joint torque for joints spanned by multiarticular muscles?" *Journal of Applied Biomechanics* 9:333–336.

the joints to an intermediate segment. However, the two forces acting on the joint centers of S_2 are equal in magnitude, parallel, and opposite in direction. Consequently, they can be represented by force couple 2 with the moment arm d_2. In general, moment arms d_1, d_2, and d_3 may be different. Thus, force couples 1, 2, and 3 may be different too. By the previous definition, the term "joint torque" collectively refers to two moments of force equal in magnitude that act on adjacent segments about the same joint rotation axis. Because the magnitudes of the moments of force acting on adjacent segments served by a biarticular muscle are different, these moments cannot be collectively referred to as joint torque if the definition is strictly followed.

The relationship among torques T_1, T_2, and T_3 is determined by the relationship among moment arms d_1, d_2, and d_3 (the same muscle force $\mathbf{F}^{mus}$ causes all three torques). Since all three moment arms are perpendicular to the line of muscle action, they run in parallel. Therefore, the moment arm of force couple

2 equals the difference between d_1 and d_3: $d_2 = d_1 - d_3$. The torque acting on S_2 about J_1 is

$$T_2 = F^{mus}d_2 = F^{mus}(d_1 - d_3) = T_1 + T_3 \qquad (2.1)$$

because $T_3 = -F^{mus}d_3$. In the particular case when $d_1 = d_3$, d_2 is zero and the resultant moment of the forces applied to the intermediate segment S_2 at J_1 and J_2 is zero.

In general, the expression "a torque (or moment) at joint X," if not defined explicitly, can be misleading. The expression "a moment of force **Y** exerted on segment Z around a joint axis X" does not lead to confusion in cases of multiarticular muscles. In experimental investigations, however, the specificity of biarticular muscles is often disregarded when joint torques are determined. The implicit assumption is that moments acting on adjacent segments at a joint are always equal in magnitude and have opposite signs. This presumes that either biarticular muscles are not active or they produce equal moments on adjacent segments whether they are attached to them or not. A typical sequence of calculations is as follows:

1. Torque T_3 acting on (distal) segment S_3 about J_2 is determined.
2. It is assumed that a torque $-T_3$ is acting on an intermediate segment S_2 about J_2 (which is not true).
3. It is assumed that torque T_3 contributes also to torque T_1 about J_1; T_1 is then determined as an algebraic sum $T_1 = -(T_3 + T_2)$.

As a result, the moments of forces acting on the intermediate segment S_2 are estimated as $-T_3$ in J_2 and $-T_1 = (T_3 + T_2)$ in J_1. The algebraic sum of these two moments is exactly T_2. Therefore, the addition of $-T_3$ to the moment acting on S_2 in J_2 and subsequent subtraction of the same value when the moment is calculated in J_1 does not change the external effects because the torque systems T_2 and $(T_2 + T_3 - T_3)$ are equivalent. Thus, when only the external effects of the forces are of concern, as in static analysis of kinematic chains consisting of rigid links, "joint torques" can be introduced. They should, however, be considered "equivalent" rather than "actual" joint torques.

The difference between actual joint torques (produced by only uniarticular muscles) and *equivalent joint torques* (calculated for the system served by two-joint muscles) is important in some situations. For instance, calculation of total joint power for joints served by one- and two-joint muscles should be performed differently (this issue is discussed in chapter **6**). When the object of interest is an internal effect of force, such as link stress and deformation or joint stability, equivalent moments cannot be immediately used. For instance, in a three-link system with two one-joint muscles (figure 2.6*a*), when $T_1 = T_3$, a bending stress acts on the intermediate link. For a similar system with one two-joint muscle (figure 2.6*b*), the bending stress is zero: only a compressive

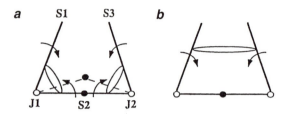

Figure 2.6 A three-link system served by (*a*) two one-joint muscles and (*b*) one two-joint muscle. An additional locked pin joint (filled circle) is located in the middle of segment S_2. If the locked joint is released, (*a*) the left and right parts of S_2 will rotate in the direction shown by the arrows (upward) or (*b*) the unlocked joint will be in a state of unstable equilibrium.

Adapted, by permission, from V.M. Zatsiorsky and M.L. Latash, 1993, "What is a joint torque for joints spanned by multiarticular muscles?" *Journal of Applied Biomechanics* 9:333-336.

force acts on the intermediate link. When the actual forces and moments acting at the joints are replaced by joint torques, this distinction is lost.

Some models of motor control are based on an assumption that the CNS plans and immediately controls joint torques. This is a debatable idea, especially when equivalent joint torques due to two-joint muscles are involved. Since equivalent joint torques are just abstract concepts, a proper understanding of the expression "the CNS controls joint torques" is important. Consider as an example another abstract concept—the center of mass (CoM) of a body. It is well known that the sum of external forces acting on a rigid body equals the mass of the body times the acceleration of the body's CoM. Does this mean that a central controller, in order to impart a required acceleration to the body, controls the force at the CoM? In a roundabout way yes, but actually no. The CoM is an imaginary point, a fictitious particle that possesses some very important features. The CoM can be outside the body, as for a bagel. One cannot actually apply a force to the CoM of a bagel since it is somewhere in the air. Hence, the expression "to control a force at the CoM" should not be taken literally. Similarly, the CNS cannot immediately control joint torques because they are just abstract concepts—like the CoM of a bagel.

2.2 TRANSFORMATION ANALYSIS

The aim of this section is to relate external forces and couples that act on an end effector with forces and torques at the joints. This relationship is established through *transformation analysis.* For brevity, the external force and couple applied at a point P is called a generalized external force, $\overline{\mathbf{F}}_P$. Generalized external force $\overline{\mathbf{F}}_P$ is a six-component vector. Its elements are three force components and three moment components. The generalized external force produces a torque at each joint, $-\mathbf{T}_i$, where the subscript i refers to an individual joint ($i = 1, 2, \ldots, n$, where n is the number of joints). To maintain equilibrium, the torque actuators must generate an equilibrating torque, $\mathbf{T}_i$, at each joint. (It is also possible that the joint torques are resisted passively, by the skeleton.) In three dimensions, $\mathbf{T}_i$ is a vector; in two dimensions, it is a scalar. When an external force is given and the task is to compute the magnitudes of the torque at each joint, the procedure is called *inverse static analysis*. Conversely, when the joint torques are known and the task is to compute the force applied to the end effector, the procedure is called *direct static analysis* (or *forward static analysis*).

We assume that the joints are ideal revolute joints powered by torque actuators—actuators that do not exert force along the links. The joint torques calculated via transformation analysis, $\mathbf{T} = [\mathbf{T}_i]$, counterbalance the force and couple exerted externally on the end effector. Such torques are called the *equilibrating joint torques* (of the endpoint force $\overline{\mathbf{F}}_p$). The equilibrating torques do not account for other forces and moments, such as the force of gravity. We are currently interested in inverse static analysis: how external contact forces and couples "propagate" along kinematic chains. Two techniques are described: a link-isolation method and the Jacobian method.

2.2.1 Link Isolation

Representative paper: Dempster (1961)

We first describe the method of link isolation and then apply the method to a simple kinematic chain.

2.2.1.1 Method

To determine the forces and moments acting at joints on body links, individual links can be separated from adjacent ones and then analyzed. This separation is usually accomplished by means of a *free-body diagram,* a schematic representation of an isolated link showing all the forces and couples acting on the link. When a free-body diagram is drawn, force calculations can be performed. Because the system is in equilibrium, the vector sum of all the forces acting on the link and the sum of all the moments are zero. Some of the forces and moments must be known. The solution involves the determination of the unknown forces and moments. Link isolation and free-body diagrams are classical techniques of statics. The technique is simple but requires strictly following two basic rules:

1. The link of interest should be completely isolated from all other contacting bodies. The external boundary of the link should be sketched.

2. All external forces and couples, both known and unknown, that act on the isolated link should be represented in the diagram by vector arrows. Note that the word "external" in this context means external to the isolated link.

According to Newton's third law, an equal and opposite reaction force exists for every acting force. External forces are those forces whose reaction forces do not act on the link under consideration. Internal forces acting between the particles of the link cancel each other and, if the link is rigid, do not change the link movement. When the diagram is completed, principles of mechanics can be applied. More details on the method can be found in any textbook on statics. An example of a free-body diagram of the forearm is presented in figure 2.7.

In the static analysis of kinematic chains, all the internal forces and couples acting on a link are reduced to forces and couples acting at the joints. Though

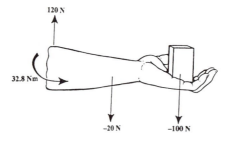

Figure 2.7 A free-body diagram of the forearm with a 100-N load held in the hand. The combined weight of the forearm and hand is 20 N. The distance from the elbow joint to the point of load application is 30 cm, and the distance from the joint to the center of gravity of the forearm-plus-hand system is 14 cm. The moment of force with respect to the elbow joint center is $100 \text{ N} \times 0.3 \text{ m} + 20 \text{ N} \times 0.14 \text{ m} = 32.8 \text{ N} \cdot \text{m}$. Note that the muscle force is considered an external force in this analysis. However, if the arm or the human body as a whole were analyzed, the muscle force would be considered an internal force.

■ ■ ■ *From the Literature* ■ ■ ■

Free-Body Diagram for Calculating Forces on a Lumbar Disk During Lifting

Source: Daggfeldt, K., and A. Thorstensson. 1997. The role of intra-abdominal pressure in spinal unloading. *J. Biomech.* 30 (11–12): 1149–55.

To construct the free-body diagram shown in figure 2.8, an imaginary cut was performed along the transverse plane through the center of the L5/S1 vertebral disk.

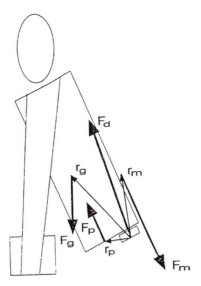

Figure 2.8 Free-body diagram for calculating loading on a lumbar disk during lifting. $\mathbf{F}_m$ is a muscle force generated by the back extensors, $\mathbf{F}_p$ is a force from intra-abdominal pressure, $\mathbf{F}_d$ is a load on the disk, and $\mathbf{r}_m$, $\mathbf{r}_g$, and $\mathbf{r}_p$ are the vector distances from the point of application of $\mathbf{F}_d$ to the points of application of $\mathbf{F}_m$, $\mathbf{F}_g$, and $\mathbf{F}_p$, respectively. The intra-abdominal pressure that results from compression of the abdominal cavity is, by assumption, created by the contraction of the transversus abdominis without producing any longitudinal tension. Because of that, the longitudinal tension in the abdominal wall is neglected in the diagram.

Reprinted from *Journal of Biomechanics*, 30 (11/12), K. Daggfeldt and A. Thorstensson, The role of intra-abdominal pressure in spinal unloading, 1149–1155, 1997, with permission from Elsevier Science.

The equations of static equilibrium for the model are

$$\mathbf{F}_g + \mathbf{F}_m + \mathbf{F}_p + \mathbf{F}_d = 0 \tag{2.2}$$
$$\mathbf{r}_g \times \mathbf{F}_g + \mathbf{r}_m \times \mathbf{F}_m + \mathbf{r}_p \times \mathbf{F}_p = 0$$

The L5/S1 disk is compressed by the back-extensor muscle force $\mathbf{F}_m$ and the force of gravity $\mathbf{F}_g$. The force generated by the intra-abdominal pressure, F_p (shown as applied on the diaphragm), acts in the opposite direction and decreases the compression. The gravitational force produces a flexor moment around L5/S1, whereas the back-extensor muscles and the intra-abdominal pressure generate an extensor moment. As a result, F_p reduces the need for back-extensor muscle force and unloads the L5/S1 disk.

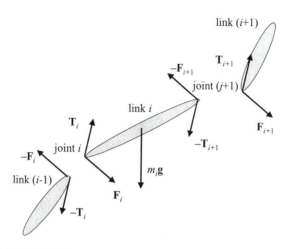

Figure 2.9 A free-body diagram of a body link i. Two resultant forces, $\mathbf{F}_i$ and $-\mathbf{F}_{i+1}$, and two resultant couples that produce the torques $\mathbf{T}_i$ and $-\mathbf{T}_{i+1}$ act on link i at the joints i and $i + 1$. Note that according to Newton's third law, equal and opposite forces and couples are applied to the neighboring links $i - 1$ and $i + 1$. For completeness, the weight $m_i\mathbf{g}$ is included in the free-body diagram but is neglected at the equilibrating torques.

some muscles have their origin and insertion quite far from the joints (for instance, the brachioradialis muscle inserts at the lower end of the radius), the far-off muscle forces are replaced by the forces and couples at the corresponding joints (for the brachioradialis, at the elbow joint). As a result, for a link with two joints, only two resultant forces acting at the joints and two couples are considered (figure 2.9). The weight of the link should also be included. One joint force and couple only are applied to distal segments (e.g., link $i + 1$ in figure 2.9). Five joint forces and couples act on the trunk: at the shoulder joints, hip joints, and the trunk–neck articulation (not shown in the figure).

To isolate individual links in a chain, we first assume that all the joints are locked and the chain is one rigid structure. This operation is sometimes called "joint freezing." Then each link is isolated. To isolate a link, constraints that keep the links together are replaced with the reaction forces and moments. After that, the equilibrium equations can be written and solved. The equations of equilibrium can be written in any appropriate system of coordinates. Usually, either a global reference frame or a local frame tied to a link is used. Both approaches are equally useful. We write the equations first in the local frames. We use the symbols $\mathbf{F}_i$ and $\mathbf{T}_i$ for the force and the moment of a couple (torque) exerted on link i by link $i - 1$ (see figure 2.9). The $\mathbf{F}_i$ and $\mathbf{T}_i$ is three-component vectors. The force exerted on link i by link $i + 1$ is $-\mathbf{F}_{i+1}$. The following equations are written in the local system of coordinates $\{i\}$. The system has its origin at the center of joint i and is fixed with link i. In this system of coordinates, the conditions for equilibrium are

$$\mathbf{F}_i - \mathbf{F}_{i+1} = 0$$
$$\mathbf{T}_i - \mathbf{T}_{i+1} - \mathbf{r}_{i+1} \times \mathbf{F}_{i+1} = 0 \tag{2.3}$$

where $\mathbf{r}_{i+1}$ is the position vector from joint i to joint $i + 1$. In equation 2.3, the joint force and moment acting at joint $i + 1$ are written in the reference frame $\{i\}$. For instance, if the angle at the joint $i + 1$ is 90° and the force acts along link $i + 1$, this force is represented in equation 2.3 as acting perpendicularly to link i. To write the equations for the forces and moments defined in their own link frames, the rotation matrix should be used:

$$^{\{i\}}\mathbf{F}_i = [R]^{\{i+1\}}\mathbf{F}_{i+1} \tag{2.4a}$$
$$^{\{i\}}\mathbf{T}_i = [R]^{\{i+1\}}\mathbf{T}_{i+1} + {}^{\{i\}}\mathbf{r}_{i+1} \times {}^{\{i\}}\mathbf{F}_{i+1} \tag{2.4b}$$

where the leading superscripts $\{i\}$ and $\{i + 1\}$ stand for the corresponding reference frames and $[R]$ is a rotation matrix that accomplishes the change from reference frame $\{i + 1\}$ into reference frame $\{i\}$. (Rotation matrices are explained in *Kinematics of Human Motion*, section **1.2.5.1**.) Hence, the symbols $^{\{i\}}\mathbf{F}_{i+1}$ and $^{\{i+1\}}\mathbf{F}_{i+1}$ designate the same force exerted on link i by link $i + 1$ but measured in different reference frames, $\{i\}$ and $\{i + 1\}$. Equation 2.4a represents two operations: the force $^{\{i+1\}}\mathbf{F}_{i+1}$ is transformed from reference frame $\{i + 1\}$ into reference frame $\{i\}$ and then moved from joint $i + 1$ to joint i. To compensate for this parallel displacement, a corresponding couple $^{\{i\}}\mathbf{r}_{i+1} \times {}^{\{i\}}\mathbf{F}_{i+1}$ is added in equation 2.4b. Equation 2.4b can also be written as

$$^{\{i\}}\mathbf{T}_i = [R]^{\{i+1\}}\mathbf{T}_{i+1} + {}^{\{i\}}\mathbf{r}_{i+1} \times [R]^{\{i+1\}}\mathbf{F}_{i+1} \tag{2.4c}$$

The expression $^{\{i\}}\mathbf{r}_{i+1} \times [R]^{\{i+1\}}\mathbf{F}_{i+1}$ in the right-hand part of the equation depicts two consecutive operations: transformation of the vector $\mathbf{F}_{i+1}$ from reference frame $\{i + 1\}$ into reference frame $\{i\}$, that is, from $^{\{i+1\}}\mathbf{F}_{i+1}$ to $^{\{i\}}\mathbf{F}_{i+1}$, and the cross product of the vectors $^{\{i\}}\mathbf{r}_{i+1}$ and $^{\{i\}}\mathbf{F}_{i+1}$. The expression $^{\{i\}}\mathbf{r}_{i+1} \times [R]$ designates this two-stage procedure. The reader is advised to compare equation 2.4a with equation 1.31a and equations 2.4b and 2.4c with equation 1.35.

In a matrix form, a force-moment transformation from frame $\{i + 1\}$ to frame $\{i\}$ is

$$^{\{i\}}\begin{bmatrix} \mathbf{F}_i \\ \mathbf{T}_i \end{bmatrix} = \begin{bmatrix} [R] & 0 \\ {}^{\{i\}}\mathbf{r}_{i+1} \times [R] & [R] \end{bmatrix}^{\{i+1\}} \begin{bmatrix} \mathbf{F}_{i+1} \\ \mathbf{T}_{i+1} \end{bmatrix} \tag{2.5}$$

which may be written as

$$^{\{i\}}\overline{\mathbf{F}}_i = [FMT]^{\{i+1\}}\overline{\mathbf{F}}_{i+1} \tag{2.6}$$

where $^{\{i\}}\overline{\mathbf{F}}_i$ and $^{\{i+1\}}\overline{\mathbf{F}}_{i+1}$ are six-component generalized force vectors and $[FMT]$ is a 6×6 *force-moment transformation matrix*.

Equations 2.4 through 2.6 are used to compute equilibrating joint torques needed to apply or resist the force and couple acting on the end effector. The equilibrating torques do not account for the weight of the links. The equations are written in the local system of coordinates (in the ith frame). Similar equations can be written in the global reference frame. The weight can also be included. Writing equations in a global reference frame fixed with the environment, we obtain

$$
\begin{aligned}
{}^{\{G\}}\mathbf{F}_i &= {}^{\{G\}}\mathbf{F}_{i+1} - m_i\mathbf{g} \\
{}^{\{G\}}\mathbf{T}_i &= {}^{\{G\}}\mathbf{T}_{i+1} + \mathbf{r}_{i,i+1} \times {}^{\{G\}}\mathbf{F}_{i+1} - \mathbf{r}_{iCM} \times m_i\mathbf{g}
\end{aligned}
\tag{2.7}
$$

The vectors in equation 2.7 are expressed in the global reference frame. The vector $\mathbf{r}_{i,i+1}$ that specifies the position of joint $i + 1$ with respect to joint i and the position vector $\mathbf{r}_{iCM}$ that defines the position of the center of mass of the ith link with respect to the joint center i are often given in the local reference frame. Before the position vectors $\mathbf{r}_i$ and $\mathbf{r}_{iCM}$ are substituted in equation 2.7, the vectors should be transformed into the global frame.

$$
\begin{aligned}
{}^{\{G\}}\mathbf{r}_{i,i+1} &= {}^{\{G\}}[R]_i \,{}^{\{i\}}\mathbf{r}_{i,i+1} \\
{}^{\{G\}}\mathbf{r}_{iCM} &= {}^{\{G\}}[R]_i \,{}^{\{i\}}\mathbf{r}_{iCM}
\end{aligned}
\tag{2.8}
$$

where $^{\{G\}}[R]_i$ is the rotation matrix that relates the ith reference frame with the global reference frame. To find joint forces and moments for all the links of the chain, a *recursive* procedure is used. Calculations start from the most distal link and then continue from link to link, one link at a time. The method is advantageous because joint forces and moments are solved for one link at a time without the need for solving many vector equations concurrently. The process is repeated until all the joint forces and moments are computed.

The equilibrating torques computed from equations 2.4 through 2.8 can be caused by muscle activity, or they can be passive, caused by the resistance provided by the skeleton and other passive tissues. For instance, when an arm is abducted 90° and supinated (the palm is up), the torque at the elbow joint is passive; elbow hyperextension is prevented by the skeleton. If the arm rotates internally, the elbow extensors become active to prevent elbow flexion. To find an active torque (i.e., a joint moment generated by the muscles), the dot product of the joint axis vector and the joint torque vector should be computed:

$$
\mathbf{T}_z = \mathbf{T}^T \cdot \mathbf{U}_z
\tag{2.9}
$$

where $\mathbf{T}_z$ is a component of the joint torque at joint i acting about joint rotation axis z, $\mathbf{T}^T$ is a transpose of the torque vector acting at joint i, and $\mathbf{U}_z$ is the joint axis vector, a unit vector along the axis of rotation. All the vectors are represented in the reference frame $\{i\}$. Equation 2.9 is essentially the same as equation 1.27. In joints with 3 DoF, in the working range of motion, all the torques are generated by torque actuators and are muscle-tendon torques.

2.2.1.2 Equilibrating Torques in a Planar Two-Link Chain

Consider a planar two-link chain (figure 2.10). We analyzed the kinematics of these chains in detail in *Kinematics of Human Motion*, section **2.2.5.1.1**. An external force $\mathbf{F}$ acts at point P. The external couple is zero. The goal is to find the equilibrating joint torques necessary to maintain the chain in equilibrium. The force that is acting on link 2, if measured in the local system of coordinates at a point P, is

$$^{\{2\}}\mathbf{F}_P = -\begin{bmatrix} F_x \\ F_y \\ 0 \end{bmatrix} \tag{2.10}$$

The force component F_x acts along the second link, and the force component F_y is normal to this link. At joint 2, force $^{\{2\}}\mathbf{F}_P$ generates a moment $-T_2 = -l_2 F_y$, which is counterbalanced by the equilibrating joint torque $T_2 = l_2 F_y$. The

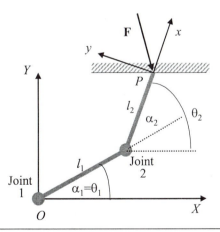

Figure 2.10 A planar two-link chain. Greek letters designate angles: α designates joint angles and θ designates the angles in the external reference frame O-XY. The numbers 1 and 2 refer to the first and second joint/link. A local reference system P-xy is fixed with the second link. A reference frame at joint 2, not shown in the figure, is parallel to the reference frame P-xy.

force component F_x is along link 2 and does not create a moment about joint 2. In vector form, the equilibrating joint torque is

$$
{}^{\{2\}}\mathbf{T}_2 = \begin{bmatrix} 0 \\ 0 \\ l_2 F_y \end{bmatrix}
\tag{2.11}
$$

Note that the external force $\mathbf{F}_p$ "propagates" to joint 2 without change, ${}^{\{2\}}\mathbf{F}_p = {}^{\{2\}}\mathbf{F}_{1,2}$, where ${}^{\{2\}}\mathbf{F}_{1,2}$ is the force exerted on link 1 by link 2. Going from link 2 to link 1 and changing the system of coordinates, we have

$$
{}^{\{1\}}\mathbf{F}_{1,2} = [R]\,{}^{\{2\}}\mathbf{F}_p = \begin{bmatrix} C_2 & -S_2 & 0 \\ S_2 & C_2 & 0 \\ 0 & 0 & 1 \end{bmatrix}\begin{bmatrix} F_x \\ F_y \\ 0 \end{bmatrix} = \begin{bmatrix} C_2 F_x - S_2 F_y \\ S_2 F_x + C_2 F_y \\ 0 \end{bmatrix}
\tag{2.12}
$$

where the abbreviations C_2 and S_2 stand for $\cos\alpha_2$ and $\sin\alpha_2$, respectively. The force ${}^{\{1\}}\mathbf{F}_{1,2}$ acts on link 1 and is measured in its reference frame. Its magnitude and line of action are the same as those of the external force ${}^{\{2\}}\mathbf{F}_p$. The equilibrating torque at joint 1 is equal to the sum of the joint moment at joint 2 (i.e., $l_2 F_y$) and the moment produced by the force $\mathbf{F}_{1,2}$ acting on link 1 at joint 2:

$$
{}^{\{1\}}\mathbf{T}_1 = \begin{bmatrix} 0 \\ 0 \\ l_2 F_y \end{bmatrix} + l_1\begin{bmatrix} 1 \\ 0 \\ 0 \end{bmatrix}\times {}^{\{1\}}\mathbf{F}_{1,2} = \begin{bmatrix} 0 \\ 0 \\ l_2 F_y + l_1 S_2 F_x + l_1 C_2 F_y \end{bmatrix}
\tag{2.13}
$$

Hence, the equilibrating joint torques in the plane of rotation are

$$
\begin{aligned}
T_1 &= l_1 S_2 F_x + \left(l_2 + l_1 C_2\right)F_y \\
T_2 &= l_2 F_y
\end{aligned}
\tag{2.14}
$$

Writing equation 2.14 in matrix form, we have

$$
\mathbf{T} = \begin{bmatrix} T_1 \\ T_2 \end{bmatrix} = \begin{bmatrix} l_1 S_2 & l_2 + l_1 C_2 \\ 0 & l_2 \end{bmatrix}\begin{bmatrix} F_x \\ F_y \end{bmatrix}
\tag{2.15}
$$

The matrix in equation 2.15 is the transpose of the *chain Jacobian* written in the local system of coordinates (recall that the force components F_x and F_y are given in the local frame P-xy). This is not a simple coincidence. The relationship among the endpoint force, equilibrating torques, and the chain Jacobian is addressed in the following section.

••• *Mechanics Refresher* •••

Virtual Work

Virtual displacement is a hypothetical, small displacement of a body or a system from an *equilibrium position*. The word "virtual" means imaginary: the displacements do not actually occur but are only theoretical.

Virtual work is the work done by a force over a virtual displacement.

Consider a system of rigid bodies interconnected by ideal joints. The following principle of virtual work is valid for a system in equilibrium: the total virtual work done on the system by all the externally applied forces is zero.

2.2.2 A General Relationship Between End Force and Joint Moments: The Jacobian Method

Representative paper: Fujikawa et al. (1997)

Consider an open kinematic chain. The joints are assumed to be frictionless. Gravity is neglected. A generalized external force $\overline{\mathbf{F}}$ is applied to the end effector. The force vector is measured in the global system of coordinates. The vector $\overline{\mathbf{F}}$ has six dimensions: three force components acting along the axes X_1, X_2, and X_3 and three moment components M_1, M_2, and M_3 about these axes. The goal is to find the equilibrating torques induced at the joints, T_j ($j = 1, 2, \ldots, N$). The following theorem is valid:

$$\mathbf{T} = \mathbf{J}^T \overline{\mathbf{F}} \qquad (2.16)$$

where $\mathbf{T}$ is the vector of the joint torques [$\mathbf{T} = (T_1, T_2, \ldots, T_N)^T$] and $\mathbf{J}^T$ is the transpose of the Jacobian matrix that relates infinitesimal joint displacement $d\boldsymbol{\alpha}$ to infinitesimal end-effector displacement $d\mathbf{P}$. The dimensionality of the vector $\mathbf{T}$ equals the number of degrees of freedom of the chain, N. The Jacobian is a $6 \times N$ matrix.

To prove the theorem, we use the principle of virtual work. According to the principle, the system is in equilibrium if and only if the virtual work equals zero. The work performed is the same regardless of whether the coordinate system used is an external system or a joint-based system. The virtual work done at the endpoint is equal to the dot product $VW_{ep} = \overline{\mathbf{F}}^T \cdot \delta\mathbf{P} = F_1\delta X_1 + F_2\delta X_2 + F_3\delta X_3 + M_1\delta\theta_1 + M_2\delta\theta_2 + M_3\delta\theta_3$, where $\delta\mathbf{P}$ is the vector of the end-effector virtual displacement, and X_1, X_2, X_3 and θ_1, θ_2, and θ_3 are the linear and angular coordinates of the end effector, respectively. Alternatively, the virtual work done at the joints equals the dot product of the joint torque vector and the vector of the joint virtual displacement, $VW_{joints} = \mathbf{T}^T \delta\boldsymbol{\alpha} = T_1 \delta\alpha_1 + T_2 \delta\alpha_2 + \ldots + T_N \delta\alpha_N$. To set off the virtual displacements from the actual ones, the Greek letter δ rather than d is used. Unlike the actual displacements, virtual displacements must satisfy only geometric constraints and are not obliged to conform to other laws of motion. For the kinematics chain under consideration, the entire virtual work is given by

$$VW = F_1 \delta X_1 + F_2 \delta X_2 + F_3 \delta X_3 + M_1 \delta\theta_1 + M_2 \delta\theta_2 +$$
$$M_3 \delta\theta_3 - T_1 \delta\alpha_1 - T_2 \delta\alpha_2 - \ldots - T_N \delta\alpha_N \qquad (2.17)$$

or

$$VW = \mathbf{T}^T \delta\boldsymbol{\alpha} - \overline{\mathbf{F}}^T \delta\mathbf{P} = 0 \qquad (2.18)$$

According to equation 2.18 and by the principle of virtual work, the virtual work done by the external generalized force $\overline{\mathbf{F}}^T \delta\mathbf{P}$ equals the work performed by the joint torques, $\mathbf{T}^T \delta\boldsymbol{\alpha}$. In other words, the virtual work done by the external force is negated by the virtual work performed at the joints, and as a result $VW = 0$. Recall that

$$\delta\mathbf{P} = \mathbf{J} \delta\boldsymbol{\alpha} \qquad (2.19)$$

Equation 2.18 can be now rewritten as

$$\mathbf{T}^T \delta\boldsymbol{\alpha} = \overline{\mathbf{F}}^T \mathbf{J} \delta\boldsymbol{\alpha} \qquad (2.20)$$

Because equation 2.20 holds for all infinitesimal joint displacements $\delta\boldsymbol{\alpha}$ and also because $\delta\boldsymbol{\alpha}$ 0,

$$\mathbf{T}^T = \overline{\mathbf{F}}^T \mathbf{J} \qquad (2.21)$$

Taking the transposes of both sides yields equation 2.16, $\mathbf{T} = \mathbf{J}^T \overline{\mathbf{F}}$. Thus, the theorem is proved. If the external couples are zero, equation 2.16 is simplified to $\mathbf{T} = \mathbf{J}^T \mathbf{F}$, where $\mathbf{F}$ is the endpoint force, a three-component vector.

Equation 2.16 expresses a general relationship between the vector of equilibrating joint torques and the external generalized force for a specific configuration of a kinematic chain. The equation allows a force defined in the external space to be transformed into the torques defined in the joint space without inverting the chain Jacobian. According to equation 2.16, the equilibrating joint torques in a given chain configuration are uniquely defined by the generalized external force. Consider the case of a subject pressing with a finger against an external object. Both the magnitude and direction of the endpoint force are constant. If the arm, hand, and finger do not move, the torques in all the joints are fixed. The subject cannot increase or decrease the torque in one joint without changing the torques in other joints. The chain is said to be overdetermined. The chain overdeterminacy is discussed in more detail in section **2.3.7**.

2.2.3 Relationships Between End Force and Joint Torque for Planar Kinematic Chains

This section deals with planar two- and three-link chains.

2.2.3.1 Two-Link Chain

Consider again the chain presented in figure 2.10. We already found the joint torques for this chain using the link-isolation method. Now we solve the same problem using the chain Jacobian. The vector **F** is given either in the global reference system, *O-XY*, or in the local system, *P-xy*. If the vector **F** is expressed in the global system, with respect to the environment, equation 2.16 can be applied immediately. The chain Jacobian (equation 3.16 in *Kinematics of Human Motion*) is

$$\mathbf{J} = \begin{bmatrix} -l_1 S_1 - l_2 S_{12} & -l_2 S_{12} \\ l_1 C_1 + l_2 C_{12} & l_2 C_{12} \end{bmatrix} \tag{2.22}$$

where the subscripts 1 and 2 refer to the angles α_1 and α_2, correspondingly, and the subscript 12 refers to the sum of the two angles, $(\alpha_1 + \alpha_2)$. Transposing the Jacobian and applying equation 2.16 gives

$$\begin{bmatrix} T_1 \\ T_2 \end{bmatrix} = \begin{bmatrix} -l_1 S_1 - l_2 S_{12} & l_1 C_1 + l_2 C_{12} \\ -l_2 S_{12} & l_2 C_{12} \end{bmatrix} \begin{bmatrix} F_X \\ F_Y \end{bmatrix} \tag{2.23}$$

The elements of the Jacobian transpose have a plain geometric meaning. They represent the moment arms of the external force **F** with respect to the individual joints 1 and 2 when the external force is written in projections on the *X* and *Y* axes (figure 2.11). For instance, the horizontal component of the external force F_X exerts a moment about joint 2 that equals $-M_2 = F_X d_2$, where the moment arm $d_2 = -l_2 S_{12}$. The moment is negative, in the clockwise direction. The moment of F_X about joint 1 is $-M_2 = F_X d_1$, where $d_1 = -(l_1 S_1 + l_2 S_{12})$. Hence, the joint torque in joint 1 is

$$T_1 = -(l_1 S_1 + l_2 S_{12})F_X + (l_1 C_1 + l_2 C_{12})F_Y \tag{2.24}$$

The reader is encouraged to trace the geometric meaning of the other elements of the Jacobian transpose.

The external force vector **F** can also be given in the local *P-xy* reference frame; for instance, the force **F** can be applied along the distal segment or

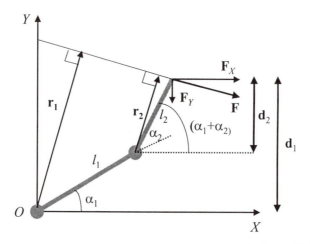

Figure 2.11 The correspondence between the moment arms of the external force **F** and the rows of the Jacobian transpose. The equivalent joint torques $\mathbf{T}_1$ and $\mathbf{T}_2$ of the external force **F** have the magnitudes $T_1 = Fr_1$ and $T_2 = Fr_2$, where r_1 and r_2 are the perpendicular distances from the corresponding joint to the line of **F**. The joint torque is normal to the plane containing the joint center and the line of **F**. Both the magnitude and the direction of $\mathbf{T}_1$ and $\mathbf{T}_2$ can be computed by using the cross product of $\mathbf{r}_i$ and **F**, where $\mathbf{r}_i$ ($i = 1, 2$) is a vector from the corresponding joint center to any point on the line of action of **F**, $\mathbf{T}_i = \mathbf{r}_i \times \mathbf{F}$. When the equations are written in scalar form, as projections on the axes of the system of coordinates, they can be conveniently represented by using the Jacobian transpose of the chain. Here the horizontal force component $\mathbf{F}_X$ is in a positive direction, and the vertical component $\mathbf{F}_Y$ is in a negative direction. Hence, both force components produce moments of force at joints 1 and 2 in the negative direction, clockwise. For instance, $T_1 = (-l_1 S_1 - l_2 S_{12})F_X + (l_1 C_1 + l_2 C_{12})(-F_Y)$ is a negative number.

perpendicularly to it, as in some strength exercise machines. When the external force is given in the local frame, the components of the vector **F** in the global frame should be obtained first by applying a rotation matrix $[R]$ to the force vector $^{(2)}\mathbf{F}_P = {}^{(2)}[F_x \ F_y]^T$. The rotation matrix is

$$[R] = \begin{bmatrix} C_{12} & -S_{12} \\ S_{12} & C_{12} \end{bmatrix} \tag{2.25}$$

where C_{12} stands for $\cos(\alpha_1 + \alpha_2)$ and S_{12} stands for $\sin(\alpha_1 + \alpha_2)$. The substitution of equation 2.25 into equation 2.16 leads to

$$\mathbf{T} = \mathbf{J}^T[R]^{(2)}\mathbf{F} \tag{2.26}$$

or, after the matrices $\mathbf{J}^T$ and $[R]$ in equation 2.26 are multiplied,

$$\begin{bmatrix} T_1 \\ T_2 \end{bmatrix} = \begin{bmatrix} l_1 S_2 & l_1 C_2 + l_2 \\ 0 & l_2 \end{bmatrix}^{\{2\}} \begin{bmatrix} F_X \\ F_Y \end{bmatrix} \tag{2.27}$$

where the components of the force vector $^{\{2\}}\mathbf{F}$ are given in the local system of coordinates $\{2\}$. Equation 2.27 is identical to equation 2.15 discussed previously in **2.2.1.2**.

If the joint torques T_1 and T_2 are known, the external force can be calculated by inverting the transposed Jacobian matrix, $\mathbf{J}^T$. For a planar two-link chain, the inverse exists in all joint configurations except singular ones. (Singular chain configurations are explained in *Kinematics of Human Motion*, section **3.1.1.1.4**.) The inverse of the transposed Jacobian is

$$\left(\mathbf{J}^T\right)^{-1} = \frac{1}{l_1 l_2 S_2} \begin{bmatrix} l_2 C_{12} & -l_1 C_1 - l_2 C_{12} \\ l_2 S_{12} & -l_1 S_1 - l_2 S_{12} \end{bmatrix} \tag{2.28}$$

Singular joint configurations are observed when α_2 equals either zero or π. In this case, $S_2 = 0$ and equation 2.28 cannot be solved. If a two-link chain represents an arm or a leg, the singular position corresponds to complete arm or leg extension when the included elbow or knee angle is 180° (complete flexion is impossible for human body links). At a singular joint configuration, the distal segment is an extension of the proximal body segment. It follows from equation 2.27 that at this particular joint posture, $T_1 = (l_1 + l_2)F_y$ and $T_2 = l_2 F_y$. Hence, at the singular chain configuration, the force component $^{\{2\}}\mathbf{F}_x$ that acts along the distal link does not influence the joint torques; $^{\{2\}}\mathbf{F}_x$ intersects the joint axes of rotation. The external load is said to be taken by the structure of the chain or, in other words, by the skeleton.

When the chain (e.g., the leg) approaches a singular joint configuration—a complete extension—the chain can bear large forces in the direction of the extension from the endpoint to the proximal joint (figure 2.12). However, at this chain configuration, the transfer of joint angular velocity into the velocity of the leg extension is minimal (see section **3.1.1.1.7** in *Kinematics of Human Motion*).

2.2.3.2 Three-Link Chain

For a planar three-link kinematic chain, the transpose of the Jacobian matrix is

$$\mathbf{J}^T = \begin{bmatrix} -l_1 S_1 - l_2 S_{12} - l_3 S_{123} & l_1 C_1 + l_2 C_{12} + l_3 C_{123} \\ -l_2 S_{12} - l_3 S_{123} & l_2 C_{12} + l_3 C_{123} \\ -l_3 S_{123} & l_3 C_{123} \end{bmatrix} \tag{2.29}$$

where, as usual, l_1, l_2, and l_3 represent the lengths of the links and subscripts 1, 12, and 123 refer to α_1, $(\alpha_1 + \alpha_2)$, and $(\alpha_1 + \alpha_2 + \alpha_3)$, respectively ($\alpha_1$, α_2, and

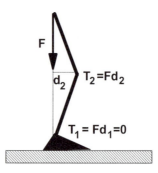

Figure 2.12 The closer the leg to full extension, the smaller is the knee joint moment required to bear the force **F**. This explains why the heaviest loads can be borne when the legs are (almost) completely extended. When the line of force action passes through the joint center, the equivalent joint moment is zero. When the leg is nearly outstretched, large external forces can be exerted with low joint moments.

■ ■ ■ *FROM THE LITERATURE* ■ ■ ■

Maximal Isometric Force During Leg Extension

Source: Hugh-Jones, P. 1947. The effect of limb position in seated subjects on their ability to utilize the maximum contractile force of the limb muscles. *J. Physiol. (Lond.)* 105: 332–44.

An experiment was performed to determine the effect of leg position on the maximal pushing force exerted on a pedal by a seated operator (figure 2.13*a*). The study was originally started for the purely practical purpose of placing controls. A foot pedal was placed in front of a seat with a vertical backrest. The pedal was mounted in different positions so that for each of five different angles (α) between the thigh and the horizontal, the relationship between pedal push and knee angle (β) was found. The subjects were asked to press on the pedal as hard as possible. The results are presented in figure 2.13*b*.

Push increased with the leg extension until a maximum was reached, before the knee joint becomes straight (figure 2.13*b*). The closer was the leg to full extension, the larger the maximal extension force. However, a well-defined limiting angle was reached when angle β was 160°, below the theoretically expected 180°. Above this angle, push suddenly decreased, the pedal subjectively felt too far away, and pushing became uncomfortable. This pattern is typical for sitting postures where the hamstring is extended. At large values of hip flexion ($\alpha > 33°$), the hamstring prevents

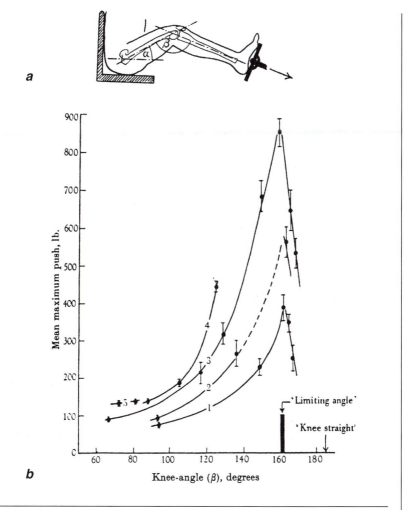

Figure 2.13 Dependence of the pushing force on limb position in seated subjects. (*a*) Experimental setup. (*b*)The mean maximum push (±2 standard deviations) exerted isometrically by six subjects on a pedal placed in different positions. For each of the five different angles between thigh and horizontal (α), the knee angle (β) varied. Curve 1 represents the data for α between −15° and −6°; curve 2, between +5° and +10°; curve 3, between +15° and +19°; curve 4, between +33° and +36°; and curve 5, between +48° and ±49°. Curves 4 and 5 necessarily stop as shown, well before the limiting angle is reached. At these thigh positions, the knee cannot be extended farther due to the limitation caused by the hamstring.

Adapted, by permission, from P. Hugh-Jones, 1947, "The effect of limb position in seated subjects on their ability to utilize the maximum contractile force of the limb muscles," *Journal of Physiology* (London) 105:332–344.

complete knee extension. At this magnitude of hip flexion, the knee cannot be extended farther. Therefore, the maximum force was exerted at much smaller knee-joint angles (curves 4 and 5). It was concluded that the limb acts as a mechanical "toggle" between the pedal and the backrest.

α_3 are the external joint angles). When a force $\mathbf{F}$ acts at the endpoint of the chain, equation 2.29 can be used immediately. When both a force and a force couple M are applied, the Jacobian transpose also contains a column corresponding to the orientation of the last link of the chain. The elements of the column are $\delta\theta/\delta\alpha_1$, $\delta\theta/\delta\alpha_2$, and $\delta\theta/\delta\alpha_3$, where θ is the angular attitude of the third link in the global reference frame. Because the orientation elements of the Jacobian equal 1 in the planar case, the solution is straightforward.

$$\begin{bmatrix} T_1 \\ T_2 \\ T_3 \end{bmatrix} = \begin{bmatrix} -l_1S_1 - l_2S_{12} - l_3S_{123} & l_1C_1 + l_2C_{12} + l_3C_{123} & 1 \\ -l_2S_{12} - l_3S_{123} & l_2C_{12} + l_3C_{123} & 1 \\ -l_3S_{123} & l_3C_{123} & 1 \end{bmatrix} \begin{bmatrix} F_x \\ F_y \\ M \end{bmatrix} \tag{2.30}$$

where M is the external couple. Note that the external couple is simply added to the joint torques.

Equation 2.30 represents a set of three linear equations with three unknowns in matrix form. We can rearrange these equations by subtracting the first scalar equation from the second and third ones (the reader who is familiar with linear algebra will easily recognize this operation as the Gaussian elimination method). This renders

$$\begin{bmatrix} T_1 \\ T_2 - T_1 \\ T_3 - T_1 \end{bmatrix} = \begin{bmatrix} -l_1S_1 - l_2S_{12} - l_3S_{123} & l_1C_1 + l_2C_{12} + l_3C_{123} & 1 \\ l_1S_1 & -l_2C_1 & 0 \\ l_1S_1 + l_2S_{12} & -l_1C_1 - l_2C_{12} & 0 \end{bmatrix} \begin{bmatrix} F_x \\ F_y \\ M \end{bmatrix} \tag{2.31}$$

The last two equations are independent from the first one and can be solved separately:

$$\begin{bmatrix} T_2 - T_1 \\ T_3 - T_1 \end{bmatrix} = \begin{bmatrix} l_1S_1 & -l_2C_1 \\ l_1S_1 + l_2S_{12} & -l_1C_1 - l_2C_{12} \end{bmatrix} \begin{bmatrix} F_x \\ F_y \end{bmatrix} \tag{2.32}$$

Equation 2.32 relates the differences in the joint torques with the positions of the first and second joints (represented by the square 2×2 matrix in the right-hand side of the equation) and the end-effector force. The reader is invited to draw a three-link kinematic chain and trace this relationship geometrically. The reader is also encouraged to explain the reason for the minus signs in some of the elements of the matrix.

For a three-link planar kinematic chain in a nonsingular configuration, the direct statics problem can be easily solved. When joint torques are known, the generalized force (i.e., the force and couple) exerted by the end effector can be calculated by inverting the Jacobian transpose:

$$\overline{\mathbf{F}} = \left(\mathbf{J}^T\right)^{-1}\mathbf{T} \tag{2.33}$$

If the Jacobian is singular, equation 2.33 cannot be solved. In the singular positions, the end-effector force can be changed in specific directions with no effect on the joint torques. For instance, when the chain is outstretched, the force acting along the chain does not influence the torques in the joints.

2.3 CONTROL OF EXTERNAL CONTACT FORCES

Representative papers: Gielen and van Ingen Schenau (1992); Prilutsky and Gregor (1997)

To move, people exert forces on the ground. To manipulate objects, people exert forces on them. In all cases, the force magnitude and direction should correspond to the requirement of the motor task. This section concentrates on the problems associated with the exertion of an intended contact force on the environment. In general, the external contact force can be generated in two ways: dynamically or statically. Dynamic force exertion involves accelerated movement of the body parts, which may be located far away from the point of force application. For instance, if a person standing in an upright posture with the legs extended performs a fast arm swing or trunk flexion, the GRF changes. In this case, the legs serve as force transmitters. They do not generate the force; they just transmit it. Force transmission is seen in many movements, especially where the body passes quickly over the supporting leg, giving only a short time for leg flexion and extension, such as during the takeoff in long jumps. In this case, the direction and magnitude of the GRF are highly influenced by the acceleration of body parts other than the grounded leg, for instance, by the swinging motion of the arms and the free leg. The supporting leg resists a thrust from the upper parts of the body and transmits it to the ground. It also exerts a force and a moment on the torso that results in the acceleration of the entire upper body. In static conditions, the situation is simpler. At a given joint configuration, the external contact force is controlled by the joint torques of the involved body limb (the arm or leg). In this section, we analyze static force exertion, starting with simple kinematic chains.

■ ■ ■ *From the Literature* ■ ■ ■

Control of the Ground Reaction Force in Squat Jumps

Source: Ridderikhoff, A., J.H. Batelaan, and M.F. Bobbert. 1999. Jumping for distance: Control of the external force in squat jumps. *Med. Sci. Sports Exerc.* 31 (8): 1196–1204.

The goal of the study was to investigate whether control in jumps for distance is similar to control in jumps for height. In the experiment, male subjects performed maximal squat jumps for height and distance with different inclination angles of the body relative to the horizontal. An inverse dynamics analysis was performed using measured kinematics and GRF. In addition, jumps were simulated with a forward dynamics model of the musculoskeletal system. To simulate the long jumps, a rotation-extension strategy was used. For the extension, the strategy repeats a pattern of coordination found for the vertical jumps with a whole-body rotation (figure 2.14). In the experiments, no significant differences in the magnitude of the peak GRF or in velocity of the body's CoM were found

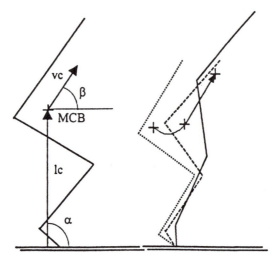

Figure 2.14 The rotation-extension model: from its starting position (dotted lines), the body is rotated until the onset of the extension phase (dashed lines), and subsequently an explosive leg extension is carried out until takeoff (solid). The cross marks the body's CoM.

Adapted, by permission, from A. Ridderikhoff, J.H. Batelaan, and M.F. Bobbert, 1999, "Control of the external force in squat jumps," *Medicine and Science in Sport and Exercise* 38 (8):1196–1197. Copyright © 1999 Lippincott Williams & Wilkens.

among jumps with different inclination levels. As the jumps were directed more forward, the net knee moment increased, whereas the hip and ankle moments decreased with the inclination. The authors concluded that a jump for distance can be achieved by applying a rotation-extension strategy to control of a vertical jump (figure 2.15).

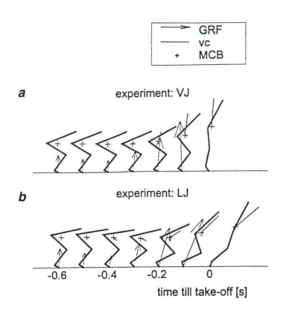

Figure 2.15 (*a*) Vertical jump (VJ) and (*b*) long jump (LJ) by one subject. The body configuration is shown with the GRF and the velocity of the body's CoM (**v**$_c$) at seven intervals. The GRF and **v**$_c$ vectors are scaled to fit the figure.

Adapted, by permission, from A. Ridderikhoff, J.H. Batelaan, and M.F. Bobbert, 1999, "Control of the external force in squat jumps," *Medicine and Science in Sport and Exercise* 38 (8):1196–1197. Copyright © 1999 Lippincott Williams & Wilkens.

2.3.1 Two-Link Chains

Consider a planar two-link chain in a nonsingular configuration (figure 2.16). The axis *X* of the global system of coordinates is along link 1. The force **F** acts at the endpoint. The magnitude of **F** is $F = T_1/r_1$ or $F = T_2/r_2$. Consequently, for a given joint configuration and a given force direction (r_1 and r_2 are constant), the magnitude of the endpoint force is determined by the joint torques T_1 and T_2. Note that in order to produce a different endpoint force in the same direction, the joint torques must change proportionally. For instance, to exert a 2*F*

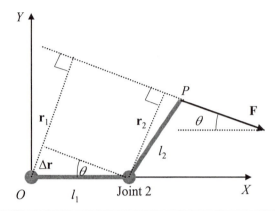

Figure 2.16 The direction of the endpoint force as a function of the difference in the moment arms r_1 and r_2. For convenience, the coordinate X axis is along the first link. Because $\Delta r = \Delta T/F$, the sine of angle θ can be computed using equation 2.35.

force, the joint torques must be $2T_1$ and $2T_2$. Hence, to alter the magnitude of the external force, the CNS should change the level of muscle activation in all the involved joints proportionally (metaphorically speaking, the CNS should multiply the muscle activity level by the same coefficient).

To characterize the force direction, consider the difference $\Delta T = T_1 - T_2$

$$\Delta T = Fr_1 - Fr_2 = F(r_1 - r_2) = F\Delta r \tag{2.34}$$

where $\Delta r = r_1 - r_2$. From figure 2.16, it follows that $\Delta r = l_1\sin\theta$, where θ is the angle of the line of force with the X axis. Therefore, the direction of the endpoint force is characterized by the trigonometric function

$$\sin\theta = \frac{1}{Fl_1}(T_1 - T_2) \tag{2.35}$$

According to equation 2.35, for an endpoint force of a constant magnitude F, the force direction is given by the difference in the joint torques. Hence, the magnitude of the endpoint force is controlled by the proportional changes of the joint torques, and if the force magnitude is constant, the force direction is controlled by maintaining the required difference in the torques. If the chain configuration and the direction of the endpoint force remain constant, either one of the joint torques is representative of the magnitude of the endpoint force.

Consider joint torques in tasks that require exerting an endpoint force of the same magnitude in various directions (figure 2.18).

It is convenient to analyze the relationship between the endpoint force and the joint torques in a system of *polar coordinates*. For a two-link model of the human

■ ■ ■ *FROM THE LITERATURE* ■ ■ ■

The Difference in the Joint Torques As an Informative Parameter in Motor Control Research

Sources: Jacobs, R., and G.J. van Ingen Schenau. 1992. Control of an external force in leg extensions in humans. *J. Physiol.* 457: 611–26.

Prilutsky, B.I., and R.J. Gregor. 1997. Strategy of coordination of two- and one-joint leg muscles in controlling an external force. *Motor Control* 1 (1): 92–116.

To study how the endpoint force in leg extension and flexion is controlled, the researchers instructed subjects to exert a constant force in five different directions. The subjects exerted an endpoint force by pressing on the ground while sitting on a chair (figure 2.17).

The difference in the joint torques at the knee and hip joints was calculated and compared with the difference in *electromyographic* (*EMG*) activity of rectus femoris and hamstrings. A linear relationship with a very high correlation coefficient was found.

a

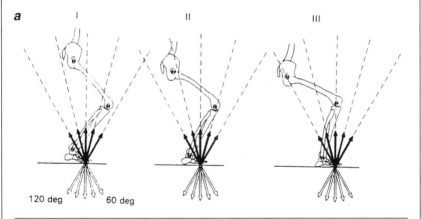

Figure 2.17 Control of external forces. (*a*) A schematic of the experiment. The subjects exerted a constant force at five different directions in three postures (I, II, and III). The pelvis is assumed to be fixed. The horizontal line represents the ground. The arrows below the horizontal line correspond to the five directions of the force exerted by the limb during pushing, from –60° to –120° in 15° increments. The arrows above the horizontal line correspond to the GRF. Angles at the hip, knee, and ankle in position I are 125°, 100°, and 78°, respectively; in position II, 107°, 100°, and 91°; and in position III, 95°, 100°, and 104°. Postural changes affect the force direction with respect to the joints.

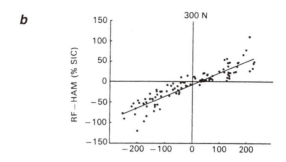

Figure 2.17 *(continued)* Control of external forces. (*b*) The difference between EMG activity of the rectus femoris and hamstring muscles (RF – HAM; SIC stands for standard isometric contraction) as a function of the joint torque difference ($M_k - M_h$; *k* for "knee" and *h* for "hip"). The endpoint force is 300 N. All individual data are shown, *r* = .96.

Reprinted, by permission, from R. Jacobs and G.J. van Ingen Schenau, 1992, "Control of external forces in leg extensions in humans," *Journal of Physiology* (London) 457:611–627.

Jacobs and van Ingen Schenau (1992) hypothesized that two-joint muscles, specifically the rectus femoris and hamstrings, control the direction of the endpoint force, whereas one-joint muscles are less sensitive to force direction, while Prilutsky and Gregor (1997) disagreed with this theory.

arm (figure 2.19), two angles with respect to the *X* axis are defined: the *polar angle θ* of the *radial axis* that connects the shoulder *S* with the endpoint and the *pointing angle ϕ* of the *pointing axis,* from the elbow *E* along the forearm.

The radial and pointing axes define four sectors, marked in figure 2.19. Flexion at the shoulder joint and extension at the elbow joint produce an endpoint force in sector 1. This effort corresponds to arm extension. In sector 2, endpoint force is due to flexion at the two involved joints. In sector 3, the force is due to shoulder extension and elbow flexion. In sector 4, the endpoint force is a result of combined extension torques in the shoulder and elbow joints.

For the leg, unlike the arm, hip flexion (or extension) is in the opposite angular direction of flexion (or extension) at the knee; that is, if flexion at the hip joint is defined in a clockwise direction, flexion is counterclockwise for the knee. Therefore, a force exertion in sector 1, which is typical for takeoffs, corresponds to simultaneous hip and knee extension. The reader is invited to find the hip and knee flexion and extension correspondence in sectors 2 through 4.

The endpoint force, like any other force, can be resolved into two or more components, for instance, into the horizontal and vertical components. In particular, the endpoint force can be resolved into two contributing force compo-

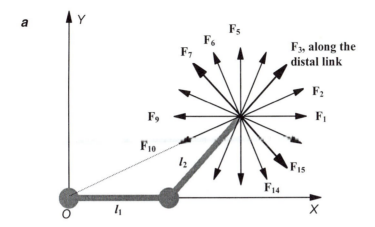

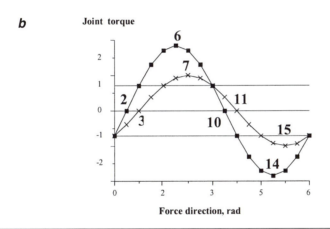

Figure 2.18 The end-effector forces and joint torques of a two-link planar chain. (*a*) An endpoint exerts a force with a magnitude of 1 on the environment in 16 directions. The X axis of the global reference system is along the proximal link. The angular distance between the neighboring forces is 22.5°. The forces are numbered in a counterclockwise sequence. The forces are due to the torques T_1 and T_2 acting at joints 1 and 2, respectively. The link lengths, l_1 and l_2, equal 1. The joint angle α_2 is 45°. Force F_{11} is along the distal link, opposite force F_3, and is not shown. (*b*) Joint torques that produce an endpoint force of unit magnitude in various directions. The numbers in the graph correspond to the force directions shown in (*a*). Black rectangles represent the torques at joint 1 and crosses represent the torques at joint 2. The joint torques have peak values when the force direction is perpendicular to the corresponding moment arms: for joint 1, the peak values are in directions 6 and 14, and for joint 2, in directions 7 and 15. The joint torques are zero when the line of force action passes through the joint center. In particular, the torque at joint 1 is zero when the endpoint force is exerted in direction 2 or 10, and the torque at joint 2 is zero when the endpoint force is in direction 3 or 11.

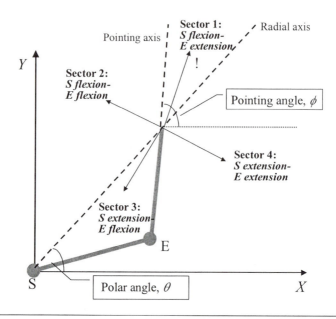

Figure 2.19 The polar and pointing angles of a two-link model of the arm and the sectors that define shoulder (*S*) and elbow (*E*) flexion or extension. Flexion is a counterclockwise rotation, and extension is a clockwise rotation of the corresponding link.

nents that are due to individual joint torques. Such a decomposition allows the relative contribution of each joint torque to the endpoint force to be determined. To find the contributing forces, consider cases when the line of force action is along either the radial or pointing axis. When the endpoint force is exerted along the radial axis, the torque at the proximal joint equals zero (figure 2.19). In the distal joint, the torque is zero when the endpoint force is exerted along the pointing axis. In both cases, the line of force action crosses the joint center, and the corresponding moment arm is zero. Thus, in the described conditions, the endpoint force is due to the activity in solely one joint. It can be said that the distal joint produces the endpoint force along the radial axis and the proximal joint produces the endpoint force along the pointing axis.

To prove that the endpoint force is a sum of the two force components generated by the individual joint torques, note that the components are concurrent at the endpoint. Therefore, they can be added according to the parallelogram rule. Inversely, an endpoint force **F** can be resolved into $\mathbf{F}_p$ acting along the pointing axis and $\mathbf{F}_r$ directed along the radial axis. The new force system is equivalent to the force **F**. Being equivalent, the two force systems, **F** and $(\mathbf{F}_p, \mathbf{F}_r)$, exert similar moments of force about all moment centers, including the proximal and distal joints of the chain. Hence, they are generated by the same joint torques.

For illustration, consider the simple model of static force exertion in leg extension presented in figure 2.20. The weight of the legs and the ankle joint

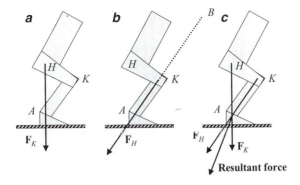

Figure 2.20 Resultant force exerted on the ground is the sum of the force components due to the torques at the hip and knee joints. The ankle-joint torque is zero. The force vectors are scaled to the figure.

torque are neglected. The hip-joint axis does not displace. In the model, the force on the ground is exerted by the combined hip and knee extension torques. The force due to knee torque, $\mathbf{F}_K$, acts along the axis from the hip, H, to the ankle joint, A (figure 2.20*a*). Hip-joint extension generates force $\mathbf{F}_H$ along the shank (figure 2.20*b*). The resultant force exerted on the ground is the vector sum of these two force components (figure 2.20*c*). Note that extension at the knee can be prevented by a cord linking H and A, and hip extension can be prevented by a cord along the line KB. The cords would behave as struts and maintain a constant direction.

■ ■ ■ *From the Literature* ■ ■ ■

Muscle Activity As a Function of Force and Movement Direction

Source: Bolhuis, B.M. van, C.C.A.M. Gielen, and G.J. van Ingen Schenau. 1998. Activation patterns of mono- and biarticular arm muscles as a function of force and movement direction of the wrist in humans. *J. Physiol. (Lond.)* 508 (1): 313–24.

The authors investigated the hypothesis that mono- and biarticular muscles have different functional roles in the control of multijoint movements. Surface EMG recordings were obtained from several arm muscles during voluntary slow movements of the wrist in a horizontal plane against an external force. The direction of force produced at the wrist and the direction of movement of the wrist were varied independently. The arm was modeled as a two-link kinematic chain (figure 2.21).

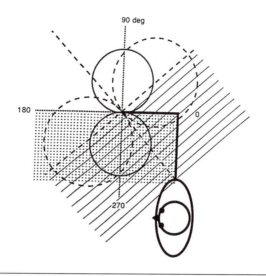

Figure 2.21 Top view of a subject in the rest position, 90° of shoulder abduction and 90° of elbow flexion. External forces of a constant magnitude F were exerted at the wrist in various directions. The forces have components that are normal to the radius of rotation. When the force direction changes, the normal force components change as a cosine function of force direction. Their magnitudes lie on the circumference of a circle whose diameter is F. The joint torques are proportional to the normal force components and hence can also be represented by similar circles. In the figure, the joint torques corresponding to different directions of the force are presented as circles: the continuous circles represent elbow torque, and dashed circles represent shoulder torque. A force in a direction of 270° causes maximal elbow flexion torque. A force in the opposite direction, 90°, causes maximal elbow extension torque. For the position shown, the directions of maximal flexion and extension torque in the shoulder are approximately 220° and 40°, respectively. The dotted area indicates the region of force directions at the wrist resulting in an elbow flexion torque. The striped area indicates the region of movement directions of the wrist corresponding to flexion of the elbow. In the region that is both dotted and striped, the direction of elbow torque is similar to the direction of change in elbow-joint angle. In this region, the one-joint muscles exert force and change their length in the same direction; they therefore produce positive work. In the regions that are either dotted or striped, the direction of elbow torque and change in elbow-joint angle have opposite signs. This implies that a monoarticular elbow flexor muscle would dissipate the work of, rather than contribute work to, the movement.

Reprinted, by permission, from B.M. van Bolhius, C.C.A.M. Gielen, and G.J. van Ingen Schenau, 1998, "Activation patterns of mono- and biarticular arm muscles as a function of force and movement direction of the wrist in humans," *Journal of Physiology* 508 (1):313–324.

The results revealed distinct differences between the activation patterns of mono- and biarticular muscles. The activation of two-joint muscles was not affected by movement direction but appeared to vary exclusively with direction of force.

From the foregoing discussion, it follows that when force is generated along the shank (in the pointing direction), we should expect zero or minimal activity of the muscles serving the knee joint. Similarly, when force is exerted in the radial direction, the activity of the muscles serving the hip joint is expected to be close to zero. The largest torques are required when force vector is orthogonal to the radius from the joint to the point of force application.

A two-joint muscle with the moment arms d_1 and d_2 at joints 1 and 2, respectively, generates at these joints the moment of force $M_1 = F_m d_1$ and $M_2 = F_m d_2$, where F_m is the magnitude of muscle force. The ratio $M_1 : M_2$ equals $d_1 : d_2$. When active alone, the muscle can counterbalance an endpoint force $\mathbf{F}$ if and only if the moment arms of $\mathbf{F}$ about joints 1 and 2 are also in the proportion $d_1 : d_2$. Otherwise, other muscles have to be active. Hence, an external force can be counterbalanced by a single two-joint muscle for only one force direction.

2.3.2 Three-Link Chains

Representative paper: Hof (2001)

For a three-link chain, the force exerted on the environment by a performer can be resolved into the components associated with the individual joint torques. However, the components are usually not concurrent at the endpoint. Consider a planar three-link chain in a nonsingular configuration (figure 2.22). An external force is exerted on the end link of the chain at a point P. It is not necessary for P to be at the endpoint of the distal link (unlike ballet dancers who can stand on their toes, most people stand on the entire plantar surface of the foot). To find the contributing forces, we introduce lines passing through the joint centers, L_{23}, L_{13}, and L_{12}, where the subscripts refer to the corresponding joint centers. Recall that when a force intersects a joint center, it does not produce a moment of force at this joint. Therefore, the line of force action that is due solely to the torque at joint 1 must intersect joint centers 2 and 3. The same is valid for other joints. The following rule exists: individually applied joint torques T_1, T_2, and T_3 cause the end effector to apply forces to the environment along the lines L_{23}, L_{13}, and L_{12}, respectively.

The forces $\mathbf{F}_1$ and $\mathbf{F}_2$ act along the lines L_{23} and L_{13}. In two-link chains, these lines would be along the radial and pointing directions, respectively. Forces $\mathbf{F}_1$ and $\mathbf{F}_2$ are concurrent at joint 3 but not at the endpoint. The three forces $\mathbf{F}_1$, $\mathbf{F}_2$, and $\mathbf{F}_3$ are coplanar and can be reduced to a single resultant force $\mathbf{F}$ and a

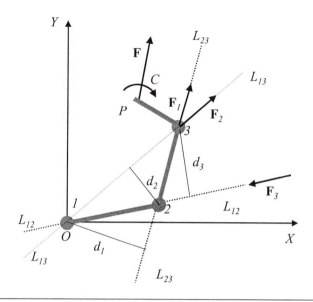

Figure 2.22 Forward static analysis of a planar three-link chain. The torque actuators at joints 1, 2, and 3 produce the joint torques that contribute to the end-effector force **F**. The torque T_1 acting at joint 1 develops a contributing force F_1 along the line L_{23}. The magnitude of **F**$_1$ is equal to the ratio T_1/d_1, where d_1 is the moment arm. The magnitudes of the contributing forces from the other joints can be computed in a similar way: $F_2 = T_2/d_2$ and $F_3 = T_3/d_3$. These forces act along the lines L_{13} and L_{12}, correspondingly. Forces **F**$_1$ and **F**$_2$ are shown originating at joint 3. Force **F**$_3$ is shown along the line of its action. A couple C, represented here by a curved arrow, is also exerted on the environment.

■ ■ ■ *FROM THE LITERATURE* ■ ■ ■

Measuring the Fingertip Forces

Sources: Valero-Cuevas, F.J., F.E. Zajac, and C.G. Burgar. 1998. Large index-fingertip forces are produced by subject-independent patterns of muscle excitation. *J. Biomech.* 31 (8): 693–703.

Li, Z.M., V.M. Zatsiorsky, and M.L. Latash. 2000. Contribution of the extrinsic and intrinsic hand muscles to the moments in finger joints. *Clin. Biomech.* 15 (3): 203–11.

The ensuing analysis of fingertip forces is limited to the sagittal plane, that is, to flexion and extension motions. In this plane, the fingers have three kinematic degrees of freedom. Thus, the distal phalange can impart a couple and a shear force independent of the normal force to an object in

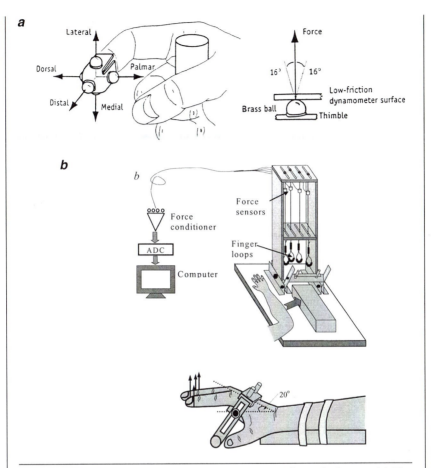

Figure 2.23 Techniques to measure well-directed fingertip forces. (*a*) The thimble technique. Subjects wrapped their hand around a fixed dowel to isolate index-finger function and generated index-finger forces in the *dorsal,* palmar, distal, *lateral,* and *medial* directions while maintaining a standard posture. (*b*) The suspension technique. The entire device is shown in the upper panel; the bottom panel represents the position of the fingers at the time of measurement. During the measurement, the subjects pressed on the loops connected to the wires with the force sensors. ADC is analog-to-digital converter.

Figure 2.23*a* reprinted from *Journal of Biomechanics*, 31, F.J. Valero-Cuevas, F.E. Zajac, and C.G. Burgar, Large index-fingertip forces are produced by subject-independent patterns of muscle excitation, 693–703, 1998, with permission from Elsevier Science.

Figure 2.23*b* reprinted from *Clinical Biomechanics*, 15, Z.M. Li, V.M. Zatsiorsky, and M.L. Latash, Contribution of the extrinsic and intrinsic hand muscle to the moments in finger joints, 203–211, 2000, with permission from Excerpta Medica Inc.

contact with the fingertip. To generate and measure a contact force in the desired direction, special techniques were developed: a thimble system and a suspension system (figure 2.23). The systems would not be necessary for measuring finger forces if the fingers were not three-link kinematic chains.

In the thimble technique, the subjects wear a thimble molded to the contour of their distal phalange with a 5-mm metal ball embedded in the location corresponding to the desired force direction. The low-friction contact of the ball with the force-sensing surface requires the subjects to both accurately direct fingertip forces (otherwise the thimble would slip) and avoid producing a force couple by the distal phalange (otherwise the thimble would rotate about the point of contact).

With the suspension technique, the subjects are constrained to produce force along the wires from which the fingers are suspended. Because the wire orientation can be changed, the forces generated in various directions can be recorded.

couple C applied to the end link of the chain. The end link transmits the force-couple system to the environment. Consequently, three-link systems allow not only push-pull forces but also rotation effects to be exerted on the environment. In particular, both a force and a couple can be exerted on work tools.

If the moment of the couple is not too large, the resultant force-couple system can be replaced by a single force acting at P. Otherwise, the situation illustrated in figure 1.12b occurs: a single resultant force can be computed but cannot really be applied to the body. The point of application of the resultant force (if it exists) can be voluntarily or involuntarily displaced along the end link of the chain. The reader can perform a simple experiment. While standing in a dependent posture (upright with arms hanging straight down at the sides of the body, head erect), perform a forceful plantar flexion. Torque about the ankle causes a difference between the vertical forces acting at the heel and the toe. Consequently, the point of force application displaces in an anterior direction.

Compare the endpoint force exertion in a two-link chain and a three-link chain (figure 2.24). In figure 2.24a, force $\mathbf{F}_{(H+K)}$ is exerted on the ground by the simultaneous torque production at the hip and knee joints (the ankle joint is statically "frozen"; the situation is similar to wearing a very tight and stiff ankle brace). This situation is the same as in figure 2.20c. When a torque is produced at the ankle joint in addition to the torques at the hip and knee joints, the force exerted on the ground $\mathbf{F}_{(H+K+A)}$ changes in magnitude and direction (figure 2.24b). The point of force application also changes. During human standing, the displacement of the point of force application by the controlled changes in the ankle joint torque plays an important role in maintaining balance.

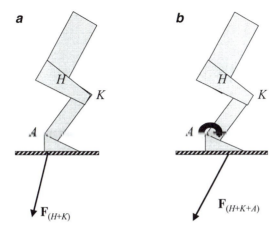

Figure 2.24 Exertion of a force on the ground. (*a*) The force is exerted by the hip and knee torques. The ankle torque is zero. (*b*) The force is exerted by the hip, knee, and ankle torques. Note the displacement of the point of force application.

> ### ▪ ▪ ▪ *FROM THE LITERATURE* ▪ ▪ ▪
>
> ## Displacing the Center of Pressure Beneath the Foot
>
> Source: King, D.L., and V.M. Zatsiorsky. 1997. Extracting gravity line displacement from stabilographic recordings. *Gait Posture* 6 (1): 27–38.
>
> During static posturography, subjects stand quietly on a force platform. Although several biomechanical variables can be measured during this task, the most common variable to measure is the center of pressure (CoP), the point of application of the vertical resultant force acting between the feet and the force platform.
>
> For quiet standing, the following assumptions are made:
>
> 1. The foot or feet do not move (heel or toe rising and steps are not permitted).
>
> 2. The feet are considered solid bodies (they do not deform, and the action of the foot and toe muscles is ignored).
>
> 3. The axis of ankle-joint rotation is fixed (this assumption was confirmed experimentally; see *Kinematics of Human Motion,* section **5.3**).
>
> 4. All of the forces and moments acting at the ankle joint can be replaced by a resultant force acting through the joint axis of rotation and a couple.

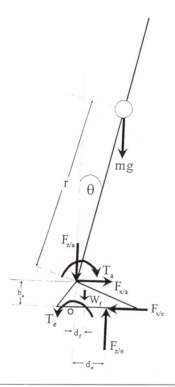

Figure 2.25 The forces and moments acting on the foot. See text for explanation of the symbols.

Reprinted from *Gait and Posture*, 6, D. King and V. Zatsiorsky, Extracting gravity line displacement from stabilographic recordings, 27–38, 1997, with permission from Excerpta Medica Inc.

Consider a free-body diagram of the foot (figure 2.25). The origin of the reference system O is at the intersection of the supporting surface and the vertical line passing through the ankle-joint center. This technical detail simplifies the calculations. Practically, it means that subjects should stand on a force platform with the malleoli (ankle-joint centers) strictly above a coordinate axis of the reference frame tied to the platform. The following forces and couples act on the foot: $\mathbf{F}_a$ and $\mathbf{T}_a$ are the force and the couple (torque) at the ankle joint, $\mathbf{W}_f$ is the weight of the foot, $\mathbf{mg}$ is the weight of the body minus the weight of the feet divided by two, $\mathbf{F}_e$ is the ground reaction force, and $\mathbf{M}_O$ is the moment of the ground reaction force about the origin of the reference system O (not shown in the figure). Because the system is in equilibrium, the sum of all the forces as well as the sum of all the moments is zero. For simplicity, we write the equations in scalar form (the sagittal projection only):

$$F_{x/e} = -F_{x/a}$$
$$F_{z/e} = -(W_f + F_{z/a}) \qquad (2.36)$$
$$F_{z/e}d_e = -(T_a + F_{x/a}h_a + W_fd_f)$$

Since the horizontal ground reaction force, $F_{x/e}$, and the vertical force acting at the ankle joint, $F_{z/a}$, pass through the origin, O, they do not contribute to the moment about this point. The moment due to the weight of the foot is evidently constant. Neglecting this moment (or after subtracting its value), we can write the following equation for the CoP location:

$$d_e = \frac{T_a + F_{x/a}h_a}{F_{z/e}} \qquad (2.37)$$

This equation states that the CoP is a function of the torque acting in the ankle joint, T_a, the horizontal force acting in the ankle joint, $F_{x/a}$, the height of the ankle joint, h_a, and the vertical component of the ground reaction force, $F_{z/e}$. During standing, the foot height is constant, and vertical force fluctuates only slightly. Hence, the ankle-joint torque and horizontal force are the primary determinants of the CoP location.

In summary, control of a multijoint chain is different from that of a single-joint chain. In the single-joint case, the endpoint force is proportional to the joint torque. In multilink chains, the endpoint force depends on both torque magnitudes and chain configuration. The same joint torques at different body postures generate different endpoint forces. Hence, when the CNS specifies joint torques to produce a desired external effect, the CNS must adjust the torques to the given posture.

• • • MATHEMATICS REFRESHER • • •

Circles, Ellipses, and Ellipsoids

The equation of a circle whose center is at the origin O and whose radius is a is

$$x^2 + y^2 = a^2, \text{ or } \frac{x^2}{a^2} + \frac{y^2}{a^2} = 1 \qquad (2.38a)$$

The equation can also be written as

$$\begin{bmatrix} x \\ y \end{bmatrix} = \begin{bmatrix} a\cos\theta \\ a\sin\theta \end{bmatrix} \qquad (2.38b)$$

where angle θ is the running parameter ($0 \leq \theta < 2\pi$).

The equation of an ellipse whose center is at O is

$$\begin{bmatrix} x \\ y \end{bmatrix} = \begin{bmatrix} a\cos\theta \\ b\sin\theta \end{bmatrix} \qquad (2.39a)$$

where a and b are semi-axes of the ellipse, $a > b$. The axis bearing the semi-axis a is called the major axis; the other axis is the minor axis. The major and minor axes are called the principal axes. The equation depicts an ellipse in so-called *standard position,* with the principal axes along the coordinates axes X and Y (figure 2.26a). The quantity $r = (x^2 + y^2)^{0.5}$ denotes the length of the radius vector $\mathbf{r}$ from the origin to a point (x, y) on the ellipse (figure 2.26a). Equation 2.39a is a parametric equation of the ellipse with the variable angle θ as a parameter. The equation represents x and y explicitly. By eliminating θ, equation 2.39a can be transformed into an implicit equation of the ellipse in a standard position:

$$\frac{x^2}{a^2} + \frac{y^2}{b^2} = 1 \qquad (2.39b)$$

Setting $y = 0$, we obtain $x = a$ or $x = -a$; setting $x = 0$, we obtain $y = b$ or $y = -b$.

For an ellipse whose center is at the origin but the principal axes do not coincide with the X and Y axes, the parametric (explicit) equation is

$$\begin{aligned} x &= a\cos\phi\cos\theta - b\sin\phi\sin\theta \\ y &= a\sin\phi\cos\theta + b\cos\phi\cos\theta \end{aligned} \qquad (2.40a)$$

where ϕ is the angle formed by the major axis of the ellipse and the positive X axis (figure 2.26b). An implicit equation for an ellipse in a slanted position is

$$Ax^2 + 2Bxy + Cy^2 = D \qquad (2.40b)$$

An expression of this type is called a *quadratic form,* or a *quadric,* specifically a quadratic form in the two variables, x and y. The equation represents an ellipse with center at O if and only if the following inequalities hold:

$$A > 0, \, C > 0, \, D > 0, \text{ and } AC > B^2 \qquad (2.40c)$$

For an ellipse whose center is at a point P, the implicit equation is

$$A(x - P_x)^2 + 2B(x - P_x)(y - P_y) + C(y - P_y)^2 = D \qquad (2.40d)$$

where P_x and P_y are coordinates of P (figure 2.26c). Equations 2.40 with fixed constants P_x, P_y, A, B, and C and variable parameter D represent a

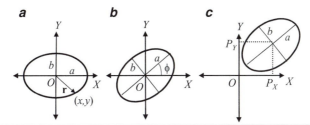

Figure 2.26 Ellipses in different positions: (*a*) in standard position, (*b*) with the center at the origin of the reference frame, and (*c*) in an arbitrary position.

Adapted, by permission, from H. Witte, F. Eckstein, and S. Recknagel, 1997, "A calculation of the forces acting on the human acetabulum during walking," *Acta Anatomica* 160:269–280. By permission of Karger, Basel.

family of geometrically similar ellipses with a common center and coinciding principal axes. All the ellipses of the family have the same ratio *a*/*b*; they differ only in size, which depends on the magnitude of *D*.

In three dimensions, the equation of an ellipsoid in standard position is

$$\frac{x^2}{a^2} + \frac{y^2}{b^2} + \frac{z^2}{c^2} = 1 \qquad (2.41a)$$

Any ellipsoid whose center is at the origin *O* can be described by a quadratic form:

$$Ax^2 + By^2 + Cz^2 + Dxy + Eyz + Fzx = G \qquad (2.41b)$$

where *A*, *B*, *C*, *D*, *E*, and *F* are appropriate constants and *G* is a positive number. A quadric describes an ellipsoid (but not, for instance, a hyperbola) when all the values *x*, *y*, and *z* that satisfy the equation are bounded (no point of the ellipsoid is at infinite distance from the origin *O*).

2.3.3 Force Ellipsoids

From the previous discussion, it follows that the endpoint force depends on the joint torques and the chain configuration. It is useful to separate these factors. To do that, we make the magnitude of the vector of joint torques **T** constant. The magnitude of a vector is described by its *norm*, also known as a *Euclidean norm*. For a vector **T**, the norm is $\sqrt{T_1^2 + T_2^2 + \ldots + T_n^2}$, where $T_1, T_2, \ldots, T_n$ are the magnitudes of the torques in the individual joints (vector components). We explore the situation when the norm of the vector **T** equals 1. Squaring gets rid of the square root, $T_1^2 + T_2^2 + \ldots + T_n^2 = 1$. According to equation 1.19, the squared value of the norm can be computed as a dot product of vector $\mathbf{T}$ with itself. Substituting equation 2.16 ($\mathbf{T} = \mathbf{J}^T\mathbf{F}$, for a zero external couple) into $\mathbf{T}^T\mathbf{T} = 1$, we obtain

$$\mathbf{F}^T\mathbf{J}\mathbf{J}^T\mathbf{F} = 1 \qquad (2.42)$$

Equation 2.42 is a quadratic form (the reader is invited to check this statement). In a planar case, equation 2.42 is a quadric in two variables, F_x and F_y, the magnitudes of the force components along the axes X and Y. Hence, the equation represents an ellipse. If the product $\mathbf{J}\mathbf{J}^T$ is a unit matrix, equation 2.42 can be reduced to $\mathbf{F}^T\mathbf{F} = 1$. This is an equation of a circle with a radius 1, $\left(F_x^2 + F_y^2 = 1\right)$. Consequently, the matrix product $\mathbf{J}\mathbf{J}^T$ transforms a circle into an ellipse.

In three dimensions, equation 2.42 represents a three-dimensional ellipsoid with the principal axes along the *eigenvectors* of the matrix $[JJ^T]$. (The eigenvectors and *eigenvalues* are explained in *Kinematics of Human Motion*, p. 58; they are also called *characteristic values* and *characteristic vectors*, respectively.) The length of a principal axis i is given by $1/\sqrt{\lambda_i}$, where λ_i is an eigenvalue associated with eigenvector i. The largest eigenvalue corresponds to the minor axis of the ellipsoid, and the smallest value corresponds to the major axis (figure 2.27). When the magnitude of the total torque vector is constant, maximal force can be exerted along the major axis. Therefore, efforts are most effective when they are directed along this axis: a unit torque is converted into the largest endpoint force. The minor axis indicates the direction in which the endpoint force is minimal.

••• MATHEMATICS REFRESHER •••

Determination of Eigenvalues and Eigenvectors

Let $[A]$ be an $n \times n$ matrix, λ be an eigenvalue of $[A]$, and $\mathbf{V}$ be an eigenvector of $[A]$. Then, by definition (see *Kinematics of Human Motion*, p. 58),

$$[A]\mathbf{V} = \lambda\mathbf{V} \qquad (2.43)$$

Hence,

$$[A]\mathbf{V} - \lambda\mathbf{V} = ([A] - \lambda[I])\mathbf{V} = \mathbf{0} \qquad (2.44)$$

where $[I]$ is an $n \times n$ unit matrix, also known as an identity matrix, and $\mathbf{0}$ is an n-component zero vector. Equation 2.44 is called the *characteristic equation* of matrix $[A]$. The eigenvalues λ can be found by solving the equation

$$\|[A] - \lambda[I]\| = 0 \qquad (2.45)$$

where the vertical lines designate the determinant. For $n = 2$,

$$\left\{ \begin{bmatrix} A_{11} & A_{12} \\ A_{21} & A_{22} \end{bmatrix} - \lambda \begin{bmatrix} 1 & 0 \\ 0 & 1 \end{bmatrix} \right\} \begin{bmatrix} v_1 \\ v_2 \end{bmatrix} = \begin{bmatrix} A_{11} - \lambda & A_{12} \\ A_{21} & A_{22} - \lambda \end{bmatrix} \begin{bmatrix} v_1 \\ v_2 \end{bmatrix} = \begin{bmatrix} 0 \\ 0 \end{bmatrix} \quad (2.46)$$

and the characteristic equation is

$$\begin{vmatrix} A_{11} - \lambda & A_{12} \\ A_{21} & A_{22} - \lambda \end{vmatrix} = (A_{11} - \lambda)(A_{22} - \lambda) - A_{12}A_{21} = 0 \quad (2.47)$$

This is a quadratic equation in λ that has two solutions, λ_1 and λ_2. For a given matrix $[A]$ and an eigenvalue λ, there exist an infinite number of eigenvectors that differ from each other by scalar multiplication $k\mathbf{V}$, where k is any number. Customarily, an eigenvector is chosen in such a way that its magnitude equals 1.

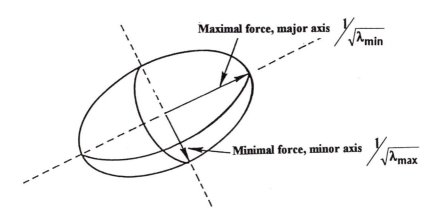

Figure 2.27 A *force ellipsoid.*

As an example, consider a planar two-link chain with $l_1 = l_2 = 1$ at the posture where $\alpha_1 = 0$ and $\alpha_2 = \pi/2$ (figure 2.28). For this chain, the Jacobian matrix is

$$\mathbf{J} = \begin{bmatrix} -S_1 - S_{12} & -S_{12} \\ C_1 + C_{12} & C_{12} \end{bmatrix} = \begin{bmatrix} -1 & -1 \\ 1 & 0 \end{bmatrix} \quad (2.48)$$

and

$$\mathbf{J}\mathbf{J}^T = \begin{bmatrix} 2 & -1 \\ -1 & 1 \end{bmatrix} \quad (2.49)$$

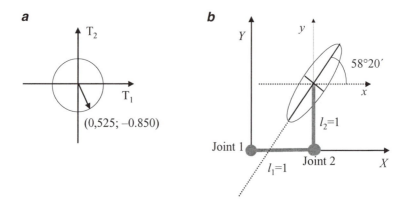

Figure 2.28 (*a*) The joint space of a planar two-link kinematic chain. A unit torque, $T_1^2 + T_1^2 = 1$, is represented by a circle of radius 1. The arrow represents a torque-torque combination corresponding to the endpoint force exertion along the major axis of the force ellipsoid, $T_1 = 0.525$ and $T_2 = -0.850$. (*b*) The force ellipse for the chain. The links are at right angles to each other. The X axis is along the first link. See the text for explanation.

Substituting equation 2.49 into equation 2.42 yields

$$2F_X^2 - 2F_XF_Y + F_Y^2 = F_X^2 + \left(F_X - F_Y\right)^2 = 1 \tag{2.50}$$

This is the equation of an ellipse whose center is at (0, 0) and whose principal axes are inclined with respect to the coordinate axes.

To find the major and minor axes of the force ellipse, we solve the characteristic equation of the matrix $[JJ^T]$. By definition of eigenvalues and eigenvectors, $[JJ^T]\mathbf{V} = \lambda\mathbf{V}$, where λ is an eigenvalue of $[JJ^T]$ and $\mathbf{V}$ is an eigenvector of $[JJ^T]$. Hence, $[JJ^T]\mathbf{V} - \lambda\mathbf{V} = \{[JJ^T] - \lambda[I]\}\mathbf{V} = 0$, where $[I]$ is a unit matrix. The equation $[JJ^T] - \lambda[I] = 0$ is a characteristic equation of $[JJ^T]$. Rewriting this equation in the elements of the matrix $[JJ^T]$ and expanding the determinant, we obtain

$$\begin{vmatrix} 2-\lambda & -1 \\ -1 & 1-\lambda \end{vmatrix} = (2-\lambda)(1-\lambda) - 1 = \lambda^2 - 3\lambda + 1 =$$

$$\left(\lambda + \frac{3+\sqrt{5}}{2}\right)\left(\lambda + \frac{3-\sqrt{5}}{2}\right) = 0 \tag{2.51}$$

Hence $\lambda_1 = \dfrac{3-\sqrt{5}}{2} = 0.382$ and $\lambda_2 = \dfrac{3+\sqrt{5}}{2} = 2.618$ are the eigenvalues of the matrix $[JJ^T]$. To compute the associated eigenvectors $\mathbf{V}_1$ and $\mathbf{V}_2$, recall that

$[JJ^T]\mathbf{V}_1 = \lambda_1\mathbf{V}_1$ and $[JJ^T]\mathbf{V}_2 = \lambda_2\mathbf{V}_2$. For λ_1, the corresponding eigenvector should satisfy the following equation:

$$\begin{bmatrix} 2 & -1 \\ -1 & 1 \end{bmatrix}\begin{bmatrix} v_{11} \\ v_{21} \end{bmatrix} = 0.382\begin{bmatrix} v_{11} \\ v_{21} \end{bmatrix} \tag{2.52}$$

Equation 2.52 represents two linearly dependent equations. Hence, we can use only one equation. Writing an equation for the first row of the matrix in equation 2.52, we obtain $2v_{11} - v_{21} = 0.382v_{11}$, or $v_{21} = 1.618v_{11}$. Normalization of the vector $\mathbf{V}_1^T = (v_{11}, v_{21})^T = (1, 1.618)^T$ by its magnitude $\left(V = \sqrt{v_{11}^2 + v_{21}^2}\right)$ yields $v_{11} = 0.525$ and $v_{21} = 0.850$. The normalized eigenvector $\mathbf{V}_1$ is chosen in such a way that its magnitude is 1. Vector $\mathbf{V}_2$ can be found in a similar way; the value $\lambda_2 = 2.618$ should be used. The normalized value of the vector $\mathbf{V}_2$ is (–0.525, 0.850). These two vectors are orthogonal to each other, and they are lined up with the principal axes of the force ellipse. The normalized eigenvectors are direction vectors. Because their magnitude equals 1, the components of these vectors are the direction cosines of the principal axes of the ellipse, that is, the cosines of the angles that the principal axes make with the coordinate axes. The major axis (the axis that corresponds to the smaller eigenvalue) is at angles $58°20'$ and $-31°40'$ with the X and Y axes. The magnitude of the force generated along the major axis by a unit torque is $1/\sqrt{\lambda_1} = 1/\sqrt{0.382} \approx 1.64$, with $F_X = (1.64 \times 0.525) = 0.861$ and $F_Y = (1.64 \times 0.850) = 1.394$. The corresponding joint torques are $T_1 = 0.525$ (in flexion) and $T_2 = -0.850$ (in extension).

For comparison, if the force is generated in the Y direction (which is a pointing direction for the given chain configuration), its magnitude equals 1, and when the endpoint force is exerted in the X direction, its magnitude is 0.707. The magnitude of the force in the radial direction is 1.414. The proof of these statements is left to the reader. (Hint: use equation 2.50 and figure 2.28). Note that the projections of the force exerted along the major axis on the X, Y, and radial directions are larger then the forces intentionally exerted in these directions (provided that a constant torque is exerted, $T_1^2 + T_2^2 = C$, where C is a constant value). Therefore, if a maximal endpoint force in a certain direction is desired, it may be advantageous to push or pull in another direction, for instance, along the major axis of the force ellipse. It may happen that the projection of the force in the desired direction is larger when this strategy is employed (this issue is covered in section **2.3.5**).

Because equation 2.42 includes the Jacobian, the shape and orientation of the force ellipsoid depend on the chain configuration. As the end effector moves from one location to another, the force ellipsoid also changes accordingly.

▪ ▪ ▪ FROM THE LITERATURE ▪ ▪ ▪

Finger Force Ellipses in Different Conditions

Source: Valero-Cuevas, F.J. 1997. Muscle coordination of the human index finger. Doctoral diss., Stanford University.

The author studied the force produced by the index finger at different angles. In the sagittal plane, the distal phalanx of the finger can exert both a force and a couple on the environment (to produce a couple, two equal forces should act in opposite directions; hence, two contacting surfaces should adhere to each other). Finger forces with zero couples at the area of contact (point forces) are better suited to the grasping of small and slippery objects. The finger was modeled as a three-link chain with 4 DoF (the metacarpophalangeal joint has 2 DoF; the proximal and distal interphalangeal joints have 1 DoF each). The finger force ellipses were computed from the finger Jacobian for two conditions: with and without production of the endpoint couple (figure 2.29).

forces in plane of finger flexion

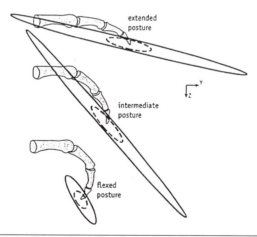

Figure 2.29 Force ellipses in the plane of finger flexion. Ellipses without couple constraint (solid lines) and ellipses constrained not to produce finger output couple (dashed lines). Introducing point-force constraints reduces the size of ellipses. The ellipses were computed under a simplifying assumption that the Euclidean norm of the torque vector was constant (more advanced approaches that are not described in this book were also used).

Adapted, by permission, from F.J. Valero-Cuevas, 1997, Muscle coordination of the human index finger. Ph.D. Dissertation. Stanford University (figure II.3, p. 35).

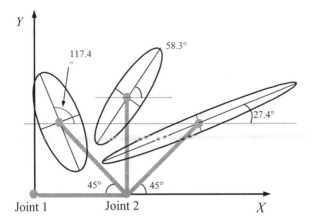

Figure 2.30 Force ellipsoids of a two-link chain at different joint configurations ($l_1 = l_2 = 1$). The ellipsoids were obtained by solving equation 2.42.

Figure 2.30 illustrates the force ellipsoids for a two-link chain with $l_1 = l_2 = 1$ at different joint configurations. When the principal axes of the force ellipsoid are of equal length, the ellipsoid reduces to a sphere. In this case, a unit joint torque produces an endpoint force of equal magnitude in all directions. Such a joint torque–endpoint force transformation is called the *isotropic transformation.*

The force ellipsoids in the preceding discussion were computed under the assumption that the Euclidean norm of the joint torques is constant. This means that the maximal joint torques in flexion and extension in all involved joints are assumed equal. In real life, the torques may differ. For instance, one joint may be stronger than another or the maximal torque in flexion may be larger than in extension. If this happens, the torque–torque relationship for the two joints is not a circle anymore, and the envelope of the endpoint force exerted in various directions may deviate from an ellipse. However, the available evidence suggests that, at least for the human arm, this deviation is relatively small (figure 2.31).

2.3.4 Chains With Holonomic Constraints

Representative paper: Woude, Veeger, and Dallmeijer (2000)

Holonomic, or geometric, *constraints* restrict movement in certain directions. Examples include pushing a cart, pedaling a bicycle, and opening a door. When the end effector is constrained, the performer may exert force in a direction different from the direction of motion and still perform the task. Many strength exercise machines prescribe the trajectory of the bar to which the athlete applies force. The component of the force along the permitted trajectory is used

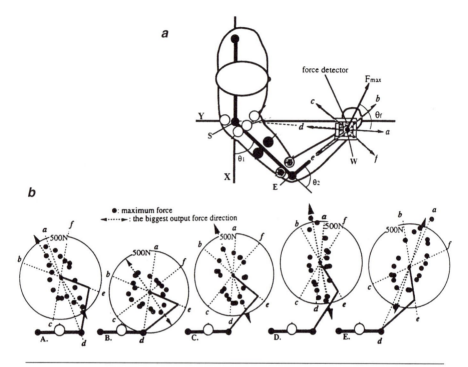

Figure 2.31   Distribution of maximal endpoint forces. (*a*) The subject's position and the measuring device. S, E, and W stand for shoulder, elbow, and wrist, respectively. Line *ad* is along the radius to the shoulder, line *be* is along the pointing axis, and line *cf* is perpendicular to the forearm. White, black, and gray circles designate the electrode placement. (The EMG data are not included here.) (*b*) Distribution of the maximal force at different arm configurations. A. $\theta_1 = 80°$, $\theta_2 = 80°$. B. $\theta_1 = 37°$, $\theta_2 = 120°$. C. $\theta_1 = 50°$, $\theta_2 = 90°$. D. $\theta_1 = 60°$, $\theta_2 = 60°$. E. $\theta_1 = 42°$, $\theta_2 = 57°$. The force envelopes resemble ellipses. (The authors, who developed a model with six muscles, suggest that the envelopes are hexagons.)

Adapted, by permission, from T. Fujikawa et al. 1997, "Functional coordination control of pairs of antagonistic muscles," *Transactions of the Japan Society of Mechanical Engineers* 63 (607):769–776.

to overcome the resistance of the apparatus. Other components are exerted on the structure of the apparatus. With such a machine, the direction of the endpoint force and the joint torques may be quite distinct from those observed in lifting or holding free weights.

Consider a planar two-link chain with hinge joints (figure 2.32). The endpoint force is exerted on a piston that slides without friction between parallel guides. The endpoint *P* is hinged, the end link of the chain can rotate around point *P* without friction, but point *P* cannot slide with respect to the piston. The chain resembles a bench press on a strength-training machine with a fixed

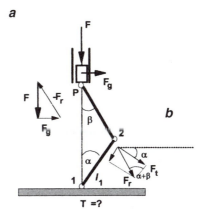

Figure 2.32   A two-link chain with a hinged contact at point *P*. (*a*) Forces acting at the endpoint *P*. (*b*) Forces acting at joint 2. The links are massless, and the hinge joints 1 and 2 are frictionless. The piston slides between the parallel guides without friction. The force **F** is exerted on the piston and is transmitted to the second link at the endpoint *P*. The guides exert a force **F**$_g$ on the piston that is transmitted to point *P*. The chain is kept in equilibrium by the equilibrating torque at joint 1. There is no torque at joint 2 and the contact point *P*. Compare with figure 2.12.

bar trajectory. The chain is similar to the chain in figure 2.12 but with one important distinction: in figure 2.12, the endpoint is not constrained. In the chain with the unconstrained endpoint, the external force is equilibrated by the torque at joint 2. The torque at joint 1 is zero. In contrast, in the chain with the constrained (hinged) endpoint, the countertorque may be at joint 1, and the torque at joint 2 may be zero.

We want to prove that the chain can be kept in equilibrium by a torque T_1 alone and to find a relation between the force **F** and the torque T_1. Consider the case when the torque at joint 2 is zero. The chain now resembles a crank mechanism that is broadly used in industry. The force **F** acts on the piston and is transmitted without change to the distal tip of the second link. Since there is no friction at point *P*, there is no couple at that endpoint. Hence, force **F** must pass through *P*. The (frictionless) guides exert a force on the piston that is normal to the guides, **F**$_g$. This force is also transmitted to point *P*. Because the second link is in equilibrium, the force acting on the second link at *P* is equilibrated by an equal and opposite force exerted on this link at joint 2. This force, −**F**$_r$, is along the second link. Therefore, three forces act at point *P*: **F**, **F**$_g$, and −**F**$_r$ (figure 2.32*a*). Their resultant is evidently zero. The compressive force **F**$_r$ = F/cosβ is transmitted to joint 2 and to the distal tip of the first link. This force can be resolved into a radial component (in which we are not interested) and a tangential component that equals $F \sin(\alpha + \beta)/\cos\beta$ (figure 2.32b). Hence, the torque at joint 1 is

$$T_1 = Fl_1 \sin(\alpha + \beta)/\cos\beta \qquad (2.53)$$

When such a torque is produced, the system is kept in equilibrium with a zero torque at joint 2. (This pattern is only an option. The torque at joint 2 can be zero but is not required to be zero.) The force exerted on the second link at joint 2, $-\mathbf{F}_r$, acts along the link. One of the force components counterbalances force $\mathbf{F}$ and the second, $-\mathbf{F}_g$, is exerted on the guides and deforms them. The torque T_1 must be balanced by a couple of the same magnitude. When the chain is outstretched, $\alpha = 0$, $\beta = 0$, and the couple becomes zero. The reader already knows that this position is singular.

Thus, actual constraints—the tangible, physical obstacles to movement—can completely change the joint torques required for equilibrium. Consequently, different muscle groups may act when body motion is free and when it is actually (physically) constrained. In human movement, the body links are often constrained in some directions. For instance, during a bench press, force components can be exerted horizontally along the bar, either extending or compressing it, and therefore when the barbell is being lifted vertically, the manual forces can act in a nonvertical direction.

■ ■ ■ FROM THE LITERATURE ■ ■ ■

Control of Pedal Force Direction

Source: Wang, X., J.P. Verriest, B. Lebreton-Gadegbeku, Y. Tessier, and J. Trasbot. 2000. Experimental investigation and biomechanical analysis of lower limb movements for clutch pedal operation. *Ergonomics* 43 (9): 1405–29.

Foot pedals are commonly used in vehicles. Researchers investigated the influence of seat height, pedal travel, pedal travel inclination, and pedal resistance on lower-limb movements for clutch-pedal operation. Fifteen subjects participated in the experiment. To explain how pedal force direction could be controlled, a biomechanical model was developed (figure 2.33) Since the pedal force reaches its maximum value at the end of travel, only the forces exerted at the posture corresponding to the end of pedal travel were analyzed. Researchers tested the hypothesis that force direction is controlled so that minimal joint moments are applied. Experiments and simulation both show that pedal force direction is close to the line connecting the hip joint and the contact point, line HP' in figure

2.33. The deviation for average men was only $0.8° \pm 5.2°$. Increasing the pedal resistance makes the pedal force direction approach the direction *HP′*.

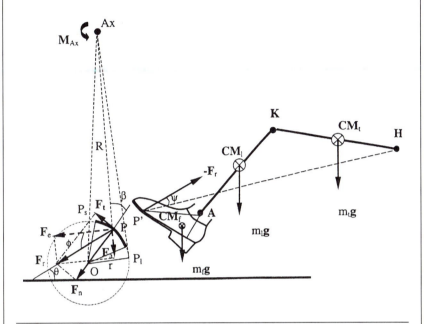

Figure 2.33 A biomechanical model of the control of pedal force direction. *A*, *K*, and *H* are the ankle, knee, and hip joint centers, respectively. $\mathbf{M}_{Ax}$ is the pedal resistance moment around the rotation axis *Ax*. $\mathbf{F}_r$ is foot force applied at the pedal; $\mathbf{F}_e$ and $\mathbf{F}_a$ are two components of $\mathbf{F}_r$ that are perpendicular and along the line *PAx* connecting the contact point *P* and the axis *Ax*. $\mathbf{F}_n$ and $\mathbf{F}_t$ are two force components that are normal and tangential to the pedal surface. *R* is the pedal rotation radius. *O* is the center of curvature of the pedal surface. The hypothesis is that the pedal force direction corresponds to the minimum of the function $\text{Min}\,(f) = k_A |M_A|^q + k_K |M_K|^q + k_H |M_H|^q$. The weighting coefficients are chosen in inverse proportion to the maximal joint moments at the joints, $k_A = 1/\left(M_{A\max}\right)^q; k_K = 1/\left(M_{K\max}\right)^q; k_H = 1/\left(M_{H\max}\right)^q$. The simulation improved when parameter *q* increased from 1 to 10. The model predicted that increasing pedal resistance makes the pedal force direction approach the direction *HP′*.

Reprinted, by permission, from X. Wang et al., 2000, "Experimental investigation and biomechanical analysis of lower limb movements for clutch pedal operation," *Ergonomics* 43 (9):1405–1429. By permission of Taylor & Francis Ltd, **http://www.tandf.co.uk/journals**

2.3.5 Preferred Directions for the Exertion of Endpoint Forces

Representative papers: Grieve and Pheasant (1981); Looze et al. (2000)

When people exert forces on objects with geometric constraints, the exerted forces may not be in exactly the direction of the desired movement. For instance, while pedaling a bicycle, even professional athletes wearing toe clips do not exert force perpendicularly to the crank throughout the complete circular cycle. In some leg positions, large force components are exerted along the crank, thus tending to compress or extend it (figure 2.34). Mechanically, these

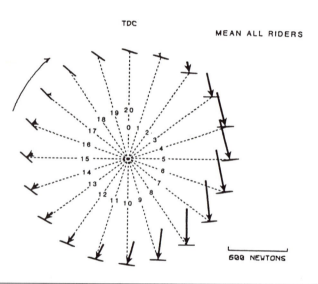

Figure 2.34   Forces applied to a pedal during bicycling. Average data for elite pursuit riders at 100 rpm and 400 W. The orientation of the pedal and the resultant force vector are shown at 20 positions of the crank. The force acting on the crank can be decomposed into tangential and normal components. Positive tangential force advances the crank, and positive normal force is directed toward the crank center. Mechanically, the normal force can be considered a loss. In position 4, the force is perpendicular to the crank and is close to being 100% effective. At the bottom center (position 10), a large force exists, but—as one can see from its orientation—it is not very effective. In positions 11 through 17, a force still pushes down on the pedals and hence produces a countertorque that opposes forward movement.

Reprinted, by permission, from P.R. Cavanagh and D.J. Sanderson, 1986, The biomechanics of cycling: Studies on the pedaling mechanics of elite pursuit riders. In *The Science of Cycling*, edited by E.R. Burke (Champaign, IL: Human Kinetics), 105.

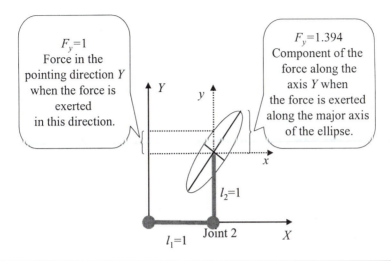

Figure 2.35 The force intentionally exerted in a given direction *Y* may be smaller than the component along *Y* of the force exerted in a more beneficial direction.

force components are losses; they do not produce useful effects. Formally, the effectiveness of force action may be computed as a projection of the force on the direction of movement. However, such a measure may be misleading.

It follows from the analysis in section **2.3.3** that the maximal force in a given direction may sometimes be achieved by deliberately directing force efforts in another direction where a larger force is produced. The projection of the exerted force on the given direction may exceed the force intentionally exerted in this direction. This assertion is illustrated in figure 2.35, which is a corollary of figure 2.28*b*.

According to experimental evidence, when large efforts are required, people are prone to exert forces in the direction of larger force production, whatever the useful direction involved. For instance, wheelchair users do not typically exert forces tangentially to the hand rims (figure 2.36). This technique, which is appropriate for hand-rim propulsion, may be detrimental in other tasks that require accurately directed exertion and are not tolerant of deviation.

To estimate the effectiveness of static force efforts, the *maximum advantage of using a component of exertion* (MACE) was introduced (Grieve and Pheasant 1981). MACE is defined as the ratio of the maximum available force component in a given direction to the manual force deliberately exerted in this direction. The following example illustrates the problem. Assume that one is asked to push horizontally a cart whose handlebar is very low, at knee level. One has two choices: either to push the handlebar strictly forward or to push it in the oblique direction, forward and down. In the first case, there will be no force loss due to pushing in an incorrect direction, but the body posture will be

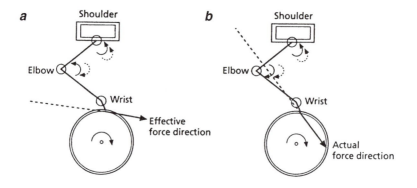

Figure 2.36 (*a*) The tangential force direction and (*b*) the actual force direction exerted by wheelchair users. The solid curved arrows indicate the joint torques at the shoulder and elbow. The dashed curved arrows indicate the rotation direction at those joints. For hand-rim rotation, the tangential direction can be viewed as the most effective mechanically. The *fraction effective force* (FEF) is defined as $FEF = (F_{tan} / F_{tot}) \times 100\%$, where F_{tan} is the magnitude of the tangential force component and F_{tot} is the magnitude of the total force exerted on the hand rim. For riding at a self-selected pace on level ground, the FEF values are only about 47% to 49%. If a wheelchair user were to propel his or her chair with a force that was directed as in (*a*), the FEF would be close to 100% but the magnitude of F_{tot} would be smaller. Because the task is not static, to follow the hand rims, the elbow has to be extended. Therefore, when directing the force tangentially, the elbow flexors have to apply force while being forcibly stretched (compare the direction of the solid and dotted arrows at the elbow joint).

Reprinted, by permission, from L.H.V. van der Woude, H.E.J. Veeger, and A.J. Dallmeijer, 2000, Manual wheelchair propulsion. In *Biomechanics of Sport*, edited by V.M. Zatsiorsky (Oxford, UK: Blackwell Science), 609-636.

awkward and the exerted force will not be very large. With the second technique, the vertical component of force will be lost to friction and cart deformation. However, the generated force will be larger. The question is which of the two techniques is better. (A similar question arises in bobsledding. After the start, the athletes push the bob. What is the optimal angle of the push?)

The method of computing the MACE in a sagittal plane is illustrated in figure 2.37. Consider a vector *OF* representing a maximal static manual force (figure 2.37*a*). The magnitude of the vector components in other directions are $F \cos\theta$, where θ is the angle formed by the vector *OF* and the direction vector. The components lie on the circumference of a circle whose diameter is *OF* (figure 2.37*b*). Suppose that the maximal force vectors (i.e., the muscular strength) are known for all directions. The envelope of the vectors, called the inner envelope, is shown in figure 2.37*c*. Each point on the inner envelope has a circle of force components, similar to the circle in figure 2.37*b*. The outer

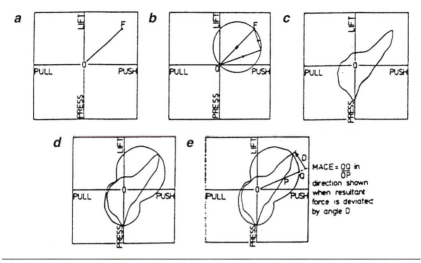

Figure 2.37   Diagram to illustrate the meaning of MACE. (*a*) Force vector *OF*. (*b*) The vector has components in other directions that lie on a circle of diameter *OF*. (*c*) An inner envelope showing the measured maximal manual forces in each direction in the sagittal plane. (*d*) The outer envelope. Each point on the inner envelope has a circle of force components associated with it. The outer envelope represents the tangents of all the circles. (*e*) In any selected direction, the ratio of the radii *OQ* to *OP* of the outer and inner envelopes represent the maximal advantage of using the components of an exertion directed at an angle *D* from the required direction.

Reprinted, by permission, from D.W. Grieve and S.T. Pheasant, 1981, "Naturally preferred directions for the exertion of maximal manual forces," *Ergonomics* 24 (9):685-693. By permission of Taylor & Francis Ltd, **http://www.tandf.co.uk/journals**

envelope for all the circles represents the maximal forces that can be attained in all directions either by exerting a force in this direction or as a component of the force exerted in another direction, figure 2.37*d*. MACE is then computed as the ratio of the radii of the outer and inner envelopes in a given direction. A MACE value greater than 1 indicates that using a component of a maximal effort is beneficial.

2.3.6 Exertion of Maximal Force: Limiting Joints

Representative paper: Hugh-Jones (1947)

According to equation 2.16, when a subject exerts a static force on the environment and the body posture and force direction do not change, the torques in the involved joints are defined uniquely. Any increase in the endpoint force requires a simultaneous, proportional increase in all the joint torques.

During maximal force exertion, the torques in some joints reach their maximal value (for the given body posture), while other joints are taxed only submaximally. The endpoint force cannot be increased further because of the insufficient torque in one or more of several joints involved in the task. Such joints are known as limiting joints (for a given performer, a given task, and a given body posture). For instance, for a boy who cannot perform a chin-up due to insufficient strength of his shoulder flexors but whose elbow flexors are strong enough to complete the task, the shoulder torque is a limiting factor. Limiting joints can differ in different athletes. For example, two athletes who can lift 100 kg from the floor but cannot lift 105 kg may have different limiting joints (e.g., knee extension in one athlete and spine extension in the other). Clearly, these athletes should be trained differently.

Some strength coaches believe that the value of maximal joint torque depends on whether the joint generates the torque or transmits it. A popular opinion among practitioners is that the joint can transmit a larger torque than it can generate. For instance, the assumption is that while testing isometric elbow flexion strength, assistive maneuvers, such as raising the shoulder, may increase the strength values. This issue is not purely a mechanical one, and it cannot be solved by application of equation 2.16. The problem should be resolved by experiments. At this time, experimental evidence is not sufficient to make a decisive conclusion. Available scientific evidence speaks against the hypothesis: it seems that the maximal values of joint torque do not depend on whether the joint generates or transmits the torque.

■ ■ ■ *FROM THE LITERATURE* ■ ■ ■

Torque Transmission Versus Torque Generation

Source: Andersson, G.B.J., and A.B. Schultz. 1979. Transmission of moments across the elbow joint and the lumbar spine. *J. Biomech.* 12:747–55.

The researchers studied the dependence of the magnitude of maximal joint torque on whether the joint generates the torque (the joint is the sole source of the endpoint force) or transmits it. One of the experiments is presented in figure 2.38. In figure 2.38a, the subject lying supine on a free-wheeling dolly extended his legs, and the elbow joint served as a torque transmitter. In figure 2.38b, the legs were already extended, and the elbow joint served as a torque generator. The measured force values were nearly the same in both cases. Hence, the maximal magnitude of the elbow-joint torque did not depend on whether or not leg extension was used.

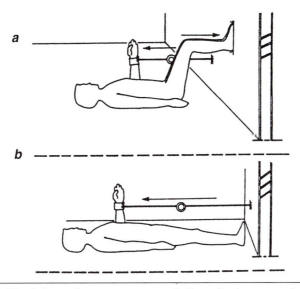

Figure 2.38 Scheme for measurement of elbow flexion strength when supine. (*a*) A typical assistive maneuver. In the assistive-maneuver test, the subjects lay on a free-wheeling dolly, which enabled the push-off force exerted with the legs to be freely transmitted to the elbow. (*b*) The unassisted test.

Reprinted from *Journal of Biomechanics*, 12, G.B.J. Andersson and A. Schultz, Transmission of moments across the elbow joint and the lumbar spine, 747–755, 1979, with permission from Elsevier Science.

▪ ▪ ▪ *FROM THE LITERATURE* ▪ ▪ ▪

While Increasing Force Magnitude, People Change the Force Direction: Linear Force Paths Emerge

Source: Gruben, K.G., and C. López-Ortiz. 2000. Characteristics of the force applied to a pedal during human pushing efforts: Emergent linearity. *J. Motor Behav.* 32 (2): 151–62.

Subjects supported by a bicycle seat pushed against a pedal in the most comfortable manner. The crank was fixed in 97 positions. The force magnitude was up to 40% of maximal. In 87% of the pushes, the measured sagittal-plane force exerted on the pedal by the foot changed magnitude and direction through time so that the path of the head of the force vector traced a straight line (figure 2.39*a*). The linearity of the foot force paths reflected proportionality in the changes of the involved joint moments (figure 2.39*b*).

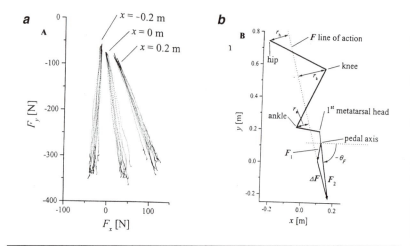

Figure 2.39   The linearity of force paths. (*a*) Sample force paths for a typical single participant in the pedal-pushing experiment. The data are from 10-repetition sets in 3 pedal axis positions (from 97 positions tested). The different locations of the upper end of the force paths reflect different initial force vectors at each of the pedal positions. (*b*) Foot force. The leg is completely extended at $x = 0$ m, $y = -0.2$ m. $\mathbf{F}_1$ and $\mathbf{F}_2$ are force vectors at the start and the end of the push, respectively. For a fixed limb position (no change in pedal inclination), vector decomposition yields a vector of the force change $\Delta\mathbf{F}$ with constant direction. The moment arms of the vector $\Delta\mathbf{F}$ at the joints (r_h, r_k, r_a) are constant. The changes in joint moment magnitudes are $\Delta M_h = r_h\Delta F$, $\Delta M_k = r_k\Delta F$, and $\Delta M_a = r_a\Delta F$. Hence, the ratios between changes of joint moments are constant, for instance, $\Delta M_h/\Delta M_k = r_h/r_k = a_1$, and pairs of joint moments relate linearly to each other, for instance, $M_h = a_0 + a_1\Delta M_k$, where a_0 and a_1 are constants.

Journal of Motor Behavior, 32, 2, 151–162, 2000. Reprinted with permission of the Helen Dwight Reid Educational Foundation. Published by Heldref Publications, 1319 Eighteenth St., NW, Washington, D.C. 20036-1802. Copyright © 2000.

2.3.7 On the Direct Problem of Statics: Motor Redundancy and Motor Overdeterminacy

Representative paper: Mussa-Ivaldi (1986)

If a kinematic chain is in equilibrium, an endpoint force $\overline{\mathbf{F}}$ uniquely determines the torques in all joints. Hence, the inverse problem of statics can always be solved. The opposite is not always true. An equilibrating endpoint force may not exist for some joint torque combinations. In this case, the direct

Sets of Equations

Underdetermined sets of equations have more unknowns than equations. The underdetermined equations cannot be uniquely solved. For instance, the equation $x + y = 10$ has an infinite number of solutions. Such sets of equations become solvable only when additional restrictions are put on the problem.

Determined sets of equations have as many equations as there are unknowns. If the coefficient matrix of a set of linear equations is not singular (its determinant is not zero), the system has a unique solution.

Overdetermined sets of equations have more equations than unknowns. If the overdetermined set of equations allows a common solution (the set is consistent), some equations duplicate other equations and can be deleted from the system with no effect.

problem of statics cannot be solved. The physical meaning of this assertion is clear: chain equilibrium cannot be maintained under arbitrarily selected joint torques.

For a single link system, the external force cannot be computed if only a joint torque is known. In such a situation, the direct statics problem cannot be solved. The reason is simple: unknown force acting at the endpoint has two characteristics, the force magnitude and direction. Hence, it is not enough to know only the joint torque; two equations are needed. If we know only the torque, we can compute only one force component: the component that is normal to the radius from the joint center to the point of force application. The component that is acting along the link remains unknown. An equation for the projection on the coordinate axes is

$$T = F_x l \sin\alpha + F_y l \cos\alpha \qquad (2.54)$$

Any combination of F_x and F_y that satisfies the constraints imposed by equation 2.54 is an appropriate solution (figure 2.40). Answers such as $F = F_x = T/l \sin\alpha$ or $F = F_y = T/l \cos\alpha$ are equally valid. This does not mean that the endpoint force is acting in all directions. It simply indicates that we do not have enough information to solve the problem. If we were to know the joint reaction force, the external force could be found easily. Equation 2.54 is evidently underdetermined; unique values of F_x and F_y cannot be computed from this equation.

In two-link planar chains, knowledge of the joint torques allows the magnitude and direction of the force acting at the endpoint to be determined. In

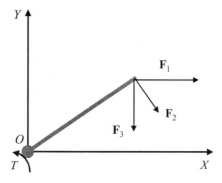

Figure 2.40 If the forces $\mathbf{F}_1$, $\mathbf{F}_2$, and $\mathbf{F}_3$ satisfy equation 2.54, any of them can counterbalance the torque T acting at the joint.

three-link chains, it allows the endpoint force and couple to be determined. For these chains, the number of available equations equals the number of unknowns. To produce a desired endpoint force—a force of a given magnitude and direction and a couple—particular joint torques must be generated in the joints. The solution is unique: there exists only one combination of joint torques that brings about the required endpoint force.

In statics, planar chains with four or more links are overconstrained. The Jacobian transpose of such a chain is not a square matrix and therefore cannot be inverted. The set of equations is overdetermined. The number of equations ($n = 4$ or more) exceeds the number of unknowns (three). Hence, for this special case of an actual physical system, the equations are linearly dependent. If we are interested in finding only the endpoint force and couple, we can use any three equations out of the n available. These equations may include the values of any three joint torques, for instance, the torques in joints 1, 2, and 3 or joints $(n - 2)$, $(n - 1)$, and n. (Note that we are speaking about the equations that correspond to the individual rows in equation 2.16; see also equation 2.24.) The remaining $(n - 3)$ equations are not necessary for the solution. If we know the values of three joint torques and we know that the chain is in equilibrium, we can delete the $(n - 3)$ other equations from the system with no effect because they duplicate other equations. However, for a chain to be in equilibrium, *all* the joint torques must correspond to the requirement of the task.

From the standpoint of motor control, overdeterminacy means that although it is sufficient for the central controller (the CNS) to set values for joint torques in only three joints in order to produce a desired endpoint force and a couple, the torques in the other $(n - 3)$ joints must also be specified precisely. If torque at one of the joints is smaller than necessary for equilibrium, this joint will yield to the load. If the torque is larger than necessary, other joints will yield.

Therefore, to control a chain with more than three joints, the central controller should have knowledge about the torques in the three joints and adjust the other torques accordingly. If it does not do this, chain equilibrium is not attained. Is it because of this complication that we have only three main segments in our extremities?

Underdetermined sets of equations (in which the number of unknowns is larger than the number of available equations) are quite common in human movement science. The number of equations usually corresponds to the number of constraints in the task, and the number of unknowns corresponds to the number of contributing elements that are used to solve the task. Examples include exerting a force of a given magnitude with several fingers, producing a given joint torque when several muscles are involved, generating a muscle force when many muscle fibers are activated, or reaching a certain point in space by a multilink chain. The equations specify the required total force, the joint torque, the muscle force, and the coordinates of the endpoint, accordingly. The unknowns are the finger forces, the muscle forces, the number and the level of activation of individual muscle fibers, and the joint angles, correspondingly. In all these situations, the central controller has the freedom to choose one solution from an infinite number of possible solutions. The underdetermined systems of equations reflect the complexity of the human motor system usually called *motor redundancy* or *motor abundance.* The problem of motor redundancy is also known as *Bernstein's problem,* after the Russian scientist Nikolai A. Bernstein (1896–1966) who first posed this problem in the early 1930s.

Overdetermined systems, or *overconstrained systems,* are also quite common in human and animal movement. For instance, during static force exertion, multilink kinematic chains represent overdetermined systems. When body posture and endpoint force are specified, the solution is unique; no freedom is allowed. The joint torques should correspond not only to the exerted endpoint force but also to each other. When torques in some joints are produced, torques in other joints should match them. Therefore, not only motor redundancy but also motor overdeterminacy reflects the intricacy of the human motor system.

In human movement science, the term "motor redundancy" is traditionally associated with underdetermined sets of equations for which the central controller should choose one solution out of an infinite number of existing solutions. In mathematics, overdetermined sets of equations are often called redundant sets. To avoid confusion, we prefer not to use the word redundant to describe sets of equations and restrict the use of this word to the description of the motor system. Hence, underdetermined sets of equations are due to motor redundancy, and overdetermined sets of equations arise due to the overdeterminacy of the human motor system, when the system is overconstrained.

> ### • • • MATHEMATICS REFRESHER • • •
>
> ## Range and Null Space: Rank of Matrices
>
> The following symbols are used: R^n and R^m stand for the vector spaces with n and m components, where n and m are real numbers. (The superscripts to R should not be interpreted as exponents of R.) X is a vector in R^n, $(X \in R^n)$, and Y is a vector in R^m, $(X \in R^n)$. Any $(m \times n)$ matrix $[M]$ defines a linear transformation (mapping) of the vector space R^n into the vector space R^m (figure 2.41).
>
>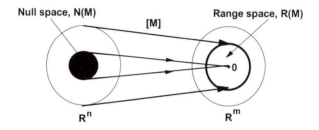
>
> **Figure 2.41**   A mapping diagram of the range and null spaces.
>
> The range space of $[M]$ is the subspace of R^m containing all vectors Y such that $Y = [M]X$ for at least one X. The range space $\mathbf{R(M)}$ is the image of R^n under the linear transformation $[M]$. The null space of $[M]$ is the subspace of R^n containing all vectors X such that $[M]X = \mathbf{0}$. The null space $\mathbf{N(M)}$ is the inverse image of the zero vector $\mathbf{0}$ in R^m under the linear transformation $[M]$. The sum of the dimensions of the range space and the null space equals n, $dim\ \mathbf{R(M)} + dim\ \mathbf{N(M)} = n$.
>
> The rank of matrix $[M]$ is the order of the largest submatrix in $[M]$ with a nonzero determinant. The rank of $[M]$ equals the number of linearly independent row (or column) vectors in $[M]$. The dimension of the range space, $dim\ \mathbf{R(M)}$, equals the rank of matrix $[M]$.

2.4 DUALITY OF STATICS
AND KINEMATICS: NULL SPACES

Representative papers: Mussa-Ivaldi (1986); Scholz and Schoner (1999)

The goal of this section is to give an intuitive view of the duality of statics and kinematics. The mathematical notions of the range space and the null space of a linear transformation are applied to the problems of kinematics and statics of multilink chains.

Earlier in the text, certain analogies between concepts in kinematics and statics were mentioned. For instance, a resemblance was demonstrated between a wrench (a system of forces) and a screw (a motion). Mathematically, such an analogy is easy to explain. Both velocity and force are vectors. Hence, they can be analyzed in a similar manner. In table 2.1 some theorems and statements from kinematics and statics are juxtaposed.

Table 2.1   Examples of the Duality of Kinematics and Statics

Kinematics	Statics
The parallelogram rule is used with . . .	
Linear and angular velocities	Forces and couples
Any angular velocity ω can be reduced to an equal angular velocity about a parallel axis and a linear velocity ωd, where d is the distance between the parallel axes.	Any force F can be reduced to an equal and parallel force and a couple Fd, where d is the distance between the lines of force action.
To define an axis of rotation, four parameters are required (e.g., two coordinates of a piercing point in any coordinate plane and two direction angles).	To define a line of force action, four parameters are required (e.g., two coordinates of a piercing point in any coordinate plane and two direction angles).
To define an angular velocity in 3-D space, five parameters must be specified (four parameters for the axis of rotation plus the magnitude of the angular velocity).	To define a force in 3-D space, five parameters must be specified (four parameters for the line of force action plus the magnitude of the force).
Any instantaneous motion can be reduced to three projections of the linear velocity on three rectangular axes and three projections of the angular velocity into these axes.	Any system of forces can be reduced to three forces acting along three rectangular axes and three couples around these axes.
Any instantaneous motion can be reduced to a screw: a linear velocity along an axis and an angular velocity around this axis.	Any system of forces can be reduced to a wrench: a resultant force and a couple that acts around the line of action of the resultant force.
Instantaneous velocity of the endpoint of a kinematic chain equals $\mathbf{v} = \mathbf{J}\dot{\boldsymbol{\alpha}}$.	A vector of joint torques equals $\mathbf{T} = \mathbf{J}^{T}\overline{\mathbf{F}}$.

Kinematically redundant systems (e.g., planar chains with 4 DoF) are overdetermined in statics. Statically redundant systems (e.g., several muscles serving one joint) are overdetermined in kinematics.

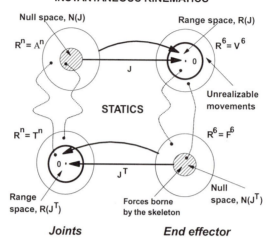

INSTANTANEOUS KINEMATICS

Joints   *End effector*

Figure 2.42 Linear mapping of velocities in instantaneous kinematics and forces in statics. The top of the figure shows a mapping from the n-dimensional vector space of joint angular velocities $\dot{\mathbf{A}}^n$ to the 6-dimensional space of end-effector velocity $\mathbf{V}^6$. The bottom of the figure shows a mapping from the 6-dimensional vector space of endpoint forces $\mathbf{F}^6$ to the n-dimensional space of joint torques $\mathbf{T}^n$.

In kinematics, the relationship between the instantaneous velocity of the endpoint and the vector of joint angular velocity is described by the equation $\mathbf{v} = \mathbf{J}\dot{\boldsymbol{\alpha}}$. In statics, the equation $\mathbf{T} = \mathbf{J}^T\overline{\mathbf{F}}$ describes the relationship between the vector of joint torques and the force at the endpoint. The same chain Jacobian is used in both cases. Both equations can be regarded as a linear mapping from one vector space into another vector space (figure 2.42).

In kinematics, the linear mapping $\mathbf{v} = \mathbf{J}\dot{\boldsymbol{\alpha}}$ represents a transformation from an n-dimensional space of joint angular velocities, $\dot{\mathbf{A}}^n$, to an m-dimensional space of end-effector velocity, $\mathbf{V}^m$. Hence, the Jacobian is a $6 \times n$ matrix and $\mathbf{V}^m = \mathbf{V}^6$. The *range space* $\mathbf{R}(\mathbf{J})$ symbolizes all the end-effector velocities that can be attained at a given joint configuration. If the rank of $\mathbf{J}$ equals 6 (or 3 in the planar case), the range space covers the entire vector space $\mathbf{V}^m$. Otherwise, the end effector cannot move in some directions. In this context, the word "direction" signifies not only linear directions but also directions of rotation. For instance, the distal link of a planar two-link chain ($n = 2$) cannot move along the X and Y axes with a prescribed linear velocity and at the same time rotate with a desired angular velocity. In this example, $m > n$ and the rank of $\mathbf{J} < 3$. In three-link planar chains, the distal links can move in all three directions, two linear and one angular. If a chain is in a singular position, the chain Jacobian is not of full rank and is not invertible. Starting from this position, the end effector cannot move in some directions; for example, it cannot penetrate the surface of the work envelope.

When $n > 6$, the chain is redundant (and the sets of equations that describe the end-effector position are underdetermined). In redundant chains, (1) the end effector can be placed at a given position in an infinite number of ways, and (2) the links of the chain can move with respect to each other without displacing the terminal points of the chain—the end effector and the center of the most proximal joint. For instance, a person can move his or her elbow while keeping a hand firmly placed on a desk and the torso stationary. The movement involves angular rotations in the shoulder and wrist joints. The explanation is evident: excluding the finger joints, the human arm has seven degrees of freedom (maneuverability of the arm equals one degree of freedom; see *Kinematics of Human Motion*, section **2.2.2**). The described arm movement, which can be represented by a vector of the joint angular velocities, results in zero endpoint velocity. The movement is possible because the angular rotations at the shoulder and wrist joints are performed about collinear joint axes that form a straight line. Therefore, rotation in the first joint is canceled by a counteracting rotation in the second joint. As a result, both the hand and the trunk position remain constant, even though the arm moves.

All the possible joint movements that are mapped into the zero vector in $\mathbf{V}^6$ form the *null space* of the linear mapping $\mathbf{J}$. The null space $\mathbf{N(J)}$ is the set of joint angular velocities that produces no movement of the end effector. For any element of the null space, $\mathbf{J}\dot{\alpha} = \mathbf{v} = 0$. The sum of the dimensionalities of the range space and the null space is always equal to n, $dim\ \mathbf{R(J)} + dim\ \mathbf{N(J)} = n$.

The concept of null spaces is employed to identify the control variables used by the CNS during human movement. To grasp the idea, consider the task of drawing on a tabletop. At each instant in time, the position of the end effector is described by the two coordinates X and Y. The position of the arm is characterized by the three joint angles at the shoulder, elbow, and wrist. Hence, the joint space has three dimensions. For a given endpoint position, the null space represents those combinations of joint angles that leave the end effector unaffected. In this particular example, the dimensionality of the null space is 1, *dim* $\mathbf{N(J)} = 3 - 2 = 1$, and the null space is just a line in the joint space. It can be found as a null space of the Jacobian matrix of the chain. During drawing, the position of the endpoint is a control variable (drawing would not be possible without controlling the endpoint). However, control variables in many other movements are up for discussion. For instance, what is the control variable in the transition from sitting to standing? Is it CoM displacement, head displacement, or something else?

The interpretation of the range and null spaces in terms of joint angular motion and end-effector motion has a counterpart in statics. In statics, the linear mapping $\mathbf{T} = \mathbf{J}^T\mathbf{F}$ represents transformation from the 6-dimensional space

■ ■ ■ *FROM THE LITERATURE* ■ ■ ■

Null Spaces in Human Movement

Source: Scholz, J.P., and G. Schoner. 1999. The uncontrolled manifold concept: Identifying control variables for a functional task. *Exp. Brain Res.* 126 (3): 289–306.

[Note. The word *space* in the following text has three different meanings: (1) the external Cartesian space with the vertical and horizontal coordinates; (2) joint space, where coordinates are joint angles; and (3) null and range space, which are mathematical abstractions where any variables can serve as the coordinates. The reader is advised to be attentive to the meaning of the word.]

The authors suggested a method of identifying the control variables in human movement and applied the method to the sitting-to-standing transition. The study was based on two postulates: (1) control variables have smaller variability than variables that are not immediately controlled, and (2) variability of the joint angles is larger in the null space than in the range space (the authors' mathematical methods allowed them to decompose the total variability into variability in the null space and in its complement).

When the control variable is known, the joint space can be divided into two subspaces, the null space (which the authors called the uncontrolled manifold) and its complement (a space perpendicular to the null space, the range space) (figure 2.43*a*). The term "uncontrolled manifold" should not be taken literally. It does not mean that the joints are not controlled. The term means that in the null space, there is no need to specify a precise joint configuration unless other constraints are brought to bear. The coordinated changes of joint configuration in the null space do not influence the control variable. For instance, if the CoM is a control variable, any changes in the body posture that leave the CoM position unaffected belong to the null space. Contrary to that, any joint changes in the range space affect the control variable.

Variability over consecutive trials of the sitting-to-standing transition was investigated. During the sitting-to-standing transition, body configuration is described by eight joint angles: the angles formed by the foot, shank, thigh, pelvis, trunk, head/neck, forearm, and arm with the horizontal. Two coordinates describe the CoM position in the sagittal plane. Therefore, the dimensionality of the range space is two, *dim* $\mathbf{R}(\mathbf{J}) = 2$, and the dimensionality of the null space (the uncontrolled manifold), *dim* $\mathbf{N}(\mathbf{J})$, is six. The dimensionality *dim* $\mathbf{R}(\mathbf{J}) = 2$ means that in order to displace the

CoM in the desired direction along the X and Y coordinates, a coupled movement in at least two joints has to be performed. In the study, the variability was estimated both in the external Cartesian space (e.g., variability of the CoM trajectory in the vertical direction) and in the joint space (the variability of the joint angles).

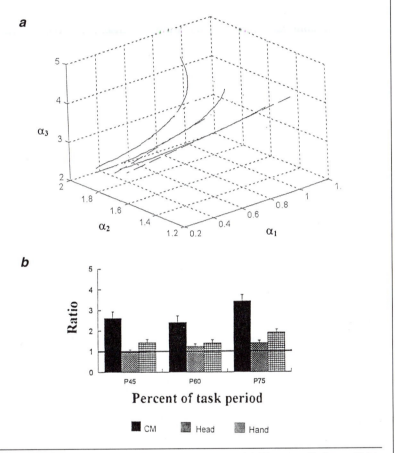

Figure 2.43 Null spaces in human movement. (*a*) Uncontrolled manifolds (null spaces) of the drawing model described in the text are illustrated in the space of the three joint angles. Each solid line represents a set of joint angles that correspond to a particular end-effector position. The null spaces in this example are simply the lines in the joint space. (*b*) The ratio of joint variability in the null space to that in the range space for the three hypothetical control variables during the sit-to-stand movement: the trajectories of the CoM, head, and hand.

The authors compared three hypothetical control variables: the trajectories of the CoM of the body, the head, and the hand. As a first step, they compared variability of these variables in the Cartesian space and found that the nervous system's priority for controlling these trajectories is CoM > head > hand. The second step addressed the variability of joint trajectories. The average variability per one degree of freedom (joint angle) was compared in the range and null spaces. For the CoM as a hypothetical control variable, the variability was found much larger in the null space than in the range space. For the head and the hand, these differences were not observed. An example of the results is presented in figure 2.43*b*. The authors concluded that during the sitting-to-standing transition, the CoM position in the sagittal plane is the primary control variable.

of the endpoint force, $\mathbf{F}^6$, to the *n*-dimensional space of the joint torques, $\mathbf{T}^n$. The Jacobian transpose is an $n \times 6$ matrix. The range space $\mathbf{R}(\mathbf{J}^T)$ is the set of all the joint torques that can equilibrate the endpoint force. The null space $\mathbf{N}(\mathbf{J}^T)$ is the set of all the endpoint forces that do not require any joint torque to resist the load. Any external force that belongs to $\mathbf{N}(\mathbf{J}^T)$ is supported entirely by the skeleton.

Due to the duality of kinematics and statics, the null and range spaces of the two linear mappings are interrelated. In particular, (1) a nonzero vector that belongs to the null space in kinematics cannot belong to the range space in statics, and (2) a nonzero vector that belongs to the null space in statics does not belong to the range space in kinematics. [For readers who are familiar with matrix algebra, these statements can be formulated differently: the spaces $\mathbf{N}(\mathbf{J})$ and $\mathbf{R}(\mathbf{J}^T)$ are orthogonal complements of each other, as are the spaces $\mathbf{N}(\mathbf{J}^T)$ and $\mathbf{R}(\mathbf{J})$.] In terms of biomechanics, the first statement implies that if a certain vector of joint velocities does not produce any endpoint velocity, none of the external endpoint forces can be equilibrated by a similar vector of joint torques. For instance, external loading does not prevent possible elbow movement while keeping the hand motionless on a support and the trunk stationary. The movement can be performed with equal ease whether the arm is loaded or not. The reader may try to perform the described elbow movement rotation while performing push-ups to prove this. As another example, imagine an athlete who is holding a barbell in a semisquatting position. The athlete slightly displaces the knees in the lateral or medial direction. This movement does not influence the leg extension and hence belongs to the null space $\mathbf{N}(\mathbf{J})$. The vector of joint velocities from this space is mapped into the zero vector of the endpoint. If the weight of the barbell increases, it does not influence the joint torques in the direction of the described movement. Hence, the external force vector is not mapped into the joint torques; it does not belong to the range space $\mathbf{R}(\mathbf{J}^T)$. As a result, if the joint torques in the $N(J)$ space increase, they cannot be balanced by the barbell weight. Hence, to maintain a stable leg configuration, the joint torques in the $N(J)$ space should be zero.

In statics, the null space $N(J^T)$ represents a set of force directions in which the external endpoint forces are mapped into the zero joint torques and counterbalanced exclusively by the skeleton. In these directions, the end effector cannot exert a force on an external object, no matter what joint torques are generated. An example is an outstretched leg with force acting along its long axis. In kinematics, in the $N(J^T)$ space, joint angular velocities do not produce endpoint movement in certain directions. The vector of joint angular velocities is not mapped into the range space $R(J)$.

2.5 SUMMARY

The chapter deals with a basic model of the human body called the stick-figure model. In this model, the body segments are represented as rigid links connected by ideal revolute joints. The joints are served by torque actuators that produce a moment of force with respect to the joint axis. The term *joint torque*, or *joint moment*, refers to two equal and opposite moments of force acting on two adjacent body links about the rotation axis of the joint. There is a difference between actual joint torques (produced by muscles that span only one joint) and equivalent torques (calculated for a system served by two-joint muscles). Several varieties of joint torques (joint moments) are introduced in table 2.2.

Table 2.2 Joint Torques

Joint torque	Function, characteristics
Equilibrating torques	Torques that resist the external endpoint force
Active joint torques	Torques generated by the force actuators
Muscle-tendon torques	Same as active torques
Passive torques	Torques borne by the passive tissues (e.g., by the bones and ligaments)
Net joint torques	Total torques exerted at the joints, without regard to their role (e.g., counterbalancing an external endpoint force or overcoming inertial resistance) or whether they are generated actively or passively

Transformation analysis (of kinematic chains) deals with determining the endpoint force and joint torques. When joint torques are known and the task is to compute the generalized force applied to the end effector, the procedure is called direct, or forward, static analysis. Inversely, when an external force is given and the task is to compute the magnitudes of the torque at each joint, the procedure is called inverse static analysis. Two methods of transformation analysis are described: the link-isolation method and the Jacobian method. In the link-isolation method, free-body diagrams are used to determine the forces and couples acting on individual links considered separately from other links of the kinematic chain, one link at a time. A recursive procedure is used to find joint forces and moments for all the links of the chain. The Jacobian method allows inverse static analysis of arbitrary chains in a single procedure. The following equation is applied: $\mathbf{T} = \mathbf{J}^T \overline{\mathbf{F}}$, where $\mathbf{T}$ is the vector of the joint torques, $\mathbf{J}^T$ is the transpose of the Jacobian matrix, and $\overline{\mathbf{F}}$ is the six-component endpoint force vector measured in the global system of coordinates. If an endpoint couple is zero, the elements of the Jacobian transpose represent the moment arms of the external force $\mathbf{F}$ with respect to the individual joints. If the Jacobian is singular, for instance, when the chain is outstretched, the direct problem of statics cannot be solved. In singular positions, the end-effector force can be changed in specific directions with no effect on the joint torques. The transformation analysis of planar two- and three-link chains is discussed in detail.

The exertion of an intended contact force on the environment in static conditions is examined. The magnitude of the endpoint force is controlled by proportional changes in the joint torques at the involved joints, and the force direction is manipulated by maintaining the required difference in torques. For a two-link chain, the radial axis, which connects the endpoint with the proximal joint, and the pointing axis, directed along the distal segment, are defined. In two-link chains, the distal joint produces an endpoint force along the radial axis, and the proximal joint produces an endpoint force along the pointing axis. Two-link chains can exert a force, but they cannot exert a torque (a couple) on the environment. Three-link chains can exert both an external force and a couple.

Endpoint force depends on joint torques and chain configuration. To study the effect of chain configuration exclusively, endpoint forces in various directions are computed at a constant magnitude of the vector of joint torques, $\mathbf{T}$. In three dimensions, the tips of the endpoint force vectors represent a three-dimensional ellipsoid, called the force ellipsoid. The principal axes of the ellipsoid are along the eigenvectors of the matrix $[JJ^T]$, where $[J]$ is a chain Jacobian. The shape and orientation of the force ellipsoid depend on the chain configuration.

Holonomic, or geometric, constraints restrict movement in certain directions. The constraints may completely change the joint torques required for equilibrium. Therefore, different muscle groups may act when body motion is free and when it is actually (physically) constrained. Maximal force in a given direction can sometimes be achieved by deliberately directing force efforts in another direction where a larger force is produced. To estimate the effectiveness of static force efforts, the ratio of the maximum available force component in a given direction to the manual force deliberately exerted in this direction is computed.

An equilibrating endpoint force may not exist for some joint torque combinations. In this case, the direct problem of statics cannot be solved. For a single link system, the external force cannot be unequivocally computed if only the joint torque is known. In this case, the number of unknowns (two force components) exceeds the number of equations (one). The set of equations is underdetermined. For two- and three-link kinematic chains, the set of equations is determined: the number of equations equals the number of unknowns. With the exception of singular chain configurations, the direct problem of statics can be solved for these chains. For chains with more than three links, the set of equations is overdetermined and $(n - 3)$ equations are linearly dependent on others. Hence, to produce a desired endpoint force and couple, it is sufficient for the central controller to set the necessary values for joint torques in only three joints. However, to maintain equilibrium, the torques in the remaining $(n - 3)$ joints must be specified in strict accordance with the torques in other joints. Due to the motor redundancy of the human motor system, the mathematical description of human motion frequently results in underdetermined sets of equations. Overdetermined sets of equations reflect the motor overdeterminacy of the human motor system.

There is a similarity (duality) between the concepts of kinematics and statics. For instance, in kinematics, the equation $\mathbf{v} = \mathbf{J}\dot{\alpha}$ describes the relationship between the instantaneous velocity of the endpoint and the vector of joint angular velocity. In statics, the equation $\mathbf{T} = \mathbf{J}^T\overline{\mathbf{F}}$ describes the relationship between the vector of joint torques and the force at the endpoint. Both equations can be regarded as a linear mapping from one vector space into another vector space. Due to the duality of kinematics and statics, (1) a nonzero vector that belongs to the null space in kinematics cannot belong to the range space in statics, and (2) a nonzero vector that belongs to the null space in statics does not belong to the range space in kinematics.

2.6 QUESTIONS FOR REVIEW

1. Give a definition of the term "joint torque."
2. Define and describe the basic model of the human body (stick-figure model).

3. Consider (a) a one-joint muscle crossing a frictionless revolute joint and (b) a two-joint muscle crossing two joints. What is the conceptual difference between the joint torque at the joint served by the one-joint muscle and the joint torque at a joint served by the two-joint muscle?

4. Define direct and inverse static analyses.

5. Write equilibrium equations for a link i of a kinematic chain in the reference frame $\{i\}$.

6. Joint forces and moments are given in a frame $\{i + 1\}$. Transform the forces and moments from frame $\{i + 1\}$ to frame $\{i\}$.

7. How is a recursive procedure used to find the joint forces and moments for the links of a kinematic chain?

8. For a joint with 1 DoF, a net joint torque is known. Find the muscle-tendon torque.

9. A generalized contact force $\overline{\mathbf{F}}$ is applied to an end effector. The goal is to find the equilibrating torques induced at the joints, $\mathbf{T}_j$. Write the necessary equation. Use the chain Jacobian.

10. Explain the geometric meaning of the elements of the Jacobian transpose.

11. Consider a planar three-link chain. Provide an example of a chain configuration for which the direct problem of statics cannot be solved.

12. Consider a planar two-link chain. How should the joint torques be altered to change the magnitude of the endpoint force (keeping the force direction constant) or to change the force direction (keeping force magnitude constant)?

13. Consider a planar three-link chain. An endpoint force is exerted on the environment. Determine the directions of the components of the endpoint force produced by the individual joint torques.

14. Discuss force ellipsoids.

15. An athlete bench-presses a barbell several times. In consecutive trials, the grip is constant, and the barbell moves along the same trajectory with the same velocity and acceleration. The joint torques and the muscle activity are different, however. Explain a possible cause of the observed variability.

16. Discuss the maximum advantage of using a component of exertion (MACE).

17. Is the set of equations that describes a human movement that illustrates motor redundancy underdetermined, overdetermined, or redundant? Is the set of equations that describes a human movement that exemplifies motor overdeterminacy underdetermined, overdetermined, or redundant?

18. Discuss the duality of kinematics and statics of kinematic chains.

2.7 BIBLIOGRAPHY

Abendroth-Smith, J., and S. Griswold. 1998. The effects of grip width on bench press performance using novice lifters. In *Proceedings of NACOB '98: The third North American Congress on Biomechanics, August 14–18, 1998,* 407–8. Waterloo, ON: University of Waterloo.

Amis, A.A., D. Dowson, and V. Wright. 1980. Elbow joint force predictions for some strenuous isometric actions. *J. Biomech.* 13:765–75.

Andersson, G.B.J., and A.B. Schultz. 1979. Transmission of moments across the elbow joint and the lumbar spine. *J. Biomech.* 12:747–55.

Bay, B.K., A.J. Hamel, S.A. Olson, and N.A. Sharkey. 1970. Statically equivalent load and support conditions produce different joint contact pressures and periacetabular strains. *J. Biomech.* 30 (2): 193–96.

Bolhuis, B.M. van, C.C.A.M. Gielen, and G.J. van Ingen Schenau. 1998. Activation patterns of mono- and biarticular arm muscles as a function of force and movement direction of the wrist in humans. *J. Physiol. (Lond.)* 508 (1): 313–24.

Carlton, L.G., and K.M. Newell. 1985. Force variability in isometric tasks. In *Biomechanics IX-A,* ed. D. Winter, R.W. Norman, R.P. Wells, K.C. Hayes, and A.E. Patla, 128–32. Champaign, IL: Human Kinetics.

Cavanagh, P.R., and D.J. Sanderson. 1986. The biomechanics of cycling: Studies on the pedaling mechanics of elite pursuit riders. In *The science of cycling,* ed. E.R. Burke, 91–122. Champaign, IL: Human Kinetics.

Chao, E.Y., J.D. Opgrande, and F.E. Axmear. 1976. Three-dimensional force analysis of finger joints in selected isometric hand functions. *J. Biomech.* 9:387–96.

Cooper, R.A., R.N. Robertson, D.P. VanSickle, M.L. Boninger, and S.D. Shimada. 1996. Projection of the point of force application onto a palmar plane of the hand during wheelchair propulsion. *IEEE Trans. Rehabil. Eng.* 4 (3): 133–42.

Daggfeldt, K., and A. Thorstensson. 1997. The role of intra-abdominal pressure in spinal unloading. *J. Biomech.* 30 (11–12): 1149–55.

Delp, S.L., A.E. Grierson, and T.S. Buchanan. 1996. Maximum isometric moments generated by the wrist muscles in flexion-extension and radial-ulnar deviation. *J. Biomech.* 29 (10): 1371–75.

Dempster, W.T. 1958. Analysis of two-handed pulls using free body diagrams. *J. Appl. Physiol.* 13 (3): 469–80.

Dempster, W.T. 1961. Free body diagrams as an approach to the mechanics of human posture and motion. In *Biomechanical studies of musculo-skeletal system,* ed. F.G. Evans, 81–135. Springfield, IL: Charles C Thomas.

Donskoy, D.D., and V.M. Zatsiorsky. 1979. *Biomechanics* (in Russian). Moscow: FiS.

Doorenbosch, C.A., J. Harlaar, and G.J. van Ingen Schenau. 1995. Stiffness control for lower leg muscles in directing external forces. *Neurosci. Lett.* 202 (1–2): 61–64.

Doorenbosch, C.A.M., and G.J. van Ingen Schenau. 1995. The role of mono- and biarticular muscles during contact control leg tasks in man. *Hum. Mov. Sci.* 14:279–300.

Dvir, Z. 1999. Coefficient of variation in maximal and feigned static and dynamic grip efforts. *Am. J. Phys. Med. Rehabil.* 78 (3): 216–21.

Fenn, W.O. 1957. The mechanics of standing on the toes. *Am. J. Phys. Med.* 36:153–56.

Fothergill, D.M., D.W. Grieve, and S.T. Pheasant. 1991. Human strength capabilities during one-handed maximum voluntary exertions in the fore and aft plane. *Ergonomics* 34 (5): 563–73.

Fowler, N.K., and A.C. Nicol. 1999. A force transducer to measure individual finger loads during activities of daily living. *J. Biomech.* 32 (7): 721–25.

Fujie, H., G.A. Livesay, M. Fujita, and S.L. Woo. 1996. Forces and moments in six-DOF at the human knee joint: Mathematical description for control. *J. Biomech.* 29 (12): 1577–85.

Fujikawa, T., T. Oshima, M. Kumamoto, and N. Yokoi. 1997. Functional coordination control of pairs of antagonistic muscles. *Trans. Jpn. Soc. Mech. Eng.* 63 (607): 769–76.

Gielen, C.C.A.M., and G.J. van Ingen Schenau. 1992. The constrained control of force and position in multi-link manipulators. *IEEE Trans. Syst. Man Cybernet.* 22:1214–19.

Gielen, C.C.A.M., G.J. van Ingen Schenau, T. Tax, and M. Theeuwen. 1990. The activation of mono- and bi-articular muscles in multi-joint movements. In *Multiple muscle systems,* ed. J.M. Winters and S.L.Y. Woo, 302–11. New York: Springer-Verlag.

Gottlieb, G.L., Q. Song, D.A. Hong, G.L. Almeida, and D. Corcos. 1996. Coordinating movement at two joints: A principle of linear covariance. *J. Neurophysiol.* 75 (4): 1760–64.

Grieve, D.W., and S.T. Pheasant. 1981. Naturally preferred directions for the exertion of maximal manual forces. *Ergonomics* 24 (9): 685–93.

Gruben, K.G., and C. López-Ortiz. 2000. Characteristics of the force applied to a pedal during human pushing efforts: Emergent linearity. *J. Motor Behav.* 32 (2): 151–62.

Herzog, W. 1998. Torque: Misuse of a misused term. *J. Manipulative Physiol. Ther.* 21 (1): 57–59.

Hof, A.L. 2001. The force resulting from the action of mono- and biarticular muscles in a limb. *J. Biomech.* 34 (8): 1085–9.

Hogan, N. 1985. The mechanics of multi-joint posture and movement control. *Biol. Cybern.* 52:315–31.

Hogan, N., E. Bizzi, F.A. Mussa-Ivaldi, and T. Flash. 1987. Controlling multi-joint motor behavior. *Exerc. Sport Sci. Rev.* 15:153–90.

Hugh-Jones, P. 1947. The effect of limb position in seated subjects on their ability to utilize the maximum contractile force of the limb muscles. *J. Physiol. (Lond.)* 105:332–44.

Hutchins, E.L., R.V. Gonzalez, and R.E. Barr. 1993. Comparison of experimental and analytical torque-angle relationship of the human elbow joint complex. *Biomed. Sci. Instrum.* 29:17–24.

Jacobs, R., and G.J. van Ingen Schenau. 1992. Control of an external force in leg extensions in humans. *J. Physiol.* 457:611–26.

Jørgenson, K., and S. Bankov. 1971. Maximum strength of elbow flexors and extensors with pronated and supinated forearm. In *Biomechanics II,* ed. J. Vredenbregt and J. Wartenweiler, 174–80. Basel: Karger.

Kaneko, Y., and F. Sato. 1999. Comparison of transformation of joint torque to ground reaction force under different running velocities. In *International Society of Biomechanics, XVIIth congress, book of abstracts,* August 8–13, 1999, ed. W. Herzog and A. Jinha, 505. Calgary, AB: University of Calgary.

King, D.L., and V.M. Zatsiorsky. 1997. Extracting gravity line displacement from stabilographic recordings. *Gait Posture* 6 (1): 27–38.

Kroemer, K.H. 1999. Assessment of human muscle strength for engineering purposes: A review of the basics. *Ergonomics* 42 (1): 74–93,

Lacquaniti, F., and C. Maioli. 1994. Independent control of limb position and contact forces in cat posture. *J. Neurophysiol.* 72 (4): 1476–95.

Lan, N., P.E. Crago, and H.J. Chizeck. 1991. Control of end-point forces of a multijoint limb by functional neuromuscular stimulation. *IEEE Trans. Biomed. Eng.* 38 (10): 953–65.

Latash, M. 2000. There is no motor redundancy in human movements. There is motor abundance. *Motor Control* 4 (3): 259–61.

Lazarus, J.A., and J.M. Haynes. 1997. Isometric pinch force control and learning in older adults. *Exp. Aging Res.* 23 (2): 179–99.

Lee, W.A., C.F. Michaels, and Y.C. Pai. 1990. The organization of torque and EMG activity during bilateral handle pulls by standing humans. *Exp. Brain Res.* 82 (2): 304–14.

Lee, W.A., and J.L. Patton. 1997. Learning of coordination during multijoint pulls: Differences and similarities with a simple model. *Biol. Cybern.* 77:197–206.

Li, Z.M., V.M. Zatsiorsky, and M.L. Latash. 2000. Contribution of the extrinsic and intrinsic hand muscles to the moments in finger joints. *Clin. Biomech.* 15 (3): 203–11.

Lindahl, O., A. Movin, and I. Ringqvist. 1969. Knee extension: Measurement of the isometric force in different positions of the knee joint. *Acta Orthop. Scand.* 40:79–85.

Looze, M.P. de, K. van Greuningen, J. Rebel, I. Kingma, and P.P. Kuijer. 2000. Force direction and physical load in dynamic pushing and pulling. *Ergonomics* 43 (3): 377–90.

Mussa-Ivaldi, F.A. 1986. Compliance. In: Morasso, P., and V. Tagliasco (eds), *Human movement understanding,* 159-212. Amsterdam: North-Holland.

Nicholas, S.C., D.D. Doxey-Gasway, and W.H. Paloski. 1998. A link-segment model of upright human posture for analysis of head–trunk coordination. *J. Vestib. Res.* 8 (3): 187–200.

Paul, J.P. 1965. Bioengineering studies of the forces transmitted by joints. In *Biomechanics and related engineering topics,* ed. R.M. Kenedi, 369–80. Oxford: Pergamon Press.

Pellegrini, J.J., and M. Flanders. 1996. Force path curvature and conserved features of muscle activation. *Exp. Brain Res.* 110 (1): 80–90.

Pheasant, S.T., and D.W. Grieve. 1981. The principal features of maximal exertion in the sagittal plane. *Ergonomics* 24 (5): 327–38.

Prilutsky, B.I., and R.J. Gregor. 1997. Strategy of coordination of two- and one-joint leg muscles in controlling an external force. *Motor Control* 1 (1): 92–116.

Ramos, C.F., and L.W. Stark. 1990. Postural maintenance during movement: Simulations of a two joint model. *Biol. Cybern.* 63:363–75.

Ridderikhoff, A., J.H. Batelaan, and M.F. Bobbert. 1999. Jumping for distance: Control of the external force in squat jumps. *Med. Sci. Sports Exerc.* 31 (8): 1196–1204.

Runge, C.F., C.L. Shupert, F.B. Horak, and F.E. Zajac. 1999. Ankle and hip postural strategies defined by joint torques. *Gait Posture* 10 (2): 161–70.

Sanderson, D.J., and P.R. Cavanagh. 1990. Use of augmented feedback for the modification of the pedaling mechanics of cyclists [comments]. *Can. J. Sport Sci.* 15 (1): 38–42.

Scholz, J.P., and G. Schoner. 1999. The uncontrolled manifold concept: Identifying control variables for a functional task. *Exp. Brain Res.* 126 (3): 289–306.

Siegler, S., G.D. Moskowitz, and W. Freedman. 1984. Passive and active components of the internal moment developed about the ankle joint during human ambulation. *J. Biomech.* 17:647–52.

Uchiyama, T., T. Bessho, and K. Akazawa. 1998. Static torque-angle relation of human elbow joint estimated with artificial neural network technique. *J. Biomech.* 31 (6): 545–54.

Valero-Cuevas, F.J. 1997. Muscle coordination of the human index finger. Doctoral diss., Stanford University.

Valero-Cuevas, F.J. 2000. Predictive modulation of muscle coordination pattern magnitude scales fingertip force magnitude over the voluntary range. *J. Neurophys.* 83: 1469-79.

Valero-Cuevas, F.J., F.E. Zajac, and C.G. Burgar. 1998. Large index-fingertip forces are produced by subject-independent patterns of muscle excitation. *J. Biomech.* 31 (8): 693–703.

van Ingen Schenau, G.J., P.J.M. Boots, G. de Groot, R.J. Snackers, and W.W.L.M. van Woensel. 1992. The constrained control of force and position in multi-joint movements. *Neuroscience* 46 (1): 197–207.

Wang, X., J.P. Verriest, B. Lebreton-Gadegbeku, Y. Tessier, and J. Trasbot. 2000. Experimental investigation and biomechanical analysis of lower limb movements for clutch pedal operation. *Ergonomics* 43 (9): 1405–29.

Wells, R., and N. Evans. 1987. Functions and recruitment patterns of one- and two-joint muscles under isometric and walking conditions. *Hum. Mov. Sci.* 6:349–72.

Woude, L.H.V. van der, H.E.J. Veeger, and A.J. Dallmeijer. 2000. Manual wheelchair propulsion. In *Biomechanics in sport,* ed. V.M. Zatsiorsky, 609–36. Oxford, UK: Blackwell Science.

Yamashita, N. 1975. The mechanism of generation and transmission of forces in leg extension. *J. Hum. Ergol. (Tokyo)* 4 (1): 43–52.

Zajac, F.E. 1993. Muscle coordination of movement: A perspective. *J. Biomech.* 26:109–24.

Zajac, F.E., and M.E. Gordon. 1989. Determining muscle force and action in multiarticular movement. In Pandolf, K. (ed), *Exercise and sports science reviews,* 187-230. Baltimore: Williams and Wilkins.

Zatsiorsky, V.M., and D.L. King. 1998. An algorithm for determining gravity line location from posturographic recordings. *J. Biomech.* 31 (2): 161–64.

Zatsiorsky, V.M., and M.L. Latash. 1993. What is a joint torque for joints spanned by multiarticular muscles? *J. Appl. Biomech.* 9:333–36.

Zuylen, E.J. van, C.C.A.M. Gielen, and J.J. Denier van der Gon. 1988. Coordination and inhomogeneous activation of human arm muscles during isometric torques. *J. Neurophysiol.* 60:1523–48.

3

CHAPTER

STATICS OF MULTILINK CHAINS: STABILITY OF EQUILIBRIUM

For every complex problem, there is a solution that is simple, neat, and wrong.

Henry Louis Mencken (1880–1956)

A kinematic chain is said to be in equilibrium when all the links of the chain are in equilibrium. If a mechanical system returns to equilibrium after being subjected to small disturbances, it is said to be in *stable equilibrium;* if not, it is in *unstable equilibrium* (both terms can be found in the glossary under *equilibrium position*). The equilibrium is *neutral* if the displacement does not cause forces acting either away or toward the original position. Examples of stable and unstable equilibrium are presented in figure 3.1.

The conditions for stable equilibrium can be expressed in multiple ways. The equilibrium of a chain or link is stable when, upon a small displacement away from the equilibrium position, (1) the forces and couples acting on the chain or link tend to decrease the deviation in position, (2) the forces and couples acting on the chain or link do negative work (the point of force application moves against the force direction), and (3) the potential energy of the chain or link increases.

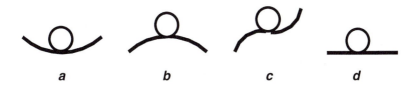

a   *b*   *c*   *d*

Figure 3.1   (*a*) Stable, (*b* and *c*) unstable, and (*d*) neutral equilibrium of a ball on a surface.

In an equilibrium position, the potential energy of the system, Q, is stationary; that is, $dQ/dx = 0$, where x is any coordinate that defines the position of the system. If the potential energy of the system is locally minimal ($Q = $ min), the equilibrium is stable. Otherwise, the equilibrium is not stable. The equilibrium is neutral when Q is constant.

• • • MECHANICS REFRESHER • • •

Statics of Deformable Bodies

Elasticity is the ability of a body or material to resist deformation and to restore its original shape and size after being deformed. Examples of elastic bodies include springs and rubber bands.

Hooke's law states that there is a linear relationship between the applied force and the amount of deformation. The law is valid only for a specific range of deformation. *Stiffness* is defined as the amount of force necessary to extend the body one unit of length; its SI unit is N/m. *Compliance* is defined as the amount of elongation per unit of force, m/N. Compliance is thus the inverse of stiffness.

Stress is the amount of force per unit of area, N/m². *Strain* is relative elongation, $\Delta l/l$, where l is the initial length of the object and Δl is its elongation. Strain is a dimensionless quantity.

Young's modulus is the stress required to elongate the body to double its original length. For the majority of materials, Young's modulus is just an abstract quantity; most bodies break before their length can be doubled.

If the restoring forces and couples acting on a kinematic chain are external to the chain, for instance, when the chain returns to the equilibrium position due to gravity, the equilibrium is *naturally stable* or *unintended*. For instance, when the arms hang straight down at the sides of the body, they are in a naturally stable equilibrium position: no muscular force or joint torque is required to maintain the equilibrium. If equilibrium is maintained because of restoring joint torques, the equilibrium is *naturally unstable* or *intended*. An upright posture is an example of naturally unstable equilibrium, while the so-called basket-hang position (suspended from a horizontal bar with the legs extended) is a naturally stable equilibrium position. In postures that are naturally unstable, the human body or its parts usually do not stay completely motionless. They oscillate around an equilibrium (reference) position. A small oscillation of body parts around an equilibrium position is called *tremor.*

■ ■ ■ **FROM THE LITERATURE** ■ ■ ■

Instant Equilibrium Point in Quiet Standing

Sources: Zatsiorsky, V.M., and M. Duarte. 1999. Instant equilibrium point and its migration in standing tasks: Rambling and trembling components of the stabilogram. *Motor Control* 3 (1): 28–38.

Zatsiorsky, V.M., and D.L. King. 1998. An algorithm for determining gravity line location from posturographic recordings. *J. Biomech.* 31 (2): 161–64.

During quiet standing, the human body does not stay motionless. It sways. The sway occurs about an equilibrium reference. When the body deviates from the reference position, a restoring force is acting. The restoring force is manifested in the horizontal component of the ground reaction force. The horizontal component of the GRF, $F_{hor,}$ equals the body mass m times the acceleration of the body's CoM in the horizontal direction, $F_{hor} = ma_{hor}$ (see equation 1.56a). When restoring forces are zero, the acceleration a_{hor} is also zero, and the body, albeit moving, is instantly in equilibrium.

Due to the body sway, the CoP of the vertical forces exerted on the support surface migrates. The position of the CoP at an instant when $F_{hor} = 0$ is called an instant equilibrium point (IEP). At this instant, the CoP momentarily coincides with the equilibrium reference (and therefore the restoring force is not generated). According to experimental observations, the IEP does not stay in the same place; rather, it displaces (see the black squares in figure 3.2). Hence, the reference position at which equilibrium is instantly maintained migrates.

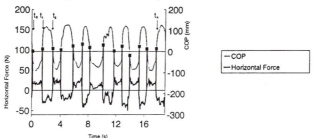

Figure 3.2 Determining instant equilibrium points. The CoP and the magnitude of the horizontal force, F_{hor}, are registered. The instances when $F_{hor} = 0$ are determined, and the CoP location at these instances (= IEP, black squares) is established.

Reprinted from *Journal of Biomechanics*, 31, V.M. Zatsiorsky and D.L. King, An algorithm for determining gravity line location from posturographic recordings, 161–164, 1998, with permission from Elsevier Science.

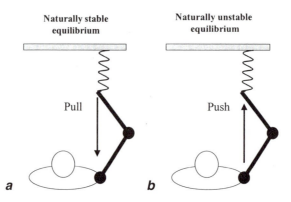

Figure 3.3 A subject is (*a*) pulling and (*b*) pushing against a spring. Pulling is a naturally stable movement, and pushing is not.

In human movements, naturally stable equilibrium is observed predominantly during pulling efforts, and naturally unstable equilibrium is observed predominantly during pushing efforts. Consider a performer who is pulling or pushing a handle that is not constrained to move along the line of force action but can deviate from this line (figure 3.3). If the handle is displaced perpendicularly to the line of force action, the end-effector force tends to return the handle to its equilibrium position during the pulling effort and tends to move the handle away from equilibrium during the pushing movement. Hence, when pushing, the performer should stabilize the handle with additional muscular effort. The stabilization is achieved mainly by *agonist-antagonist* muscle co-contraction.

■ ■ ■ *From the Literature* ■ ■ ■

Arm Stabilization During Pushing Efforts

Sources: Bober, T., S. Kornecki, R.P. Lehr, Jr., and J. Zawadski. 1982. Biomechanical analysis of human arm stabilization during force production. *J. Biomech.* 15 (11): 825–30.

Kornecki, S., W. Kulig, and J. Zawadzki. 1985. Biomechanical implications of the artificial and natural strengthening of wrist joint structures. In *Biomechanics IX-A,* ed. D. Winter, R.W. Norman, R.P. Wells, K.C. Hayes, and A.E. Patla, 118–22. Champaign, IL: Human Kinetics.

The subjects pushed maximally a handle that was either fixed or free to rotate with respect to one or two axes (figure 3.4). In some experiments, the wrist joint was fixed with a plastic cast.

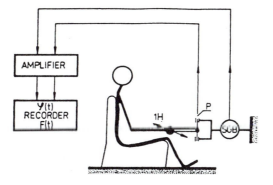

Figure 3.4 A schematic of the experimental device. P = potentiometer used for measuring angular displacement; SGB = strain gauge bridge used to measure the pushing force; 1H = one degree of freedom in the horizontal plane.

Reprinted, by permission, from S. Kornecki, W. Kulig, and J. Zawadzki, 1985, Biomechanical implications of the artificial and natural strengthening of wrist and joint structures. In *Biomechanics IX-A*, edited by D. Winter et al. (Champaign, IL: Human Kinetics), 119.

When the handle was not fixed, the peak force was on average equal to 76% of the force exerted against the stationary handle. Hence, arm stabilization required a significant percentage of the possible maximal force ($\Delta F = 24\%$). The percentage did not change after special training and did not depend on athletic mastership (athletes, including world champions in various sports, were compared with students who did not practice sports). The force loss was a price that all the subjects paid to stabilize the arm and the handle. Mechanical stabilization of the wrist joint increased the force production. When the wrist joint was fixed with a plastic cast, the force exerted on the stationary handle increased 1.9%, and the force applied to the movable handle increased 16.4%.

To analyze the equilibrium of kinematic chains that are naturally unstable, the notions of joint stiffness and *joint compliance* are commonly used. For the joints with one degree of freedom (DoF) and a zero resting torque, joint stiffness k is defined as angular stiffness:

$$k = \frac{T}{\alpha - \alpha_0} \tag{3.1}$$

where T is torque around the joint, α is joint angle, and α_0 is the resting angle of the joint. For equation 3.1 to be valid, the angular deflection, $\Delta\alpha = \alpha - \alpha_0$, should be small. If the resting torque is not equal to zero and the displacement is infinitesimal, the equation is

$$k = \frac{dT}{d\alpha} \qquad (3.2)$$

Joint stiffness equals the amount of torque increment per unit of joint angular deflection. The units of coefficient k are (N · m)/rad or, because radians are dimensionless, simply N · m. Joint compliance is the inverse of joint stiffness. Describing the mechanical behavior of a joint by equation 3.1 or 3.2 is analogous to replacing all the muscles and passive tissues at the joint, the reflexes, and the higher-order neural control by a simple spring. Whether such a simplification is reasonable or not depends on the circumstances.

■ ■ ■ *FROM THE LITERATURE* ■ ■ ■

Ankle Joint Stiffness Required for Maintaining Upright Posture

Source: Bergmark, A. 1989. Stability of the lumbar spine: A study in mechanical engineering. *Acta Orthop. Scand.* 60 (Suppl. 230): 1–230.

The body in upright posture was modeled as an inverted pendulum with the axis of rotation at the ankle joints (figure 3.5). The body deviates from the vertical in anteroposterior direction.
    The following assumptions were made:

1. The body above the feet is a single, rigid body.
2. The resistance at each ankle joint is provided by the torque (T) that is proportional to the joint angular displacement (α) and acts without a time delay.
3. The body sway is small.
4. The angular acceleration of the body is small and can be neglected. Hence, the inertial effects were ignored, and the problem was reduced to a problem of statics.

Under these assumptions, the model can be described by the equation

$$2T = k\alpha = Wh \sin\alpha \qquad (3.3)$$

where W is the body weight above the feet and h is the height of the center of gravity above the ankle joints. Because the deviation is small, $\sin\alpha$ is approximately equal to α. The critical joint stiffness k_{crit} that is sufficient

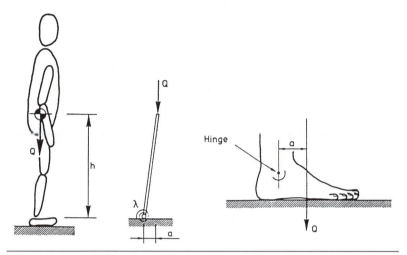

Figure 3.5 A simple pendulum model of upright standing. **Q** is gravity force, *a* is the distance from the hinge to gravity line, *h* is vertical height to the center of gravity of the body, and λ is the angular stiffness of the ankle joint.

Reprinted, by permission, from A. Bergmark, 1989, "Stability of the lumbar spine. A study in mechanical engineering," *Acta Orthopaedica Scandinavica* 60 (Suppl. 230): 1–230.

to maintain the posture, that is, to generate the ankle joint torque adequate to return the body to the equilibrium position, is

$$k_{crit} = \frac{W}{2} h \qquad (3.4)$$

Equation 3.4 allows estimating the critical value of the stiffness at the ankle joints, that is, the stiffness sufficient to maintain the equilibrium. For instance, if $h = 0.9$ m and the body weight (minus the weight of the feet) is 600 N, the critical value of the ankle joint stiffness is $k_{crit} = 0.5 \times 600 \times 0.9 = 270$ (N · m)/rad. To maintain stability, ankle joint stiffness should be equal to or larger than k_{crit}. According to previously published data, the gravity line passes 63 mm anterior to the lateral malleolus on average. These results enable computation of the ankle joint torque, $T_a = 0.5 \times 600$ N $\times 0.063$ m $= 18.9$ N · m. The angle of body inclination is $a/\sin\alpha \approx 0.063/0.9 = 0.07$ rad. Hence, the ankle joint stiffness necessary to maintain the posture is $k_{crit} = 18.9$ N · m/0.07 rad $= 270$ N · m.

This chapter deals with equilibrium of naturally unstable kinematic chains. In these chains, restoring joint torques is required to retain the equilibrium. If a chain or its link deflects from an equilibrium position, joint torque or torques that act opposite to the deflection should be generated to restore the equilibrium. Section **3.1** discusses the difference between the stiffness of passive and active objects. Apparent endpoint stiffness is addressed in section **3.2**. In section **3.3**, apparent joint stiffness is described. Section **3.4** deals with the transformation analysis of the apparent endpoint and joint stiffness. The Jacobian matrices of the kinematic chains are used for this purpose. This section also considers the main factors that influence apparent endpoint stiffness. They are (1) the chain Jacobian matrix (i.e., the chain configuration), (2) apparent joint stiffness, and (3) external force. Experimental data on the apparent stiffness of the human arm during posture maintenance are briefly discussed. Section **3.5** examines the mechanical and physiological factors that complicate the measurement of joint stiffness. Section **3.6** concentrates on the problems associated with the measurement and interpretation of the apparent endpoint and joint stiffness.

3.1 STIFFNESS OF PASSIVE AND ACTIVE OBJECTS

Representative papers: Mussa-Ivaldi (1986); Latash and Zatsiorsky (1993)

The musculoskeletal system contains *passive* and *active elements*. Passive bodies deform under the influence of an external force, resist the deformation, and can store elastic energy. If passive elements are elastic, they restore their original shape when all forces are removed. The passive elements of the musculoskeletal system include tendons, ligaments, fascia, cartilage, bones, skin, and relaxed (not activated) muscles. The word "passive" when applied to biological objects should not be taken literally. In living tissue, the response always includes biological reactions. For instance, when a passive object is overstressed, it may become inflamed. Still, from a purely mechanical standpoint, these objects can be considered passive. The passive components deform only when an external force is applied. In the absence of external forces, these bodies maintain constant shape. There is a one-to-one relationship between exerted force and deformation of a given passive object. Hence, to uniquely determine the mechanical state of a passive object (e.g., a tendon), either the deformation or the force should be specified. The notion of the stiffness was introduced in classical mechanics solely for passive bodies.

The notion of stiffness can be applied to the study of passive biomechanical objects without conceptual difficulties. For instance, when the muscles serving a joint are completely relaxed, the passive structures at the joint provide resistance to the joint movement. In the clinical literature, such a resistance is often called the *joint stiffness* (the term "passive joint stiffness" would be more appropriate in this case). To determine the passive joint stiffness, an external torque is applied at the joint and the angular deflection is measured (alternatively, a given joint deflection may be induced and the resistive torque measured). Passive stiffness of joints is higher in older populations (old-age stiffness). It is also higher in some patients with rheumatic disease (rheumatic stiffness). The joints are usually stiffer in the morning (morning stiffness). Some people with rheumatoid arthritis wake up in the morning with their joints "locked." To restore joint mobility, they have to perform several movements. This passive joint resistance may be due to many factors, such as friction, adhesion of structures that are normally free to move with respect to each other, fluid viscosity (how easily the fluid moves around in the tissues), and elasticity. Passive joint stiffness may increase after surgery due to scar-tissue formation, inflammation, and adhesions. We do not analyze passive joint stiffness in detail in this book.

The active elements of the musculoskeletal system (active muscles and joints) can change their length (or angle) without external forces. Therefore, a one-to-one relationship between the joint torque and joint angle does not exist. Knowledge of the joint angle does not reveal much about the torque at the joint, and knowledge of the torque does not provide information about the joint angle. To specify the mechanical state of a muscle (or joint) in a static condition, both the force (torque) and length (joint angle) must be known. Also, joints are not bodies; rather, they are connections between adjacent segments of the human body that are under complex neural control. Hence, the situation is rather different from that assumed in classical mechanics when stiffness is used to study passive deformable bodies.

Because torque (force) and angle (length) in active objects can be changed independently, the force–length relationship of the active objects is not unique. When the level of activation varies, the entire force–length relationship changes. The measurements make sense only if the level of activation and its time course is specified, for instance, under conditions of maximal voluntary contraction. However, even in this case, at least two options exist. In the first approach, the derivative of the joint torque–angle curve, $dT/d\alpha$, or the force–length curve, dF/dx, is measured. In the second, the reaction to a deflection from the equilibrium position is registered (*incremental stiffness*).

When subjects produce maximal voluntary contractions at various joint angles, the force values are different. The corresponding curves are called *joint strength curves*. The joint strength curve results from the interaction between two factors: changes in the force generated by individual muscles and changes in the moment arms of these muscles (figure 3.6).

For a hypothetical hinge joint served by only one muscle, the equation is $T = F(\alpha)r(\alpha)$, where T is joint torque (the moment of muscle force), $F(\alpha)$ is the force produced by the muscle at joint angle α, and $r(\alpha)$ is the muscle moment

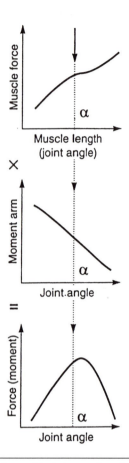

Figure 3.6   Joint torque registered in joint angle α is the product of muscle force and muscle moment arm in this joint configuration.

Reprinted, by permission, from V.M. Zatsiorsky, 1995, *Science and Practice of Strength Training* (Champaign, IL: Human Kinetics), 54.

arm at this angle. A derivative of the joint strength curve with respect to the joint angle can easily be computed:

$$\frac{dT}{d\alpha} = \frac{dF(\alpha)}{d\alpha}r(\alpha) + \frac{dr(\alpha)}{d\alpha}F(\alpha) \qquad (3.5)$$

The derivative has two terms. The first term includes the derivative of the force–length curve of the muscle (the change in the muscle force due to the infinitesimal joint displacement), and the second contains the derivative of moment arm with respect to the joint angle. Note that the experiment employed to compute the expression $dT/d\alpha$ in equation 3.5 differs from the experiment employed previously for determining joint stiffness, equation 3.2. Although the derivatives $dT/d\alpha$ in both equations appear identical, they were obtained using different operations. One cannot expect that the two experimental approaches—recording a torque-angle curve and measuring the incremental stiffness—would lead to the same results. In fact, they do not. Muscles produce larger forces when extended than can be expected from their force–length curves (this issue will be elaborated in the next companion volume of this book). To avoid confusion, it is recommended that the derivative $dT/d\alpha$ in equation 3.5 not be called "joint stiffness." We reserve the term "joint stiffness" exclusively for incremental stiffness, the stiffness obtained from equations 3.1 and 3.2, as a reaction to a deflection from the equilibrium position.

In biomechanics, the concept of joint stiffness is borrowed from robotics. When an external force is applied to a robot arm, the arm deflects. The deflection comes from two sources: the deformation of the links and the joint displacement. In robots, each joint torque actuator is controlled by a controller. If there is a discrepancy between the desired joint configuration and the actual joint position, the joint controller generates a restoring torque. The restoring torque is proportional to the difference between the reference position and the actual position. Thus, robots really behave as if they have springs at their joints. However, even in robots, the "pure" stiffness analysis describes only the steady-state response (if deflection velocity is neglected). It does not describe the transient response, that is, the robot behavior during the transition from one state or position to another.

Unlike robots, humans can react in different ways to an external load applied to a biokinematic chain such as the arm. The reaction depends on the motor task, for example, on the instruction given to the subjects (e.g., "resist" or "do not resist" the perturbation) and the subject's will to follow the instruction. In humans, the reaction to limb deflection and thus the application of the concept of joint stiffness are much more complex than in robots or passive objects. When a joint angle or a kinematic chain is perturbed, the resistive force depends on many factors, such as the background torque, the amplitude and speed of the stretch, the time elapsed after the stretch, the co-contraction of antagonistic muscles, and so on.

If not all of the details of the task and measurement procedure are specified, the expression "joint stiffness" is imprecise. Measured stiffness values at the same joint can be very different. In active objects, the stiffness (i.e., the resistive force per unit of deflection) is always motor-task-specific and time-dependent. It is unfortunate that the same word "stiffness" is used in the literature to designate a quality of both passive objects, such as tendons and ligaments, and active objects, such as muscles and joints. Despite this ambiguity, we continue the practice simply because a better term has not been suggested. To differentiate between the stiffness-like parameters of active objects and the authentic stiffness of passive objects, we use the term *apparent stiffness*. The apparent stiffness of active objects only seems like the stiffness of passive objects; it has different—and much more complex—mechanisms.

Measurement of the apparent joint stiffness, if performed properly, is potentially valuable. It enriches our understanding of human movement and allows us to estimate the amount of elastic potential energy that can be stored in the muscles and connective tissues during the eccentric phases of movement.

3.2 APPARENT ENDPOINT STIFFNESS

Representative papers: Hogan (1985b); Mussa-Ivaldi, Hogan, and Bizzi (1985); Lacquaniti, Carrozzo, and Borghese (1993)

In this section we analyze open kinematic chains with hinge joints. The joints are considered to be ideal, that is, without friction and backlash. All of the links are rigid, and their mass is neglected. An external force is applied to the endpoint of the chain. The external force causes a deflection of the endpoint and displacements at the joints. We consider only small displacements. Because the deflections are small, the force–displacement relationships are approximately linear. Hence, equations 3.1 and 3.2 can be applied. We limit the discussion to a two-link planar chain.

3.2.1 Matrices of Endpoint Stiffness

In multilink chains, the direction of the endpoint deflection is generally not coincident with that of the line of force, even when the external couple is zero. More often than not, the force direction differs from the direction of deflection, as shown in figure 3.7. Therefore, the *endpoint stiffness* is not a scalar quantity; the force–displacement relationship cannot be reduced to a simple ratio as it was for unidimensional objects discussed at the beginning of this chapter.

If a restoring force at the endpoint $\mathbf{F} = (F_x, F_y)^T$ is a differentiable function of the endpoint position, the relationship between the force differential increment $d\mathbf{F}$ and the linear differential displacement $d\mathbf{P} = (dX, dY)^T$ can be expressed as

$$dF_X = -\left(\frac{\partial F_X}{\partial X} dX + \frac{\partial F_X}{\partial Y} dY\right) = -\left(S_{XX} dX + S_{XY} dY\right)$$

$$dF_Y = -\left(\frac{\partial F_Y}{\partial X} dX + \frac{\partial F_Y}{\partial Y} dY\right) = -\left(S_{YX} dX + S_{YY} dY\right)$$

$$(3.6)$$

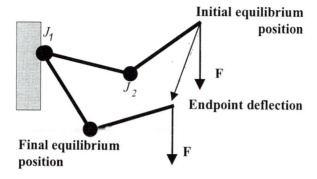

Figure 3.7 A vertical force **F** is applied to the endpoint of a two-link chain connected to a wall by a hinge joint J_1. The direction of the endpoint deflection is different from the force direction. The angular changes (deflections) occur in both joints, J_1 and J_2.

This relationship can be described by the endpoint *stiffness matrix* [S]

$$\begin{bmatrix} dF_X \\ dF_Y \end{bmatrix} = -\begin{bmatrix} \dfrac{\partial F_X}{\partial X} & \dfrac{\partial F_X}{\partial Y} \\ \dfrac{\partial F_Y}{\partial X} & \dfrac{\partial F_Y}{\partial Y} \end{bmatrix} \begin{bmatrix} dX \\ dY \end{bmatrix} = -\begin{bmatrix} S_{XX} & S_{XY} \\ S_{YX} & S_{YY} \end{bmatrix} \begin{bmatrix} dX \\ dY \end{bmatrix} \tag{3.7}$$

or

$$d\mathbf{F} = -[S]d\mathbf{P} \tag{3.8}$$

The minus sign in the right sides of equations 3.6 through 3.8 indicates that the restoring force is opposite to the displacement. Equation 3.8 is exact only for infinitesimal displacements but is approximately valid for small deflections from equilibrium. For instance, for a small displacement in the horizontal direction, the restoring force is

$$\begin{aligned} F_X &= -S_{XX}dX \\ F_Y &= -S_{YX}dX \end{aligned} \tag{3.9}$$

If the chain does not sustain any external load before the displacement ($\mathbf{F}_0 = 0$), the left side of equation 3.8 can be written as $\mathbf{F} = (F_X, F_Y)^T$. This convention is typical in the literature. We use both the designations **F** and *d***F**. The reader is advised to compare the matrix [S] with the Jacobian (equation 3.10 on p. 154 in *Kinematics of Human Motion*). Both matrices are similar; they contain partial derivatives that relate small incremental changes in one vector ($d\mathbf{P}$ or $d\boldsymbol{\alpha}$) with the small variations in another vector ($d\mathbf{F}$ or $d\mathbf{P}$).

The inverse of the stiffness matrix $[S]$ is called the *compliance matrix, $[C]$*. The infinitesimal deflection at the endpoint $d\mathbf{P}$ is related to the endpoint force $\mathbf{F}$ by the matrix $[C]$:

$$d\mathbf{P} = [C]\mathbf{F} = [S]^{-1}\mathbf{F} \tag{3.10}$$

Note that $\mathbf{F}$ in this equation is not a restoring force (the force opposite to the direction of the deflection) but the force that is exerted by an external agent in the direction of the displacement. Because of that, the minus sign is omitted in the equation. To compute the compliance matrix $[C]$ from a stiffness matrix $[S]$, the stiffness matrix must be invertible. Because $[S]$ is a square matrix, this requirement means that the determinant of $[S]$ must not be equal to zero. This requirement is not satisfied when the chain is in a singular position, for instance, when the arm is completely extended. This issue is elucidated later in the text.

The stiffness matrix $[S]$ is not diagonal; it contains off-diagonal terms. This means that a restoring force in the X direction depends on the displacement in both the X and Y directions. For the chain to be maintained in a stable equilibrium by restoring forces, the matrix $[S]$ must be a symmetric matrix, $S_{XY} = S_{YX}$. In this case, the deflected endpoint returns back to its initial position along the route of deflection. Otherwise, the endpoint is involved in a circular motion. The requirement for symmetry of $[S]$ can also be formulated in terms of vector fields.

• • • VECTOR ALGEBRA REFRESHER • • •

Vector and Scalar Fields

A *vector field* is a function that assigns a vector to each point in some region of space. A function that assigns a number to each point in a region is called a *scalar field*.

Consider a function $z = f(x, y)$ that has continuous partial derivatives. The largest directional derivative of f at point (a, b) is called the *gradient* of f at this point and is denoted ∇_f. The symbol ∇ is the Greek letter delta printed upside down; it is called "del" or "nabla." The directional derivative at (a, b) has its largest value in the direction of ∇_f.

Let $\int_C F \cdot V ds$ be the integral of the tangential component V of the vector field F along a curve C (the *line integral*). A vector field F is called the *conservative field* if the line integrals depend only on the endpoints of the curve C and not on the particular path between them. For instance, the field of gravity is conservative: the work of gravitational force depends only on the initial and final positions of the body and not on the route traveled. Any gradient field is a conservative field.

The *curl* of a vector field F, usually denoted as *curl F*, is a measure of circulation (the tendency to rotate) at a given point. *Curl F* signifies the magnitude and direction of maximum circulation at a point. The curl of a vector field in a plane is characterized by the quantity $\dfrac{\partial F_y}{\partial x} - \dfrac{\partial F_x}{\partial y}$. The curl of a conservative field F is zero.

3.2.2 Vector Field Representation

The magnitude and direction of the restoring force acting at the endpoint depend on the magnitude and direction of the deflection and are different at various endpoint locations. The restoring forces at the endpoint can be represented by a vector field. The field is called the *postural force field*. A necessary and sufficient condition for a postural force field to be conservative is

$$curl\ F = \frac{\partial F_Y}{\partial X} - \frac{\partial F_X}{\partial Y} = S_{YX} - S_{XY} = 0 \tag{3.11}$$

Hence, for the curl to be zero, the matrix $[S]$ must be a *symmetric matrix*, $S_{XY} = S_{YX}$. In experimental research, the recorded matrix $[S]$ is usually separated into a *symmetric component* $[S_s]$ and an *antisymmetric component* $[S_a]$; $[S] = [S_s] + [S_a]$.

$$[S_s] = \begin{bmatrix} S_{XX} & \dfrac{S_{XY} + S_{YX}}{2} \\ \dfrac{S_{YX} + S_{XY}}{2} & S_{YY} \end{bmatrix}$$

$$[S_a] = \begin{bmatrix} 0 & \dfrac{S_{XY} - S_{YX}}{2} \\ \dfrac{S_{YX} - S_{XY}}{2} & 0 \end{bmatrix} \tag{3.12}$$

The decomposition allows the separate study of the *conservative* and *rotational (curl) components* of postural force fields. The restoring force that is due to the antisymmetric component is at right angles to the displacement; its magnitude is proportional to that of the displacement. Two force fields that correspond to the symmetric and antisymmetric components of apparent endpoint stiffness are shown in figure 3.8.

It has been experimentally observed that the matrix $[S]$ of a human arm in equilibrium is nearly symmetric (Mussa-Ivaldi, Hogan, and Bizzi 1985). This finding has been confirmed by several investigations (Flash and Mussa-Ivaldi

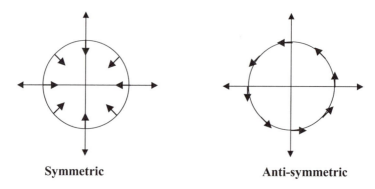

Symmetric   Anti-symmetric

Figure 3.8 Force vectors corresponding to the symmetric and antisymmetric components of apparent endpoint stiffness. The circles represent displacement of constant magnitude in all directions. The tails of the force vectors are drawn at various directions of the displacement. For symmetric components, the curl is zero. The nonzero curl associated with the antisymmetric components causes rotation. (For symmetric components, the presented example corresponds to equal eigenvalues of the stiffness matrix. This is elucidated later in the text.)

1990; Dolan et al. 1993; Shadmehr, Mussa-Ivaldi, and Bizzi 1993; McIntyre, Mussa-Ivaldi, and Bizzi 1996; Gomi and Kawato 1997). This result is not trivial.

■ ■ ■ *From the Literature* ■ ■ ■

Apparent Stiffness
While Maintaining an Arm Posture

Source: Mussa-Ivaldi, F.A., N. Hogan, and E. Bizzi. 1985. Neural, mechanical, and geometric factors subserving arm posture in humans. *J. Neurosci.* 10:2732–43.

This influential paper started an important direction of research.

In the experiments, the subjects gripped the handle of a planar two-link manipulandum (a device similar to a robot arm). The handle was placed in five different locations. At each location, the handle was displaced 5 to 8 mm in various directions 45° apart, from 0° to 325°. To prevent the early occurrence of voluntary reactions, the subjects were given several instructions: first, to concentrate on perceiving the direction of the displacement; second, to say "one, two" aloud; and third, to move rapidly in the direction opposite the imposed displacement. A holding phase was observed after a transient phase of the handle displacement and prior to the voluntary reaction. During this period, the handle was at rest; nevertheless, the subjects exerted a force on the handle. This force represented the static restoring force that was driving the hand toward the original equilibrium position. The authors especially emphasized that the measurements of the apparent endpoint stiffness were made at zero hand velocity, between the end of the hand movement and the beginning of the voluntary reaction.

The stiffness represented by the matrix $[S]$ was decomposed into symmetric and antisymmetric components, $[S] = [S_s] + [S_a]$. The results were presented in both numerical and graphical forms (figure 3.9).

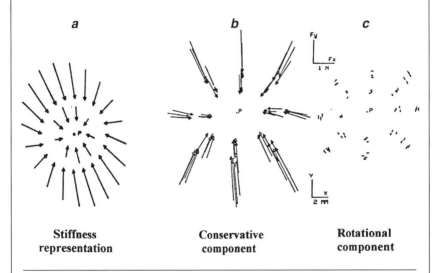

a	**b**	**c**
Stiffness representation	**Conservative component**	**Rotational component**

Figure 3.9 Endpoint stiffness of a multijoint arm. (*a*) Restoring forces (arrows) and the endpoint displacements. The distance between the equilibrium position P and the tip of the arrow represents the displacement. (*b*) Conservative (symmetric) components of the endpoint stiffness. The stiffness matrix $[S_s]$ was used to derive the force vectors represented by arrows. (*c*) Rotational (antisymmetric) components of the endpoint stiffness. They were calculated from the antisymmetric component of the stiffness matrix, $[S_a]$. The forces that were due to the conservative components of stiffness were much larger than the forces due to the rotational components. In the majority of cases, the curl was responsible for less than 10% of the restoring forces generated by the symmetric stiffness component.

Adapted, by permission, from F.A. Mussa-Ivaldi, N. Hogan, and E. Bizzi, 1985, "Neural, mechanical, and geometric factors subserving arm posture in humans," *The Journal of Neuroscience* 5 (10):2732–2743. Copyright © 1985 by the Society of Neuroscience.

3.2.3 The Stiffness Ellipse and the Principal Transformation of the Endpoint Stiffness Matrix

Representative papers: Dolan et al. (1993); Katayama and Kawato (1993)

Suppose that an endpoint of an open kinematic chain undergoes a displacement of amplitude 1 in various directions ϕ such that $0 < \phi < 2\pi$. The excursion can be described by a function:

$$dX = \cos\phi$$
$$dY = \sin\phi$$

(3.13)

which is the formula of a circle with a radius of 1. Using equation 3.8 ($\mathbf{F} = -[S]d\mathbf{P}$) and multiplying the rotating unit excursion by the symmetric stiffness matrix $[S_s]$, we obtain the output force vectors for a given displacement direction ϕ.

$$F_X = -\left(S_{XX}\cos\phi + S_{XY}\sin\phi\right)$$
$$F_Y = -\left(S_{YX}\cos\phi + S_{YY}\sin\phi\right)$$

(3.14)

When $S_{XY} = S_{YX}$ and the direction of displacement ϕ varies from 0 to 2π, equation 3.14 represents the *stiffness ellipse* (figure 3.10). The distance from a point on the

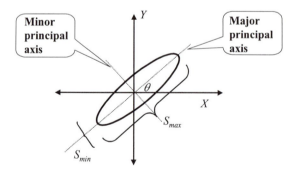

Figure 3.10 The stiffness ellipse. The perimeter of the ellipse is the locus of the force vectors for a unit displacement in various directions, $0 < \phi < 2\pi$. The perimeter can be obtained by multiplying the endpoint displacement of unit amplitude by the symmetric matrix $[S_s]$ (equation 3.14). The deflection and the force are aligned only along the major and minor axes. The angle θ between the major principal axis and the X axis defines the apparent stiffness orientation. The shape of the ellipse is given by the ratio S_{max}/S_{min}. The size of the stiffness ellipse is represented by its area, $\pi S_{max}S_{min}/4$.

stiffness ellipse to its center represents the magnitude of the restoring force generated in response to a deflection of unit magnitude from the equilibrium position.

A stiffness ellipse is characterized by three parameters: (1) the *ellipse orientation* is defined by the angle between the major axis and the *X* axis of the fixed reference system; (2) the *ellipse shape* is given by the ratio of the major axis to the minor axis (the ratio of the maximal to minimal stiffness); (3) the *ellipse size*, or magnitude, equals the ellipse area. The area is proportional to the determinant of the matrix [*S*], that is, to the expression $\det[S] = S_{XX}S_{YY} - S_{XY}S_{YX}$. (The reader who has forgotten the concept of determinants can look at the "Mathematics Refresher" in *Kinematics of Human Motion*, p. 29.)

In general, the deflection and the restoring force are not collinear except along the major and minor axes of the ellipse. The major axis is oriented along the direction of maximal stiffness, and the minor axis is along the direction of minimal stiffness. The major and minor axes are orthogonal to each other. When an endpoint force acts along a principal axis, the endpoint deflects in the same principal direction and the displacement takes an extreme value. In any other direction, the force and the deflection are not collinear and the deflection is not extreme (figure 3.11).

The stiffness in the directions of the major and minor axes of the stiffness ellipse is equal to the maximum and minimum eigenvalues of the stiffness matrix, λ_2 and λ_1, correspondingly. The larger and smaller eigenvalues equal the forces along the major and minor axes against a unit displacement in these

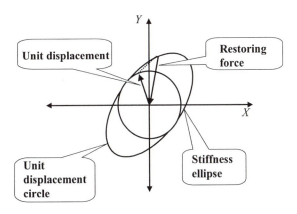

Figure 3.11 The endpoint deflection and the restoring force. The deflection and the force are collinear if the displacement takes place along a principal direction. They are also collinear if the eigenvalues equal each other, that is, if the contour is a circle rather than an ellipse. Note that the force vectors are plotted with their tips at the center to obtain the stiffness ellipse. If the disturbing force **F** were plotted instead of the restoring force –**F**, the graph would represent the relationship between the force and the resulting displacement.

directions. (Note the difference from the force ellipsoids introduced in **2.3.3**, for which the length of a principal axis i was given by $1/\sqrt{\lambda_i}$ of the matrix $[JJ^T]$. The reader is invited to explain the reason for this difference.) The direction of the axes is given by the eigenvectors corresponding to the maximum and minimum eigenvalues. For a symmetric (2×2) stiffness matrix $[S_s]$ with the elements S_{XX}, S_{YY}, and $_{XY}S = (S_{XY} + S_{YX})/2$, the characteristic equation is

$$\begin{vmatrix} S_{XX} - \lambda & S_{XY} \\ S_{YX} & S_{YY} - \lambda \end{vmatrix} = \left(S_{XX} - \lambda\right)\left(S_{YY} - \lambda\right) - _{XY}S^2 = 0 \qquad (3.15)$$

and the eigenvalues are

$$\lambda_{2,1} = \tfrac{1}{2}\left[\left(S_{XX} + S_{YY}\right) \pm \sqrt{\left(S_{XX} + S_{YY}\right)^2 + 4\left(_{XY}S^2 - S_{XX}S_{YY}\right)}\right] \qquad (3.16)$$

where $\lambda_{2,1}$ is the eigenvalues $\lambda_2 \geq \lambda_1$, $\lambda_2 = S_{max}$ and $\lambda_1 = S_{min}$. Both eigenvalues are positive. Unless the eigenvalues of the stiffness matrix are equal, only displacement along the eigenvectors of the stiffness matrix induces a restoring force that acts in precisely the opposite direction. If the endpoint is displaced in any other direction, the restoring force acts at an angle to the direction of the displacement.

As mentioned previously, the orientation θ of the ellipse is the direction of S_{max}. The area (A) and shape (aspect ratio, B) can be computed as follows:

$$A = \pi \lambda_1 \lambda_2$$
$$B = \frac{\lambda_2}{\lambda_1} \qquad (3.17)$$

Sometimes, it is convenient to choose coordinate axes in the principal directions. The stiffness matrix expressed in the principal directions becomes diagonal. This transformation of the reference axes is called the *principal transformation* of the endpoint stiffness matrix.

• • • CLASSICAL MECHANICS REFRESHER • • •

Potential Function

Consider a force **F** in a conservative *force field.* The work done by **F** during a displacement of its point of application depends only on the final location of the point and does not depend on the path followed. The scalar function that determines the work accomplished by **F** when the point of its application moves from a given location to the origin is called a *potential function,* or simply *potential.* The potential function $-Q$ assigns a certain amount of potential energy to any point in the field. The minus sign is arbitrary and introduced for convenience. As an example, consider a simple spring. An elastic force exerted by a spring on a body that stretches it is $F = -k(x - x_0)$, and the potential function is $-Q = \frac{1}{2}k(x - x_0)^2$. The potential function is defined everywhere except at the origin. The force field is the negative of the gradient of the potential.

When the second derivative of the potential function is positive, $d^2Q/dx^2 > 0$, the potential energy is minimal, and consequently the equilibrium is stable. When $d^2Q/dx^2 < 0$, the equilibrium is unstable. If the second derivative equals zero, the stability of equilibrium is characterized by higher derivatives of the potential function. If a lowest order of the nonzero derivative is odd, the equilibrium is stable in one direction of x and unstable in the opposite direction, similar to the condition in figure 3.1c.

3.2.4 Potential Functions of Kinematic Chains

When the postural force field is conservative, a deflection of a limb is associated with a certain potential function: a given amount of work is spent displacing the arm, and the same amount of potential energy is stored. If the restoring force is proportional to the endpoint displacement (the force–displacement relation is linear), the potential function is a quadratic function:

$$-Q = \frac{1}{2}\mathbf{F}^T d\mathbf{P} = \frac{1}{2}d\mathbf{P}^T[S]d\mathbf{P} =$$
$$\frac{1}{2}\left(S_{xx}dX^2 + S_{xy}dXdY + S_{yx}dXdY + S_{yy}dY^2\right) \tag{3.18}$$

If the matrix of apparent endpoint stiffness is symmetric, $S_{xy} = S_{yx}$, equation 3.18 can be written as

$$-Q = \frac{1}{2}\left(S_{xx}dX^2 + 2S_{xy}dXdY + S_{yy}dY^2\right) \tag{3.19}$$

Because the force field $\mathbf{F}(X, Y)$—that is, the restoring force as a function of the endpoint coordinates—is the negative of the gradient of the potential, we can write

$$\mathbf{F}(X,Y) = -grad\, Q = \begin{bmatrix} -\dfrac{\partial Q}{\partial X} \\ -\dfrac{\partial Q}{\partial Y} \end{bmatrix} = -\begin{bmatrix} F_X(X,Y) \\ F_Y(X,Y) \end{bmatrix} \tag{3.20}$$

Differentiating equation 3.20 with respect to X and Y yields

$$F_X = -\left(S_{XX}dX + S_{XY}dY\right)$$
$$F_Y = -\left(S_{YX}dX + S_{YY}dY\right)$$

(3.21)

which is essentially equation 3.7 for the case when the chain does not support an external load before the deflection.

Hence, the information about the apparent limb stiffness can be presented as a force field, a stiffness matrix, a stiffness ellipse, and a limb potential function.

■ ■ ■ *FROM THE LITERATURE* ■ ■ ■

Postural Force Fields of the Human Arm

Source: Shadmehr, R., F.A. Mussa-Ivaldi, and E. Bizzi. 1993. Postural force fields of the human arm and their role in generating multi-joint movements. *J. Neurosci.* 13 (1): 45–62.

Subjects were instructed to grip the handle of a planar manipulandum with their right hand (figure 3.12*a*), to place the handle at a specified location, and to close their eyes. The manipulandum was a two-joint planar robot with a force transducer mounted at the handle. The measurements were performed at two positions: "right" and "left." The handle was slowly displaced; the displacement was 5.0 cm over a period of 8.5 s. The subjects were told "not to intervene voluntarily." The restoring forces generated by the subjects were measured. The measurements were repeated for 24 directions spanning 360°. [Note the difference in the methods of this study and the study of Mussa-Ivaldi, Hogan, and Bizzi (1985) described previously. In this study, the magnitudes of hand displacement and the restoring forces were much larger.]

The measured postural fields for one subject are presented in figure 3.12*b*. Note that

1. the direction of the restoring force did not, as a rule, coincide with the direction of the displacement;
2. the magnitude of the restoring forces was not equal for equal displacements from equilibrium in various directions; and
3. the shape and orientation of the fields were different for the two configurations of the arm.

The force field data were converted into matrices of apparent endpoint stiffness and presented graphically as stiffness ellipses (figure 3.12*c*). Note that the stiffness was not constant. It depended on the magnitude of the

displacement: the largest stiffness was observed for the smallest perturbations; then it gradually decreased. The potential function represented these
changes (figure 3.12*d*).

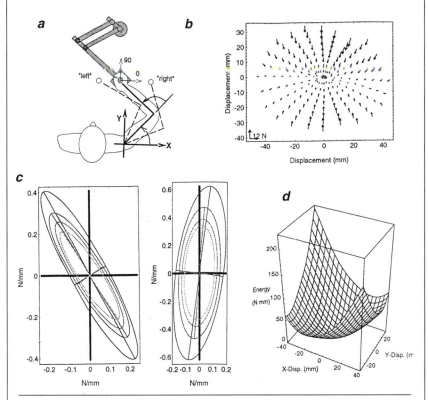

Figure 3.12   Measuring the postural force fields of the human arm. (*a*) The
experimental setup. (*b*) Postural force fields for one subject with the arm in
the "right" position. The arm was displaced from the center to the periphery.
Restoring force vectors are shown at 1-s intervals. (*c*) Stiffness ellipses for
the displacements 0–1, 1–2, 2–3, and 3–4 cm. The largest ellipses represent
the stiffness for the smallest perturbations, 0–1 cm. Stiffness becomes progressively smaller as the deflection gets larger. The left and right ellipses
correspond to the left and right positions of the arm, respectively. (*d*) The
potential function for the left arm position. The steepest region of the potential function corresponds to the direction of maximum stiffness of the arm.
The equilibrium point is at the bottom of the "bowl."

3.3 APPARENT JOINT STIFFNESS

The reaction of a kinematic chain to a perturbation is a consequence of the reactions in individual joints. We limit our examination to the simplest case, a planar two-link chain. Consider a planar two-link chain that undergoes a small displacement from an equilibrium position. This chain is similar to the chains analyzed in chapters **2** and **3** of the *Kinematics of Human Motion* and chapter **2** of this book. We call the proximal joint "joint 1," or the shoulder joint, and the distal joint "joint 2," or the elbow joint. For simplicity, we assume that the proximal and distal links (the upper arm and the forearm) are of the same length. For the human arm, this is a valid approximation if a person holds a handle in the hand, the wrist is fixed, and the length of the distal link is measured from the elbow to the handle.

A differential increment in the joint torque vector $d\mathbf{T}$ is related to the angular differential displacement vector $d\boldsymbol{\alpha}$ by the matrix of apparent joint stiffness [K]:

$$\begin{bmatrix} dT_1 \\ dT_2 \end{bmatrix} = -\begin{bmatrix} \dfrac{\partial T_1}{\partial \alpha_1} & \dfrac{\partial T_1}{\partial \alpha_2} \\ \dfrac{\partial T_2}{\partial \alpha_2} & \dfrac{\partial T_2}{\partial \alpha_2} \end{bmatrix}\begin{bmatrix} d\alpha_1 \\ d\alpha_2 \end{bmatrix} = -\begin{bmatrix} K_{11} & K_{12} \\ K_{21} & K_{22} \end{bmatrix}\begin{bmatrix} d\alpha_1 \\ d\alpha_2 \end{bmatrix} \tag{3.22}$$

where the subscripts 1 and 2 refer to the shoulder and elbow joint, respectively, or

$$d\mathbf{T} = -[K]d\boldsymbol{\alpha} \tag{3.23}$$

In general, the matrix [K] is not diagonal. Diagonal elements of the matrix, called *direct terms,* relate changes in the torque at a given joint to the angular deflection at this joint, for example, a change in the elbow torque that is due to a change in elbow angle. Off-diagonal terms, called *cross-coupling terms*, relate changes in the torque at one joint to the angular displacements at another joint, for example, a change in the elbow torque due to the shoulder-joint angular displacement. In humans and animals, the cross-coupling terms are due to two factors: (1) the biarticular muscles that span both joints and (2) the *heterogenic, or nonautogenic, reflexes* that occur when the stretching of one muscle changes the activity of other muscles. Mechanically, cross coupling means that every joint deflection affects the torques at all joints. Equation 3.23 is a matrix equation that is equivalent to a set of two linear equations in four unknowns, the joint stiffness values. This set is underdetermined. Hence, if one obtains the values of $d\mathbf{T}$ and $d\boldsymbol{\alpha}$ from one set of measurements, one cannot determine all four elements of matrix [K]. Several measurements or experiments are necessary to find the matrix of the apparent joint stiffness. The same is true of the matrix of apparent endpoint stiffness.

If a matrix of apparent joint stiffness [K] is diagonal (no two-joint muscles and no reflex interaction between the joints), each joint behaves as an independent spring. However, without coupling between the joints, control of endpoint stiffness sharply deteriorates (Hogan 1985b). Perhaps this is an additional reason for the existence of two-joint muscles.

3.4 TRANSFORMATION ANALYSIS

Endpoint stiffness depends on the apparent stiffness of the involved joints and the chain configuration. In turn, apparent joint stiffness depends on the muscle stiffness and the joint angles. Section **3.4.1** deals with the transformation analysis of endpoint stiffness and joint stiffness. In section **3.4.2**, the muscle Jacobian is introduced, and the relationship between joint stiffness and muscle stiffness is discussed. This relationship is analogous to the relationship between endpoint and joint stiffness; the transformation equations look similar. Section **3.4.3** discusses the factors that influence apparent endpoint stiffness. Section **3.4.4** looks at the specific example of the apparent stiffness of the human arm during posture maintenance.

3.4.1 Apparent Endpoint Stiffness Versus Joint Stiffness

Representative papers: Lacquaniti, Carrozzo, and Borghese (1993); McIntyre, Mussa-Ivaldi, and Bizzi (1996); Gomi and Kawato (1997)

We are going to establish a relationship between the apparent stiffness of the endpoint and the apparent stiffness of the individual joints. First, consider a situation where the endpoint couple is zero and the endpoint force is zero before the perturbation. To establish a relationship between the apparent stiffness at the endpoint and at the joints, we combine three equations that were presented previously:

$$\mathbf{T} = \mathbf{J}^T\mathbf{F} \qquad (2.16)$$

$$\mathbf{F} = -[S]d\mathbf{P} \qquad (3.8)$$

and

$$d\mathbf{P} = \mathbf{J}d\alpha \qquad \text{(equation 3.9 from } Kinematics \\ of Human Motion, \text{ p. 154)}$$

Consequently, we have

$$\mathbf{T} = -\mathbf{J}^T[S]d\mathbf{P} = -\mathbf{J}^T[S]\mathbf{J}d\alpha \qquad (3.24)$$

where **J** stands for the Jacobian matrix that relates joint angular velocities to the hand velocity and **J**^{T} stands for its transpose. Because **T** = −[K]$d\boldsymbol{\alpha}$ (equation 3.23) the transformation from the apparent endpoint stiffness to the apparent joint stiffness is

$$[K] = \mathbf{J}^T[S]\mathbf{J} \qquad (3.25)$$

and consequently, the transformation from the joint stiffness to the endpoint stiffness is

$$[S] = (\mathbf{J}^T)^{-1}[K]\mathbf{J}^{-1} \qquad (3.26)$$

Note that both [K] and [S] depend on the chain Jacobian, that is, on the chain configuration. In experiments in which [K] and [S] are determined, the Jacobian is usually assumed to be constant. This assumption is valid only if the limb deflection is small. Therefore, equations 3.25 and 3.26 cannot be used to study large-range arm and leg movements without making the necessary corrections to the equations (the Jacobians are variable).

If the endpoint force is not equal to zero before the perturbation, for instance, when a load is maintained in the hand, one should use the equation $d\mathbf{F} = -[S]d\mathbf{P}$ rather than $\mathbf{F} = -[S]d\mathbf{P}$. Differentiating equation 2.16 (**T** = **J**^{T}**F**) yields

$$[K] = \frac{d\mathbf{T}}{d\boldsymbol{\alpha}} = \frac{d(\mathbf{J}^T\mathbf{F})}{d\boldsymbol{\alpha}} = \mathbf{J}^T\frac{d\mathbf{F}}{d\boldsymbol{\alpha}} + \frac{d\mathbf{J}^T}{d\boldsymbol{\alpha}}\mathbf{F} \qquad (3.27)$$

The equation includes two terms $\mathbf{J}^T\dfrac{d\mathbf{F}}{d\boldsymbol{\alpha}}$ and $\dfrac{d\mathbf{J}^T}{d\boldsymbol{\alpha}}\mathbf{F}$. It follows that the stiffness matrix [K] depends on the changes in the external load, its initial magnitude, and the chain configuration (the chain Jacobian). (Compare equation 3.27 with equations 3.22 and 3.23.) Because $[S] = d\mathbf{F}/d\mathbf{P}$ and $d\mathbf{P} = \mathbf{J}d\boldsymbol{\alpha}$, the derivative of the endpoint force **F** with respect to the joint angles is

$$\frac{d\mathbf{F}}{d\boldsymbol{\alpha}} = [S]\mathbf{J} \qquad (3.28)$$

Therefore, equation 3.27 can be rewritten as

$$[K] = \mathbf{J}^T\frac{d\mathbf{F}}{d\boldsymbol{\alpha}} + \frac{d\mathbf{J}^T}{d\boldsymbol{\alpha}}\mathbf{F} = \mathbf{J}^T[S]\mathbf{J} + \frac{d\mathbf{J}^T}{d\boldsymbol{\alpha}}\mathbf{F} \qquad (3.29)$$

The endpoint stiffness matrix is equal to

$$[S] = (\mathbf{J}^T)^{-1}\left\{[K] - \frac{d\mathbf{J}^T}{d\boldsymbol{\alpha}}\mathbf{F}\right\}\mathbf{J}^{-1} \qquad (3.30)$$

If the endpoint force $\mathbf{F}$ is zero, equation 3.30 can be transformed into equation 3.26. Under certain conditions (a large $\mathbf{F}$ directed toward a joint at a particular chain configuration), the term $\dfrac{d\mathbf{J}^T}{d\boldsymbol{\alpha}}\mathbf{F}$ can be very close to $[K]$. In this case, the kinematic chain becomes unstable, even though the joint stiffness is large. As an example, imagine an athlete who is trying to hold a very heavy barbell above the head with the arms bent. This is likely a risky situation.

3.4.2 Muscle Jacobian: Apparent Joint Stiffness Versus Muscle Stiffness

Representative papers: Tsuji (1997); Prilutsky (2000a)

Whatever the physiological mechanisms of the apparent joint stiffness are, resistance to the joint deflection from an equilibrium position is provided by the muscles (presuming that the joint is not in an extreme angular location). We limit our discussion to a highly idealized case of the basic model with muscles (BMM). In the BMM, the muscles are considered force actuators that exert a force along the line between the *origin* and *insertion*. When muscles are forcibly stretched, they provide resistance to the deformation. Apparent muscle stiffness (or simply muscle stiffness) S_i^m of a muscle i is defined as the derivative

$$S_i^m = dF_i / dl_i \qquad (3.31)$$

For a chain, muscle stiffness can be written as a diagonal matrix $[S^m]$. $[S^m]$ is an $(m \times m)$ matrix, where m is the number of muscles. The length of a given muscle may depend on several joint angular values. Some muscles cross two or more joints, and angular position in joints with more than one degree of freedom is described by two or three angular values. The muscle length vector $\mathbf{L}$ depends on the joint angle vector $\boldsymbol{\alpha}$, $\mathbf{L} = f(\boldsymbol{\alpha})$. Considering small displacements around a posture $\boldsymbol{\alpha}$, we can write

$$d\mathbf{L} = \mathbf{G}(\boldsymbol{\alpha})d\boldsymbol{\alpha} \qquad (3.32)$$

where $\mathbf{G}(\boldsymbol{\alpha})$, or simply $\mathbf{G}$ or $[G]$, is a matrix of partial derivatives that relates small muscle-length changes with the small joint-angle changes:

$$
[G] = \left.\begin{bmatrix}
\dfrac{\partial l_1}{\partial \alpha_1} & \dfrac{\partial l_1}{\partial \alpha_2} & \cdots & \dfrac{\partial l_1}{\partial \alpha_k} \\[2ex]
\dfrac{\partial l_2}{\partial \alpha_1} & \dfrac{\partial l_2}{\partial \alpha_2} & \cdots & \dfrac{\partial l_2}{\partial \alpha_k} \\[2ex]
\cdots & \cdots & \cdots & \cdots \\[2ex]
\dfrac{\partial l_{lm}}{\partial \alpha_1} & \dfrac{\partial l_{lm}}{\partial \alpha_2} & \cdots & \dfrac{\partial l_{lm}}{\partial \alpha_k}
\end{bmatrix}\right\} \text{Muscles } (1, 2, \ldots, m) \qquad (3.33)
$$

where the top bracket spans Joint angles $(1, 2, \ldots, k)$.

If the angle change does not influence the length of a certain muscle, the corresponding element in the matrix equals zero. Matrix $[G]$ is evidently a Jacobian matrix. It is called the *muscle Jacobian*. The muscle Jacobian is similar to the chain Jacobian used extensively in previous chapters. The elements in both Jacobians are partial derivatives with respect to the joint angles. The chain Jacobian relates the endpoint displacement with the small shifts in the joint angles, and the muscle Jacobian relates the changes in the muscle length with these shifts. In simple cases when a muscle can be modeled as a straight line from origin to insertion, the analogy between the chain Jacobian and muscle Jacobian is especially evident.

Like the relationship between the endpoint force and the joint torque vector that was derived previously (equation 2.16), a relationship exists between the joint torques and the muscle forces:

$$\mathbf{T} = \mathbf{G}^T\mathbf{F}_m \qquad (3.34)$$

where $\mathbf{F}_m$ is the vector of muscle forces and $\mathbf{G}^T$ is the muscle Jacobian transpose, which is the matrix of muscle moment arms. If we were to repeat the derivation of equations 3.24 and 3.25, replacing the endpoint force stiffness $[S]$ by the muscle stiffness $[S^m]$ and chain Jacobian $\mathbf{J}$ by the muscle Jacobian $\mathbf{G}$, we would arrive at the equation that expresses joint stiffness $[K]$ as a function of muscle stiffness $[S^m]$:

$$[K] = \mathbf{G}^T[S^m]\mathbf{G} \qquad (3.35)$$

which is analogous to equation 3.25.

■ ■ ■ *FROM THE LITERATURE* ■ ■ ■

Endpoint Stiffness As a Function of Muscle Stiffness

Source: Prilutsky, B.I. 2000a. Coordination of two- and one-joint muscles: Functional consequences and implications for motor control. *Motor Control* 4 (1): 1–44.

In this study, the arm was represented as a planar two-link chain with six muscles acting about the shoulder and elbow joints in the horizontal plane (figure 3.13). Because only four muscles (two flexors and two extensors) are necessary to control two hinge joints, the number of muscles is redundant. The arm response to perturbation of the endpoint position was studied. The changes in muscle length and joint torques were computed from previously published values of joint stiffness for the studied arm postures. Individual muscle forces were estimated using various assumptions. A hypothesis was tested that the muscle forces are distributed in such a way as to minimize the sum $\sum \left(F_i / \text{PCSA}_i \right)^3$, where F_i is the force exerted by muscle i and PCSA_i is the physiological cross-sectional area of this muscle (the criterion was suggested by Crowninshield and Brand 1981).

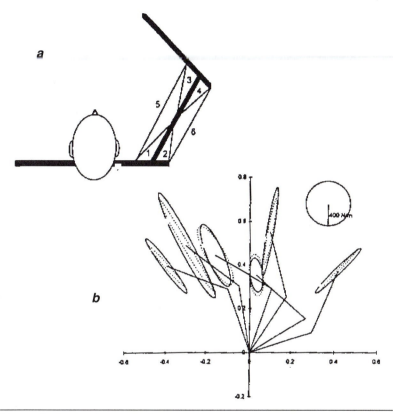

Figure 3.13 Prediction of the endpoint stiffness from muscle stiffness. (*a*) The model. Muscles 1 and 2 are one-joint shoulder flexor and extensor, respectively; muscles 3 and 4 are one-joint elbow flexor and extensor, respectively; and muscles 5 and 6 are two-joint heads of biceps and triceps, respectively. (*b*) Measured (solid lines) and predicted (dotted lines) stiffness ellipses at the endpoint of the arm. The measured data are from Flash and Mussa-Ivaldi (1990). For the prediction, the criterion $\sum \left(F_i / \text{PCSA}_i \right)^3 \rightarrow \text{min}$ was used. The endpoint stiffness was computed from the calculated muscle stiffness. The arm coordinates were measured in meters (abscissa and ordinate axes), and stiffness was measured in N/m (scale shown in the circle).

Reprinted, by permission, from B.I. Prilutsky, 2000a, "Coordination of two- and one-joint muscles: Functional consequences and implications for motor control," *Motor Control* 4 (1): 1–44.

Muscle stiffness depends on the force exerted by the muscle. Relaxed muscles can be extended more easily than contracted muscles. The larger the muscle force, the greater the resistance to the stretch that the muscle provides. This is not a law of mechanics; it is just an experimental fact. Because muscle stiffness depends on the level of muscle activation, joint stiffness can be controlled by co-contracting the antagonistic muscles around the joint. In this case, the level of joint stiffness increases without changing the joint position.

3.4.3 Factors Influencing Apparent Endpoint Stiffness

Representative papers: Flash (1987); Flash and Mussa-Ivaldi (1990)

Consider again a planar two-link system with an endpoint P (figure 3.14). Two systems of coordinates, Cartesian and polar, are located at the proximal joint. If the Cartesian coordinates are X, Y, the coordinates of P in the polar system are

$$r = \sqrt{X^2 + Y^2} \text{ and } \theta = \tan^{-1}(Y / X) \tag{3.36}$$

The differential transformation from polar to Cartesian coordinates is expressed as

$$\begin{bmatrix} dX \\ dY \end{bmatrix} = \begin{bmatrix} \dfrac{\partial X}{\partial r} & \dfrac{\partial X}{\partial \theta} \\ \dfrac{\partial Y}{\partial r} & \dfrac{\partial Y}{\partial \theta} \end{bmatrix} \begin{bmatrix} dr \\ d\theta \end{bmatrix} = \begin{bmatrix} \cos\theta & -\sin\theta \\ \sin\theta & \cos\theta \end{bmatrix} \begin{bmatrix} dr \\ rd\theta \end{bmatrix} = [J] \begin{bmatrix} dr \\ rd\theta \end{bmatrix} \tag{3.37}$$

where $[J]$ is the Jacobian transformation matrix. The differential displacement of P in the X and Y directions, dX and dY, depends on the polar angle θ, the displacement along the radial axis dr, and the product of the differential change in the polar angle θ and the distance from P to the origin, $rd\theta$. Note that for differential transformation from polar to Cartesian coordinates, the Jacobian matrix $[J]$ is simply a rotation matrix (see equation 1.9 in *Kinematics of Human Motion*).

From equation 3.30, it follows that apparent endpoint stiffness depends on the chain Jacobian matrix (i.e., the chain configuration), joint apparent stiffness, and external force. In this section, we consider the influence of the first two factors on apparent endpoint stiffness. It is convenient to perform this analysis in a polar system of coordinates.

The configuration of a chain has significant effects on the stiffness ellipse. As the hand moves toward proximal locations, the shape ratio (λ_2/λ_1) decreases and approaches 1. The ellipse becomes more like a circle. When the arm extends and the hand moves to the working range boundary—the set of boundary points that can be reached by the end effector—the stiffness ellipse becomes more elongated.

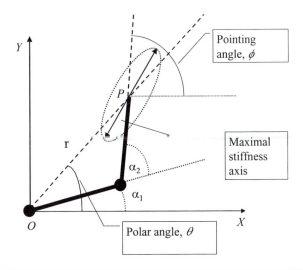

Figure 3.14   A planar two-link model of the human arm. Axis *X* is the *polar axis,* the axis with respect to which coordinate angles are computed. The polar angle *θ* is between the ray from the shoulder to the endpoint (the radial axis) and the polar axis *X*. The pointing angle *φ* is between the line along the forearm (the pointing line) and the polar axis.

The maximal value of stiffness in the direction of the major axis increases. The orientation angles of the stiffness ellipse that are usually between the polar angle and pointing angle for a human arm change to a more polar orientation.

If the arm is completely extended, the chain is in a singular position, and there exists a null space $N(J^T)$ of the external forces. Forces that belong to $N(J^T)$ do not require any joint torque to resist them. Endpoint stiffness becomes infinite in at least one direction for a chain in a singular position. Any endpoint force that acts in this direction is resisted by the skeleton; no joint deflection occurs, and no joint torques are generated. As a result, there is no endpoint deflection, and the endpoint stiffness is infinite. For such joint configurations, the determinant of the Jacobian matrix is zero, and the inverse Jacobian does not exist. Hence, equation 3.26 cannot be applied.

When the endpoint stiffness ellipse has a polar orientation, its matrix expressed in polar coordinates is diagonal. Hence, the transformation from a Cartesian to a polar reference system is a principal transformation of the endpoint stiffness matrix. The endpoint stiffness ellipse has a polar orientation if and only if the stiffness at the shoulder joint due to the angular changes in this joint (K_{11}) equals the sum of the cross-coupling (off-diagonal) terms of the joint stiffness matrix ($K_{11} = 2K_{12}$ and $K_{12} = K_{21}$). The elbow joint stiffness K_{22} may assume any value. The orientation of the major axis when various apparent joint stiffness values are zero is as follows (see also figure 3.15):

Table 3.1 Dependence of the Orientation of the Ellipse of the Apparent Joint Stiffness on the Joint Stiffness Values

| Case | Stiffness coefficients | | | Major axis direction |
	K_{11}	K_{22}	$K_{12} = K_{21}$	
1	Nonzero	Zero	Zero	Along the pointing axis (the forearm)
2	Zero	Nonzero	Zero	Along the radial axis
3	Zero	Zero	Nonzero	Parallel to the upper arm

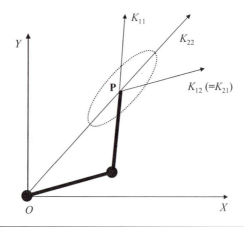

Figure 3.15 The endpoint stiffness resulting from the contribution of K_{11}, K_{22}, and K_{12} (which is equal to K_{21}). When $K_{11} = 2K_{12}$, the major axis of the ellipse is aligned with the radial axis and the orientation angle of the endpoint stiffness ellipse equals the polar angle.

The shape of the stiffness ellipse depends on the stiffness in the radial direction S_r and the stiffness in the polar direction S_θ. For the human arm, movement in the radial direction is described as an arm flexion or extension or *reaching movement,* and movement in the polar direction is an arm rotation or *whipping movement* (see *Kinematics of Human Motion,* pp. 117–18). The shape of the stiffness ellipse is

$$\frac{S_{rr}}{S_{\theta\theta}} = \left(1 + \frac{2K_{22}}{K_{12}}\right)\cot^2\frac{\alpha_2}{2} \tag{3.38}$$

Hence, the shape depends on the ratio of the stiffness in the distal (elbow) joint to the cross-coupling term of the stiffness matrix and the angle at the elbow joint. The smaller the angle, the larger the ratio (remember that we are using external, or anatomic, angles) and the more elongated the stiffness ellipse is.

3.4.4 Apparent Stiffness of the Human Arm During Posture Maintenance

Representative papers: Mussa-Ivaldi, Hogan, and Bizzi (1985); Flash and Mussa-Ivaldi (1990); Dolan et al. (1993); Lacquaniti, Carrozzo, and Borghese (1993); Tsuji et al. (1995); Tsuji (1997); Gomi and Kawato (1997)

The methods described in the previous sections have been widely used in numerous experiments to study the behavior of the human arm during posture maintenance and point-to-point movement. During the experiments, a small perturbation was applied to the arm. This section is not intended to be a comprehensive review of these studies. The discourse is limited to highlighting the main directions of research and basic discoveries. One of these facts was already mentioned: when perturbations are applied in the horizontal plane, the matrix of the apparent endpoint stiffness is (almost) symmetrical. Consequently, the postural force fields of the human arm are conservative (Mussa-Ivaldi, Hogan, and Bizzi 1985; Flash and Mussa-Ivaldi 1990; Tsuji et al. 1995). These studies investigated the shoulder–elbow responses. However, the symmetry of the endpoint stiffness matrix was not confirmed when perturbations were applied in a vertical plane and the elbow–wrist responses were examined (Lacquaniti, Carrozzo, and Borghese 1993).

For a given arm configuration, the shape and orientation of the stiffness ellipse are similar for different subjects (figure 3.16). The values of these parameters in a subject are highly reproducible over time. In contrast, the size of the apparent stiffness ellipse varies among different subjects and is not reproducible for the same subject when measurements are performed on different days (figure 3.17).

The shape and orientation of the stiffness ellipse depend heavily on arm configuration. In particular, (1) the major principal axis intersects the shoulder joint or passes close to it (hence, K_{11} is approximately equal to $2K_{12}$; see figure 3.13 again); (2) the ellipse becomes more elongated when the arm is extended and the hand moves farther from the shoulder; and (3) the ellipse becomes more isotropic (circlelike) when the hand assumes a more proximal location.

Because the apparent stiffness ellipse depends on both the arm configuration (chain Jacobian, see equations 3.26 and 3.30) and the values of the apparent stiffness at the joints, the question arises about the relative contributions of geometric and biomechanical factors in the stiffness ellipse. It has been

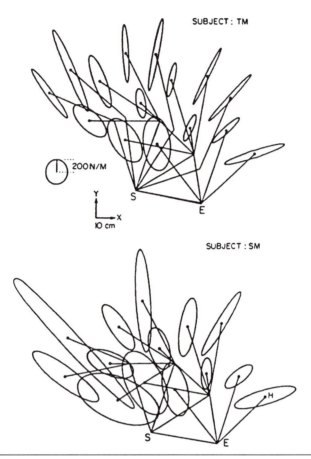

Figure 3.16 Hand stiffness ellipses obtained from two subjects during a postural task. Each ellipse was derived by a regression on about 60 force and displacement vectors. Two line segments (SE) and (EH) represent the upper arm and forearm, respectively. The stiffness ellipses are placed on the hand (H). The circle on the left represents an isotropic endpoint stiffness of 200 N/m.

Reprinted, by permission, from *Experimental Brain Research*, Human arm stiffness characteristics during the maintenance of posture, T. Flash and F. Mussa-Ivaldi, 82, 315–326, 1990, © of Springer-Verlag.

reported that geometric factors alone cannot explain the changes observed in the orientation and shape of the stiffness ellipse (figure 3.18). Hence, joint stiffness values vary when the arm changes its position.

When experimental subjects are informed about the direction of future perturbations, they are not able to increase the stiffness in the direction of impending perturbation exclusively. Instead, the entire size of the stiffness ellipse is altered: the stiffness changes in all directions. Therefore, people cannot change

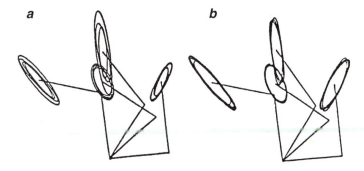

Figure 3.17 Stiffness variability. (*a*) Stiffness ellipses measured in a subject on four different days over a period of six months. (*b*) The same ellipses normalized to cover the same area.

Reprinted, by permission, from F.A. Mussa-Ivaldi, N. Hogan, and E. Bizzi, 1985, "Neural, mechanical, and geometric factors subserving arm posture in humans," *The Journal of Neuroscience* 5 (10): 2732–2743. Copyright © 1985 by the Society of Neuroscience.

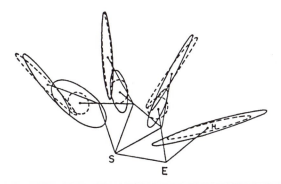

Figure 3.18 The stiffness ellipse varies with arm configuration. Measured ellipses are marked by solid lines. The dashed lines represent simulated ellipses. During the simulation, joint stiffness values corresponding to a single arm configuration were computed. Then, with the stiffness terms kept constant, the model of the arm was displaced to six locations.

Reprinted, by permission, from *Experimental Brain Research*, Human arm stiffness characteristics during the maintenance of posture, T. Flash and F. Mussa-Ivaldi, 82, 315–326, 1990, © of Springer-Verlag.

arm stiffness in a given direction without proportional stiffness changes in other directions. They can, however, change the magnitude of endpoint force in any chosen direction. Hence, the control mechanisms of endpoint force and stiffness are different. Apparent endpoint stiffness of the arm increases with grip force.

3.5 WHY IS IT SO DIFFICULT TO DETERMINE JOINT STIFFNESS?

Representative papers: Kearney and Hunter (1990); Latash and Zatsiorsky (1993)

Although the ratio of joint torque to angular displacement, or force to linear displacement, can always be computed, it does not mean that this computation always makes sense or provides an advantage over a separate analysis of the force and displacement. In addition, it does not mean that the ratio represents stiffness in any meaningful way. The ratio can be interpreted in different ways. Let us consider mechanical and physiological factors that influence the interpretation of the force-to-displacement ratio.

3.5.1 Mechanical Factors

Although stiffness is measured as a force-to-displacement ratio, not every force-to-displacement ratio or derivative dF/dx refers to stiffness. Consider, for instance, the movement of a material particle on a horizontal surface without friction. The equation of motion is $F = ma$. The derivative $dF/dx = m\ddot{x}/\dot{x}$ can easily be calculated (x stands for displacement, $\ddot{x}$ for jerk, and $\dot{x}$ for velocity).

▪ ▪ ▪ *From the Literature* ▪ ▪ ▪

Dependence of the Apparent Stiffness Ellipse on Force Direction

Source: Tsuji, T. 1997. Human arm impedance in multi-joint movement. In *Self-organization, computational maps, and motor control,* ed. P. Morasso and V. Sanguineti, 357–82. Amsterdam: North-Holland, Elsevier Science.

The subjects were seated in front of a robot in a manner similar to the experimental method developed by Mussa-Ivaldi, Hogan, and Bizzi (1985); see From the Literature, p. 214. The elbow of the right arm was supported in the horizontal plane by a chain attached to the ceiling (the shoulder angle was 1.04 rad, and the elbow angle was 1.57 rad). The subjects were instructed to exert a force vector of prescribed magnitude and direction on a handle on the robot. The required force was set to 0.5, 1.0, and 2.0 (× 9.8 N) along the eight directions shown in figure 3.19. Hand deflections of 5-mm amplitude were induced by the robot in the eight directions in a random sequence. The hand force was measured, and the stiffness ellipses computed. The area, orientation, and shape of the stiffness ellipses depended on the magnitude and direction of the target force.

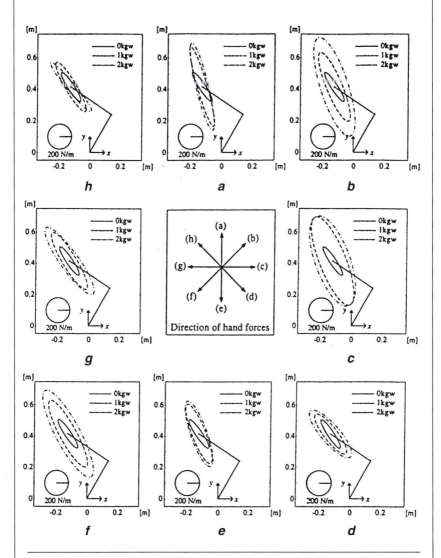

Figure 3.19 Changes of the apparent endpoint stiffness ellipses with the direction and amplitude of target hand forces. The data are from one subject.

Reprinted from Self-Organization, Computational Maps, and Motor Control, P. Morasso and V. Sanguineti, Human arm impedance in multijoint movements, 357–381, with permission from Excerpta Medica Inc.

However, this expression is unusable as a rule and has nothing in common with stiffness as it is understood in classical mechanics.

The joints are occasionally modeled as viscoelastic hinges where resistance is provided by elastic forces and damping forces. The elastic forces depend on the amount of joint angular displacement, and the damping forces depend on the joint angular velocity. For the purpose of our discussion, we substitute the joint angular motion for the rectilinear deformation. If a model includes a damping element and a spring along with inertia, the equation of motion is

$$F(t) = m\ddot{x} + b\dot{x} + k(x - x_0) \qquad (3.39)$$

where m stands for mass, b is a damping coefficient, k is an elastic (or spring) constant, $\ddot{x} = d^2x / dt^2$, and $\dot{x} = dx / dt$. All three coefficients—m, b, and k—are scalars. Even when m, b, and k are not time dependent, the derivative dF/dx is a rather complex function. Let us differentiate both sides of equation 3.39 by t and then divide both sides by dx/dt:

$$\frac{dF}{dx} = m\frac{\dddot{x}}{\dot{x}} + b\frac{\ddot{x}}{\dot{x}} + k \qquad (3.40)$$

The right side of the equation is quite different from the spring constant k. The magnitude of dF/dx depends not only on the spring constant k and the magnitude of the deviation from an equilibrium position, x, but also on the acceleration and velocity at the instant of the measurement. It equals k only when the measurements are performed at equilibria and both the acceleration and velocity are equal to zero. Therefore, using the term "stiffness" for the derivative dF/dx or for the ratio $\Delta F/\Delta x$ when a body of interest is moving may be misleading. This name is reserved for the spring constant k.

For human joints, coefficients k and b in equation 3.39 can be determined experimentally, particularly if k and b are constant during the measurement, provided that several assumptions are satisfied. If a model that includes a single torsional spring and a single torsional damping element (dashpot) is valid, the moment produced by the spring should be exactly in phase with the joint angular displacement, and the damping moment should be in phase with the joint angular velocity. In experiments, a body limb is subjected to oscillation at various frequencies. The exerted torque or force and the limb displacement are measured, and the values of k and b are computed from these data. (These methods are beyond the scope of this book.) If values of k and b happen not to be constant (e.g., if the joint stiffness depends on the test frequency), equation 3.39 cannot be used.

The main challenge in employing the concept of joint stiffness is, however, not in the measurement techniques but in the biological interpretation of the obtained data.

3.5.2 Physiological Factors

When only the apparent joint stiffness (i.e., the part of the total resistance that is proportional to the link displacement and that does not depend on time, velocity, and acceleration) is considered, three constituents of the reaction can be singled out:

1. Purely peripheral factors: muscles and tendons oppose the length change. This element of the apparent joint stiffness is known as the *intrinsic stiffness,* or *preflex stiffness.* Intrinsic stiffness is a complex phenomenon. In a muscle–tendon complex, the intrinsic stiffness depends on the stiffness of the muscle and the stiffness of the tendon. The intrinsic muscle stiffness is not constant. It depends on the level of muscle excitation; during forcible stretching, cross bridges are bent and broken. When experiments are conducted in situ on human or animal subjects, it is frequently assumed that there is no time delay between the link deviation from an equilibrium position and the force rise due to the nonreflex stiffness. With such an assumption, the nonreflex stiffness is comparable to genuine stiffness of passive objects. Real muscles, however, behave in a much more complex way than passive mechanical objects; in particular, their behavior is usually time dependent.

2. Reflex contribution (*reflex stiffness*). There may be an increase in the level of muscle activation due to various reflexes that act with different delays, particularly the *stretch reflex.* The reflex contribution to apparent joint stiffness is always time dependent. It starts after a certain latent period and then changes with time. To distinguish the preflex and reflex contributions to the apparent stiffness, complex experimental methods have been developed. An example of the contributions of preflex (intrinsic) and reflex components to total joint apparent stiffness is presented in figure 3.20. Note that the stiffness values were measured 200 ms after a stretch.

3. Preprogrammed postural reactions that are instruction dependent.

If the problem of joint stiffness appears difficult after this discussion, rest assured that it is.

3.6 STIFFNESS IN MOTOR CONTROL MODELS

The problem of endpoint and joint stiffness becomes even more involved when the stiffness is measured or its existence is implied during movement rather than during posture maintenance. To discuss this issue further, we first review the notion of mechanical modeling.

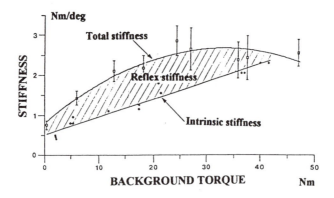

Figure 3.20 Total, intrinsic (preflex), and reflex stiffness in a human ankle joint plotted against the background torque after a plantar stretch. The shaded area indicates the reflex-mediated stiffness. Bars indicate one standard deviation. The intercept of the linear regression line reflects the passive nonreflex stiffness. The goal of the study from which these data were obtained was to evaluate the mechanical response to stretch in normal human dorsiflexors at different levels of voluntary contraction. An electrical stimulation of the deep peroneal nerve that eliminates the stretch reflex was used to activate the muscles and measure the nonreflex component of the apparent stiffness. The total stiffness was defined as the ratio between the torque increment and the amplitude of the stretch and was decomposed into two components: (1) the nonreflex response from the contractile apparatus (intrinsic stiffness) and (2) the reflex-mediated response (reflex stiffness).

Reprinted, by permission, from T. Sinkjær et al., 1988, "Muscle stiffness in human ankle dorsiflexions: Intrinsic and reflex components," *Journal of Neurophysiology* 60:1110–1121.

3.6.1 Mechanical Models and Mechanical Analogies

In human movement science, two types of mechanical models are used. In models of the first kind, a mechanical object of great complexity is replaced by a simpler mechanical object. For instance, the human body is modeled as a multilink chain connected by ideal joints powered by torque actuators (the basic model). This approach is widespread in mechanics, where complex mechanical objects, such as an aircraft, are commonly analyzed as solid bodies or even as material particles. Whether a particular model is valid or not depends on the situation, in particular, on the required accuracy. If the size of an aircraft is much smaller than the distance it travels, the aircraft can be considered a point. If displacement of the human body segments is much larger than their deformation, the deformation can be neglected and the body segments can be considered rigid. Such modeling is the bedrock on which the science of mechanics is based.

In models of the second type, a biological object is replaced by a simple mechanical object (such models are sometimes called *mechanical analogies*). For instance, a joint or a muscle is represented by a spring (with or without damping). Mechanical analogies are not intended to describe the structure of the modeled biological objects or their internal functioning (nobody assumes that the joint is built as a spring). Their usual goal is to describe the manifested behavior of the modeled object in certain circumstances. The description is frequently not literal; it is metaphorical. For instance, the same word, "stiffness," can be used to characterize muscle-tendon deformation, the outcome of a certain spinal reflex, the combined effect of tissue deformation and spinal reflexes, or even the result of a voluntary reaction. Only in the case of muscle-tendon deformation is the word "stiffness" used in its original meaning. The metaphoric use of scientific terms requires special attention from the reader: the same term can have different meanings in various studies. When it comes to mechanical analogy, accuracy alone is not sufficient. Understanding what exactly is represented by the analogy is crucial. Also, the extent of applicability of the analogy should be described and understood. In mechanical analogies, the problem under investigation is essentially nonmechanical. Methods of mechanics have their own limits. They are appropriate for exploring the constructed model (analogy) but not the modeled object itself. Whether the analogy fits reality or not is another issue that is beyond the scope of mechanics. When the goal is to describe a system's performance under certain circumstances (a common goal in engineering), mechanical analogies are frequently useful. Their usefulness for describing internal motor control processes is much less evident.

An example of an "engineering" model of the human leg is presented in figure 3.21. An immediate goal for using this model was to design a running track of optimal compliance. If the surface compliance is too large (as on a trampoline or a thick, soft pillowlike mat), running is almost impossible. If the stiffness is too large (as on concrete), performance deteriorates and injury risk increases. The idea was to find the optimal track compliance by comparing leg stiffness with track stiffness. The model consists of a rack and pinion and a serially connected, damped spring. During posture maintenance, the rack and pinion is locked and external perturbations are resisted by the spring. The model is based on two postulates of motor control:

1. The leg behavior during the support period in running represents the outcome of three events: (a) a central command, (b) spinal reflexes, and (c) muscle-tendon properties.
2. The spinal reflexes together with the muscles and tendons can be considered one lumped object.

According to the model, extension and flexion of the unloaded leg is determined by a central command that cranks the rack and pinion to a new set point

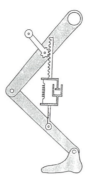

Figure 3.21 Schematic representing the separate roles of descending commands (rack and pinion) and muscle properties plus local reflexes (damped spring). To estimate leg stiffness during support periods in running, an assumption was made that the rack and pinion is locked. The vertical motion of the runner was assumed to be due to the mass–spring system.

Reprinted from *Journal of Biomechanics*, 12, T.A. McMahon and P.R. Green, The influence of track compliance on running, 893–904, 1979, with permission from Elsevier Science.

for the joint angle. When the leg is loaded, its length depends on both the position of the rack and pinion (the central preprogrammed command) and the spring deformation (the muscle-tendon properties and spinal reflexes combined). However, to determine leg stiffness, a simplifying assumption was made that during the support period, the rack and pinion is locked. The model was successful in predicting an optimal range of track stiffness. Running on tracks whose stiffness is less than the leg's stiffness decreases running speed.

The model possesses an important property. The model's spring deformation depends on the exerted force and does not depend on the initial joint angle or leg length (which are determined by the rack and pinion). Consequently, if the spring is not linear, its stiffness depends on the magnitude of the acting force but not on the joint angle. The independence of the incremental stiffness of an active object from its initial length is a general feature of all "serial" models, in which a force-generating nonextensible element is connected in series with an elastic element. It is valid, for instance, for a two-element model of muscle that contains a contractile component and a series elastic component (this model will be described in the next volume of this series).

The model under discussion has clear limitations. It is limited even from a purely mechanical standpoint: among other limitations, the leg stiffness is represented as a scalar instead of a matrix, and the model neglects that leg deformation is not small and that consequently the leg Jacobian is not constant. It is limited in application. The model does not work, for instance, when landing and takeoff are performed as two separate movements rather than as one unin-

---

• • • *PHYSIOLOGY REFRESHER* • • •

Stretch Reflexes

Stretch reflexes are induced by stretching spindle receptors within a muscle. The stretching initiates an *afferent* signal along a sensory neuron to the spinal cord. The sensory neurons relay the impulse to other neurons. In *short-latency stretch reflexes,* which are induced by the fast stretching of the muscle, the impulse goes directly to the motoneurons that innervate the stretched muscle. Because only two neurons and one central synapse are involved in the pathway, this stretch reflex is also called the *monosynaptic stretch reflex.* This reflex causes the stretched muscle to contract.

The *tonic stretch reflex* is induced by slow stretching or maintenance of a muscle at a new, longer length. The pathways of the tonic reflex are presently not known. The tonic stretch reflex causes the stretched muscle to maintain sustained contraction.

---

terrupted movement or when the legs bend and extend slowly so that the leg length changes but the exerted force is practically constant, among other tasks. In these movements, leg stiffness cannot be computed; it "disappears." With the exception of the previously mentioned assumptions, the model does not address the underlying physiological and motor control processes that define the observed behavior. It does not allow, for instance, separating the contribution to leg deformation of the rack and pinion (preprogrammed changes of leg length) and the damped spring (peripheral processes). Because the model was designed for a very specific purpose (it is an engineering model), the mentioned limitations do not decrease its value. However, the application of the model in motor control studies or in biomechanics of sport technique is limited. In the engineering approach, the assumed locking of the rack and pinion during the support period is simply a small technical detail. In motor control, it is the cornerstone of the problem: what is the role of the central command and the peripheral mechanisms in the control of movement? One of the most popular approaches to solving this problem is the *equilibrium-point hypothesis.*

3.6.2 The Equilibrium-Point Hypothesis

Representative papers: Feldman and Levin (1995), Latash and Gottlieb (1991)

The equilibrium-point hypothesis, also known as the λ-model, was suggested by A. Feldman in 1966, and since then it has become an object of active research and heated debate. The hypothesis has its proponents and critics. The goal of this section is to familiarize readers with the basics of this hypothesis.

Formally, the equilibrium-point hypothesis has a certain similarity with the mechanical model discussed in the previous section. At least, it is based on the same presumptions: human movement is determined by an interaction of a central command, peripheral reflexes (plus intrinsic muscle-tendon properties), and external forces. In addition, in both models, the force–displacement curve does not depend on the joint position prior to the displacement. The similarity ends at this point. The basic quest of the λ-hypothesis is to understand the interaction among the central command, peripheral reflexes (plus intrinsic muscle-tendon properties), and the external forces—an issue that is neglected in engineering models, which are mainly interested in the description of mani-fested behavior rather than its mechanisms.

3.6.2.1 Control of Muscle Length and Force

In the first experiments that led to the λ-hypothesis, the subjects assumed a joint position while resisting an applied load. The load was suddenly decreased, and the joint displacement was measured. It was found that the shape of the force–displacement curve did not depend on the initial joint angle. The force–displacement curves were called "invariant characteristics" or "joint compliant characteristics" (JCC; figure 3.22a). The independence of the force–deformation curve from the initial length of an active object may be expected from any "serial" system analogous to the rack-and-pinion-plus-spring model described previously. These models, however, do not tell anything about physiological mechanisms that preceded the observed upshot. The equilibrium-point hypothesis is trying to do exactly that. The hypothesis has two names, the equilibrium-point hypothesis or λ-hypothesis, because it is based on two main ideas:

1. The central command does not immediately prescribe the activity of the α motoneuron pools that innervate the muscles. Instead, the descending central com-mand dictates the parameters of muscle reflexes, specifically, the tonic stretch reflex. To control both the muscle length and muscle force, the central motor command specifies only one variable, the threshold of the tonic stretch reflex, λ. For a given λ, the JCC represents the force–length relationship for a muscle length exceeding the threshold value (to the right of the corresponding λ, figure 3.22a).

2. For a given λ and a fixed external load, the muscle-load system attains equilibrium at a certain muscle length. The equilibrium point is the muscle length (a point along the JCC) at which the muscle force equals the load and the two forces cancel each other (muscle force + load = 0).

A motor command dictates neither a unique muscle force nor a unique muscle length. It defines only λ, the muscle length or joint angle at which the tonic

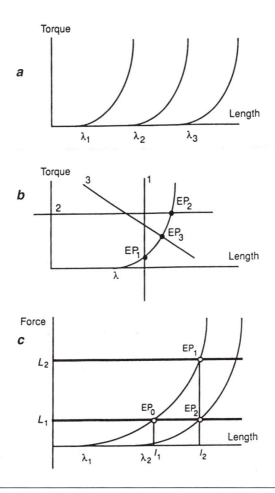

Figure 3.22 Basic ideas of the equilibrium-point hypothesis. (*a*) The force–deformation curves are similar for any initial muscle length or joint angle; they are invariant characteristics. The descending central commands specify only the threshold of the tonic stretch reflex, λ. (*b*) For a given λ, the muscle length at equilibrium is determined by the interplay of the muscle force and the external load. The equilibrium point (EP), that is, the muscle length at a given load and a fixed λ, can be seen as the point of intersection of the JCC and the load characteristic curve. Three examples of load characteristic curves are presented: (1) static, (2) constant load not depending on the joint angle, and (3) elastic. (*c*) The muscle force or length changes due to change in the central motor command, the load characteristics, or both. For example, the initial state EP_0 corresponds to a central command λ_1 and load L_1; a change in the muscle length from l_1 to l_2 can result from a change in the load (L_2, EP_1) or from a change in the central command (λ_2, EP_2).

Adapted, by permission, from M.L. Latash, 1993, *Control of Human Movement* (Champaign, IL: Human Kinetics), 27–28.

stretch reflex is activated. Muscle force or muscle length is modified either by shifting λ (changing the central command) or by changing the external load (figure 3.22c). If the load alters, the muscle length also changes. These deviations change the afferent signals and modulate the level of muscle activation via the tonic stretch reflex. The descending central command may remain constant in this case. Changes in the level of muscle activation—as evidenced by EMG, for instance—do not immediately represent the central command; they represent the intensity of the tonic stretch reflex.

The JCCs are similar in appearance to the force–deformation curves known in the mechanics of deformable bodies. However, the mechanisms are different. The JCCs are caused by the tonic stretch reflex, which is more comparable to a feedback control mechanism used in a technical device (e.g., a room thermostat, cruise control in a car) than to a spring.

3.6.2.2 Control of Joint Angle and Torque

Representative paper: Feldman (1986)

According to the equilibrium-point hypothesis, to change the muscle's length or force, the CNS specifies only one parameter: λ. Joints are controlled by at least two muscles (we refer to these as the flexor and the extensor). Hence, at least two commands are necessary to control the joint position or torque, λ_{fl} for the flexor and λ_{ext} for the extensor. When both λ_{fl} and λ_{ext} shift in one direction, the level of activation of one muscle increases and the level of activation of the second muscle decreases (figure 3.23). This is similar to the reciprocal activation of the antagonists known in physiology. Coactivation of the antagonists can be modeled by shifts of λ_{fl} and λ_{ext} in opposite directions. The larger (or smaller) the difference between λ_{fl} and λ_{ext}, the larger (or smaller) the simultaneous activation of the flexor and extensor. It is suggested that the central regulation of a joint is realized by two commands, the reciprocal command r and the coactivation command c, defined by equation 3.41.

$$r = \frac{1}{2}\left(\lambda_{fl} + \lambda_{ext}\right)$$
$$c = \frac{1}{2}\left(\lambda_{fl} - \lambda_{ext}\right)$$

(3.41)

The caption to figure 3.23 explains the details.

For a joint at equilibrium, the reciprocal command r defines the joint configuration (for a given set of external forces), while the coactivation command c prescribes the level of co-contraction of the muscles serving the joint and hence the joint apparent stiffness. Note that if the external forces change and the reciprocal command does not vary, the equilibrium position changes. The joint equilibrium position for a zero load (and a given r) is known as the *virtual position*. The virtual joint angle is a characteristic of the reciprocal command

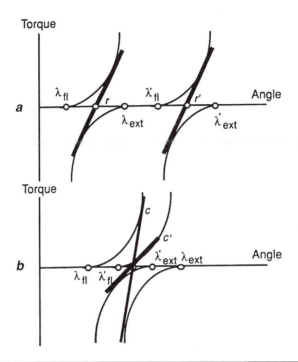

Figure 3.23 Central commands to a joint are described by λ shifts for the muscles (λ_{fl} and λ_{ext}) or by shifts of the reciprocal and coactivation commands (r and c). (a) The r command corresponds to a shift of λ_{fl} and λ_{ext} in the same direction and leads to the shift of the JCC along the angle axis without changing its slope. (b) The c command corresponds to shifts of λ_{fl} and λ_{ext} in opposite directions and leads to a change in the slope of the JCC (apparent joint stiffness).

Reprinted, by permission, from M.L. Latash, 1993, *Control of Human Movement* (Champaign, IL: Human Kinetics), 30.

r. The virtual position can be thought of as the position toward which the muscular activity generated by the tonic stretch reflex is driving the limb. The virtual position can be beyond the reach of the limb. In this case, the limb either is at equilibrium, exerting a force on the environment, or is moving toward the virtual position.

3.6.2.3 Virtual Trajectories and Their Reconstruction

Representative papers: Hogan (1985b); Latash and Gottlieb (1991); Latash et al. (1999); Domen et al. (1999)

According to the equilibrium-point hypothesis, limb movement can be generated by displacing the virtual position. A central command for a single-joint movement can be described by a pair of time functions, $r(t)$ and $c(t)$. The successive positions of $r(t)$ define the *virtual trajectory*. The virtual trajectory is a trajectory toward which the muscular activity generated by the tonic stretch reflex is driving the limb at any instant in time. Due to the mechanical resistance to the movement (limb inertia, damping, variable external forces, etc.), the real limb trajectory differs from the virtual trajectory.

The virtual trajectory, which represents displacement of the JCC, can be estimated (in other words, reconstructed) from experimental recordings. Instant values of r and c at a certain time point after the start of the movement are obtained by linear regression analysis of the joint torque and joint angle values for a set of trials (figure 3.25). The slope of the regression line represents the apparent joint stiffness (and hence the c command), and the intercept (i.e., the joint angle corresponding to zero torque) portrays the JCC position (r command). Consecutive positions of the intercept stand for the virtual trajectory.

■ ■ ■ *From the Literature* ■ ■ ■

Virtual Trajectory in Reaching Movements

Source: Won, J., and N. Hogan. 1995. Stability properties of human reaching movements. *Exp. Brain Res.* 107: 125–36.

Subjects were seated in front of a computer-controlled robot that was equipped with electromagnetic brakes. A handle containing a force transducer was situated at the end effector of the robot. Subjects grasped this handle at shoulder level. The brakes were capable of independently locking the shoulder or the elbow joint of the robot.

While holding the handle, subjects performed specified point-to-point planar arm movements. The shoulder or the elbow joint of the robot was locked for some trials, and the movement trajectory was constrained. The target points were selected so that the subjects were able to complete the movement along the constrained trajectory. Three situations were tested:

1. Unconstrained throughout the movement
2. Constrained throughout the entire movement
3. Initially constrained and then released during the movement

The robotic constraints evoked from the subjects significant forces oriented to restore the unconstrained hand path (figure 3.24). When released from the constraint, the forces tended to return the hand to the unconstrained path before the end of the movement was reached.

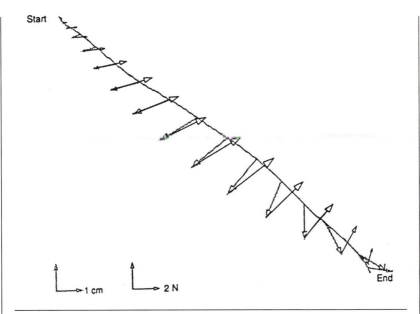

Start

1 cm   2 N

End

Figure 3.24 A sample force–displacement profile of the unconstrained and fully constrained trajectory. The line from "start" to "end" is the unconstrained trajectory, and the vectors with the tails on the line represent the displacement of a similar point in time on the constrained trajectory. The vectors directed toward the line represent the vector difference of the forces between the two trajectories.

Reprinted, by permission, from *Experimental Brain Research*, Stability properties of human reaching, J. Won and N. Hogan, 107:125–136, 1995, © Springer-Verlag.

The results support the equilibrium-point hypothesis and the existence of a virtual trajectory for multijoint arm movements.

When a fast arm movement is performed, an endpoint of the virtual trajectory beyond the target is initially specified. During the movement, the endpoint of the virtual trajectory is adjusted to the actual position of the target. As a result, the virtual trajectory in fast movements is shaped like an N.

3.6.2.4 Rambling-Trembling Hypothesis

Representative papers: Zatsiorsky and Duarte (1999, 2000)

The rambling-trembling hypothesis is concerned with maintaining balance in a standing posture. In particular, the hypothesis addresses the mechanics of the CoP migration during quiet standing. Similar to the equilibrium-point

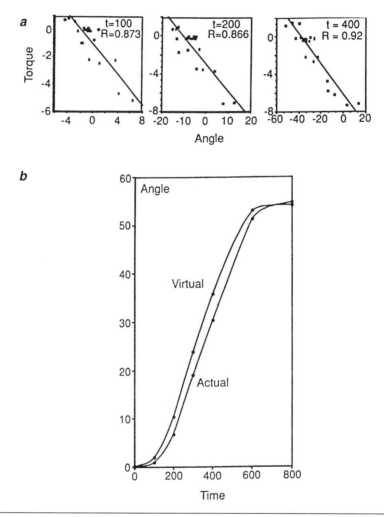

Figure 3.25 Reconstruction of virtual trajectories. The data are for one subject. Subjects were trained to perform a standardized and consistent elbow flexion. They were instructed to reproduce the same time profile of the learned voluntary command while ignoring changes in the externally applied load. During a movement, the torque could unexpectedly change in either direction (loading or unloading) or could stay constant. After sufficient practice, the subjects generated consistently repeatable kinematic profiles. In the analysis, the time 0 corresponded to the beginning of the acceleration. (*a*) Joint torque versus joint angle relationships for different time points after the start of slow elbow flexion. (*b*) Virtual and actual trajectories. Note the time delay between the virtual and actual trajectory.

Adapted from *Neuroscience*, 43, M.L. Latash and G.L. Gottlieb, Reconstruction of joint compliant characteristics during fast and slow movements, 697–712, 1991, with permission from Excerpta Medica Inc.

hypothesis, the rambling-trembling hypothesis is based on the idea that motor control processes are hierarchical and include at least two levels. The highest level determines the reference position with respect to which equilibrium is maintained, and the lower level actually maintains the balance. The reference position is characterized by a reference, or set, point on a supporting surface. During standing, the set point migrates (figure 3.26*a*). This migration is called *rambling*. While standing, the CoP moves along the rambling trajectory and oscillates around it. The oscillation of the CoP around the rambling trajectory is called *trembling*. According to the rambling-trembling hypothesis, the movement of the CoP along the rambling trajectory does not give rise to the restoring forces, but the deviation from the rambling trajectory (trembling) does. This contention is illustrated in figure 3.26.

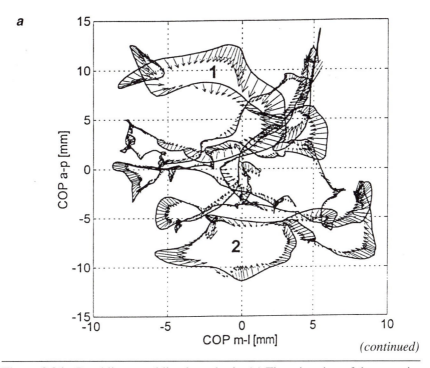

(continued)

Figure 3.26 Rambling-trembling hypothesis. (*a*) The migration of the set point during quiet standing. The graph shows the CoP migration with superimposed horizontal force vectors during quiet standing (for 30 s by a young, healthy man). The force vectors are directed to a migrating set point (not shown). Over short intervals of time, consecutive force vectors converge to (approximately) one pole, and over long intervals of time, they converge to different poles. Hence, the set point migrates. Areas of the graph labeled 1 and 2 are examples of the time intervals during which the consecutive force vectors converge to a small region.

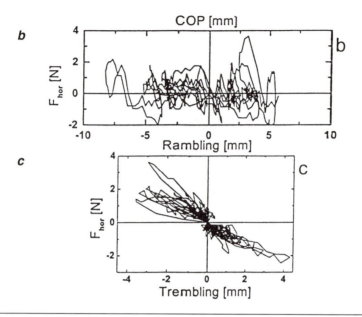

Figure 3.26 *(continued)* (*b*) Rambling and (*c*) trembling values versus the horizontal GRF in quiet standing by one subject for 30 s. The coefficients of correlation are −0.2 and −0.9, respectively. The analysis was performed for the anteroposterior direction. The rambling was estimated by an interpolation of the instant equilibrium points (see From the Literature, p. 201). The trembling values were obtained as the deviation of the CoP from the rambling trajectory. The existence of a large negative correlation between trembling (relative CoP position) and F_{hor} is evident. This means that a deviation of the CoP from the rambling trajectory is always associated with a horizontal force, which tends to reduce the deviation. The GRF is the only external force acting on the body in the horizontal direction. The horizontal component of the GRF is proportional to the acceleration of the body's CoM, $F_{hor} = ma_{hor}$, where m is the mass of the body.

Adapted, by permission, from V.M. Zatsiorsky and M. Duarte, 2000, "Rambling and trembling in quiet standing," *Motor Control* 4(2):185–200.

In figure 3.26*c*, the restoring force is proportional to the deviation and acts with no time lag. These features are typical for elastic forces that tend to return a system to equilibrium. Hence, small deviations from the reference point trajectory during quiet standing are counterbalanced by the apparent intrinsic stiffness at the ankle joints. The restoration of equilibrium in this case does not require a reflex intervention.

3.7 SUMMARY

This chapter is intended to familiarize readers with the equilibrium of naturally unstable kinematic chains. To this end, the concepts of apparent endpoint stiffness and apparent joint stiffness are introduced. The adjective "apparent" emphasizes the difference between the stiffness of passive objects, such as tendons and ligaments, and the stiffness-like property of active objects, that is, the muscles and joints.

Apparent endpoint stiffness cannot be reduced to a scalar value. It is conveniently represented by a matrix. Restoring forces at the endpoint are represented by a postural force field. Alternatively, they can be represented by a stiffness ellipse. The ellipse depicts the magnitude of the restoring force generated in response to a deflection of unit magnitude from the equilibrium position. When the postural force field is conservative, a potential function can be used to describe the chain behavior under deflection from the equilibrium position. The endpoint stiffness depends on (1) the chain Jacobian matrix (i.e., the chain configuration), (2) the joint apparent stiffness, and (3) the external force.

Joint apparent stiffness is described by a matrix that includes direct terms, which relate changes in the torque at a given joint to the angular deflection at this joint, and cross-coupling terms, which relate changes in the torque at one joint to the angular displacements at another joint.

By using the chain Jacobians, it is possible to transform the values of endpoint stiffness into values of joint stiffness, and vice versa. The transformation agrees well with the transformation analysis described previously in kinematics (chapter 3 in *Kinematics of Human Motion*) and statics (chapter 2 in this book). Table 3.2 and figure 3.27 summarize the main equations of the transformation analysis.

The muscle Jacobian is similar to the chain Jacobian. The elements of the muscle Jacobian are partial derivatives of the small changes in muscle length over the small changes in joint angles. The elements of the muscle Jacobian transpose are the moment arms of the muscles. The relationship between muscle stiffness and joint stiffness is similar to the relationship between the endpoint and joint stiffness.

The ratio of joint torque to angular displacement does not always represent joint stiffness. The ratio can be interpreted differently. Mechanical and physiological factors that influence the interpretation of the force-to-displacement ratio are briefly discussed.

Two types of mechanical models are discussed. In one kind of model, a complex mechanical object is replaced by a simpler mechanical object. In the second type, called mechanical analogies, a biological object is replaced by a simple mechanical object. In mechanical analogies, the problem under investigation

Table 3.2   Primary Transformations of the Biomechanical Parameters in Kinematic Chains

From	To	Symbols	Equation
Transformation			
Differential of joint angles	Differential of endpoint coordinates	$d\boldsymbol{\alpha} \rightarrow d\mathbf{P}$	$d\mathbf{P} = \mathbf{J}\, d\boldsymbol{\alpha}$
Differential of endpoint coordinates	Endpoint force or differential of endpoint force	$d\mathbf{P} \rightarrow \mathbf{F}$ or $d\mathbf{P} \rightarrow d\mathbf{F}$	$\mathbf{F} = -[S]d\mathbf{P}$ or $d\mathbf{F} = -[S]d\mathbf{P}$
Endpoint force	Vector of joint torques	$\mathbf{F} \rightarrow \mathbf{T}$	$\mathbf{T} = \mathbf{J}^T\mathbf{F}$
Differential of joint angles	Vector of joint torques or differential of joint torques	$d\boldsymbol{\alpha} \rightarrow \mathbf{T}$ or $d\boldsymbol{\alpha} \rightarrow d\mathbf{T}$	$\mathbf{T} = -[K]d\boldsymbol{\alpha}$ or $d\mathbf{T} = -[K]d\boldsymbol{\alpha}$
Endpoint stiffness	Joint stiffness	$[S] \rightarrow [K]$	$[K] = \mathbf{J}^T[S]\mathbf{J}$
Joint stiffness	Endpoint stiffness	$[K] \rightarrow [S]$	$[S] = (\mathbf{J}^T)^{-1}[K]\mathbf{J}^{-1}$

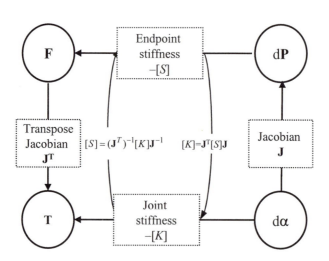

Figure 3.27   The main transformations of biomechanical parameters in kinematic chains. $\mathbf{F}$ is the vector of the endpoint force, $\mathbf{T}$ is the vector of the joint torques, $d\mathbf{P}$ is the differential displacement of the endpoint, $d\boldsymbol{\alpha}$ is the differential displacement of the joint angles, $[S]$ is the matrix of the endpoint stiffness, $[K]$ is the matrix of joint stiffness, and $\mathbf{J}$ is the Jacobian matrix of the chain. The weight of the links is neglected in this analysis.

is essentially nonmechanical. The methods of mechanics have limits. They are appropriate for exploring the constructed model (or analogy) but not the modeled object itself. Whether the analogy fits cannot be proved by the methods of mechanics alone. The mechanical concept of stiffness is used in several models (hypotheses) of motor control.

According to the equilibrium-point hypothesis, to change the muscle length or force, the CNS specifies only the threshold of the tonic stretch reflex, λ. The central regulation of a joint is realized by two commands: the reciprocal command r and the coactivation command c. The reciprocal command defines the position of the joint compliant characteristics (JCC), while the co-activation command defines the slope of the JCC (the joint apparent stiffness).

The rambling-trembling hypothesis explains the CoP migration during quiet standing as a superimposition of two processes. The rambling component reveals the motion of a moving reference point with respect to which the body's equilibrium is instantly maintained. The trembling component reflects the body's oscillation around the reference point trajectory. Trembling is strongly correlated with a restoring force that possesses the qualities of elastic forces; the force is proportional to the trembling amplitude and acts without time delay.

3.8 QUESTIONS FOR REVIEW

1. What is required for equilibrium to be stable? Is equilibrium stable in an upright standing posture?

2. What is the dimensionality of joint compliance?

3. Discuss the difference between the force–length curves of passive and active objects.

4. Explain the meaning of the individual elements of the endpoint stiffness matrix.

5. Some authors estimate leg stiffness during a takeoff in running or jumping as the ratio of the maximal vertical GRF to the changes in "leg length" (a distance between the point of GRF application and the hip joint or the center of mass of the body). Compare this method of estimating leg stiffness with the endpoint stiffness discussed in the chapter. Discuss the problems associated with the described technique of measuring leg stiffness.

6. What is a postural force field? Compare conservative and nonconservative force fields.

7. Explain the meaning of a stiffness ellipse.

8. Name the main parameters used to describe the stiffness ellipse.

9. A subject is maintaining a constant arm posture. Suggest a method to determine a potential function of the arm.

10. Discuss diagonal and off-diagonal elements of the joint stiffness matrix.

11. Write the equations that relate endpoint stiffness to joint stiffness.

12. For a planar two-link chain, compare the stiffness ellipse discussed in this chapter with the force ellipse described in chapter **2**. Explain why these ellipses are not identical.

13. Name the main factors that influence the endpoint stiffness of a kinematic chain.

14. What is intrinsic muscle stiffness?

15. Discuss the equilibrium-point hypothesis.

16. Explain the notion of virtual trajectory.

17. Discuss the rambling-trembling hypothesis.

3.9 BIBLIOGRAPHY

Agarwal, G.C., and G.L. Gottlieb. 1977. Compliance of the human ankle joint. *J. Biomech. Eng.* 99: 166–170.

Akazawa, K., T.E. Milner, and R.B. Stein. 1998. Modulation of reflex EMG and stiffness in response to stretch of human finger muscle. *J. Neurophysiol.* 49: 16–27.

Alexander, R.M. 1988. *Elastic mechanisms in animal locomotion.* Cambridge, UK: Cambridge University Press.

Allum, J.H.J., and K.-H. Mauritz. 1984. Compensation for extrinsic muscle stiffness by short-latency reflexes in human triceps surae muscles. *J. Neurophysiol.* 52: 797–818.

Almeida-Silveira, M.I., D. Lambertz, C. Perot, and F. Goubel. 2000. Changes in stiffness induced by hindlimb suspension in rat Achilles tendon. *Eur. J. Appl. Physiol.* 81 (3): 252–57.

Aruin, A.S., and V.M. Zatsiorsky. 1984. Biomechanical characteristics of human ankle-joint muscles. *Eur. J. Appl. Physiol.* 52 (4): 400–406.

Bach, T.M., A.E. Chapman, and T.W. Calvert. 1983. Mechanical resonance of the human body during voluntary oscillations about the ankle joint. *J. Biomech.* 16 (1): 85–90.

Baratta, R., and M. Solomonow. 1991. The effect of tendon viscoelastic stiffness on the dynamic performance of skeletal muscle. *J. Biomech.* 24: 109–16.

Batman, M., and R. Seliktar. 1993. Characterization of human joint impedance during impulsive motion. *J. Electromyogr. Kinesiol.* 3 (4): 221–30.

Beek, P.J., R.C. Schmidt, A.W. Morris, M.-Y. Sim, and M.T. Turvey. 1995. Linear and nonlinear stiffness and friction in biological rhythmic movements. *Biol. Cybern.* 73: 499–507.

Bennett D.J., M. Gorassini, and A. Prochazk. 1994. Catching a ball: Contributions of intrinsic muscle stiffness, reflexes, and higher order responses. *Can. J. Physiol. Pharmacol.* 72: 525–34.

Bennett, D.J., J.M. Hollerbach, Y. Xu, and I.W. Hunter. 1992. Time-varying stiffness of human elbow joint during cycling voluntary movement. *Exp. Brain Res.* 88: 433–42.

Bergmark, A. 1989. Stability of the lumbar spine: A study in mechanical engineering. *Acta Orthop. Scand.* 60 (Suppl. 230): 1–230.

Biryukova, E.V., V.Y. Roschin, A.A. Frolov, M.E. Ioffe, J. Massion, and M. Dufosse. 1999. Forearm postural control during unloading: Anticipatory changes in elbow stiffness. *Exp. Brain Res.* 124 (1): 107–17.

Bizzi, E., N. Accornero, W. Chapple, and N. Hogan. 1984. Posture control and trajectory formation during arm movement. *J. Neurosci.* 4: 2738–44.

Bizzi, E., and F.A. Mussa-Ivaldi. 1989. Geometrical and mechanical issues in movement planning and control. In *Foundations of cognitive science,* ed. M.I. Posner. Cambridge, MA: MIT Press.

Bizzi E., F.A. Mussa-Ivaldi, and S.F. Giszter. 1991. Computations underlying the execution of movement: A biological perspective. *Science* 253: 287–91.

Bloem, B.R., D.J. Beckley, M.P. Remler, R.A. Roos, and J.G. van Dijk. 1995. Postural reflexes in Parkinson's disease during "resist" and "yield" tasks. *J. Neurol. Sci.* 129 (2): 109–19.

Bober, T., S. Kornecki, R.P. Lehr, Jr., and J. Zawadski. 1982. Biomechanical analysis of human arm stabilization during force production. *J. Biomech.* 15 (11): 825–30.

Brand, P.W. 1995. Mechanical factors in joint stiffness and tissue growth. *J. Hand Ther.* 8 (2): 91–96.

Brown, M., J.S. Fisher, and G. Salsich. 1999. Stiffness and muscle function with age and reduced muscle use. *J. Orthop. Res.* 17 (3): 409–14.

Burguete, R.L. 1992. Methods to calculate joint stiffness: A review and classification of joint types. *Proc. Int. Offshore Mech. Arctic Eng. Symp.* 3 (Pt. B): 347–54.

Buyruk, H.M. 1995. Measurements of sacroiliac joint stiffness with colour Doppler imaging: A study on healthy subjects. *Eur. J. Radiol.* 21 (2): 117–22.

Buyruk, H.M., H.J. Stam, C.J. Snijders, J.S. Lameris, W.P. Holland, and T.H. Stijnen. 1999. Measurement of sacroiliac joint stiffness in peripartum pelvic pain patients with Doppler imaging of vibrations (DIV). *Eur. J. Obstet. Gynecol. Reprod. Biol.* 83 (2): 159–63.

Canon, F., and F. Goubel. 1995. Changes in stiffness induced by hindlimb suspension in rat soleus muscle. *Pflugers Arch.* 429 (3): 332–37.

Capaday, C., R. Forget, and T. Milner. 1994. A re-examination of the effects of instruction on the long-latency stretch reflex response of the flexor pollicis longus muscle. *Exp. Brain Res.* 100: 515–21.

Carpenter, M.G., J.S. Frank, and C.P. Silcher. 1999. Surface height effects on postural control: A hypothesis for a stiffness strategy for stance. *J. Vestib. Res.* 9 (4): 277–86.

Carter, R.R., P.E. Crago, and M.W. Keith. 1990. Stiffness regulation by reflex action in the normal human hand. *J. Neurophysiol.* 64: 105–18.

Certer, R., P. Crago, and P. Gorman. 1993. Nonlinear stretch reflex interaction during cocontraction. *J. Neurophysiol.* 69: 943–52.

Chapman, E.A., H.A. deVries, and R. Swezey. 1972. Joint stiffness: Effects of exercise on young and old men. *J. Gerontol.* 27 (2): 218–21.

Cholewicki, J., and S.M. McGill. 1995. Relationship between muscle force and stiffness in the whole mammalian muscle: A simulation study. *J. Biomech. Eng.* 117: 339–42.

Chou, P.C., and B. Hannaford. 1997. Study of human forearm posture maintenance with a physiologically based robotic arm and spinal level neural controller. *Biol. Cybern.* 76 (4): 285–98.

Cook, C.S., and M.J. McDonagh. 1996. Measurement of muscle and tendon stiffness in man. *Eur. J. Appl. Physiol.* 72 (4): 380–82.

Cornu, C., M.I. Almeida-Silveira, and F. Goubel. 1997. Influence of plyometric training on the mechanical impedance of the human ankle joint. *Eur. J. Appl. Physiol.* 76 (3): 282–88.

Cornu, C., F. Goubel, and M. Fardeau. 1998. Stiffness of knee extensors in Duchenne muscular dystrophy. *Muscle Nerve* 21 (12): 1772–74.

Crowninshield, R.D., and R.A. Brand. 1981. A physiologically based criterion of muscle force prediction in locomotion. *J. Biomech.* 14: 793-801.

Dalleau, G., A. Belli, M. Bourdin, and J.R. Lacour. 1998. The spring-mass model and the energy cost of treadmill running. *Eur. J. Appl. Physiol.* 77 (3): 257–63.

Davis, R.B., and C.J. DeLuca. 1995. Second rocker ankle joint stiffness during gait. *Gait Posture* 3: 79.

Davis, W.E., and W.E. Sinning. 1987. Muscle stiffness in Down syndrome and other mentally handicapped subjects: A research note. *J. Motor Behav.* 19: 130–44.

De Serres, S.J., and T.E. Milner. 1991. Wrist muscle activation patterns and stiffness associated with stable and unstable mechanical loads. *Exp. Brain Res.* 86: 451–58.

Dobrunz, L.E., D.G. Pelletier, and T.A. McMahon. 1990. Muscle stiffness measured under conditions simulating natural sound production. *Biophys. J.* 58 (2): 557–65.

Dolan, J.M., M.B. Friedman, and M.L. Nagurka. 1993. Dynamic and loaded impedance components in the maintenance of human arm posture. *IEEE Trans. Sys. Man. Cybern.* 23 (3): 698-709.

Domen, K., M.L. Latash, and V.M. Zatsiorsky. 1999. Reconstruction of equilibrium trajectories during whole-body movements. *Biol. Cybern.* 80 (3): 195-204.

Doorenbosch, C.A., J. Harlaar, and G.J. van Ingen Schenau. 1995. Stiffness control for lower leg muscles in directing external forces. *Neurosci. Lett.* 202 (1–2): 61–64.

Dornay, M., F.A. Mussa-Ivaldi, J. McIntyre, and E. Bizzi. 1993. Stability constraints for the distributed control of motor behavior. *Neural Netw.* 6: 1045–59.

Farley, C.T., and O. González. 1996. Leg stiffness and stride frequency in human running. *J. Biomech.* 29 (2): 181–86.

Farley, C.T., H.H. Houdijk, C. Van Strien, and M. Louie. 1998. Mechanism of leg stiffness adjustment for hopping on surfaces of different stiffnesses. *J. Appl. Physiol.* 85 (3): 1044–55.

Feldman, A.G. 1974. Control of length of a muscle. *Biophysics* 19: 771–76.

Feldman, A.G. 1986. Once more on the equilibrium-point hypothesis (λ model) for motor control. *J. Mot. Behav.* 18: 17-54.

Feldman, A.G., and M.F. Levin. 1995. Positional frames of reference in motor control: Their origin and use. *Behav. Brain Sci.* 18: 723–806.

Fenn, W.O., and P.H. Garvey. 1934. The measurement of the elasticity and viscosity of skeletal muscle in normal and pathological case: A study of so-called "muscle tonus." *J. Clin. Invest.* 13: 383–97.

Fitzpatrick, R.C., J.L. Taylor, and D.I. McCloskey. 1992. Ankle stiffness of standing humans in response to imperceptible perturbation: Reflex and task-dependent components. *J. Physiol. (Lond.)* 454: 533–47.

Flash, T. 1987. The control of hand equilibrium trajectories in multi-joint arm movements. *Biol. Cybern.* 57: 257–74.

Flash, T. 1989. Generation of reaching movements: Plausibility and implications of the equilibrium trajectory hypothesis. *Brain Behav. Evol.* 33 (2–3): 63–68.

Flash, T., and I. Gurevich. 1997. Models of human adaptations and impedance control in human arm movements. In *Self-organization, computational maps, and motor control,* ed. P. Morasso and V. Sanguineti, 423–81. Amsterdam: North-Holland, Elsevier Science.

Flash, T., and F.A. Mussa-Ivaldi. 1990. Human arm stiffness characteristics during the maintenance of posture. *Exp. Brain Res.* 82: 315–26.

Forcinito, M., M. Epstein, and W. Herzog. 1998. A numerical study of the stiffness of a sarcomere. *J. Electromyogr. Kinesiol.* 8 (2): 133–38.

Gielen, C.C., E.J. Vrijenhoek, T. Flash, and S.F. Neggers. 1997. Arm position constraints during pointing and reaching in 3-D space. *J. Neurophysiol.* 78 (2): 660–73.

Gielo-Perczak, K., A.E. Patla, and D.A. Winter. 1999. Analysis of the combined effects of stiffness and damping of body system on the strategy of the control during quiet standing. In *International Society of Biomechanics, XVIIth congress, book of abstracts,* ed. W. Herzog and A. Jinha, 832. Calgary, AB: University of Calgary.

Gilbert, J.A., J. Eylers, and A.J. Banes. 1985. A new instrument to assess animal joint stiffness in vitro. *J. Biomed. Mater. Res.* 19 (5): 601–5.

Giszter S., F. Mussa-Ivaldi, and E. Bizzi. 1993. Convergent force fields organized in the frog's spinal cord. *J. Neurosci.* 13: 467–91.

Goddard, R., D. Dowson, M.D. Longfield, and V. Wright. 1970. Stiffness of the knee in normal and osteoarthrosic subjects. *Ann. Rheum. Dis.* 29 (2): 194–200.

Goddard, R., D. Dowson, and V. Wright. 1970. A study in vivo of the stiffness of the human knee. *Ann. Rheum. Dis.* 29 (6): 689–94.

Gomi, H., and M. Kawato. 1996. Equilibrium-point control hypothesis examined by measured arm stiffness during multijoint movement. *Science* 272 (5258): 117–20.

Gomi, H., and M. Kawato. 1997. Human arm stiffness and equilibrium-point trajectory during multi-joint movement. *Biol. Cybern.* 76 (3): 163–71.

Gomi, H., and R. Osu. 1998. Task-dependent viscoelasticity of human multijoint arm and its spatial characteristics for interaction with environments. *J. Neurosci.* 18 (21): 8965–78.

Gottlieb, G.L. 1996. Muscle compliance: Implications for the control of movement. *Exerc. Sport Sci. Rev.* 24: 1–34.

Gottlieb, G.L. 1998. Rejecting the equilibrium-point hypothesis. *Motor Control* 2 (1): 10–12.

Gottlieb, G.L., and G.C. Agarwal. 1978. Dependence of human ankle compliance on joint angle. *J. Biomech.* 11 (4): 177–81.

Gottlieb, G.L., and G.C. Agarwal. 1988. Compliance of single joints: Elastic and plastic characteristics. *J. Neurophysiol.* 59 (3): 937–51.

Goubel, F., S. Boisset, and F. Lestienne. 1971. Determination of muscular compliance in the course of movement. In *Biomechanics II,* ed. J. Vredenbregt and J. Wartenweiler, 154–58. Basel: Karger.

Greene, P.R., and T.A. McMahon. 1979. Reflex stiffness of man's anti-gravity muscles during kneebends while carrying extra weights. *J. Biomech.* 12 (12): 881–91.

Grey, M., and T. Milner. 1999. Viscoelastic properties of the human wrist during the stabilization phase of a targeted movement. *International Society of Biomechanics, XVIIth congress, book of abstracts,* ed. W. Herzog and A. Jinha, 589. Calgary, AB: University of Calgary.

Grillner, S. 1972. The role of muscle stiffness in meeting the changing postural and locomotor requirements for force development by the ankle extensors. *Acta Physiol. Scand.* 86: 92–108.

Hagbarth, K.E., M. Nordin, and L.G. Bongiovanni. 1995. After-effects on stiffness and stretch reflexes on human finger flexor muscles attributed to muscle thixotropy. *J. Physiol.* 482: 215–23.

Hasan, Z. 1986. Optimized movement trajectories and joint stiffness in unperturbed, inertially loaded movements. *Biol. Cybern.* 53: 373–82.

Hasan, Z. 1992. Is stiffness the mainspring of posture and movement? *Behav. Brain Sci.* 15: 756–58.

Hayes, K.C., and H. Hatze. 1977. Passive visco-elastic properties of structures spanning the human elbow joint. *Eur. J. Appl. Physiol.* 37: 265–74.

Hazes, J.M., R. Hayton, and A.J. Silman. 1993. A reevaluation of the symptom of morning stiffness. *J. Rheumatol.* 20 (7): 1138–42.

Helliwell, P.S. 1997. Use of an objective measure of articular stiffness to record changes in finger joints after intra-articular injection of corticosteroid. *Ann. Rheum. Dis.* 56 (1): 71–73.

Helliwell, P.S., J.E. Smeathers, and V. Wright. 1994. The contribution of different tissues to stiffness of the wrist joint. *Proc. Inst. Mech. Eng. Pt. H. J. Eng. Med.* 208: 223–28.

Hicklin, J.A., R.J. Wighton, and F.J. Robinson. 1968. Measurement of finger stiffness. *Ann. Phys. Med.* 9 (6): 234–42.

Hills, B.A., and K. Thomas. 1998. Joint stiffness and "articular gelling": Inhibition of the fusion of articular surfaces by surfactant. *Br. J. Rheumatol.* 37 (5): 532–38.

Hoffer, J.A., and S. Andreassen. 1981. Regulation of soleus muscle stiffness in premammillary cats: Intrinsic and reflex components. *J. Neurophysiol.* 45: 267–85.

Hogan, N. 1985a. Impedance control: An approach to manipulation. *ASME J. Dyn. Syst. Meas. Control* 107: 1–24.

Hogan, N. 1985b. The mechanics of multi-joint posture and movement control. *Biol. Cybern.* 52: 315–31.

Hogan, N. 1990. Mechanical impedance of single- and multi-articular systems. In *Multiple muscle systems,* ed. J.M. Winters and S.L.Y. Woo, 149–64. New York: Springer-Verlag.

Hogan, N., E. Bizzi, F.A. Mussa-Ivaldi, and T. Flash. 1987. Controlling multi-joint motor behavior. *Exerc. Sports Sci. Rev.* 15: 153–90.

Hogan, N., and F. Mussa-Ivaldi. 1992. Muscle behavior may solve motor coordination problem. In *The head-neck sensory motor system,* ed. A. Berthoz, W. Graf, and P.P. Vidal, 153–57. New York: Oxford University Press.

Houk, J.C. 1979. Regulation of stiffness by skeletomotor reflexes. *Annu. Rev. Physiol.* 41: 99–114.

Hunter, I.W., and R.E. Kearney. 1982. Dynamics of human ankle stiffness: Variation with mean ankle torque. *J. Biomech.* 15: 747–52.

Johns, R.J., and V. Wright. 1962. Relative importance of various tissues in joint stiffness. *J. Appl. Physiol.* 17: 824–28.

Joyce, G.C., P.M.H. Rack, and H.F. Ross. 1974. The forces generated in the human elbow joint in response to imposed sinusoidal movements of the forearm. *J. Physiol.* 240: 351–74.

Katayama, M., and M. Kawato. 1993. Virtual trajectory and stiffness ellipse during multijoint arm movement predicted by neural inverse models. *Biol. Cybern.* 69 (5–6): 353–62.

Kearney, R.E., and I.W. Hunter. 1982. Dynamics of human ankle stiffness: Variation with displacement amplitude. *J. Biomech.* 15: 753–56.

Kearney, R.E., and I.W. Hunter. 1990. System identification of human joint dynamics. *Crit. Rev. Biomed. Eng.* 18 (1): 55–87.

Kearney, R.E., R.B. Stein, and L. Parameswaran. 1997. Identification of intrinsic and reflex contributions to human ankle stiffness dynamics. *IEEE Trans. Biomed. Eng.* 44: 493–504.

King, D.L., and V.M. Zatsiorsky. 1997. Extracting gravity line displacement from stabilographic recordings. *Gait Posture* 6 (1): 27–38.

Kodama, T., K. Narasaki, Y. Ogino, M. Takatori, and Y. Oka. 1966. Dynamics of rheumatoid joint. *Acta Med. Okayama* 20 (2): 53–89.

Konczak, J., K. Brommann, and K.T. Kalveram. 1999. Identification of time-varying stiffness, damping, and equilibrium position in human forearm movements. *Motor Control* 3 (4): 394–413.

Kornecki, S. 1992. Mechanism of muscular stabilization process in joints. *J. Biomech.* 25 (3): 235–45.

Kornecki, S., A. Janura, and A. Piotrowska. 1998. Stabilizing functions of muscles and their electromyographic shares. In *The problem of muscle synergism,* ed. S. Kornecki, 23–33. Wroclaw, Poland: Wydawnictwo AWF.

Kornecki, S., W. Kulig, and J. Zawadzki. 1985. Biomechanical implications of the artificial and natural strengthening of wrist joint structures. In *Biomechanics IX-A,* ed. D. Winter, R.W. Norman, R.P. Wells, K.C. Hayes, and A.E. Patla, 118–22. Champaign, IL: Human Kinetics.

Kornccki, S., and V. Zschorlich. 1994. The nature of the stabilizing functions of skeletal muscles. *J. Biomech.* 27 (2): 215–25.

Lacquaniti, F., N.A. Borghese, and M. Carrozzo. 1992. Coordinate transformations in the control of limb stiffness. *Exp. Brain Res.* 22: 17–25.

Lacquaniti, F., M. Carrozzo, and N.A. Borghese. 1993. Time-varying mechanical behavior of multijoined arm in man. *J. Neurophysiol.* 69 (5): 1443–64.

Lafortune, M.A., E.M. Hennig, and M.J. Lake. 1996. Dominant role of interface over knee angle for cushioning impact loading and regulating initial leg stiffness. *J. Biomech.* 29 (12): 1523–29.

Lakie, M., and L.G. Robson. 1988. Thixotropic changes in human muscle stiffness and the effects of fatigue. *Q. J. Exp. Physiol.* 73: 487–500.

Lakie, M., E.G. Walsh, and G.W. Wright. 1981. Measurement of inertia of the hand and the stiffness of the forearm using resonant frequency methods with added inertia or position feedback. *J. Physiol. (Lond.)* 310: 3P–4P.

Lan, N. 1997. Analysis of an optimal control model of multi-joint arm movements. *Biol. Cybern.* 76 (2): 107–17.

Lan, N., and P.E. Crago. 1994. Optimal control of antagonistic muscle stiffness during voluntary movements. *Biol. Cybern.* 71 (2): 123–35.

Latash, M.L. 1993. *Control of human movement.* Champaign, IL: Human Kinetics.

Latash, M.L., A.S. Aruin, and V.M. Zatsiorsky. 1999. The basis of a simple synergy: Reconstruction of joint equilibrium trajectories during unrestrained arm movements. *Hum. Mov. Sci.* 18: 3-30.

Latash, M.L., and G.L. Gottlieb. 1991. Reconstruction of joint compliant characteristics during fast and slow movements. *Neuroscience* 43: 697–712.

Latash, M., and V.M. Zatsiorsky. 1993. Joint stiffness: Myth or reality? *Hum. Mov. Sci.* 12: 653–92.

Lauk, M., C.C. Chow, L.A. Lipsitz, S.L. Mitchell, and J.J. Collins. 1999. Assessing muscle stiffness from quiet stance in Parkinson's disease. *Muscle Nerve* 22 (5): 635–39.

Leger, A.B., and T.E. Milner. 2000a. The effect of eccentric exercise on intrinsic and reflex stiffness in the human hand. *Clin. Biomech.* 15 (8): 574–82.

Leger, A.B., and T.E. Milner. 2000b. Passive and active wrist joint stiffness following eccentric exercise. *Eur. J. Appl. Physiol.* 82 (5–6): 472–79.

Levinson, S.F., M. Shinagawa, and T. Sato. 1995. Sonoelastic determination of human skeletal muscle elasticity. *J. Biomech.* 28 (10): 1145–54.

MacKay, W.A., D.J. Crammond, H.C. Kwan, and J.T. Murphy. 1986. Measurements of human forearm viscoelasticity. *J. Biomech.* 19: 231–38.

Malamud, J.G., R.E. Godt, and T.R. Nichols. 1996. Relationship between short-range stiffness and yielding in type-identified, chemically skinned muscle fibers from the cat triceps surae muscles. *J. Neurophysiol.* 76 (4): 2280–89.

McIntyre, J., F.A. Mussa-Ivaldi, and E. Bizzi. 1996. The control of stable postures in the multijoint arm. *Exp. Brain Res.* 110: 248–64.

McMahon, T.A. 1990. Spring-like properties of muscles and reflexes in running. In *Multiple muscle systems,* ed. J.M. Winters and S.L.-Y. Woo, 578–90. New York: Springer-Verlag.

McMahon, T.A., and P.R. Greene. 1979. The influence of track compliance on running. *J. Biomech.* 12 (12): 893–904.

McQuade, K.J., I. Shelley, and J. Cvitkovic. 1999. Patterns of stiffness during clinical examination of the glenohumeral joint. *Clin. Biomech.* 14 (9): 620–27.

Milner, T.E., C. Cloutier, A.B. Leger, and D.W. Franklin. 1995. Inability to activate muscles maximally during cocontraction and the effect on joint stiffness. *Exp. Brain Res.* 107 (2): 293–305.

Milner, T.E., and D.W. Franklin. 1998. Characterization of multijoint finger stiffness: Dependence on finger posture and force direction. *IEEE Trans. Biomed. Eng.* 45 (11): 1363–75.

Morasso, P.G., and M. Schieppati. 1999. Can muscle stiffness alone stabilize upright standing? *J. Neurophysiol.* 82 (3): 1622–26.

Morgan, D.L. 1977. Separation of active and passive components of short-range stiffness of muscle. *Am. J. Physiol.* 232: C45–C49.

Morgan, D.L., U. Proske, and D. Warren. 1978. Measurement of muscle stiffness and the mechanism of elastic storage of energy in hopping kangaroos. *J. Physiol. (Lond.)* 282: 253–61.

Mussa-Ivaldi, F.A. 1986. Compliance. In *Human movement understanding,* ed. P. Morasso and V. Tagliasco, 159–212. Amsterdam: North-Holland.

Mussa-Ivaldi, F.A., and E. Bizzi. 1997. Learning Newtonian mechanics. In *Self-organization, computational maps, and motor control,* ed. P. Morasso and V. Sanguineti, 191–238. Amsterdam: North-Holland, Elsevier Science.

Mussa-Ivaldi, F.A., N. Hogan, and E. Bizzi. 1985. Neural, mechanical, and geometric factors subserving arm posture in humans. *J. Neurosci.* 10: 2732–43.

Myers, D.B., K. Wilson, and D.G. Palmer. 1981. An objective measurement of change in morning stiffness. *Rheumatol. Int.* 1 (3): 135–37.

Nichols, T.R., and J.C. Houk. 1973. Reflex compensation for variations in the mechanical properties of a muscle. *Science* 181 (95): 182–84.

Nichols, T.R., and J.C. Houk. 1976. Improvement in linearity and regulation of stiffness from actions of stretch reflex. *J. Neurophysiol.* 39: 119–42.

Nichols, T.R., and J.D. Steeves. 1986. Resetting of resultant stiffness in ankle flexor and extensor muscles in the decerebrate cat. *Exp. Brain Res.* 62 (2): 401–10.

Nielsen, J., T. Sinkjaer, E. Toft, and Y. Kagamihara. 1994. Segmental reflexes and ankle joint stiffness during co-contraction of antagonistic ankle muscles in man. *Exp. Brain Res.* 102 (2): 350–58.

Nigg, B.M., and W. Liu. 1999. The effect of muscle stiffness and damping on simulated impact force peaks during running. *J. Biomech.* 32 (8): 849–56.

Oshima, T., and M. Kumamoto. 1995. Robot arm constructed with bi-articular muscles (stiffness properties of bi-articular muscles and its effect). *Trans. Jpn. Soc. Mech. Eng.* 61 (592): 4696–703.

Osu, R., and H. Gomi. 1999. Multijoint muscle regulation mechanisms examined by measured human arm stiffness and EMG signals. *J. Neurophysiol.* 81 (4): 1458–68.

Perreault, E.J., R.F. Kirsch, and A.M. Acosta. 1999. Multiple-input, multiple-output system identification for characterization of limb stiffness dynamics. *Biol. Cybern.* 80 (5): 327–37.

Piziali, R.L., J.C. Rastegar, and D.A. Nagel. 1977. Measurement of the nonlinear, coupled stiffness characteristics of the human knee. *J. Biomech.* 10: 45–51.

Prilutsky, B.I. 2000a. Coordination of two- and one-joint muscles: Functional consequences and implications for motor control. *Motor Control* 4 (1): 1–44.

Prilutsky, B.I. 2000b. Muscle coordination: The discussion continues. *Motor Control* 4 (1): 97–116.

Proske, U., and D.L. Morgan. 1987. Tendon stiffness methods of measurement and significance for the control of movement: A review. *J. Biomech.* 20: 75–82.

Rack, P.M.H., and D.R. Westbury. 1974. The short-range stiffness of active mammalian muscle and its effects on mechanical properties. *J. Physiol.* 240: 331–50.

Rice, M.H. 1967. A simple method of measuring joint stiffness in the hand. *J. Physiol. (Lond.)* 188 (2): 1P–2P.

Roberson, L., and D.J. Giurintano. 1995. Objective measures of joint stiffness. *J. Hand Ther.* 8 (2): 163–66.

Roeleveld, K., R. Baratta, M. Solomonow, and P. Huijing. 1994. Role of tendon stiffness in the dynamic performance of three load-moving muscles. *Ann. Biomed. Eng.* 22: 682–91.

Rostedt, M., L. Ekström, H. Broman, and T. Hansson. 1998. Axial stiffness of human lumbar motion segments, force dependence. *J. Biomech.* 31 (6): 503–9.

Savelberg, H.H.C.M. 1992. Stiffness of the ligaments of the human wrist joint. *J. Biomech.* 25 (4): 369–76.

Scheiner, A. 1994. Effect of joint stiffness on simulation of the complete gait cycle. *Ann. Int. Conf. IEEE Eng. Med. Biol. Soc. Proc.* 16 (Pt. 1): 386–87.

Scholten, P.J.M., A.G. Veldhuizen, and J.C. Sterk. 1987. The bending stiffness of the human trunk in vivo. In *Biomechanics X-A,* ed. B. Johnson, 213–17. Champaign, IL: Human Kinetics.

Shadmehr, R. 1993. Control of equilibrium position and stiffness through postural modules. *J. Motor Behav.* 25 (3): 228–41.

Shadmehr, R., and M.A. Arbib. 1992. A mathematical analysis of the force-stiffness characteristics of muscles in control of a single joint system. *Biol. Cybern.* 66 (6): 463–77.

Shadmehr, R., and F.A. Mussa-Ivaldi. 1994. Adaptive representation of dynamics during learning of a motor task. *J. Neurosci.* 14 (5 Pt. 2): 3208–24.

Shadmehr, R., F.A. Mussa-Ivaldi, and E. Bizzi. 1993. Postural force fields of the human arm and their role in generating multi-joint movements. *J. Neurosci.* 13 (1): 45–62.

Siegler, S., G.D. Moskowitz, and W. Freedman. 1984. Passive and active components of the internal moment developed about the ankle joint during human ambulation. *J. Biomech.* 17: 647–52.

Sinkjaer, T. 1997. Muscle, reflex and central components in the control of the ankle joint in healthy and spastic man. *Acta Neurol. Scand.* 170 (Suppl.): 1–28.

Sinkjaer, T., and R. Hayashi. 1989. Regulation of wrist stiffness by the stretch reflex. *J. Biomech.* 22 (11–12): 1133–40.

Sinkjaer, T., and I. Magnussen. 1994. Passive, intrinsic and reflex-mediated stiffness in the ankle extensors of hemiparetic patients. *Brain* 117 (Pt. 2): 355–63.

Sinkjaer, T., E. Toft, S. Andreassen, and B.C. Hornemann. 1988. Muscle stiffness in human ankle dorsiflexors: Intrinsic and reflex components. *J. Neurophysiol.* 60 (3): 1110–21.

Sinkjaer, T., E. Toft, K. Larsen, S. Andreassen, and H.J. Hansen. 1993. Non-reflex and reflex mediated ankle joint stiffness in multiple sclerosis patients with spasticity. *Muscle Nerve* 16 (1): 69–76.

Stefanyshyn, D.J., and B.M. Nigg. 1998. Dynamic angular stiffness of the ankle joint during running and sprinting. *J. Appl. Biomech.* 14 (3): 292–99.

Steinberg, A.D. 1978. On morning stiffness. *J. Rheumatol.* 5 (1): 3–6.

Svantesson, U., U. Carlsson, H. Takahashi, R. Thomee, and G. Grimby. 1998. Comparison of muscle and tendon stiffness, jumping ability, muscle strength and fatigue in the plantar flexors. *Scand. J. Med. Sci. Sports* 8 (5 Pt. 1): 252–56.

Thys, H., T. Farragiana, and R. Margaria. 1972. Utilization of human elasticity in exercise. *J. Appl. Physiol.* 32 (4): 491–94.

Tiselius, P. 1969. Studies on joint temperature, joint stiffness and muscle weakness in rheumatoid arthritis: An experimental and clinical investigation. *Acta Rheumatol. Scand.* 14 (Suppl. 14): 1.

Tsuji, T. 1997. Human arm impedance in multi-joint movement. In *Self-organization, computational maps, and motor control,* ed. P. Morasso and V. Sanguineti, 357–82. Amsterdam: North-Holland, Elsevier Science.

Tsuji, T., P.G. Morasso, K. Goto, and K. Ito. 1995. Human arm impedance characteristics during maintained posture. *Biol. Cybern.* 72: 475–85.

Van Doren, C.L. 1998. Grasp stiffness as a function of grasp force and finger span. *Motor Control* 2 (4): 352–78.

van Roermund, P.M., A.A. van Valburg, E. Duivemann, J. van Melkebeek, F.P. Lafeber, J.W. Bijlsma, and A.J. Verbout. 1998. Function of stiff joints may be restored by Ilizarov joint distraction. *Clin. Orthop.* 348: 220–27.

Viale, F., G. Dalleau, P. Freychat, J.R. Lacour, and A. Belli. 1998. Leg stiffness and foot orientations during running. *Foot Ankle Int.* 19 (11): 761–65.

Vincken, M.H., C.C.A.M. Gielen, and J.J. Denier van der Gon. 1984. Intrinsic and afferent components in apparent muscle stiffness in man. *Neuroscience* 9: 529–34.

Walsh, E.G., and G.W. Wright. 1987. Inertia, resonant frequency, stiffness and kinetic energy of the human forearm. *Q. J. Exp. Physiol.* 72: 161–70.

Weiss, P.L., I.W. Hunter, and R.E. Kearney. 1988. Human ankle joint stiffness over the full range of muscle activation levels. *J. Biomech.* 21: 539–40.

Weiss, P.L., R.E. Kearney, and I.W. Hunter. 1986a. Position dependence of ankle joint dynamics: I. Passive mechanics. *J. Biomech.* 19: 727–35.

Weiss, P.L., R.E. Kearney, and I.W. Hunter. 1986b. Position dependence of ankle joint dynamics: II. Active mechanics. *J. Biomech.* 19: 737–51.

Winter, D.A., A.E. Patla, F. Prince, M. Ishac, and K. Gielo-Perczak. 1998. Stiffness control of balance in quiet standing. *J. Neurophysiol.* 80 (3): 1211–21.

Winters, J.M., L. Stark, and A.H. Seif-Naraghi. 1988. An analysis of the sources of muscle-joint impedance. *J. Biomech.* 21: 1011–25.

Won, J., and N. Hogan. 1995. Stability properties of human reaching movements. *Exp. Brain Res.* 107: 125–36.

Wright, V. 1973. Stiffness: A review of its measurement and physiological importance. *Physiotherapy* 59 (4): 107–11.

Wright, V., D. Dowson, and M.D. Longfield. 1969. Joint stiffness: Its characterisation and significance. *Biomed. Eng.* 4 (1): 8–14.

Wright, V., D. Dowson, and A. Unsworth. 1971. The lubrication and stiffness of joints. *Mod. Trends Rheumatol.* 2: 30–45.

Wright, V., and R.J. Johns. 1960. Physical factors concerned with the stiffness of normal and diseased joints. *Bull. Johns Hopkins Hosp.* 106: 215–31.

Wright, V., and R.J. Johns. 1961. Quantitative and qualitative analysis of joint stiffness in normal subjects and in patients with connective tissue diseases. *Ann. Rheum. Dis.* 20: 36–46.

Yung, P. 1986. Measurement of stiffness in the metacarpophalangeal joint: The effects of physiotherapy. *Clin. Phys. Physiol. Meas.* 7 (2): 147–56.

Zatsiorsky, V.M. 1995. *Science and practice of strength training.* Champaign, IL: Human Kinetics.

Zatsiorsky, V.M., and M. Duarte. 1999. Instant equilibrium point and its migration in standing tasks: Rambling and trembling components of the stabilogram. *Motor Control* 3 (1): 28–38.

Zatsiorsky, V.M., and M. Duarte. 2000. Rambling and trembling in quiet standing. *Motor Control* 4 (2): 185–200.

Zatsiorsky, V.M., and D.L. King. 1998. An algorithm for determining gravity line location from posturographic recordings. *J. Biomech.* 31 (2): 161–64.

4
CHAPTER

INERTIAL PROPERTIES OF THE HUMAN BODY

La metrologia non é scienza, é un incubo.
(Metrology is not a science; it's a nightmare.)

Gaetano de Sanctis (1870–1957)

Nothing is as easy as it looks.

A corollary to Murphy's law

We have now completed our study of statics. We have not yet discussed the relationship between forces and motion. To relate forces to motion, we need to know the inertial properties of body segments—such as mass, tensors of inertia, and locations of the centers of mass (CoM)—also known as the *body segment parameters.* This chapter is devoted to this topic and can be considered an intermediate step to studying the *dynamics* of human motion, a field of biomechanics that deals with the cause–effect relationship between forces and motion (introduced in chapter **5**). The inertial properties of human body segments determine the relationship between the forces and moments acting on these segments and their acceleration.

Body segment parameters are difficult to measure. Although studies of body-segment parameters have more than a 300-year history, we are still quite far from characterization of the inertial properties of body segments in various populations that differ in sex, age, race, body type, sport, and occupation. Experimental data on body segment parameters are usually presented in large tables or as lengthy lists of empirical equations. Neither the tables nor the empirical equations are well suited for a textbook whose goal is to provide the basic knowledge necessary for understanding the scientific literature and for working in human movement science. Thus, this chapter is restricted to basic mechanics theory and examples of experimental facts and their interpretations. Experimental data are by necessity limited to a bare minimum. More detailed information on body segment parameters appears in the appendices of this book, and an extensive list of publications is provided at the end of this chapter.

The chapter includes five sections. Section **4.1** is intended to familiarize readers with basic mechanics theory. Section **4.2** concentrates on the inertial properties of the whole body, while section **4.3** is concerned with the inertial properties of individual body segments. Section **4.4** discusses subject-specific inertial characteristics, and section **4.5** deals with the control of body inertia in human movement. The chapter is complemented by appendices at the end of the book that contain quantitative data on inertial properties of the human body segments.

••• CALCULUS REFRESHER •••

Definite Integral Over a Region

Let f be a function that assigns a number $f(P)$ to each point P in a region R. The region can lie either in a plane or in three-dimensional space. If the region lies in a plane, its area is A, and if the region is three-dimensional, its volume is V. The definite integrals of f over R in two and three dimensions are defined as $\int_A f(P)dA$ and $\int_V f(P)dV$, or $\int_R f(P)dA$ and $\int_R f(P)dV$.

In Cartesian coordinates, the definite integral over a region in a plane can be computed as a double integral, $\int_R f(P)dA = \iint_R f(x,y)dxdy$, where x and y are the coordinates of point P. The definite integral over a region in three-dimensional space can be computed as the triple integral $\int_R f(P)dV = \iiint_R f(x,y,z)dxdydz$ throughout the region. The double and triple integrals, which are integrals of integrals, are also called *iterated integrals* or *repeated integrals*.

In polar coordinates, the definite integral $\int_R f(P)dA$ equals the iterated integral $\iint_R f(r,\theta)rdrd\theta$. Note the extra r in the integrand.

4.1 BASIC MECHANICS THEORY

In this section, we discuss the center of mass, moments and products of inertia, radius of gyration, and the inertia matrix of a body and its transformation.

4.1.1 Center of Mass

Although center of mass (CoM) was already mentioned in chapter **1**, we now reintroduce this concept more formally.

Mass is the fundamental property of matter to resist a change in velocity. Mass is also manifest in gravitational attraction between bodies. The SI unit of mass is the kilogram (kg). *Density,* ρ, is the amount of mass, m, per unit of volume, V: $\rho = m/V$. If the masses contained in any two equal volumes of a given body are equal, the body is said to be a *homogeneous body.* Otherwise, it is a nonhomogeneous, or *heterogeneous, body.* Because bones, muscles, internal organs, and fat tissue have different densities, the human body is nonhomogeneous. Due to the air-filled lungs, the average density of the upper part of the trunk is much lower than the density of the other parts of the body.

The density at a point P is equal to the derivative of mass with respect to volume at this point:

$$\rho = \frac{dm}{dV} \tag{4.1}$$

where dV is a differential volume. If the density at every point in a body is known, the total mass of the body m_b can be found as the integral over the region:

$$m_b = \int_V \rho dV \tag{4.2}$$

The product of a mass m_i concentrated at a point P_i by the distance x_i from a given plane yz is $M_i^{(1)} = m_i x_i$. This product is termed the *moment of the first order* of a material particle with respect to the yz plane. Note that x_i and the corresponding first-order moment can be either positive or negative numbers. A continuous body is an aggregate of material particles, and hence the first moment of the entire body can be found by integration. For a continuous mass, we can omit the subscript i and replace mass (m) with density (ρ). The integral

$$M^{(1)} = \int_V mxdV = \int_V \rho xdV \tag{4.3}$$

is known as the *mass moment of the first order* of the body with respect to the yz plane. Mass moments of the body with respect to other planes, points, or lines can be computed in a similar manner.

There always exists a point with respect to which the mass moment of first order in any plane equals zero. When the mass moment of the body is computed

■ ■ ■ **FROM THE LITERATURE** ■ ■ ■

Center of Mass of Helmets

Source: Aruin, A., and V. Zatsiorsky. 1988. *Biomechanics in ergonomics.* Moscow: Mashinostroenie.

The vertical line through the CoM of a helmet should be close to the vertical line through the CoM of the head or slightly behind it. During upright standing or sitting, the CoM of the head is located in front of the atlantoaxial articulation. Therefore, the vertical position of the head is maintained because of the moment of force produced by the cervical spine extensors. If the CoM of the helmet is anterior to the CoM of the head, an additional moment of force due to gravity arises (figure 4.1). Wearing a helmet with an incorrect CoM location induces fatigue of the neck muscles and may lead to headache.

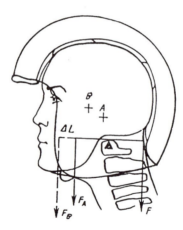

Figure 4.1 Effect of a helmet on the moment of gravity force about the atlantoaxial articulation. *A* is the CoM of the head, *B* is the CoM of the head + helmet system, and ΔL is the change in the moment arm.

Adapted from A. Aruin and V. Zatsiorsky, 1988, *Biomechanics in Ergonomics* (Moscow: Mashinostroenie).

with respect to this point, the integral of the positive *mx* products in the integrand of equation 4.3 equals the integral of the negative products, and they cancel each other. This point is called the center of mass (CoM) of the body. The Cartesian coordinates of the CoM are given by these equations:

$$x_c = \frac{1}{m_b} \int_V \rho x \, dV, \ y_c = \frac{1}{m_b} \int_V \rho y \, dV, \ z_c = \frac{1}{m_b} \int_V \rho z \, dV, \tag{4.4}$$

or

$$\mathbf{r}_c = \frac{1}{m_b} \int_V \mathbf{r} \rho \, dV \tag{4.5}$$

where $\mathbf{r}_c$ is the position vector of the CoM. If the body is homogeneous, the CoM coincides with its center of volume, the *centroid*.

When the masses (m_i) and the location of the CoM of the individual body parts along a given coordinate (d_i) are known, the location of the total CoM can be easily found from $d_{CoM} = \sum m_i d_i / \sum m_i$, which is essentially equation 4.4.

The attractive force of gravity exerted on a body is proportional to its mass. Also, the gravity forces acting on the parts of a body are parallel forces. Therefore, the unique resultant force can be computed, and Varignon's theorem is valid: the moment of the resultant force about any point is equal to the sum of the moments of the components (see **1.1.6**). The point of application of the resultant force is called the center of gravity. The vertical line through the center of gravity is called the *gravity line*. For the human body, the center of gravity coincides with the CoM.

4.1.2 Moment of Inertia and Radius of Gyration

The moment of inertia characterizes the resistance of a body to a change in the magnitude of angular velocity. Consider a material particle of mass m mounted on a massless rod that can rotate freely about a center O (figure 4.2). Force F is exerted on the particle perpendicularly to the rod. According to Newton's second law, $F = ma$, where a is the linear acceleration of the particle. Let us multiply the left and the right parts of this equation by r and replace linear acceleration a by the product of the angular acceleration $\ddot{\theta}$ and the radius r. We obtain $Fr = (mr^2)\ddot{\theta}$. Replacing Fr with M and replacing mr^2 with I yields

$$M = I\ddot{\theta} \tag{4.6}$$

where M is the moment of force F about O and $I = mr^2 = m(x^2 + y^2)$ is the *moment of inertia* of the particle–rod system about O. The moment of inertia

of a material particle with respect to a given axis depends on the distance from that axis and is independent of the direction from the axis. Mass moments of inertia are always positive numbers. The moment of inertia is also known as the *mass moment of the second order*, or simply the *second moment*. The SI unit for mass moment of inertia is kg · m². The equations $F = ma$ and $M = I\ddot{\theta}$

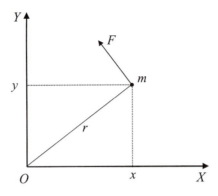

Figure 4.2   Material particle *m* connected by a rod *r* to the center *O*. See explanation in the text.

are similar. The moment of inertia plays the same role in angular dynamics that mass plays in linear dynamics.

The moment of inertia of a body with respect to a given axis *O-O* equals the integral of the second moment about the axis:

$$I = \int_V r^2 \rho dV, \text{ or simply } I = \int r^2 dm \tag{4.7}$$

where *r* is the perpendicular distance from the axis *O-O* to the arbitrary element *dV* and *dm* is the differential mass. The moments of inertia of a body with respect to Cartesian coordinate axes are

$$I_{XX} = \int (y^2 + z^2) dm$$
$$I_{YY} = \int (x^2 + z^2) dm$$
$$I_{ZZ} = \int (x^2 + y^2) dm \tag{4.8}$$

where I_{XX}, I_{YY}, and I_{ZZ} are the moments of inertia of the body with respect to the axes *X*, *Y*, and *Z*, respectively. Because the distances in equations 4.7 and 4.8 are measured from the coordinate axes, the moments of inertia I_{XX}, I_{YY}, and I_{ZZ} are termed the *axial moments of inertia*. The moment of inertia with regard to the origin is called the *polar moment of inertia* and is computed as

$$I_O = \int (x^2 + y^2 + z^2) dm = \int r_O^2 dm \tag{4.9}$$

where r_O is the shortest distance from the origin *O* (the pole) to the arbitrary element *dm*. Note that I_O is independent of the direction of the reference axes;

hence, it is invariant with respect to the rotation of the reference system. In a planar case, the polar moment coincides with the axial moment. In three dimensions, the polar moment has no physical meaning (because the body does not rotate about a pole; it rotates about an axis) and is used as an auxiliary quantity in the calculation of the axial moments. In particular, from equations 4.8 and 4.9, the following property of the moments of inertia can be obtained: the sum of the axial moments of inertia A, B, and C about any three rectangular axes meeting at a given point O equals twice the polar moment of inertia with respect to this point and is constant:

$$A + B + C = 2I_o \tag{4.10}$$

To prove this statement, take the given point as an origin and combine equations 4.8 and 4.9. The sum of any two moments of inertia is always larger than the third moment of inertia. For instance, the expression

$$I_{xx} + I_{yy} - I_{zz} = 2\int z^2 dm \tag{4.11}$$

is always positive. Equation 4.11 is used for checking whether the moments of inertia are computed correctly.

If the density of a homogeneous body is 1 $(\rho = 1)$, the moment of inertia defines a pure geometric property of the body. Instead of the integral $\int (x^2 + y^2)dm$, the integral $\int (x^2 + y^2)dA$, where dA is an element of area, can be computed. Although this integral has nothing in common with inertia because no mass is considered, when computed for a lamina or a cross section of a body, it is called the *area moment of inertia*. The area moments can be computed for each cross section of the body. They are expressed in meters to the fourth power, m^4. The area moments of inertia define the mechanical strength of the bodies, in particular, their resistance to bending and torsion.

The moment of inertia of a body depends on its mass and on the distribution of mass within the body with respect to the axis of rotation. Two bodies of the same mass may possess different moments of inertia. It is useful to have a measure that depends only on the mass distribution, not on the amount of mass. Let I denote the moment of inertia of the body about a specified axis and m the mass of the body. The ratio I/m has the dimension of the square of the length. The length

$$k = \sqrt{\frac{I}{m}} \tag{4.12}$$

is called the *radius of gyration* of the body with respect to the given axis. The radius of gyration specifies the distance from the axis at which a particle of mass m has the same moment of inertia as the original body. If the body is

■ ■ ■ **FROM THE LITERATURE** ■ ■ ■

Area Moment of Inertia and the Risk of Injury

Source: Asang, E. 1978. Injury thresholds of the leg: Ten years of research on safety in skiing. In *Skiing safety II*, ed. J.M. Figueras, 103–29. Baltimore: University Park Press.

To prevent ski trauma, the release level of the ski bindings should be below the injury threshold. In experiments, injury thresholds of the human tibia were established by different loading experiments. The diagram of ventral bending—an imitation of a forward fall in skiing—is presented in figure 4.3*a*. The fracture limits depend on the area and the shape of the cross sections at the fraction sites of the tibia. To characterize the cross sections, the area moments of inertia were computed (figure 4.3*b*). A large correlation was observed between the area moments of inertia and the threshold value of bending moment of force (figure 4.3*c*). The larger the area moment of inertia, the larger load the tibia can sustain without an injury.

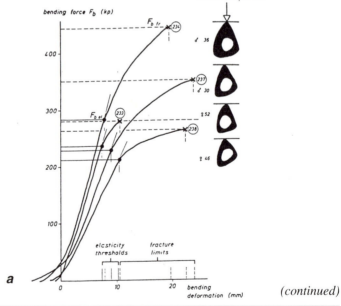

(continued)

Figure 4.3   Area moments of inertia and resistance to injury of the tibia. (*a*) Forward bending diagram and respective cross sections at the fracture sites. a.p. = anterioposterior; kp = kilopond (a nonsystemic unit of force equal to 9.8 N); $F_{b\,el}$ = bending force at the elasticity threshold; $F_{b\,fr}$ = bending force at the fracture limit.

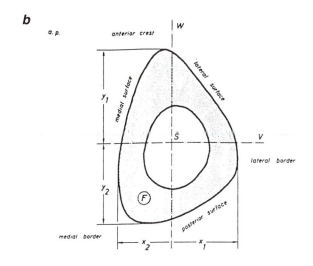

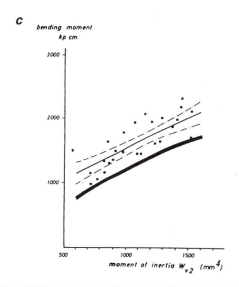

Figure 4.3 *(continued)* Area moments of inertia and resistance to injury of the tibia. (*b*) Cross section of the tibia at a fracture site, the basis for computing the area moments of inertia. (*c*) The relationship between the threshold value of the bending moment and the area moment of inertia. The thick line represents a recommended threshold value for the release of ski bindings.

Adapted, 1978, Injury thresholds of the leg: Ten years of research on safety in skiing. In *Skiing Safety II*, edited by J.M. Figueras (Baltimore: University Park Press), 103–129.

replaced by a particle of equal mass located at a distance from the axis of rotation equal to the radius of gyration, the moment of inertia of the body about the axis does not change. The radius of gyration is independent of the total mass of the body and is a measure of distribution of mass about the given axis. In the literature, the moment of inertia of a body about a given axis is often reported by specifying the radius of gyration and the mass of the body.

Moments of inertia of nonhomogeneous bodies and bodies with an irregular shape are difficult to find by integration. For these bodies, equations 4.7 and 4.8 are still valid, but their application is limited. A practical solution is to measure the moments of inertia rather than to compute them. This can be accomplished, for instance, by a pendulum method.

Moments of inertia specify the body's resistance to angular acceleration (equation 4.6). If the body rotates at a constant angular velocity, its angular acceleration is zero and therefore the *tangential acceleration* of the material particles in the body is also zero. The magnitude of the linear velocity of the individual particles does not change in this case. However, the direction of the velocity changes; the particles of the body experience *normal (centripetal) acceleration*. The resistance to the centripetal acceleration is specified by the *products of inertia*.

■ ■ ■ *FROM THE LITERATURE* ■ ■ ■

Determining the Moments of Inertia by the Pendulum Method

Source: Chandler, R.F., C.E. Clauser, J.T. McConville, H.M. Reynolds, and J.W. Young. 1975. *Investigation of the inertial properties of the human body.* Technical report AMRL-TR-74-137, AD-A016-485, DOT-HS-801-430. Ohio: Aerospace Medical Research Laboratories, Wright-Patterson Air Force Base.

The study was performed on cadavers that were embalmed, frozen, and then dissected. It is known from mechanics that the period of oscillation of a rigid body about an axis *O-O* equals $T = 2\pi\sqrt{I / mgl}$, where *I* is the moment inertia of the body about *O-O*, *m* is the mass, *l* is the distance from the CoM to the axis *O-O*, and *g* is the acceleration due to gravity. The authors determined the periods of oscillation of the total body and the individual segments about six different axes. The principal moments of inertia were computed.

4.1.3 Products of Inertia

If a stone is fastened to the end of a string and whirled around horizontally fast enough, the stone will eventually rotate in the horizontal plane. During the transition, in addition to rotating in the horizontal plane, the stone rotates about the point of suspension in the vertical plane. This principle is used in speed governors employed in steam engines (figure 4.4). Whenever the speed of the engine increases above the predetermined value, the masses m_1 and m_2 ($m_1 = m_2 = m$) move up and the governor closes a valve in the steam supply. This causes the engine to slow down. The suspended masses rotate together with the axle of the engine (around the Y axis), and they also rotate around the point of suspension O in the XY plane. We are interested in determining the inertial resistance to this movement.

The centripetal acceleration of m is $a_c = v^2 / x = \omega^2 x$, where v is the linear velocity and ω is the angular velocity (see *Kinematics of Human Motion,* p. 196). The centripetal force is $F_c = mx\omega^2$. The force F_c generates a moment of force at O that is equal to $M_O = mxy\omega^2$. The vector of the moment is normal to the plane XY containing the centripetal force. The product mxy is called the product of inertia.

Consider a perfectly symmetric, flat plate rotating uniformly at an angular velocity ω about an axis X (figure 4.5). An elemental mass a with coordinates (x, y) is under the action of a centripetal force $F_c = mx\omega^2$ that produces a

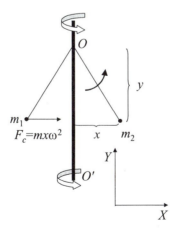

Figure 4.4   An engine speed governor (schematic). The governor mechanism consists of two masses m_1 and m_2 ($m_1 = m_2 = m$) connected at a point O to a shaft by means of two rods. The shaft is rotating, and so are the rods and the masses. The connection at O is flexible and allows the masses to rotate around O in the XY plane.

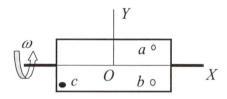

Figure 4.5 A flat plate rotating around the X axis. The plate is shown in its instantaneous position in the plane of drawing. The particles a and b are symmetric about the axis of rotation. An additional particle c disturbs the symmetry. See explanation in the text.

moment of force about an origin $M_o = mxy\omega^2$. Because the plate is symmetric, there is another elemental mass b with the coordinates $(x, -y)$ that is under an equal and opposite centripetal force. These two forces cancel each other, and their total moment is zero. However, if we add an additional mass element c to one of the corners of the plate, while rotating about X, this element will generate a moment about O that will result in nonzero bearing reaction forces. The product of inertia numerically equals the moment of force about the center O caused by unit angular velocity about the X axis. The vector of this moment is normal to the XY plane.

The product of inertia of a body in the XY plane, or with respect to the axes X and Y, is defined as the integral

$$I_{XY} = I_{YX} = \int_m xy\,dm \tag{4.13}$$

The products of inertia with respect to the YZ axes and ZX axes are defined in a similar way. Because the two coordinates that define the product of inertia can vary independently from each other, the product of inertia can be positive, negative, or zero. The product of inertia is zero if the axis of rotation is an axis of symmetry for the mass. The two products of inertia with respect to the same coordinate axes are equal; that is, $I_{XY} = I_{YX}$, $I_{YZ} = I_{ZY}$, and $I_{ZX} = I_{XZ}$. For instance, in the XY plane, the products of inertia are the same for rotation about the X axis and the Y axis.

4.1.4 Moments of Inertia With Respect to Parallel and Rotated Axes

Suppose that the moments and products of inertia of a body about three orthogonal axes are known. We want to determine the moments of inertia of the same body with respect to other axes, either parallel to the original axes or at an angle to them. The term *centroidal axis* denotes a line through the CoM.

4.1.4.1 Parallel Axis Theorem

According to the *parallel axis theorem,* the moment of inertia about any axis equals the moment with respect to the parallel centroidal axis plus the product of the mass and the square of distance between the axes:

$$I_X = I_c + ml^2 \qquad (4.14)$$

where I_X is the moment of inertia of the body about an axis X, I_c is the moment about a parallel axis through the centroid, l is the shortest distance between the two axes, and m is the mass of the body. The first term in equation 4.14 is known as the local term and the second as the remote, or transfer, term. Note that the theorem cannot be applied unless one axis passes through the CoM and the axes are parallel. From the parallel axis theorem, it follows that the moment of inertia about a centroidal axis is smaller than the moment about any parallel axis. Let us prove this theorem for a planar object when the axis of the interest X is also in the plane. In this case, only one coordinate, y, is important. The distance of a differential element dm to the axis X is $y = l + y_c$, where l is the shortest distance between X and the parallel centroidal line and y_c is the distance from the element to the centroidal line. The moment of inertia of the element is $dm(l + y_c)^2 = dml^2 + 2ly_c dm + dmy_c^2$. After integration, the first and the third terms yield ml^2 and I_c, correspondingly. The integral of the second term $2l \int y_c dm$ represents the moment of the first order about the centroidal line and hence equals zero (see **4.1.1**). This concludes the proof.

The parallel axis theorem for the products of inertia is

$$I_{XY} = (I_{XY})_c + mx_c y_c \qquad (4.15)$$

where I_{XY} is the product of inertia with respect to the axes X and Y, $(I_{XY})_c$ is the product of inertia with respect to the centroidal axes that are parallel to the axes X and Y, and x_c and y_c are the coordinates of the CoM. The remaining moments and products of inertia can be computed by permuting the x, y, and z values in equations 4.14 and 4.15.

The parallel axis theorem is routinely used in biomechanics research. In the literature, the moments of inertia of body segments with respect to centroidal axes are published. To compute the moments of inertia with respect to the joint axes, equation 4.14 is used. In this case, l is the shortest distance from the CoM of the segment to the axis of rotation at the joint, and m is the mass of the body segment. If a body consists of several parts, the total moment of inertia about a given axis can be obtained by adding together the moments of inertia of the individual parts. For instance, the moments of inertia of the upper, middle, and bottom parts of the trunk are published in the literature as individual entities. To compute the moment of inertia of the whole trunk about a given axis, the distances from the CoM of the individual parts of the trunk to the axis of

■ ■ ■ **FROM THE LITERATURE** ■ ■ ■

Determining the Moment of Inertia
of the Human Body About the Ankle Joints

Source: Smith, J.W. 1957. The forces operating at the human ankle during standing. *J. Anat.* 91: 545–64.

To determine the moments of inertia of the human body about the ankle joints, the author used the pendulum method. Direct determination of the moment of inertia of the living body about the ankle axis by this method would require suspension of the subject upside down about this axis. Use of the parallel axis theorem makes this procedure unnecessary. In the experiment, each subject was suspended from a beam by ropes passing beneath the axillae and allowed to swing through a small arc, as in figure 4.6. The period of oscillation was measured. The value of the moment of inertia about the axis of the beam I_b was then determined. The moment of inertia about the parallel axis through the CoM is $I_c = I_b - mD^2$. Therefore, the moment of inertia about the ankle axis is $I_a = I_b + m(L^2 - D^2)$, where D is the distance from the CoM to the beam and L is the distance from the CoM to the ankle axis. This method neglects the contribution of the feet to the moment of inertia about the beam.

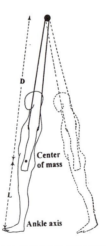

Figure 4.6   Determination of the moment of inertia of the human body by the pendulum method.

Reprinted with the permission of Cambridge University Press, J.W. Smith, 1957, "The forces operating at the human ankle during standing," *Journal of Anatomy* 91:545–564.

rotation should be determined, and then the parallel axis theorem should be applied. Note that this procedure is not valid for the radius of gyration. The radii of gyration of the individual parts cannot be added.

In what follows, unless otherwise specified, the expression "moment of inertia" means the "moment of inertia about a centroidal axis."

4.1.4.2 Moment of Inertia About a Rotated Axis

If the moments and products of inertia of a body about the three coordinate axes are known, the moment with respect to any axis through the origin can be computed. Consider the body shown in figure 4.7. We wish to compute the moment of inertia of the body about an axis OA through the origin. The orientation of the line is given by the unit vector $\mathbf{u} = u_x\mathbf{i} + u_y\mathbf{j} + u_z\mathbf{k}$, where u_x, u_y, and u_z are the direction cosines of the line OA.

The moment of inertia about OA is represented by the integral $I_{OA} = \int l^2 dm$, where l is the shortest distance from a differential element dm to OA. From figure 4.7, it follows that $l = r\sin\theta$, and hence l equals the magnitude of the cross product $\mathbf{u} \times \mathbf{r}$. Therefore, the moment of inertia about OA can be written as

$$I_{OA} = \int_m l^2\,dm = \int_m (\mathbf{u}\times\mathbf{r})^2\,dm = \int_m (\mathbf{u}\times\mathbf{r})\cdot(\mathbf{u}\times\mathbf{r})dm \qquad (4.16)$$

where $\mathbf{u} = u_x\mathbf{i} + u_y\mathbf{j} + u_z\mathbf{k}$ and $\mathbf{r} = x\mathbf{i} + y\mathbf{j} + z\mathbf{k}$. The cross product $\mathbf{u} \times \mathbf{r}$ is

$$\mathbf{u} \times \mathbf{r} = (u_y z - u_z y)\mathbf{i} + (u_z x - u_x z)\mathbf{j} + (u_x y - u_y x)\mathbf{k} \qquad (4.17)$$

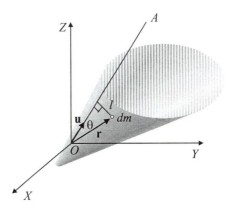

Figure 4.7 Determining a moment of inertia about an axis OA through the origin. Vector $\mathbf{r}$ is the position vector of a differential element dm; $\mathbf{u}$ is the unit vector along the line of interest, OA; and l is the shortest distance from dm to OA, $l = r\sin\theta$.

After computing the dot products, we obtain

$$I_{OA} = \int_m \left[\left(u_y z - u_z y \right)^2 + \left(u_z x - u_x z \right)^2 + \left(u_x y - u_y x \right)^2 \right] dm \qquad (4.18)$$

After expanding the squares, rearranging the terms, and recognizing the moments and products of inertia, equation 4.18 can be reduced to the following expression:

$$I_{OA} = I_x u_x^2 + I_y u_y^2 + I_z u_z^2 - 2I_{xy} u_x u_y - 2I_{yz} u_y u_z - 2I_{zx} u_z u_x \qquad (4.19)$$

where I_x, I_y, and I_z are the moments of inertia about the three coordinate axes; u_x, u_y, and u_z are the direction cosines of the axis OA; and I_{xy}, I_{yz}, and I_{zx} are the products of inertia with respect to the X and Y axes, the Y and Z axes, and the Z and X axes, correspondingly. Equation 4.19 is known as the *theorem of the six constants of a body*. It defines the moment of inertia of the body with respect to any axis through O. Because an infinite number of axes can pass through the center of mass, a rigid body has an infinite number of moments of inertia. Fortunately, their values are related in a regular manner. Equation 4.19 is a quadric that specifies the *ellipsoid of inertia* of the body at O. Ellipsoids of inertia are discussed in section **4.1.5**.

When the moments and products of inertia about any set of rectangular axes through the CoM are known, the moment of inertia of the body about any line L can be computed. This is done in two steps. In the first step, the theorem of the six constants is applied to find the moment of inertia about the line through the CoM parallel to L. In the second step, the parallel axis theorem is used.

4.1.5 Tensor of Inertia and Principal Moments of Inertia

The moments and the products of inertia of a body with respect to three orthogonal coordinate axes and for a given location of the origin O can be conveniently written as a matrix:

$$[I] = \begin{pmatrix} I_{XX} & -I_{XY} & -I_{XZ} \\ -I_{YX} & I_{YY} & -I_{YZ} \\ -I_{ZX} & -I_{ZY} & I_{ZZ} \end{pmatrix} \qquad (4.20)$$

This matrix is called the *inertia matrix*. The diagonal terms of the inertia matrix are the moments of inertia, and the off-diagonal elements are the products of inertia. Note that to comply with equation 4.19, the negative values of the products of inertia are included in the inertia matrix. Because the products of inertia in a given plane are similar ($I_{XY} = I_{YX}$, $I_{YZ} = I_{ZY}$, and $I_{ZX} = I_{XZ}$), the matrix is symmetric, and only six elements are independent. Due to the symmetry, the inertia matrix is equal to its transpose, and hence, when an inertia matrix is multiplied by a vector, the order of multiplication is immaterial. The inertia matrix can also be written as $\mathbf{I}_{ij}$ ($i, j = x, y, z$), where the sign of the element is positive when $i = j$ and negative when $i \neq j$. The values of the elements in the inertia matrix are unique for a given location of the origin O and a given orientation of the coordinate axes. The values change when the origin or orientation varies.

Heretofore, we worked mainly with two types of mathematical objects: scalars, such as mass or length, and vectors, such as force, velocity, and acceleration. Because the elements of the inertia matrix change when the axes vary, they are not scalars in the same sense as mass or length. Nor are they components of a vector. The inertia matrix belongs to a class of mathematical objects called *tensors* and is also known as the *tensor of inertia*. Tensors undergo transformation with changes of the coordinate system. In what follows, we limit the discourse to an intuitive geometric explanation of the problems under discussion. Mathematical proofs are not provided.

Let O-XYZ and O-xyz be two reference systems with a common origin at O that coincides with the CoM of the body. For brevity, we use the letter G to denote the O-XYZ system and the letter L to denote the O-xyz system. Let $[R]$ be the rotation matrix that relates the reference frame L to the frame G. If a tensor of inertia is defined about the coordinate axes x, y, and z and we want to find the value of this tensor with respect to the O-XYZ system, the following relationship is valid:

$$[I]_G = [R][I]_L[R]^T \tag{4.21}$$

where the subscripts G and L specify the reference frames and the superscript T denotes, as usual, the matrix transposition. Equation 4.21 transforms a tensor of inertia expressed in the reference frame L into the reference frame G. If the reference frame L is fixed with a body segment and the reference axes O-XYZ maintain a constant orientation in space, the elements of $[I]_L$ are constant and the elements of $[I]_G$ depend on the segment orientation. If both reference systems are fixed with the segment, equation 4.21 provides a tool to compute the moments and the products of inertia about the different axes.

There always exists a reference frame O-XYZ such that the products of inertia about its axes vanish. Such axes are called the *principal axes of inertia*. A moment of force about a principal axis produces rotation about this axis alone. Due to the nonzero products of inertia, a moment of force about any axis other than a principal axis produces simultaneous rotation about three coordinate axes. The principal axes of inertia are always orthogonal with respect to each other. The planes defined by the principal axes are known as the principal planes; the products of inertia vanish for any axis along the intersection of the principal planes. The moments of inertia about the principal axes are termed the *principal moments of inertia.* When the moments of inertia are taken with respect to the principal axes, the tensor of inertia assumes a simple diagonal form:

$$^d[I] = \begin{pmatrix} I_X & 0 & 0 \\ 0 & I_Y & 0 \\ 0 & 0 & I_Z \end{pmatrix}$$ (4.22)

where the leading superscript d indicates that the inertia matrix is in a diagonal form. With the tensor of inertia in a diagonal form, equation 4.19 that specifies the moment of inertia about any axis through the origin O reduces to

$$I_{OA} = I_X u_x^2 + I_Y u_y^2 + I_Z u_z^2$$ (4.23)

where I_X, I_Y, and I_Z are the principal moments of inertia taken about the principal axes of inertia X, Y, and Z, correspondingly, and u_x, u_y, and u_z are the direction cosines of the line OA with these axes. For instance, during baseball pitching, the upper arm performs an internal–external rotation at the shoulder joint while the elbow joint is flexed (figure 4.8). If the principal moment of inertia

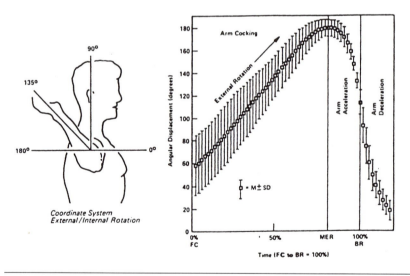

Figure 4.8 Shoulder external and internal rotation of the arm in baseball pitching. To find the moment of inertia of the forearm + hand + baseball system about the longitudinal axis of the upper arm, equation 4.23 and the parallel axis theorem should be used. Note that if the elbow-joint angle differs from 90°, the products of inertia cannot be neglected. FC is foot contact; MER is maximal external rotation; BR is ball release.

Reprinted, by permission, from C.J. Dillman, 1994, "Biomechanical contributions to the science of rehabilitation in sports," *Sport Science Review* 3(2):70–78.

of the forearm + hand + baseball system about the transverse axes is 10 arbitrary units, if the principal moment about the longitudinal axis is 1 arbitrary unit, and if the angle of elbow flexion is 45°, the moment of inertia of the forearm + hand + baseball system about an axis parallel to the upper arm is $I = I_x \cos^2 0°$ + $I_y \cos^2 45° + I_z \cos^2 45° = 5.5$ arbitrary units. To find the moment of inertia about the axis along the upper arm, the parallel axis theorem should be used.

For a homogeneous body that exhibits symmetry, the principal axes of inertia can be determined by inspection. Let a body be symmetric about the plane *XY*. Then for every element on one side of the plane whose coordinates are (x, y, z), there is another element of equal mass on the other side with coordinates $(x, y, -z)$. For such a body, the integrals $\int xz\,dm = 0$ and $\int yz\,dm = 0$. If the body has two planes of symmetry, then all the products of inertia vanish. For such a body, the coordinate axes in the planes of symmetry are the principal axes. The human body is approximately symmetric about the *midsagittal plane*. Hence, we may expect that for a diver or gymnast in a layout posture, the principal axes of inertia are close to the *cardinal axes* through the CoM. The moment of inertia of the body about the anteroposterior axis has the maximum value.

Equation 4.22 essentially says that when the axis of rotation coincides with a principal axis, the tensor of inertia can be replaced by a single scalar. In such a case, the angular momentum of the body (the product of the moment of inertia I times the angular velocity, see *Kinematics of Human Motion*, p. 16) equals $\mathbf{H} = I\boldsymbol{\omega}$, where I is a scalar value of the moment of inertia with respect to the axis of rotation and $\boldsymbol{\omega}$ is the angular velocity vector that is along a principal axis of inertia. When the moment of inertia is expressed with respect to another system of coordinates $O\text{-}XYZ$, the equation is $\mathbf{H} = [I]\boldsymbol{\omega}$. Because the angular momentum is the same in both coordinate systems, $[I]\boldsymbol{\omega} = I\boldsymbol{\omega}$. This is an equation that defines the eigenvalues and eigenvectors of the inertia matrix $[I]$; see equation 2.43 for comparison. The principal axes of inertia of a body are the eigenvectors of its tensor of inertia, and the principal moments of inertia are the eigenvalues. To find the principal moments of inertia, the characteristic equation of the inertia matrix should be solved.

$$\begin{vmatrix} I_{XX} - \lambda & -I_{XY} & -I_{XZ} \\ -I_{YX} & I_{YY} - \lambda & -I_{YZ} \\ -I_{ZX} & -I_{ZY} & I_{ZZ} - \lambda \end{vmatrix} = 0 \qquad (4.24)$$

This is a cubic equation in λ that has three solutions, λ_1, λ_2, and λ_3, representing the principal moments of inertia, I_1, I_2, and I_3, respectively. The solutions are always positive real numbers. The minimal and maximal principal moments of inertia are also the minimum and maximum among the moments computed with respect to any axis.

■ ■ ■ *From the Literature* ■ ■ ■

Perceiving the Orientation of Handheld Objects Is Based on Perceiving the Eigenvectors of the Inertia Tensor

Source: Pagano, C.C., and M.T. Turvey. 1998. Eigenvectors of the inertia tensor and perceiving the orientations of limbs and objects. *J. Appl. Biomech.* 14 (4): 331–59.

People regularly manipulate various objects, such as cups, utensils, hammers, or rackets. Subjects' ability to perceive the orientation of handheld objects that they could wield but could not see was investigated. It was found that to perceive the object's orientation, the subjects relied on determining the principal axes of inertia of the object, that is, the eigenvectors of the inertia tensor.

In one of the experiments, the subjects wielded an occluded object and adjusted the position of a visible marker to point to the felt direction of the object. The objects consisted of a stem with either one additional branch (figure 4.9*a*) or two branches forming a V-like shape (figure 4.9*b*). By loading one branch with an attached mass, it was possible to construct V-shaped objects such that similar geometric orientations were characterized by different orientations of the principal axes of inertia. Because the mass distribution was asymmetric, the principal axes of inertia were not along the stem, and the eigenvector of the inertia tensor pointed from the proximal end of the stem through the loaded branch. Subjects selected an object orientation with zero products of inertia. The perceived orientation of the object corresponded to the direction of the principal axis of inertia of the object. It was concluded that perceived orientation of handheld objects is a function of the eigenvectors of the inertia tensor rather than the object's geometry.

a

(continued)

Figure 4.9 Perceiving the orientation of handheld objects. (*a*) The apparatus used in the experiment.

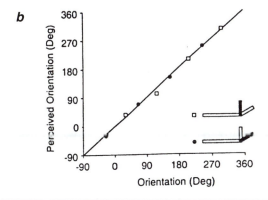

Figure 4.9 *(continued)*   *(b)* Relationship between the direction of the eigenvector of the inertia tensor and the perceived orientation of the object.

Figure 4.9*a* adapted, by permission, from C.C. Pagano and M.T. Turvey, 1998, "Inertial and perceiving eigenvectors orientations of hand-held object by dynamic touch," *Journal of Applied Biomechanics* 14(4):331–59. Figure 4.9*b* adapted, by permission, from C.C. Pagano and M.T. Turvey, "Eigenvectors of the inertia tensor and perceiving the orientation of a hand-held object by dynamic touch," *Perception & Psychophysics I* 52:617–624.

In another set of experiments, the authors investigated whether another mechanical variable, the moment of gravity force rather than the inertia tensor, is used as the main source of haptic (touch) perception. These two mechanical characteristics correspond to the mass moments of the first and the second order ($\sum mr$ and $\sum mr^2$, respectively), correspondingly. It was found that the inertia tensors of the objects seem to provide the basis for haptic abilities.

Any inertia matrix can be transformed to a diagonal form ("diagonalized") according to equation 4.21. To do this one should choose the rotation matrix

$$[R] = [R_1\ R_2\ R_3] \tag{4.25}$$

where the column matrices R_1, R_2, and R_3 are the direction cosines of the three eigenvectors of the inertia matrix. Then the transformation equation 4.24 reduces the inertia matrix to its diagonal form:

$$^{d}[I] = [R][I][R]^{T} = \begin{pmatrix} I_X & 0 & 0 \\ 0 & I_Y & 0 \\ 0 & 0 & I_Z \end{pmatrix} = \begin{pmatrix} \lambda_1 & 0 & 0 \\ 0 & \lambda_2 & 0 \\ 0 & 0 & \lambda_3 \end{pmatrix} \tag{4.26}$$

where $^{d}[I]$ is the inertia matrix in the diagonal form.

The tensor of inertia, as well as equation 4.19, allows a very useful geometric interpretation. Let an axis through the fixed center O change its orientation so that it assumes all possible angular positions. The moments of inertia about the axis are then determined for every angular orientation. Let the radius vector OQ along the axis be equal to the inverse square of the moment of inertia about the axis, $OQ = 1 / \sqrt{I}$. It is known that the locus of the points Q just obtained forms an ellipsoid called the ellipsoid of inertia. The ellipsoid defines the moment of inertia of the body about any axis through O. The shape of the ellipsoid depends on the distribution of the mass in the body and does not depend on the orientation of the rotation axes. When the orientation of the axes varies, the magnitude of the moments of inertia changes but the ellipsoid itself remains unchanged. Ellipsoids of inertia exist for all bodies, including those with irregular shapes. In some sports—for instance, in diving—athletes change the moments of inertia of the body by changing the joint configuration. The inertia of the entire body in any possible posture is described by a certain inertia ellipsoid.

4.1.6 Tensors of Inertia With Respect to Remote Axes

We so far have analyzed the tensors of inertia with respect to the axes that pass through the CoM of an object, whether it is a body segment or the entire human body. To analyze complex human movements, one needs to know the inertia of the body parts with respect to remote axes of rotation. For instance, to analyze a throwing movement, information on the inertia of the hand with respect to the shoulder may be necessary. In a planar case, the moment of inertia with regard to any axis can be easily computed by using the parallel axis theorem, equation 4.14. In three dimensions, in addition to the moments of inertia, the products of inertia should be taken into account.

Let the local reference frame L_i be along the principal axes of inertia of an individual body segment i. A rotation matrix $[R]_i$ relates the orientation of the frame L_i to a global frame G. The global frame is not obliged to be fixed with the environment; it can be anywhere, for instance, at a proximal joint. An inertia matrix $[I]_i$ is calculated with respect to the local principal axes and hence is a diagonal matrix. The coordinates of the origin of the local reference frame L_i in the global frame G are x_i, y_i, and z_i. After application of equation 4.21, we obtain a local term of the tensor of inertia of the segment i with respect to the global reference frame G, and application of the parallel axis theorem (equations 4.14 and 4.15) yields the remote term:

$$
{}^{G}[I]_i = \overbrace{[R]_i [I]_i [R]_i^{T}}^{\text{Local term}} + m_i \overbrace{\begin{bmatrix} y_i^2 + z_i^2 & -x_i y_i & -x_i z_i \\ -y_i x_i & x_i^2 + z_i^2 & -y_i z_i \\ -z_i x_i & -z_i y_i & x_i^2 + y_i^2 \end{bmatrix}}^{\text{Remote term}} \tag{4.27}
$$

The local term in equation 4.27 can be seen as an inertia tensor of the segment expressed with respect to a frame having an origin at the CoM of the segment and oriented the same as the global reference frame *G*.

When the inertia of the body segments is computed with respect to the joint axes, the contribution from the remote terms usually exceeds the contribution from the local terms.

4.2 INERTIAL PROPERTIES OF THE ENTIRE HUMAN BODY

Inertial properties of the entire body can be either computed from the known properties of individual body segments or determined experimentally.

4.2.1 Total Center of Mass

In 1679 Giovanni Borelli (1608–1679) determined the position of the body's CoM along the longitudinal axis by placing naked subjects supine on a platform balanced on a narrow ridge. This was one of the first experiments in human biomechanics. Since then, many researchers have determined the CoM position. Some experimental results are presented in table 4.1.

On average, the relative height of the CoM in women is 0.5% to 2.0% lower than in men. The location of the transverse plane of the CoM is different in experienced athletes from various sports (compare the speed skaters and rowers in the study by Zatsiorsky, Aruin, and Seluyanov 1981). The difference in the location of the transverse plane of the CoM in standing and supine postures does not exceed 1% (Page 1974). A convenient landmark for the approximate estimation of the CoM position is the anterior superior iliac spine (figure 4.10). In children of different ages, the transverse plane of the CoM lies in a different section of the body, but the height percentage is constant (Palmer 1944; figure 4.11).

When body posture changes, so does the position of the CoM with respect to anatomic reference points. For instance, during walking, the CoM is displaced by about 1% of the body height. In some body postures, the CoM may be located outside the body.

4.2.2 Moments of Inertia

The moment of inertia of the body about a centroidal axis is equal to the sum of the moments of inertia of its individual segments about this axis. Therefore, the moment of inertia of the entire body depends on body posture and hence can be controlled by the performer.

Table 4.1 Position of the Transverse Plane of the CoM in Anatomic Position

Study	Sample, n	% of body height or a regression equation	Anatomic reference
Males			
Borelli 1679	—	—	Between the buttocks and the os pubis
Palmer 1944	596 boys from birth to 20 years of age	H (from soles of feet) = 0.557 height + 1.4 cm	
Ivanitsky 1965	650 adults	54.5–58.0	Between the first and fifth sacral vertebrae
Kozyrev 1962	300 adults	57.1 ± 1.2	Between the fifth sacral and fourth lumbar vertebrae
Zatsiorsky, Aruin, and Seluyanov 1981 (gamma-scanner method)	100 adults, including 12 speed skaters 6 rowers	56.4 ± 0.9 55.8 ± 0.9 56.9 ± 0.8	
Females			
Hellebrandt et al. 1938	357 college students	Mean: 55.17 Range: 53–59	
Palmer 1944	576 girls from birth to 20 years	H (from soles of feet) = 0.557 height + 1.4 cm	
Kozyrev 1962	125 adults	55.9 ± 1.1	From the *coccyx* to the fifth lumbar vertebra
Raitsina 1976	47 cyclists	55.4 ± 1.5	Between the first and second sacral vertebrae
	90 swimmers	56.1 ± 0.9	Between the first sacral and the fifth lumbar vertebrae
	80 tennis players	55.7 ± 1.6	The first sacral vertebra

4.2.2.1 Moments of Inertia in a Standing Posture

The moments of inertia of the human body in an upright posture with respect to an axis through the CoM of the body are presented in table 4.2. To find the

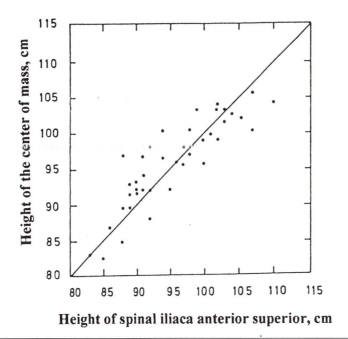

Height of spinal iliaca anterior superior, cm

Figure 4.10   Correlation between the vertical location of the CoM and the height of the anterior superior iliac spine. The 45° line is not a regression line. The individual data points deviate from the line up to 6 cm in the *cranial* direction and as much as 5 cm in the downward direction. The average location of the CoM was at 57% of the body height (±5 cm). The regression equation was $h_{CoM} = 0.63H - 0.1$, where H is the body height in meters. The standard error of the equation was ±2.5 cm. The data are from 45 subjects: 29 men and 16 women.

Reprinted, by permission, from G. Beier and R. Doschl, 1974, "Height of the center of gravity and momentum of inertia in the pedestrian," *Beitrage Gerichtl Med.* 32:60–65.

moment of inertia with respect to an axis at the level of support, the parallel axis theorem should be used.

The maximal moment of inertia is about the anteroposterior axis, I_{xx}, and the minimal one is about the longitudinal axis, I_{zz}. The moment about the mediolateral axis, I_{yy}, has an intermediate value. The ratio I_{yy}/I_{xx} is around 93% for all subject groups (Matsuo et al. 1995). It is known from classical mechanics that free rotation about a central axis with an intermediate moment of inertia is naturally unstable. Even a small perturbation, such as a deviation from a perfectly symmetric posture, induces a body twist. This facilitates forward and backward somersaults with twists, routinely used in such sports as diving and gymnastics. Performing twists is impossible when an athlete rotates in the air sideways around the anteroposterior axis (as in the so-called Arabian somersault).

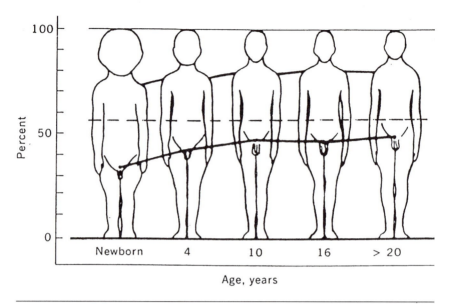

Figure 4.11 Location of the transverse plane of the center of mass in children of different ages. Body heights are scaled to the same size. The figure is based on data by Palmer (1944).

Table 4.2 Moments of Inertia of the Human Body When Standing Erect With Arms at the Sides (mean ± standard deviation, kg · m²)

Study	Sample, n	I_{xx} Anteroposterior axis	I_{yy} Frontal axis	I_{zz} Longitudinal axis
Santschi et al. 1963	66 males in U.S. Air Force	13.0 ± 2.1	11.6 + 2.0	1.2 ± 0.2
Whitsett 1963	U.S. Air Force mean man, 175.54 cm, 74.138 kg	12.84	12.27	0.91
Ignazi et al. 1972	11 males	12.3 ± 2.0	11.5 ± 1.9	1.1 ± 0.2
McConville et al. 1980	31 males, 177.4 ± 9.6 cm, 77.1 ± 13.0 kg	13.5 ± 3.4	12.6 ± 3.1	1.4 ± 0.4
Matsuo and Fukunaga 1995	Adult females: 36 sedentary, 158.6 ± 5.3 cm, 53.6 ± 6.8 kg;	7.97 ± 1.44	7.46 ± 1.39	—

Study	Sample, n	I_{xx} Anteroposterior axis	I_{yy} Frontal axis	I_{zz} Longitudinal axis
	47 athletes, 165.7 ± 7.6 cm, 61.7 ± 6.3 kg	9.87 ± 1.83	9.18 ± 1.61	—
Matsuo et al. 1995	117 Japanese boys, 13–18 years	5.6 – 14.0	4.2 – 13.5	—

In an upright posture, the remote terms contribute 86.5% and the local terms only 13.5% to the total moment of inertia about the mediolateral axis (Whitsett 1963). This allows the local moments of inertia for the hands and feet to be neglected because their contribution to the total moment of the body is less than the usual experimental error. In the sitting posture, the moments of inertia about the anteroposterior and mediolateral axes are very close, 6.73 and 6.55 kg · m², correspondingly (Chandler et al. 1975; the data are averages from three cadavers).

The moments of inertia strongly depend on body size (figure 4.12). Because the body mass is approximately proportional to a linear dimension cubed, the moments of inertia should be approximately proportional to the height of the body to the fifth power. Hence, even small changes in body size should result in large changes in the moments of inertia. A few empirical regression equations for estimating the moments of inertia of the entire body are given in table 4.3.

4.2.2.2 Moments of Inertia in Arbitrary Postures

In three dimensions, the tensor of inertia of the entire body can be computed by summing the tensors of inertia of the individual body segments with respect to an axis of rotation (see equation 4.27). During airborne movements, the axis of rotation is always through the CoM of the body. The expression for the tensor of inertia of the whole body is

$$[I]_{\text{body}} = \sum_i {}^G[I]_i = \overbrace{\sum_i [R]_i [I]_i [R]_i^T}^{\text{Local terms}} +$$

$$\underbrace{\begin{bmatrix} \sum_i m_i (y_i^2 + z_i^2) & -\sum_i m_i x_i y_i & -\sum_i m_i x_i z_i \\ -\sum_i m_i y_i x_i & \sum_i m_i (x_i^2 + z_i^2) & -\sum_i m_i y_i z_i \\ -\sum_i m_i z_i x_i & -\sum_i m_i z_i y_i & \sum_i m_i (x_i^2 + y_i^2) \end{bmatrix}}_{\text{Remote terms}} \quad (4.28)$$

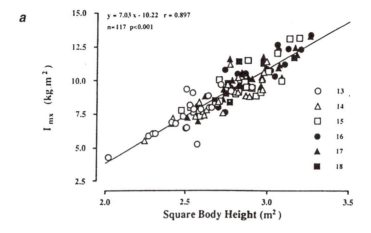

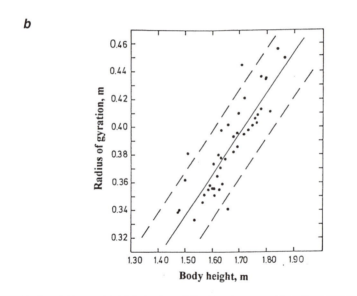

Figure 4.12 (a) Relationship between body height squared and moment of inertia about the central anteroposterior axis in 13- to 18-year-old adolescent boys. Symbols for each age group are presented. (b) Relationship between body height and radius of gyration (R) about the central frontal axis in 29 male and 16 female adults. The regression equation is $R = 0.297H - 0.11$ (SD = ±0.016 m), where H is the body height in meters.

Figure 4.12a reprinted from *Journal of Biomechanics*, 28 (2), A. Matsuo et al., Moment of inertia of whole body using an oscillating table in adolescent boys, 219–223, 1995, with permission of Elsevier Science.

Figure 4.12b reprinted, by permission, from G. Beier and R. Doschl, 1974, "Height of the center of gravity and momentum of inertia in the pedestrian," *Beitrage Gerichtl Med.* 32:60–65.

Table 4.3   Regression Equations for Estimating the Moment of Inertia of the Human Body

Study	Subjects	Axis of rotation	Equation (W is body mass, kg; H is body height, m; R is multiple coefficient of correlation)
Bolonkin 1973 (cited in Zatsiorsky, Aruin, and Seluyanov 1981)	Adult men	Frontal	$I = 0.043WH^2$
Matsuo et al. 1995	Adolescent boys	Anteroposterior	$I_{xx} = 3.44H^2 + 0.144W - 8.04$ ($R = 0.973$)
		Frontal	$I_{yy} = 3.52H^2 + 0.125W - 7.778$ ($R = 0.972$)
Matsuo and Fukunaga 1995	Female athletes	Anteroposterior	$I_{xx} = 4.04H^2 + 0.142W - 10.0$ ($R = 0.981$)
		Frontal	$I_{yy} = 4.01H^2 + 0.13W - 8.8$ ($R = 0.979$)

Computation of the tensor of inertia for asymmetric body postures, such as those seen in certain phases of pole vaulting or during a takeoff in high jumping, is a cumbersome task. Approximate estimates of the moments of inertia for some typical gymnastic activities are shown in figure 4.13.

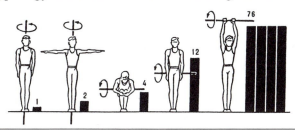

Figure 4.13   Comparative values of the moments of inertia of the human body in different postures and about different axes of rotation. Moment of inertia about the longitudinal axis of the body is used as a unit. In the tucked position, for example, the moment of inertia about the central frontal axis is about one third of the moment in the upright posture (a layout position). Note the huge moment of inertia when the rotation is performed about an axis through the hands with the arms up, such as during giant swings on a high bar.

Reprinted from D. Donskoi and V. Zatsiorsky, 1979, *Biomechanics* (Moscow: FiS Publishers).

4.3 INERTIAL PROPERTIES OF BODY PARTS

To study the inertial properties of body parts, the parts should be somehow separated from each other; or in other words, the body should be segmented. We first describe segmentation procedures, then provide some representative data, discuss the factors that influence their values, and finally look at the example of the inertial properties of the trunk.

4.3.1 Segmentation

The goal of segmentation is to approximately represent the body as a kinematic multilink chain with links of known mass-inertial characteristics. An immediate task is to divide the body in such a way that the individual parts move similarly to rigid bodies rotating around the joint axes. This task can be solved only with limited accuracy. The human body is continuous, and any separation of it into individual segments is subject to judgment. The body parts are composite; they include rigid and soft-tissue components and liquids. The muscles and other soft tissues cross from one body part to the next, so only discretionary boundary lines can be drawn. For instance, it is not clear how to divide the gluteus muscle into the portions belonging to the trunk and the thigh. Unfortunately, various authors used different segmentation protocols in their

---

▪ ▪ ▪ *From the Literature* ▪ ▪ ▪

Total Body Moments of Inertia in Young Gymnasts

Source: Jensen, R.K. 1986b. The growth of children's moment of inertia. *Med. Sci. Sports Exerc.* 18: 440–45.

Data were collected in a longitudinal study on 12 young gymnasts. The subjects were originally 4, 6, 9, and 12 years of age, with three subjects at each age, each of a different body type (ectomorphic, mesomorphic, or endomorphic). Body-segment parameters were estimated from stereophotographs. Moments of inertia for the entire body were computed for two reference configurations typical for the back handspring and back somersault maneuvers. The best predictor of total moment of inertia for both body positions was the product of body mass and height squared $(M \cdot H^2)$; see figure 4.14. During the fast-growth period associated with puberty, the rate of change of moment of inertia was approximately 30% a year. Hence, to generate the same angular acceleration as before, young gymnasts have to increase the moment of force by 30% each year.

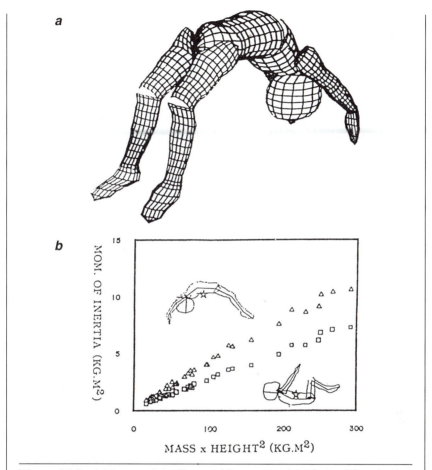

Figure 4.14   (*a*) Image of the 5-year-old mesomorph in the back handspring position. (*b*) The relationship between the product of mass and height squared ($M \cdot H^2$) and the total body moment of inertia for the back handspring (triangles) and the back somersault (rectangles). The corresponding coefficients of correlation were .995 and .977.

Reprinted, by permission, from R.K. Jensen, 1986, "The growth of children's moment of inertia," *Medicine and Science in Sport and Exercise* 18 (4):440–445. Copyright © 1986 Lippincott Williams & Wilkens.

research (figure 4.15). Hence, comparison of the data from different studies has to be done with due caution.

There is a certain advantage in separating the body parts along the joint axes of rotation. There are, however, two problems associated with this approach: (1) in some joints (e.g., hip, glenohumeral) the joint axes are not at the end of

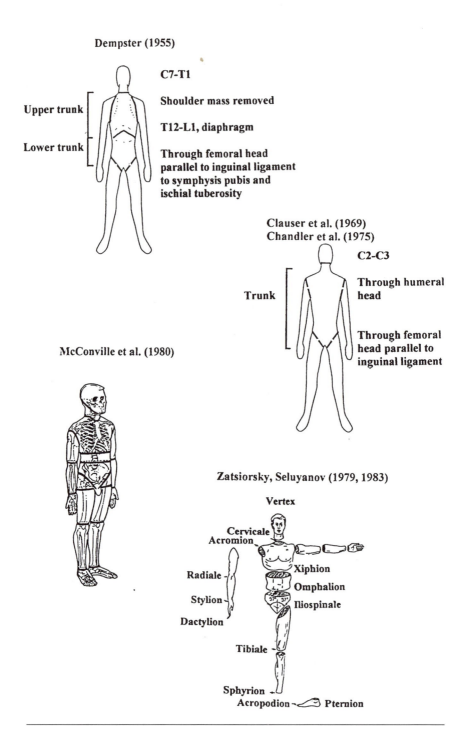

Figure 4.15 Segmentation procedures used in various studies.

the segment, and (2) the joint axes are difficult to locate. For instance, determination of the axis of rotation in the hip joint is a topic of serious research; several rather complex methods have been suggested. In other joints, for instance, in the knee joints, the axes are not fixed. When so-called joint nominal axes are used, the axes are determined only approximately and their position is a matter of convention (the joint axes are discussed in detail in chapters **4** and **5** in *Kinematics of Human Motion*). Therefore, segmenting the body from joint axis to joint axis is difficult to do. Even in cadavers, dismembering the limb segments by disarticulating the joints does not provide the desired result. It is much more precise to divide the body into individual parts according to anatomic landmarks rather than by the joint axes and then to adjust the known inertial properties from one reference system to another.

When a body is segmented according to the bony landmarks, the individual body parts are called *body segments*. When a (hypothetical) separation is performed with respect to the joint axes and the parts are considered rigid, the parts are called the *body links*. By definition, a body link is the central straight line that extends longitudinally through a body segment and terminates at both ends in axes about which the adjacent members rotate (Dempster 1955). The basic model of human body introduced in chapter **2** is based on the link representation. A body link is an ideal representation of a body segment. Although the segment mass is distributed along the body link, in some applications the mass is regarded as concentrated at the CoM of the link.

The dimensions, especially length, of the body segments and the corresponding body links may be different. Consider as an example the upper arm. In anthropology, the length of the upper-arm segment is usually measured as a projected distance from *acromion* (the lateral point on the lateral margin of the acromial process of the scapula) to *radiale* (the superior point on the proximal head of the radius) on a subject standing erect with arms at the sides. However, the glenohumeral joint center, and the corresponding joint axis, is located several centimeters below the acromion. Therefore, the length of the upper arm link is shorter than the length of the upper arm segment. Note that the part of the arm that is above the center of the glenohumeral joint and below the acromion moves together with the upper arm and should be included in its mass. The length of a body segment, when measured as the distance between predetermined anatomic landmarks, is known as the *anatomic length* of the segment. The distance between the joint axes of a body member is called the *biomechanical length* (figure 4.16). Dempster (1955) estimated the biomechanical length of the upper arm to be 89% of the anatomic length (in his study, the anatomic length was measured as the length of humerus, smaller than the distance from acromion to radiale). Hence, when the length of the upper arm is measured from acromion to radiale in the standard *anatomic position,* the biomechanical length is much shorter than the anatomic length. However, when

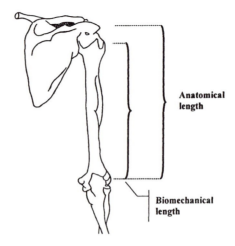

Figure 4.16 Anatomic and biomechanical length of the upper arm.

measurement is performed with the arm abducted 90°, the measured length provides a good approximation of the biomechanical length of the upper arm; in this position, the acromion is much closer to the radiale.

The separation of the thigh from the trunk poses a special problem. The lateral projection of the hip joint center is approximately at the superior aspect of the greater trochanter (for a more detailed discussion, see *Kinematics of Human Motion,* section **5.5**). When a person flexes the hip, a large bulk of flesh that is located above the greater trochanter moves together with the femur. If the thigh is dissected from the trunk along the transverse plane through the greater trochanter, the mass above the trochanter is mistakenly included in the trunk mass, and the lowest part of the trunk is mistakenly included in the thigh mass. Ideally, the thigh should be separated from the trunk along the plane through the femoral head. Such a plane goes through the *anterior iliospinale* (the inferior point of the anterior superior iliac spine) at an angle approximately 37° to the midsagittal plane. The thigh length in this case is measured from anterior iliospinale to the *tibiale laterale.*

When body-segment parameters are determined with respect to bony landmarks, in biomechanics research they should be adjusted to the joint axes of rotation (see table A2.11 and figure A2.3 in appendix **2**).

4.3.2 Principal Mechanical Axes Versus Principal Anatomic Axes

In anatomy, the expression *principal axes* (or cardinal axes) refers to the axes at the intersection of the sagittal, transverse, and frontal planes of the body. In mechanics, the principal axes of inertia are the axes with respect to which the products of inertia are zero. It is quite possible that the principal axes of inertia of human body segments do not coincide with the principal anatomic axes. In practical applications, this possibility is usually disregarded. The reason is that the trunk and the head are approximately symmetric about the midsagittal plane, and the upper arm, forearm, thigh, and shank are approximately symmetric in two planes, sagittal and frontal. Therefore, the moments of inertia of these body segments about the frontal and anteroposterior axes are close, and precise orientation of the two principal axes in the transverse plane is not important. The third principal axis is expected to be very close to the longitudinal axis of the segment. Also, due to the symmetry of the trunk and head about the midsagittal plane, it is customarily assumed that the principal axes of inertia of these body members are close to the cardinal anatomic axes through the corresponding centers of mass. These assumptions may or may not be correct, but at this time, there is insufficient experimental evidence to either accept or reject

■ ■ ■ *From the Literature* ■ ■ ■

Location of the Principal Axes of Inertia About the Cardinal Anatomic Axes

Source: McConville, J.T., T.D. Churchill, I. Kaleps, C.E. Clauser, and J. Cuzzi. 1980. *Anthropometric relationships of body and body segment moments of inertia.* Report no. AFAMRL-TR-80-119. Ohio: Aerospace Medical Research Laboratories, Wright-Patterson Air Force Base.

Investigators combined stereophotogrammetric and anthropometric techniques to measure 31 male subjects. Bodies were photographically segmented into 24 parts and their volume, centers of volume, and principal axes of inertia were computed. Density data for the individual body parts were taken from studies performed on cadavers. The body segments were assumed to be homogeneous. Because the gas contents of cadaver torsos vary, the data on the density of the trunk and its parts were not considered reliable. The authors found that for the majority of body parts, the principal axes of inertia deviate substantially ($>10°$) from the cardinal anatomic axes (figure 4.17). However the difference between the two largest moments of inertia was usually small and thus can be disregarded. For instance, for the left upper arm, the principal axes of inertia were inclined up to $42.05°$ with respect to the anatomic axes. However, the two largest principal moments of inertia were equal to 123.9 ± 42.7 and 129.9 ± 46.2 kg $\cdot$ cm^2. The difference is statistically insignificant and is comparable to the error of the estimates.

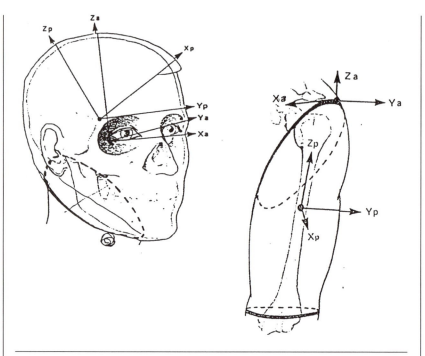

Figure 4.17 The principal axes of inertia (X_p, Y_p, and Z_p) and the anatomic axes (X_a, Y_a, and Z_a) for the head and the left upper arm. Orientation of the principal axes of inertia with respect to anatomic axes is described by the rotation matrix expressed in degrees:

	X_a	Y_a	Z_a		X_a	Y_a	Z_a
X_p	36.13	86.95	54.04	X_p	41.50	48.82	94.25
Y_p	93.75	3.75	90.02	Y_p	131.49	42.05	95.62
Z_p	125.88	92.19	36.96	Z_p	90.54	82.97	7.05

Head | Left upper arm

For the head (the left matrix), the first row indicates that the mean principal axis of inertia X_p lies 36.13° from the anatomic X_a axis, 86.95° from the anatomic Y_a axis, and 54.04° from the anatomic Z_a axis. The principal moments of inertia for the head were $I_x = 204.1 \pm 25.4$ kg $\cdot$ cm^2, $I_y = 232 \pm 31.8$ kg $\cdot$ cm^2, and $I_z = 150.8 \pm 15.6$ kg $\cdot$ cm^2. Note that the head was dissected from the neck along an inclined plane and that with such a dissection the mass distribution of the head about the longitudinal axis of the body is clearly not symmetric.

Adapted from McConville, J.T., T.D. Churchill, I. Kaleps, C.E. Clauser, and J. Cuzzi. 1980. *Anthropometric relationships of body and body segment moments of inertia.* Report no. AFAMRL-TR-80-119. Ohio: Aerospace Medical Research Laboratories, Wright-Patterson Air Force Base.

them. In an overwhelming majority of biomechanics research, the moments of inertia are taken with respect to the cardinal anatomic axes.

4.3.3 Some Representative Data

Comprehensive data on inertial properties of human body segments in populations that vary by sex, age, race, health, physical activity, and body build are presently not available. The available data are printed in the literature in large tables that cannot be reproduced here due to space limitations. Therefore, in table 4.4, only emblematic data on body-segment parameters are presented. Figures 4.18 and 4.19 illustrate the location of the CoM on the long axis of the

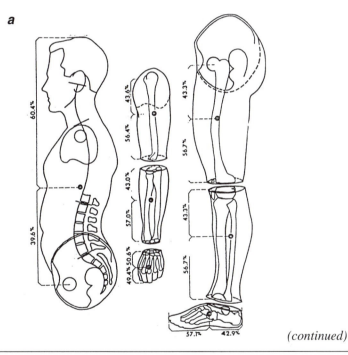

(continued)

Figure 4.18 Position of the CoM on the longitudinal axis of various body segments. Note that different anatomic landmarks for some segments were used in the studies by (*a*) Dempster (1955) and (*b*) Zatsiorsky, Aruin, and Seluyanov (1981).

Figure 4.18*a* adapted from Dempster, W.T. 1955. *Space requirements of the seated operator.* Technical report WADC-TR-55-159. Ohio: Wright-Patterson Air Force Base.

Figure 4.18*b* adapted from Zatsiorsky, V.M., A.S. Aruin, and W.N. Seluyanov. 1981. *Biomechanics of musculoskeletal system* (in Russian). Moscow: FiS. Also published in German translation as *Biomechanik des menschlichen Bewegungsapparates.* Berlin: Sportverlag, 1984.

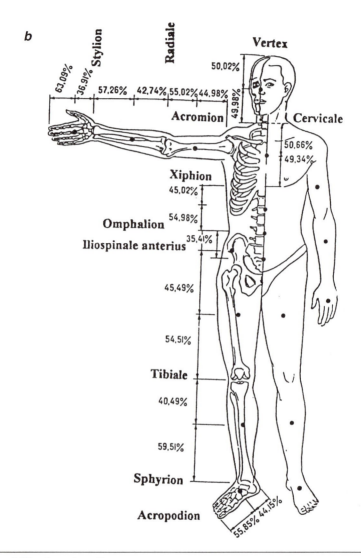

Figure 4.18 *(continued)*

segments and in the frontal plane, respectively. Some additional data on body-segment parameters are presented in the appendices at the end of the book.

4.3.4 Factors Influencing Inertial Properties of Human Body Segments

The inertial properties of a human body segment depend on the local density (the mass distribution with respect to the axis of rotation) and the size of the

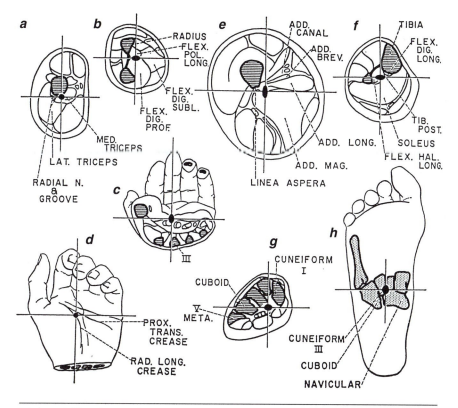

Figure 4.19   Anatomic location of CoM in the cross sections of various segments. The black spots indicate the average location of the CoM; the relative height and width of the black spot reflects the mean deviation of the position of the CoM. The bones in the section are shaded, and the important muscles and other landmarks are labeled. (*a*) Arm. (*b*) Forearm. (*c* and *d*) Two sections through the hand. (*e*) Thigh. (*f*) Leg. (*g* and *h*) Foot.

Reprinted from Dempster, W.T. 1955. *Space requirements of the seated operator.* Technical report WADC-TR-55-159. Ohio: Wright-Patterson Air Force Base.

segment. This section addresses body density and the relationship between total body mass and the mass of individual body segments.

4.3.4.1 Body Density

The human body is nonhomogeneous; it is a composite of biological tissues that have different densities. These are the approximate densities of various body tissues: cortical (compact) bone, 1.8 g/cm³; cancellous bone, 1.1 g/cm³; muscle, 1.06 to 1.08 g/cm³; visceral organs and body fluids such as blood, 1.06 g/cm³; fat, 0.94 to 0.96 g/cm³. The density of the lungs varies from 0.15 g/cm³ during inhalation to 0.25 g/cm³ during exhalation.

People have different percentages of fat and muscles in the body; therefore, the total body density of different people varies. The relationship between the body density and the body composition is used for estimating the fat percentage in the body. A "gold standard" method for measuring the percentage of body fat is *underwater weighing,* also called the *hydrodensitometric method.* Being submerged in water, the body experiences the *force of buoyancy* (the Archimedes force), which is directed upward and equals the weight of the displaced fluid. The density of water is close to 1 g/cm^3 but depends on the water temperature. The difference between the body weight in the air, W_a, and in the water, W_w, divided by the density of water at a standardized temperature

Table 4.4 The Inertial Characteristics of Human Body Segments of 100 Male Subjects

Segment	m, kg	CoM, %	m, %	I_{ap}, kg·cm^2	I_{ml}, kg·cm^2	I_{lg}, kg·cm^2	R_{ap}, %	R_{ml}, %	R_{lg}, %
Foot[a]	0.997 (0.141)	55.85 (3.65)	1.370 (0.155)	44.0 (10.1)	40.0 (9.0)	10.3 (3.2)	25.70 (0.95)	24.50 (0.80)	12.40 (1.21)
Shank	3.160 (0.439)	40.47 (2.81)	4.330 (0.305)	385.0 (89.4)	371.0 (90.6)	64.6 (25.0)	28.10 (0.63)	27.50 (0.61)	11.40 (1.96)
Thigh	10.360 (1.568)	45.49 (1.94)	14.165 (0.998)	1,997.8 (449.5)	1,999.4 (453.0)	413.4 (106.9)	26.70 (0.99)	26.70 (0.99)	12.10 (0.93)
Hand	0.447 (0.072)	63.09 (4.85)	0.614 (0.083)	13.2 (4.2)	8.76 (2.9)	5.37 (2.0)	28.50 (2.16)	23.30 (1.71)	18.20 (2.30)
Forearm	1.177 (0.161)	57.26 (3.26)	1.625 (0.140)	64.7 (14.3)	60.2 (13.50)	12.6 (3.8)	29.50 (0.86)	28.40 (0.65)	13.00 (1.51)
Upper arm	1.980 (0.319)	55.02 (4.19)	2.707 (0.243)	127.3 (35.0)	114.4 (33.5)	38.95 (13.9)	32.80 (1.61)	31.00 (1.25)	18.20 (3.27)
Head and neck	5.018 (0.393)	50.02 (2.23)	6.940 (0.707)	272.1 (45.8)	293.9 (44.8)	202.4 (38.9)	30.30 (1.29)	31.50 (1.34)	26.10 (2.10)
Upper part of the trunk (thorax)	11.654 (1.873)	50.66 (2.24)	15.955 (1.529)	1,725.6 (429.0)	705.2 (224.2)	1,454.5 (359.0)	50.50 (3.74)	31.99 (1.71)	36.50 (4.64)
Middle part of the trunk (abdomen)	11.953 (2.176)	45.02 (2.12)	16.327 (1.725)	1,280.8 (394.2)	819.1 (275.6)	1,203.1 (377.3)	48.20 (4.24)	38.30 (2.43)	46.80 (5.60)
Lower part of the trunk (pelvis)	8.164 (1.492)	35.41 (3.01)	11.174 (1.428)	656.8 (211.4)	525.0 (168.4)	592.4 (181.8)	35.60 (2.05)	31.90 (2.37)	34.00 (3.00)

Segment	m, kg	CoM, %	m, %	I_{ap}, kg $\cdot$ cm^2	I_{ml}, kg $\cdot$ cm^2	I_{lg}, kg $\cdot$ cm^2	R_{ap}, %	R_{ml}, %	R_{lg}, %
Entire trunk	31.771	43.70	43.456	—	—	—	26.70	24.80	14.40
	(3.329)	(7.37)	(4.560)				(5.11)	(2.83)	(6.85)
Middle and lower parts of trunk combined	20.117	39.20	27.501	—	—	—	30.50	28.20	20.50
	(2.292)	(3.14)	(3.140)				(4.62)	(2.42)	(6.34)

Notes. Standard deviations are given in parentheses. Symbols: m (kg) is the mass of the segment; CoM (%) is the location of the CoM of the segment along its longitudinal axis as a percentage of the segment length; m (%) is the relative mass of the segment, the ratio of the mass of the segment to the total body mass; I_{ap}, I_{ml}, and I_{lg} are the moments of inertia about the anteroposterior, mediolateral, and longitudinal axes of the segments, correspondingly; R_{ap}, R_{ml}, and R_{lg} are the radii of gyration as percentages of the segment length about the anteroposterior, mediolateral, and longitudinal axes of the segment, respectively. The details of the measurement protocol are described in appendix **2**.

[a] For the foot, the axes are given for an imaginary position with the foot rotated outward, that is, an axis that is parallel to the flexion axis at the ankle joint is considered an antero-posterior axis. The axis from the heel to the tip of the second toe is the longitudinal axis.

From Zatsiorsky and Seluyanov (1979, 1983); Zatsiorsky, Seluyanov, and Chugunova (1990a)

estimates the total body volume. During underwater weighing, the body should be completely under the surface of the water, and the subject should not move. Corrections should be made for air in the lungs and respiratory passages (which has to be measured) and in the gastrointestinal tract (around 150 cm^3). The body density, D_b, is then computed from the formula

$$D_b = \frac{W_a}{\left(W_a - W_w\right)/D_w - V_l}$$ (4.29)

where D_w is the density of the water at a given temperature and V_l is the volume of the air in the lungs, respiratory airways, and the gastrointestinal tract.

The total density of the body measured with the hydrodensitometric method in physically active young men is about 1.06 g/cm^3 on average. It varies with the fat percentage of the body and other factors such as bone density from roughly 1.03 to 1.11 (Brozek and Keys 1952). In overweight patients, body density can be as low as 1.01.

Body density has an immediate impact on the floating ability of the body. If gross body density (without correction for the air in the lungs and in the gastrointestinal tract) exceeds the water density, the body sinks. Otherwise, the body floats. Body densities in college women after full inspiration and maximal expiration were found to be 0.9812 and 1.018, respectively (Rork and

Hellebrandt 1937). Hence, after full inspiration, an average female swimmer can stay afloat motionless. The center of volume of the human body, also known as the *center of buoyancy,* and the CoM do not coincide; the center of buoyancy is 1.5 to 2.0 cm more *caudal* than the CoM. Its location depends on the lung volume.

Underwater weighing does not provide information about the density of individual body segments. Density is not uniform throughout a segment. For instance, the wrist zone is denser than the more proximal parts of the forearm. The center of volume of an individual body segment does not coincide, as a rule, with its CoM. However, the difference is small (with the exception of the trunk) and customarily neglected. In the trunk, the density of the thorax is much smaller than the density of the other parts (figure 4.20). Trunk density at the T6 level was found to be 0.74 g/cm^3, while at the level of L5, it was 1.04 g/

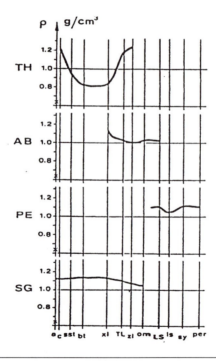

Figure 4.20 Density profiles for parts of the trunk: TH = thorax, AB = abdomen, PE = pelvis, SG = shoulder girdle. The following anthropometric landmarks are indicated along the horizontal axis: a is acromion, c = *cervicale* (C7), sst = *suprasternale,* bt = bifurcatio tracheae, xi = xiphoidale, TL = T12/L1, zl = zonale radiale, om = *omphalion* (umbilicus), LS = L5/S1, is = iliospinale, sy = syphision, per = perianale.

Reprinted from *Journal of Biomechanics*, 30 (7), W.S. Erdmann, Geometric and inertia of the trunk in adult males, 679–688, 1997, with permission from Elsevier Science.

cm³ (Pearsall, Reid, and Livingston 1996). The density profiles were similar among subjects for the lower vertebral levels but varied at the thoracic region; at the T10 level, the density values ranged from 0.65 to 1.15 g/cm³ among subjects. The density of body segments decreases with aging as bone and muscle mass are lost and fat is gained. Several estimates of the density of the individual body parts are presented in table 4.5. Note that the estimates were ob-

Table 4.5 Mean Density of Body Segments in Male Cadavers

	Dempster 1955, $n = 8$	Clauser, McConville, and Young 1969, $n = 13$	Chandler et al. 1975, $n = 6$; r = right, l = left
Total body		1.042 ± 0.018	
Head		1.071	1.056 ± 0.020
Head and neck	1.11 ± 0.120		
Neck and trunk		1.023 ± 0.032	0.853 ± 0.039
Trunk minus limbs (same as for trunk minus shoulders)	1.03 ± 0.065		
Shoulders	1.04 ± 0.006		
Thorax	0.92 ± 0.056		
Abdomen and pelvis	1.01 ± 0.014		
Thigh	1.05 ± 0.008	1.045 ± 0.017	1.018 ± 0.012 r 1.021 ± 0.012 l
Shank	1.09 ± 0.015	1.085 ± 0.014	1.059 ± 0.007 r 1.071 ± 0.021 l
Foot	1.10 ± 0.056	1.085 ± 0.014	1.073 ± 0.024 r 1.069 ± 0.024 l
Upper arm	1.07 ± 0.027	1.058 ± 0.025	0.997 ± 0.012 r 1.012 ± 0.015 l
Forearm	1.13 ± 0.037	1.099 ± 0.018	1.043 ± 0.018 r 1.061 ± 0.017 l
Hand	1.16 ± 0.110	1.108 ± 0.019	1.079 ± 0.017 r 1.081 ± 0.011 l

tained from cadavers. (To the best of my knowledge, at the time of this writing, no single study has measured the density of all the individual body segments in living people.)

The density of tissues in cadavers may differ from the density of living tissues. If the separation of segments of the body is performed on fresh cadavers that are not frozen and not embalmed, there are unavoidable decreases in mass due to loss of body fluids (the loss is between 5% and 6% of total body weight). Freezing changes the body fluid to ice and consequently increases the volume and decreases the density. Some studies used embalmed cadavers; the density of the preservation solution chosen was close to 1.060 g/cm³. The main *post-mortem* changes occur in lung tissue. In living people, its density is between 0.15 and 0.25 g/cm³, but soon after the death the lungs shrink, and its density increases to between 0.563 and 0.850 g/cm³ (Erdmann and Gos 1990) and even to 1.05 g/cm³ (Woodard and White 1986). At the same time, owing to the collapse of organs, the trunks of cadavers may contain large amounts of air in the thoracic and abdominal cavities (Chandler et al. 1975). On the whole, the densities of cadaver trunks may differ from the densities of the trunks of living people. The data on trunk density obtained from living subjects are presented in table 4.6.

4.3.4.2 Total Body Mass

It is trivial knowledge that the mass and moments of inertia of body parts are different in people of different body dimensions. The average *relative masses*

Table 4.6   Density of the Trunk in Living People

Segment	Erdmann 1997, 15 men, 20–40 years old, computer tomography	Pearsall, Reid, and Livingston 1996, 2 men and 2 women, 46–68 years old, computer tomography	Pearsall, Reid, and Ross 1994, 26 men, 40.5 ± 14.4 years old, magnetic resonance imaging
Upper part of the trunk (thorax)	0.908	0.87	0.82 ± 0.04
Middle part of the trunk (abdomen)	1.043	1.02	1.01 ± 0.01
Lower part of the trunk (pelvis)	1.077	1.07	1.02 ± 0.01
Whole trunk	1.030	0.96	0.94 ± 0.02

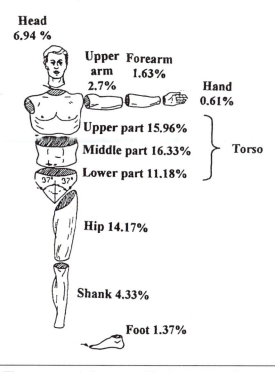

Figure 4.21   The average relative mass of body segments (as a percentage of the total body mass) in 100 physically fit young men.

Reprinted, by permission, from V. Zatsiorsky, V. Selyanov, and L. Chugunova, 1990, In vivo body segment inertial parameters determination using a gamma-scanner method. In *Biomechanics of human movement: applications in rehabilitation, sports and ergonomics*, edited by N. Berme and A. Cappozzo (Worthington, OH: Bertec Corporation), 186–202.

of body segments (segment mass as a percentage of body mass) are presented in figure 4.21 and in table A1.2 (see appendix 1).

Although the relative mass of body segments is broadly cited in the literature, this measure can be used only when the subject's data are close to the average values of the corresponding sample. The percentage mass may be misleading because, according to experimental observations, the intercept a in the regression equation, segment mass = $a + b$(body mass), does not equal zero. This fact is easy to accept: we do not expect that the mass of the head of a 100-kg person is twice the mass of the head of a 50-kg individual. There is also a slight tendency to a curvilinear relationship between total body mass and the mass of some body segments (figure 4.22). The tendency, however, is small and can be neglected. The regression equations for predicting the mass of the individual body segments from the total body mass are presented in table 4.7.

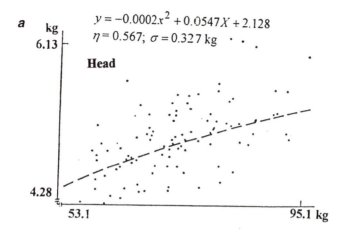

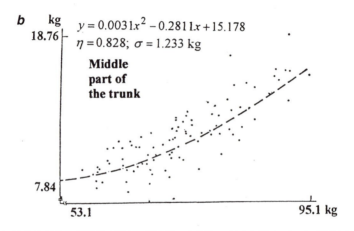

Figure 4.22 Relationship between body mass and mass of (*a*) the head and (*b*) the middle part of the trunk, measured using the gamma-scanner method in 100 young men. Although a tendency toward curvilinearity seems to exist, the linear regression equations were almost as good as the parabolic ones. For instance, the mass of the middle part of the trunk was predicted with a standard error of ±1.233 kg by a parabolic equation and ±1.268 kg by a linear equation. In the equations, η is a correlation ratio, σ is the standard error of the parabolic fit.

Reprinted from V. Zatsiorsky and V. Selyanov, 1979, Mass-inertia characteristics of the human body segments and their relationship with anthropometric measures. In *Questions of anthropology* (Moscow, Russia: Moscow University), 95, 96.

The relative mass of the body segments (as a percentage of the total body mass) is not constant but varies with the total body mass. For instance, the relative mass of the abdomen is larger and the relative mass of the head,

Table 4.7   Regression Equations for Predicting the Mass of Body Parts (Y, kg) From Total Body Mass (X, kg) of Adult Men

Segment	Equation	Coefficient of correlation	Standard error, kg
Foot	$Y = 0.259 + 0.01X$ (ZS)	.66	0.107
Calf	$Y = 0.141 + 0.041X$ (ZS)	.86	0.226
Thigh	$Y = -0.799 + 0.153X$ (ZS)	.89	0.717
Hand	$Y = 0.109 + 0.0046X$ (ZS)	.58	0.060
Forearm	$Y = 0.165 + 0.0139X$ (ZS)	.78	0.102
Upper arm	$Y = -0.142 + 0.029X$ (ZS)	.83	0.178
Head and neck	$Y = 3.243 + 0.024X$ (ZS)	.57	0.327
Upper part of the trunk	$Y = -0.078 + 0.161X$ (ZS) $Y = 1.625 + 0.11X$ (PRR)	.78 .79	1.176 —
Middle part of the trunk	$Y = -2.222 + 0.194X$ (ZS) $Y = -5.249 + 0.22X$ (PRR)	.82 .87	1.268 —
Lower part of the trunk	$Y = -0.348 + 0.117X$ (ZS) $Y = -0.222 + 0.14X$ (PRR)	.71 .90	1.051 —

Note. Equations marked (ZS) are from Zatsiorsky and Seluyanov (1979), $n = 100$, and those marked (PRR) are from Pearsall, Reid, and Ross (1994), $n = 26$.

From Zatsiorsky, V.M., and V.N. Seluyanov. 1979. Mass-inertia characteristics of human body segments and their relationship with anthropometric parameters (in Russian). *Voprosy Anthropologii (Problems in Anthropology)* 62: 91–103.

Pearsall, D.J., J.G. Reid, and R. Ross. 1994. Inertial properties of the human trunk of males determined from magnetic resonance imaging. *Ann. Biomed. Eng.* 22 (6): 692–706.

feet, and hands is smaller in heavier people than in lighter people (figure 4.23).

To compare the body shapes of people of different mass, the ratios of the two people's body mass and their segment mass can be computed. These ratios are known as proportionality coefficients. In the case of proportional changes, the body-mass and segment-mass ratios should be equal. For instance, for two subjects with body mass of 100 kg and 50 kg, respectively, the ratio of total body mass equals 2.0. The corresponding ratio for the mass of the head is only 1.27, and it is 2.30 for the mass of the abdomen (table 4.8). Therefore, the

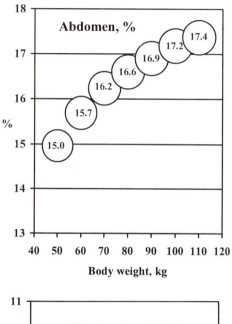

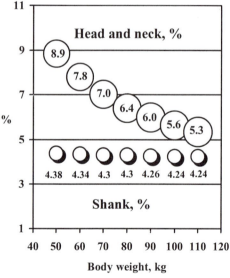

Figure 4.23 The estimated relative mass of the abdomen, head, and shank in men of different body mass (from 50 kg to 110 kg). For instance, the mass of the head and neck is 8.9% (4.45 kg) in a 50-kg man, but it is only 5.3% (5.87 kg) in a 110-kg man. If it was a 40-kg man, the estimated relative mass of the head and neck would be 10.5% (4.2 kg), which is almost twice the relative mass of the head in a 110-kg man. Note that all the estimates are from a relatively homogeneous sample of healthy, physically fit young men. The data are from Zatsiorsky and Seluyanov (1983).

Table 4.8   Mass and Proportionality of Body Segments in 50-kg and 100-kg Physically Fit Young Men

Body segment	Total body mass, kg		Proportionality coefficient m_{100}/m_{50}
	50	100	
Foot	0.759	1.259	1.66
Shank	2.191	4.241	1.94
Thigh	6.851	14.503	2.12
Hand	0.349	0.579	1.66
Forearm .	0.860	1.555	1.81
Upper arm	1.308	2.758	2.11
Head and neck	4.443	5.643	1.27
Thorax	7.972	16.022	2.01
Abdomen	7.478	17.178	2.30
Pelvis	5.502	11.352	2.06

average values of the relative mass presented in figure 4.21 and table A1.2 (in appendix **1**) should be used with caution.

4.3.5 Inertial Properties of the Trunk

Modeling the inertial properties of the trunk is a challenging task. There are several reasons for this. The trunk can barely be considered a single, rigid body. In many movements, the parts of the trunk move with respect to each other. Due to the lungs, the density of the upper portion of the trunk is much lower than the density of the middle and lower portions. The inertial properties of the trunk fluctuate with inspiration and expiration; the rib cage is displaced, and the lung density changes with the phases of breathing. The internal organs are displaced within the trunk when the body changes its orientation in space (e.g., during a handstand, the internal organs shift in the direction of the head). The internal organs may also be displaced when the trunk moves with a large acceleration. The trunk also is difficult to separate from other parts of the body.

■ ■ ■ *FROM THE LITERATURE* ■ ■ ■

Visceral Mass Displacement

Source: Minetti, A.E., and G. Belli. 1994. A model for the estimation of visceral mass displacement in periodic movements. *J. Biomech.* 27 (1): 97–101.

A method was suggested to detect the displacement of the visceral part of the body, provided that its mass is known and the movement is periodic (figure 4.24). The mass of the visceral part of the body was assumed to be 9 kg. During hopping in place, the internal visceral mass oscillates out of phase with respect to the body frame.

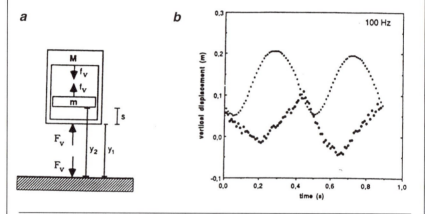

Figure 4.24 (*a*) The model. *M* is the container mass, *m* is the internal (hidden) mass, and y_1 and y_2 are distances from ground level ($s = y_2 - y_1$). The whole system oscillates vertically and exerts a vertical GRF of F_v, while the internal mass exerts a force f_v on the container. (*b*) Results from an experiment with a hopping subject (two cycles are shown). The larger dots represent the relative vertical displacement of the internal visceral mass with respect to the container, while the smaller dots show the vertical pattern of the CoM as measured by cameras (i.e., with the head–trunk segment lightened of the internal mass).

Reprinted from *Journal of Biomechanics*, 27 (1), A.E. Minetti and G. Belly, A model for the estimation of visceral mass displacement in periodic movements, 97–101, 1994, with permission from Elsevier Science.

The relative trunk mass varies substantially among people. The data from different studies are difficult to compare due to dissimilar methods of segmentation. The published mean data vary from 42.2% (Pearsall, Reid, and Ross 1994) to 52.4% (Chandler et al. 1975). However, even when the segmentation is consistent and the sample is more or less homogeneous (e.g., adult men), the range can be as large as 12.2 percentage points, from 35.8% to 48.0% (Pearsall, Reid, and Ross 1994). The abdominal region of the trunk is where body fat is most actively accumulated. Consequently, changes in body weight modulate the relative mass of this body segment, which can range from 10.4% to 21.6% even in a relatively homogeneous sample (26 men of mean age 40.5 ± 14.4 years; Pearsall, Reid, and Ross 1994).

■ ■ ■ *From the Literature* ■ ■ ■

Changes in the Mass of Cosmonauts' Body Parts After 6 Months of Hypokinesia

Source: Tischler, V.A., V.M. Zatsiorsky, and V.N. Seluyanov. 1981. Study of the mass-inertial characteristics of human body by the gamma-scanning method during 6-month hypokinesia (in Russian). *Kosmicheskaja Biologija i Aviakosmicheskaja Medicina (Space Biology and Aerospace Medicine)* 15 (1): 36–42.

The study was a part of a space research project designed to study the effects of limited movement (hypokinesia) on Russian cosmonauts. Space flights of long duration were simulated. In the experiment, 18 volunteers, men 36 to 40 years of age, laid for 182 days in special beds inclined at an angle of –4.5° to the horizontal plane with the legs higher than the head (antiorthostatic position). The subjects were divided into three groups with six subjects per group. The first group (control) did not exercise. The second and the third groups used different training protocols. The exercises were performed in the horizontal position twice a day. The duration of training sessions was 1 h.

In the first group, the average gain of body mass was around 4 kg, of which 1.0 kg was accumulated in the abdominal region of the trunk and 0.9 kg was accumulated in the thighs. The effects in the exercise groups were different in various subjects. The authors attributed these differences to the standard exercise program used in the study; for some of the subjects the program was too easy, and for others it was too demanding. The subjects with the largest gain of total mass also showed the largest gain in the mass of the middle portion of the trunk. For instance, in a subject with a total body mass gain of 7.5 kg, the mass of the abdominal region increased by 3.1 kg; in another subject, the total mass increased 7.08 kg, and the mass of the middle portion of the trunk increased by 2.7 kg.

Data on the distribution of mass and density along the vertebral trunk segments and the location of segmental CoMs are presented in figure 4.25. The lowest density, 0.74 g/cm^3, was observed at the level of T6. Most CoMs of the individual vertebral trunk segments lie anterior to the centers of their respective vertebrae. The largest distance is in the lower thoracic region. Therefore, the trunk's line of gravity is anterior to the vertebral column, and gravity force produces a flexion moment with regard to the spine. To maintain an upright posture, this moment must be counterbalanced by the spine erectors and ligaments.

4.4 SUBJECT-SPECIFIC INERTIAL CHARACTERISTICS

Representative papers: Hatze (1980); Hinrichs (1985); Zatsiorsky, Seluyanov, and Chugunova (1990a, b); Yeadon and Morlock (1989)

Researchers and health care professionals often need to know the inertial properties of the body segments of a certain person. Ideally, these properties should be measured. However, the measurement is technically complex and presently cannot be performed on a routine basis. The solution is to adjust the average data available in the literature to the individual subject or patient. This is a two-step procedure. The researcher should first make an educated selection from the data available in the literature and, second, execute the necessary measurements on the subjects and adjust the published data appropriately. During the first step, the main points to consider are (1) the sample characteristics, particularly the sex, age, and physical status of the subjects; (2) the method used in the original study, for example, whether the measurements were performed on cadavers or living people; and (3) the complexity of the recommended scaling procedure as compared with the desired accuracy of the estimates. The data obtained on a certain population group, such as physically fit young men, cannot be immediately applied to subjects whose anthropometric characteristics differ from the experimental population, for instance, to elderly men or toddlers. Unfortunately, for some populations (e.g., elderly women), appropriate data are presently not available.

Tuning the body-segment parameters for individual subjects is always a trade-off between the accuracy of the estimates and the amount of work in making the anthropometric measurements. If one limits the measurements to body weight and height, one cannot expect the same accuracy as if 200 or more measurements were made. However, hundreds of anthropometric measurements require several hours of work. Not all researchers and subjects are prepared to spend this length of time, so there is a need for compromise.

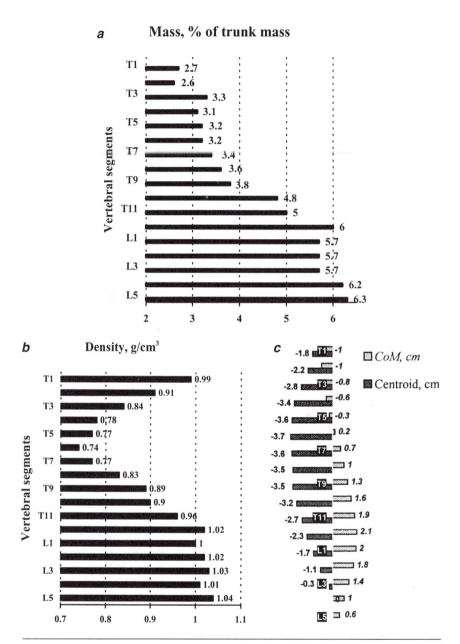

Figure 4.25 Inertial parameters of the individual vertebral trunk segments (based on data by Pearsall, Reid, and Livingston 1996). (*a*) Mass of each vertebral trunk segment as a percentage of total trunk mass. (*b*) Density of the vertebral trunk segments. (*c*) Position of the centers of volume (centroids) of the vertebrae and position of the centers of mass of the corresponding vertebral trunk segments with respect to the frontal plane at the trunk.

The following sections provide a brief overview of the existing methods of adjustment of published data for specific subjects. The geometric modeling described in **4.4.1** requires from the literature only data on the density of body parts; all other methods are based on adjusting the inertial properties obtained on a group of subjects to an individual person. To help with the literature search, a rather lengthy literature list is provided at the end of this chapter. Fortunately, several comprehensive reviews on the topic are available (Reid and Jensen 1990; Pearsall and Reid 1994). The old literature is nicely reviewed by Hay (1973, 1974) and by Miller and Nelson (1973). Tabulated data from different studies can be found in H.M. Reynolds (1978).

4.4.1 Geometric Modeling

The idea of the method of geometric modeling is to model body segments or their parts as homogeneous solids with simple geometric shapes. Density data are taken from the literature. For homogeneous solids, the main inertial parameters can easily be computed.

As an example, let us compute the moments of inertia of a homogeneous circular cylinder about the coordinate axes with the origin at the centroid (figure 4.26a). Consider first a flat, uniform, circular disk of radius R and total mass m (figure 4.26b). We are to find the value of the moment of inertia about a central axis perpendicular to the disk. For integration, we choose a thin ring (shell element) of thickness dr, of which all elements are at an equal distance r from the axis. The mass of the ring is

$$dm = m\frac{dA}{A} = m\frac{2\pi r dr}{\pi R^2} = \frac{2m}{R^2} r dr \qquad (4.30)$$

and the moment of inertia of the disk about its center O is

$$I_O = \int_0^R r^2 dm = \int_0^R r^2 \frac{2m}{R^2} r dr = \frac{2m}{R^2} \int_0^R r^3 dr = \frac{2m}{R^2} \cdot \frac{R^4}{4} = \frac{mR^2}{2} \qquad (4.31)$$

The moment I_O is a polar moment, and at the same time, it is an axial moment about the central axis of the disk. Two other axial moments of the flat disk are about the axes along the disk diameters. These two moments are clearly equal. Hence, from equations 4.10 and 4.31, it follows that the moment of a disk about its diametral axis is $mR^2/4$.

Now consider the homogeneous cylinder of radius R, length h, and mass m (figure 4.26a). The moment of inertia of the cylinder about its central longitudinal axis is the same as the moment of inertia of the flat disk that we just analyzed, $mR^2/2$. The moment of inertia about a diametral axis through the centroid can be found from the following consideration: a moment of inertia of

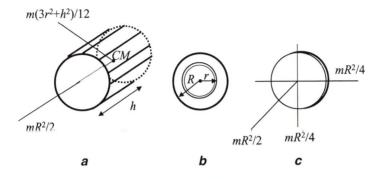

Figure 4.26 (*a*) Determining the moments of inertia of a solid, homogeneous cylinder. (*b* and *c*) The thin disk is obtained by squashing the cylinder down while preserving the radius *R* and mass *m*.

any mass element is $(x^2 + y^2)dm$. The coordinate x is measured along the central line, and coordinate y is normal to the plane containing the diameter and the central line. The moment of inertia of the cylinder is then

$$I = \int \left(x^2 + y^2 \right) dm = \int x^2 dm + \int y^2 dm \qquad (4.32)$$

The first term is the moment of inertia of a slender rod of mass m, length h, and negligible sideways dimensions. The second is the moment of inertia of the flat disk that can be obtained by squashing the cylinder down to a thin disk of the same radius and mass as the cylinder. This term equals $mR^2/4$. The moment of a slender rod with regard to an axis through its centroid is $mh^2/12$. Thus, for the moment of inertia about an axis along the diameter of the cylinder and through its centroid, we obtain

$$I = \frac{mh^2}{12} + \frac{mR^2}{4} \qquad (4.33)$$

The simple geometric forms that are most commonly used for modeling human body segments are shown in appendix **3** (see figure A3.1) along with the necessary equations (table A3.1). Geometric models of human body segments of different complexity have been developed. Some of them are presented in figure 4.27.

The method of geometric modeling is potentially promising, especially if body shape can be determined automatically with digital video equipment. At present, however, the method is limited. It neglects individual differences in segment density. The density data used in the geometric models were obtained on a limited sample of elderly cadavers (see table 4.5) of which the body weight was very low for the body height. Unfortunately, better data are currently not available.

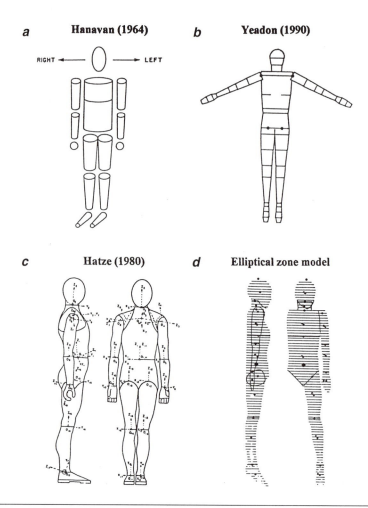

Figure 4.27 Examples of geometric models of the human body. (*a*) The simplest model was suggested by Hanavan (1964), who divided the body in 15 main segments of simple geometric shape and uniform density. The trunk was modeled as an elliptical cylinder; the upper arm, forearm, thigh, shank, and feet as frustums of circular cones; the head as a circular ellipsoid; and the hand as a sphere. To define the parameters of the model, 25 anthropometric dimensions are needed. (*b*) The model by Yeadon (1990) divides the body into 40 solids. A total of 95 measurements are necessary. The solids representing the torso, hands, and feet are stadium solids. The solids representing the head, arms, and legs are assumed to have a circular cross section. (*c*) The model developed by Hatze (1980) is more detailed and requires 242 anthropometric measurements. (*d*) In the elliptical zone method (Weinbach 1938; Jensen 1978), stereophotogrammetry is used to estimate the shape of the body segments. The body is sectioned into elliptical disks of a small width (usually 20 mm).

All methods other than geometric modeling are based on tuning the mean body-segment parameter data obtained on a sample to a particular individual subject.

4.4.2 Geometric Scaling: Isometry and Allometry

The simplest model of geometric scaling, known as the isometry model, assumes that human bodies are geometrically similar and differ only in size. The word *isometry,* meaning geometric similarity, is used in mathematics and theoretical biology. In the isometry model, complete proportionality among all linear dimensions is assumed, and differences in body proportions are entirely neglected. In the model, the masses of individual body parts are proportional to the mass of the body, and the CoMs of the body segments are always located at the same percentage of the length of the segments. The moments of inertia are proportional to the mass times a linear dimension squared. If body isometry is assumed, the scaling of segmental moments of inertia for individual subjects can be performed as a two-stage procedure. First, a scaling factor S is computed as

$$S = \frac{m_s h_s^2}{m_g h_g^2} \tag{4.34}$$

where m_s and m_g are the body mass of the subject and the mean mass of the group, correspondingly, and h_s and h_g are the standing height of the subject and the mean height of the group, correspondingly. To obtain a subject-specific value, the mean value of the moment of inertia of a given body segment for the group is multiplied by a scaling factor S. In the isometry model, the scaling factor S is the same for all body segments. When the difference between the body sizes of subjects is very large, as in children of different ages, the model provides close estimates for the moments of inertia of the entire body (see figure 4.14 again).

The main assumption of the isometry model that all people are geometrically similar is an evident oversimplification. People usually have different body proportions; for instance, leg length often constitutes a different percentage of body height in different people. In part, these differences are associated with body size. It was already mentioned (section **4.3.4.2**) that the mass of an individual body part is not a fixed percentage of total body mass. Hence, body isometry can be accepted only as a rather crude approximation. It is interesting that Galileo Galilei (1564–1642) understood that. He was the first to introduce the *square-cube law:* volume increases as the cube of linear dimensions, but strength increases only as the square. For that reason, Galileo wrote, "It would be impossible to build up the bony structures of men, horses, or other animals so as to hold together and perform their normal functions if these animals were to be increased enormously in height." Large animals require proportionally stronger supports than small ones. A cat expanded to the size of a lion and kept in exact proportion would collapse.

The *allometry* model is based on the presumption that body proportions change systematically with body size. For instance, during the growth period in children, the body height increases, and the relative leg length (as a percentage of body height) also changes (see figure 4.11 again). The relationship can be described by an allometric equation. Following is the most commonly used allometric equation:

$$Y = aX^b \tag{4.35}$$

where a is the scaling coefficient and b is the scaling exponent. Other empirical equations can also be employed. In this model, the dimensions of body segments do not represent a fixed percentage of the whole-body dimension; rather the percentage itself depends on body size. Either *ontogenetic scaling* or *static scaling* is used for allometric analysis. In ontogenetic scaling, the data on children of different ages (and hence different sizes) are analyzed. In static scaling, the measurements are performed on adults of different sizes. Allometric equations are just regular regression equations obtained from empirical statistical analysis and should be used as such. For instance, ontogenetic allometric equations can be employed for estimating the magnitude of a morphological variable (e.g., the leg length in a 5-year-old boy) when another variable (e.g., body height) is known. In the allometry model, equation 4.34 has limited application. The following equation, which uses the mass and length of individual body segments instead of total body mass and height, provides better results:

$$S_i = \frac{m_{is} l_{is}^2}{m_{ig} l_{ig}^2} \tag{4.36}$$

In this equation, S_i stands for the scaling factor for segment i, m is segment mass, l is segment length, the subscripts s and g refer to a specific subject and the group mean, respectively. Note that all the parameters are segment specific. Equation 4.36 is based on the assumption that even though the relationship between total body size and the size of individual body segments is allometric (nonproportional), individual segments are proportional (local isometry). To offset possible deviations from the ideal local isometry, the following scaling formula has been suggested for scaling the longitudinal moments of inertia (Dapena 1978; Hinrichs 1985):

$$S_l = \frac{m_s^2 h_g}{m_g^2 h_s} \tag{4.37}$$

where the subscript *l* stands for *longitudinal.* The formula is based on the assumptions that a tall subject is likely to have thinner segments than a shorter subject of the same mass and that a heavy subject is likely to have thicker segments than a lighter subject of the same height. In the case of ideal local isometry, equations 4.36 and 4.37 are identical; they both scale the moments of inertia in proportion to a linear dimension to the fifth power.

People's bodies vary not only in total size and in the size of individual body segments but also in the orientation and spatial location of some anatomic objects. For instance, orientation of the subtalar joint axis varies in different subjects from 4° to 47°, and the axis can be located differently with respect to the bony landmarks (for a review, see *Kinematics of Human Motion,* figure 5.10 on p. 297). A more compete geometric scaling should consider not only the difference in size ("deformation") but also rotation and translation. Such scaling is called *affine scaling.* In biomechanics, affine scaling is used mainly to scale the spatial locations of the origins and insertions of various muscles and ligaments. Simple length and width measurements, even if they are combined with allometric scaling, cannot provide this information.

Anatomic objects are not the only items that can be scaled with respect to body size. Biomechanical variables (such as step length and step frequency in locomotion) and physiological measures (such as energy expenditure or maximal oxygen consumption) can also be scaled. The general aspects of biological scaling are not covered in this book.

4.4.3 Regression Equations

Regression equations, especially linear regression equations, are convenient and simple tools for estimating subject-specific inertial parameters. They are

---

■ ■ ■ *From the Literature* ■ ■ ■

Affine Scaling of Skeletal Segments

Source: Sommer, H.J., III, N.R. Miller, and G.J. Pijanowski. 1982. Three-dimensional osteometric scaling and normative modeling of skeletal segments. *J. Biomech.* 15 (3): 171–80.

Three-dimensional coordinates of the muscle attachments are at times necessary for surgeons and researchers. However, the desired landmarks are inaccessible for direct measurement. They must be scaled from other specimens. The affine scaling used for this purpose is visualized in figure 4.28. It includes a deformation (change in dimensions), rotation, and translation of the scaled object.

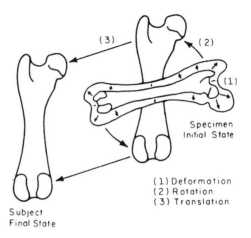

(1) Deformation
(2) Rotation
(3) Translation

Specimen
Initial State

Subject
Final State

Figure 4.28 The affine scaling of an anatomic object (a bone). The scaling is a combination of a deformation, rotation, and displacement.

Reprinted from *Journal of Biomechanics*, 15 (3), H.J. Sommer III, N.R. Miller, and G.J. Pijanowski, Three-dimensional osteometric scaling and normative modeling of skeletal segments, 171–180, 1982, with permission from Elsevier Science.

not based on mechanics theory but are just curve-fitting procedures. For instance, a regression equation can have the form I (moment of inertia) $= B_0 + B_1$ (body weight) $+ B_2$ (body height). The form of this equation implies that the moment is the weighted sum of the body weight and body height, but it is not, of course. The equation, however, allows estimating the value of the moment with certain accuracy. If the accuracy is sufficient, a researcher does not bother about the theory behind the equation but simply uses it.

Several sets of multiple regression equations are presented in appendices **1** and **2**. As input variables, the equations use either the mass and height of the body or numerous other anthropometric measurements.

4.4.4 Combined Methods (Nonlinear Regression Equations)

Representative papers: Zatsiorsky, Seluyanov, and Chugunova (1990a, 1990b)

The nonlinear regression method is based on a *gamma-scanner method* that allows the determination of the mass of individual body segments in living people (see appendix **2**). The method combines geometric scaling with statistical (regression) analysis. In this method, all segments are modeled as cylinders. The quasi volume V of a segment is calculated from its length L and the circumference (perimeter) C as for a cylinder but with a correction factor k':

$$V = \frac{LC^2}{4\pi} k' = kLC^2 \qquad (4.38)$$

where $k = k'/4\pi$. Dividing the mass of the segment m (measured with the gamma-scanner method) by the quasi volume V, we obtain the estimate of the quasi density δ of the segment:

$$\delta = m/V \qquad (4.39)$$

Combining equations yields

$$K_m = k\delta = \frac{m}{LC^2} \qquad (4.40)$$

The coefficient K_m takes into account the differences in shape and density between the segments and their models (cylinders). The moment of inertia I of a segment about an axis through its CoM is $I = mR^2$, where R is the radius of gyration. The radii of gyration about the sagittal (anteroposterior) and frontal axes depend on the length L of the segment, while the radius of gyration about the longitudinal axis correlates with the circumference C. To estimate the moments of inertia of each segment, the following coefficients were introduced:

$$K_s = \frac{I_s}{mL^2}$$

$$K_f = \frac{I_f}{mL^2} \qquad (4.41)$$

$$K_l = \frac{I_l}{mC^2}$$

where I_s, I_f, and I_l are the moments of inertia relative to the sagittal, frontal, and longitudinal axes, respectively. The values of all four coefficients were computed from experimental measurements.

The following regression equations are used for estimating subject-specific values of the inertial parameters:

$$m_i = K_{mi} L_i C_i^2$$

$$I_{si} = K_{si} m_i L_i^2$$

$$I_{fi} = K_{fi} m_i L_i^2 \qquad (4.42)$$

$$I_{li} = K_{li} m_i C_i^2$$

where the subscript i refers to the ith segment and m_i estimates the masses of individual segments. The sum of the calculated masses should be equal to the real (measured) mass of the whole body. If a difference between the estimated and the real mass is observed, it can be used to correct the mean body density. To apply equation 4.42, the data on the circumference and the length of body segments of the subjects must be known. Note that the biomechanical and not the anthropometric values of lengths must be used. If the anthropometric length of a segment is known, the biomechanical length can be estimated using the values of coefficient K^b given in table A2.9 in appendix **2**. The necessary anthropometric measurements are explained in detail in table A2.10 and figure A2.2 (in appendix **2**).

Use these steps to compute the subject-specific inertial parameters:

1. Measure lengths and circumferences of segments.
2. Calculate the biomechanical lengths of segments by multiplying the measured anthropometric lengths by the coefficients K^b.
3. Find the values of segment masses m_i.
4. Add the values of segment masses m_i and compare the calculated total sum with the known mass of the body. In case of discrepancies, introduce coefficients for correction.
5. Calculate the moments of inertia.

▪ ▪ ▪ *From the Literature* ▪ ▪ ▪

Inertial Properties of Body Segments During Pregnancy

Source: Jensen, R.K., T. Treitz, and S. Doucet. 1996. Prediction of human segment inertias during pregnancy. *J. Appl. Biomech.* 12 (1): 15–30.

Changes in body-segment parameters over the last two trimesters of pregnancy were investigated longitudinally in 15 women. At the start of testing, mean body weight was 66.3 ± 8.6 kg; by the final testing session, mean weight had increased to 77 ± 8.8 kg. The geometric model of the body consisted of 16 segments with stacked elliptical cylinders sectioned at regular 2.0-cm intervals. The density data were taken from studies performed on cadavers (see table 4.5). Regression equations similar to equation 4.42 were developed:

$$m = kp^2h$$
$$I_z = k_1 p^4 h$$
$$I_t = \frac{1}{2} I_z + k_2 p^2 h^3 \tag{4.43}$$

where k, k_1, and k_2 are constants; m is segment mass; p is perimeter; h is length of the segment; I_z is the longitudinal moment of inertia; and I_t is the "transverse" moment of inertia averaged from I_x and I_y, the anteroposterior and mediolateral moments. The constants k, k_1, and k_2 were computed from experimental data. An example of nonlinear regression is presented in figure 4.29.

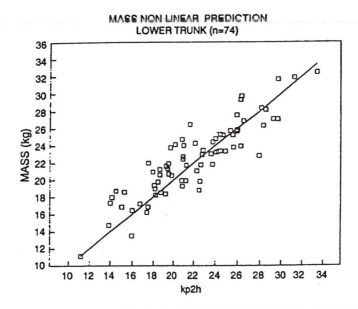

Figure 4.29 Prediction of lower trunk mass by a nonlinear regression equation. Regression equations were computed from a cross-sectional study on 15 women in middle to later pregnancy and then applied to 74 measurements representing the longitudinal aspect of the investigation. The standard error of fit is 5.22%. Note that the values of the term kp^2h are plotted along the abscissa axis.

Reprinted, by permission, from R.K. Jensen, T. Treitz, and S. Doucet, 1996, "Prediction of human segment inertias during pregnancy," *Journal of Biomechanics* 12 (1):15–30.

4.5 CONTROL OF BODY INERTIA IN HUMAN MOVEMENT

Inertial properties of individual body segments are not subject to voluntary control. On the other hand, inertial properties of the body limbs and the entire body can be controlled voluntarily. This can be achieved by changing the posture or orientation of the body with respect to external forces and by stiffening or relaxing the joints.

4.5.1 Moments and Products of Inertia

Moments of inertia of the human body and limbs depend on the distance of the rotating body parts from the axis of rotation. Therefore, the magnitude of the moment can be controlled by extending or flexing the extremities. This opportunity is commonly used by athletes who perform whole-body rotations (somersaults in gymnastics and diving, twists in figure skating, etc.). By flexing their extremities, the athletes decrease the moment of inertia of the body and correspondingly increase the angular velocity of rotation (recall that the angular momentum during airborne movements is constant; see the Mechanics Refresher "Linear and Angular Momenta" in chapter **1**, p. 50). For instance, the moment of inertia of a gymnast in a tucked position equals one third of the moment of inertia in the layout position (see figure 4.13 again). Hence, when an athlete starts a somersault in a layout position and then assumes a tucked posture, his or her angular velocity triples. The "tighter" the tucked position (the closer the limbs to the trunk), the smaller the moment of inertia and the larger the velocity of body rotation. Similar reasoning is valid for the extremities: the more they are flexed, the smaller their moment of inertia with respect to the proximal joint. For instance, to decrease the moment of inertia of a swinging leg during the recovery phase in running, the runner flexes the leg (figure 4.30).

Repositioning the body segments can also change the products of inertia. Any deviation of a body segment from a mass symmetry with respect to the axis of rotation causes nonzero products of inertia to appear. This phenomenon is used by gymnasts and divers to induce body twist during somersaulting movements. Consider a diver who initiates a somersault and is in the layout position with the arms abducted 90° (figure 4.31*a*). If the diver raises one arm above the head and moves the second arm down to the side as in figure 4.31*b*, the inertia ellipsoid inclines with respect to the axis of the angular momentum vector (which remains constant while the diver is in the air). The simple rotation about the frontal axis of the body changes to a complex rotation in space that includes the somersaulting rotation about the horizontal axis through the diver's CoM and a twisting rotation about the longitudinal axis of the diver's body.

4.5.2 Inertia Matrix of the Chain

The inertial properties of multilink chains are not limited to the masses and moments of inertia of the individual body segments. Additional properties emerge. They require a generalization of the very notion of inertia. Recall that the concept of inertia is based on Newton's laws, in particular, the second:

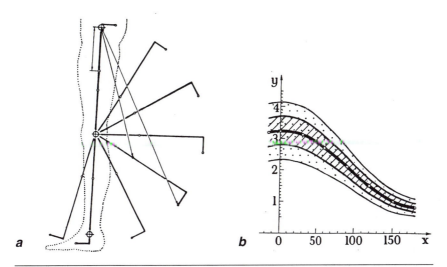

Figure 4.30 Moment of inertia of the leg about the hip joint in 100 male subjects, 21 to 25 years of age. (*a*) The range of angles between the thigh and calf used in the model: −15°, 0°, 30°, 60°, 90°, 120°, 150°, and 180°. The distances between the hip and the centers of mass of the thigh, calf, and foot are indicated. (*b*) Mean data on moments of inertia as a function of the knee angle with one and two standard deviations.

Reprinted, by permission, from W.S. Erdmann, 1987, Individual moment of inertia of the lower extremity of the living young adult male. In *Biomechanics X-B*, edited by B. Johnson (Champaign, IL: Human Kinetics), 1158–9.

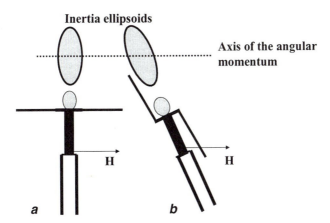

Figure 4.31 Initiating a twist during a somersault. When body posture changes from (*a*) symmetric to (*b*) asymmetric, nonzero products of inertia appear. The off-diagonal terms of the inertia tensor are not zero anymore, and the inertia ellipsoid is inclined with respect to the axis of the angular momentum vector **H**.

■ ■ ■ *From the Literature* ■ ■ ■

Inducing Aerial Twists
by Changing the Orientation of the Inertia Tensor

Source: Yeadon, M.R. 1993. The biomechanics of twisting somersaults: III. Aerial twist. *J. Sports Sci.* 11: 209–18.

A computer simulation model was developed to evaluate aerial twisting techniques in gymnastics and diving. In the absence of a somersault (the angular momentum of the body is zero), asymmetric arm movement produces a body tilt without any twisting movement (figure 4.32a). When performed during a straightforward somersault, the same arm movement induces a body twist, the rotation of the athlete's body around its longitudinal axis (figure 4.32b).

a

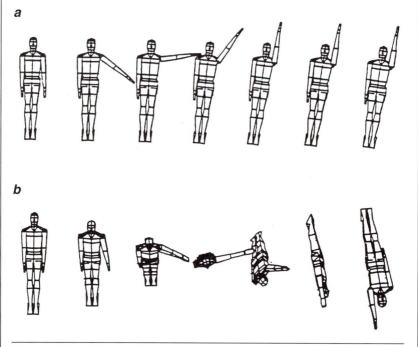

b

Figure 4.32 Effects of an asymmetric arm movement on body motion in the air. (*a*) Straight jump. (*b*) Straight forward somersault.

Reprinted, by permission, from M.R. Yeadon, 1993, "The biomechanics of twisting somersaults," *Journal of Sports Sciences* 11:209-218. By permission of Taylor & Francis Ltd, **http://www.tandf.co.uk/journals**

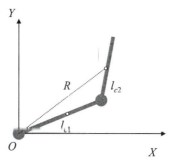

Figure 4.33 A planar two-link chain. The chain rotates at the proximal joint (joint 1) and the distal joint (joint 2) with angular accelerations $\left(\ddot{\alpha}_1 \text{ and } \ddot{\alpha}_2 \right)$, respectively. The masses of the links are m_1 and m_2, and their moments of inertia with respect to the CoM are I_1 and I_2; l_{c1} is the distance from joint 1 to the CoM of the first link, and l_{c2} is the distance from joint 2 to the CoM of the second link; l_1 (not shown in the figure) is the length of link 1. R is the radius from the proximal joint to the CoM of the second link.

$F = m\ddot{x}$ for a linear translation and $M = I\ddot{\theta}$ for an angular rotation. In these equations, inertia is a coefficient of proportionality that relates acceleration, $\left(\ddot{x} \text{ or } \ddot{\theta} \right)$, with a force or a moment of force, F or M. Hence, if we obtain the equation Force = Something × Acceleration, we can regard the "something" as inertia. From this point of view, consider a planar two-link chain (figure 4.33).

We limit ourselves to the analysis of some inertial properties of the chain (a detailed analysis of dynamics of the chain is in chapter **5**). Application of the parallel axis theorem yields

$$I_{1,1} = \underbrace{\left(m_1 l_{c1}^2 + I_1 \right)}_{\substack{\text{Moment of inertia} \\ \text{of link 1 about joint 1}}} + \underbrace{\left(m_2 R^2 + I_2 \right)}_{\substack{\text{Moment of inertia} \\ \text{of link 2 about joint 1}}} \tag{4.44}$$

and

$$I_{2,2} = m_2 l_{c2}^2 + I_2 \tag{4.45}$$

where $I_{1,1}$ is the moment of inertia of the entire chain about joint 1 (the inertial resistance at joint 1 associated with angular acceleration at that joint) and $I_{2,2}$ is the moment of inertia of link 2 about joint 2. In these two-digit subscripts, the second numeral refers to the joint where acceleration occurs, and the first numeral refers to the joint where inertial effect of the acceleration is felt. The corresponding dynamic equations are $\tau_{1,1} = I_{1,1} \ddot{\alpha}_1$ and $\tau_{2,2} = I_{2,2} \ddot{\alpha}_2$, where $\tau_{1,1}$ and $\tau_{2,2}$ are the torques at joints 1 and 2, respectively, that are associated with the angular acceleration in these joints. These torques are just mechanical abstractions. When movement is performed in both joints, $\tau_{1,1}$ and $\tau_{2,2}$ do *not* represent the entire torque in the joints. The reason is that, in kinematic chains, acceleration in one joint is associated with torque in other joints. In particular, when link 1 rotates nonuniformly at joint 1 and rotation in joint 2 is absent, a certain torque is necessary to maintain the constant angular position at joint 2. For this particular case, we can write the equation $\tau_{2,1} = I_{2,1}\ddot{\alpha}_1$, where $\tau_{2,1}$ is the torque in joint 2 that is related to the acceleration in joint 1 and $I_{2,1}$ is known as the *coupling inertia* or *interactive inertia*. Similarly, angular acceleration at joint 2 is linked with a torque at joint 1. The corresponding equation is $\tau_{1,2} = I_{1,2}\ddot{\alpha}_2$, where $\tau_{1,2}$ is the torque in joint 1 that is associated with the acceleration in joint 2 and $I_{1,2}$ is the coupling inertia. We defer the itemized analysis of the coupling inertia terms $I_{2,1}$ and $I_{1,2}$ to chapter **5**; here we mention only that

$$I_{2,1} = I_{1,2} = m_2 l_1 l_{c2} C_2 + m_2 l_{c2}^2 + I_2 \qquad (4.46)$$

where $C_2 = \cos\alpha_2$. Note that $I_{2,1}$ and $I_{1,2}$ are *not* moments of inertia. They are inertial properties that account for the effect of the motion of one link on the other joint.

For the chain under consideration, the inertial parameters can be written as a (2×2) *chain inertia matrix:*

$$[I^c] = \begin{bmatrix} I_{1,1} & I_{1,2} \\ I_{2,1} & I_{2,2} \end{bmatrix} \qquad (4.47)$$

The corresponding dynamic equations are

$$\begin{bmatrix} \tau_1 \\ \tau_2 \end{bmatrix} = \begin{bmatrix} I_{1,1} & I_{1,2} \\ I_{2,1} & I_{2,2} \end{bmatrix} \begin{bmatrix} \ddot{\alpha}_1 \\ \ddot{\alpha}_2 \end{bmatrix}, \text{ or } \boldsymbol{\tau} = [I^c]\ddot{\boldsymbol{\alpha}} \qquad (4.48)$$

where $\boldsymbol{\tau}$ is the vector of the torque components associated with the angular accelerations at the joints, $[I^c]$ is the inertia matrix of the chain, and $\ddot{\boldsymbol{\alpha}}$ is the vector of joint angular accelerations. Equation 4.48 represents the components of the joint torque that are related solely to the angular acceleration in the joints. The torque components associated with the centripetal and *Coriolis acceleration* and gravity force are *not* included in the equation (see chapter **5**).

4.5.3 Endpoint Inertia

Representative papers: Hogan (1985); Tsuji et al. (1995)

Consider a solid rod that can rotate around a hinge O (figure 4.34). Gravity and friction are neglected. The moment of inertia of the rod with respect to O is I.

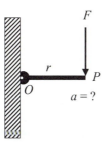

Figure 4.34   Force F is acting on a slender rod at the endpoint P. The acceleration a of the endpoint is sought. Gravity and friction are neglected. See text.

A force F is applied perpendicularly to the bar at the endpoint. The distance from O to the endpoint is r. We are interested in acceleration of the endpoint. Force F generates the moment Fr that induces angular acceleration Fr/I. The linear acceleration of the endpoint is then equal to $a = Fr^2 / I$, and consequently $F = (I / r^2)a$. The expression in parentheses signifies the inertial resistance to force F. It plays the same role as mass plays in the equation $F = ma$ and is known as *apparent mass*. The acceleration in the equation $F = ma$ is experienced by either a material particle or the CoM of the body on which force F is being exerted. The acceleration in the equation $F = (I / r^2)a$ is the acceleration of the point of force application. It is evident that the apparent mass of the rod under consideration is not constant. It depends on the point of force application and the force direction. (For instance, if the force acts along the rod, the effective mass is infinite.)

In multilink chains, the apparent mass affects the *endpoint inertia,* the inertial resistance to the external force exerted on the end of the chain. The endpoint inertia depends on (1) the mass and moments of inertia of individual links; (2) the chain geometry, in particular, the chain Jacobian; and (3) the apparent joint stiffness, the resistance at the joints to the applied perturbation. If the joint stiffness is infinite (whatever the mechanism of stiffness), the chain behaves as one solid body, and analysis of its endpoint inertia is not much different from the analysis performed earlier for a solid rod. We concentrate on the opposite case: zero stiffness. In a chain with zero joint stiffness, the joint torques are always zero. Analysis of such a chain allows estimating the contribution of inertia to the chain's total resistance to the external forces. From the discussion in section **3.2**, it follows that direction of the endpoint deflection does not usually coincide with the direction of the external force. Hence, like endpoint stiffness, the endpoint inertial behavior has a directional property. The apparent endpoint mass (inertia) of a kinematic chain is described by a matrix [M].

$$[M] = \begin{bmatrix} M_{XX} & M_{XY} \\ M_{YX} & M_{YY} \end{bmatrix} \qquad (4.49)$$

where the individual matrix elements represent inertial resistance in the X and Y directions to the force components in these directions. The relationship between the chain inertia matrix (equation 4.47) and the apparent endpoint mass (equation 4.49) resembles the transformation of joint to endpoint stiffness (section **3.4**). We provide the transformation equation without proof:

$$[M] = \left(J[I^c]^{-1} J^T \right)^{-1} = \left(J^T \right)^{-1} [I^c] J^{-1} \qquad (4.50)$$

where [M] is the matrix of the endpoint apparent mass, **J** is the chain Jacobian, and [I^c] is the chain inertia matrix. The chain inertia matrix is symmetric. Its inverse always exists. Note that both [I^c] and **J** depend on the chain configuration. Like endpoint stiffness, apparent endpoint inertia can be represented by an inertia ellipse. The major axis of the inertia ellipse represents the direction of maximal inertial resistance; its minor axis represents the direction of least apparent inertia. The inertia ellipses differ in shape and orientation from the stiffness ellipses (figure 4.35).

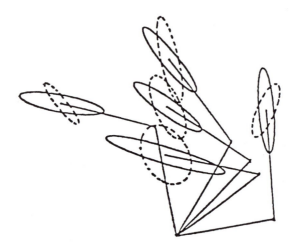

Figure 4.35 Inertia ellipses (solid lines) and stiffness ellipses (dashed lines) obtained at different arm configurations on one subject. The forearm and hand are combined in one segment. The wrist position is constant. The major axis of the inertia ellipses is always along the forearm–hand segment.

Reprinted, by permission, from F.A. Mussa-Ivaldi, N. Hogan, and E. Bizzi, 1985, "Neural, mechanical, and geometric factors subserving arm posture in humans," *The Journal of Neuroscience* 5 (10):2732–2743. Copyright © 1985 by the Society of Neuroscience.

Equation 4.50 can be simplified by using the inverse of the inertia matrix, called the *mobility tensor,* instead of the inertia matrix itself. In an elementary case, this is equivalent to changing the classical equation $F = ma$ to the form $a = F \cdot (1/m) = Fq$, where q is the *mobility,* the inverse of mass. With this designation, acceleration equals force times mobility. In kinematic chains, an acceleration vector **a** produced by an endpoint force **F** equals the product of the endpoint mobility tensor $[Q_{end}]$ and the force vector:

$$\mathbf{a} = [Q_{end}]\mathbf{F} \tag{4.51}$$

The correspondence between the endpoint mobility $[Q_{end}] = [M]^{-1}$ and the chain mobility in joint coordinates $[Q_{joint}] = [I^c]^{-1}$ is

$$[Q_{end}] = \mathbf{J}[Q_{joint}]\mathbf{J}^T \tag{4.52}$$

Equation 4.52 describes the relationship between two different representations of the chain mobility tensor: in external Cartesian coordinates and in joint coordinates. Note that equation 4.52 does not include an inverse Jacobian. Therefore, the equation is valid even when the Jacobian does not have an inverse, such as when the Jacobian is not square or when the link assumes a singular position. Like endpoint inertia, endpoint mobility can be represented by a *mobility ellipse.* The eigenvectors of the mobility ellipse correspond to the directions in which a force and acceleration induced by it are collinear. By

---

■ ■ ■ *FROM THE LITERATURE* ■ ■ ■

Comparing Inertia and Stiffness Ellipses

Source: Tsuji, T. 1997. Human arm impedance in multi-joint movements. In *Self-organization, computational maps and motor control,* ed. P. Morasso and V. Sanguinetti, 357–72. Amsterdam: Elsevier Science.

A two-joint planar robot was used to apply external displacements to the hand of a subject (figure 4.36*a*). The force vector between the hand and the handle was measured, and the arm displacement was registered by a stereo camera. The mass-inertial characteristics of the arm segments were used in the mechanical model that was developed (figure 4.36*b*). The EMG of the biceps brachii caput longum and triceps brachii caput longum muscles was displayed on the oscilloscope to provide information about muscle activation levels. Subjects were asked to maintain the arm in the initial position while keeping the muscle activation level at a target value. Among other parameters, stiffness and inertia ellipses were computed (figure 4.36*c* and *d*).

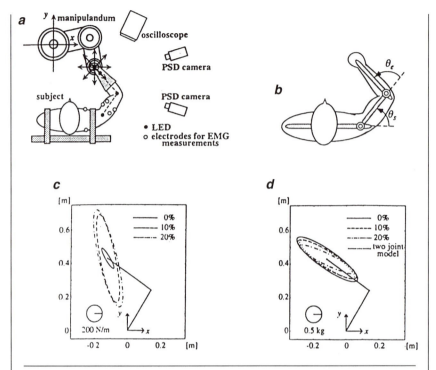

Figure 4.36 Inertia and stiffness of the human arm at different postures. (*a*) Subject and manipulandum. (*b*) Upper-limb model. (*c*) The stiffness ellipses and (*d*) the inertia ellipses at different muscle activation levels: 0%, 10%, and 20% of the maximal force. The inertia ellipses do not depend on the muscle activation levels, while the stiffness ellipses do. The major axes of the inertia ellipses are along the forearm, while the major axes of the stiffness ellipses at nonzero activation levels are close to the shoulder joint.

Adapted from T. Tsuji, *Self-organization, computational maps and motor control*, 1997, 357–372, with permission from Elsevier Science.

changing the body position, the CNS can control the endpoint inertia and mobility of the limbs.

4.5.4 Apparent Mass During Impacts and Striking

Representative papers: Elliott (2000); Tsaousidis and Zatsiorsky (1996)

Control of apparent body inertia is especially important in fast, transient interaction with the environment, such as during the collision phase in striking, kicking, and landing. The apparent mass depends on (1) the point of application

The Effect of Wrist Position on the Apparent Mass of the Arm at the Endpoint

Source: Hogan, N. 1985. The mechanics of multi-joint posture and movement control. *Biol. Cybern.* 52: 315–31.

The effects of different wrist positions on the inertial resistance of the arm at the endpoint were computed. The location of the endpoint was fixed. For a given position of the endpoint, as the wrist rotated, the apparent inertia in any direction underwent a substantial change (figure 4.37 and table 4.9).

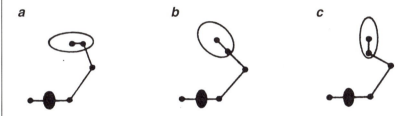

Figure 4.37   Inertia ellipses of a model of the human arm with the endpoint in a constant position and the wrist in different positions. The apparent mass changes by a factor of six between configurations *a* and *c*.

Reprinted, by permission, from *Biological Cybernetics*, The mechanics of multi-joint posture and movement control, N. Hogan, 52: 323, 1985, © 1985 Springer-Verlag.

Table 4.9   Apparent Endpoint Mass of a Planar Model of the Human Arm at Various Wrist Positions

Hand orientation (°)	Apparent endpoint mass (kg)	
	X direction	Y direction
(*a*) 180	1.823	0.322
(*b*) 135	0.568	0.568
(*c*) 90	0.322	1.823

For orientations of the wrist shown in figure 4.37*a*, a perturbing force directed toward the shoulder joint is resisted primarily by the rotational inertia of the hand. For this orientation, the apparent inertia of the end-point is small. For other orientations of the wrist (figure 4.37*b* and *c*), the motion of the endpoint in response to the same perturbing force is resisted by the combined inertia of the hand, the forearm, and the upper arm. For the hand position in figure 4.37*c*, the apparent endpoint inertia is large. As the distal link rotates through 90°, the apparent mass in each of the two orthogonal directions changes by a factor of six.

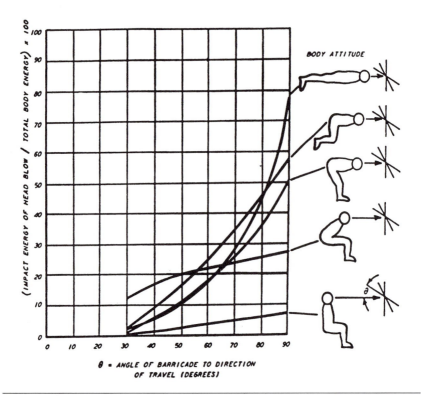

Figure 4.38 Energy of impact as a function of body posture (apparent mass). Data were obtained from the collision of a dummy with a wooden barricade. Body postures ranged from a swan-dive attitude (the top figure on the right) to a seated position (the bottom figure). The ordinate of the chart is the ratio of the energy of the head impact to the energy of the whole body. The abscissa is the angle between the surface of the barricade and the line of the flight of the head at the time of impact. A collision in the swan-dive attitude against the surface at a 90° angle produces the most energy for a given crash velocity.

Reprinted, by permission, from E.R. Dye, 1956, "Kinematics of the human body under crash conditions," *Clinical Orthopaedics and Related Research* 8:305–309.

and direction of the impact force, (2) body posture, and (3) the level of apparent stiffness in joints (body *rigidity*), that is, whether the muscles are contracted or relaxed. The apparent mass strongly influences the impact force and energy (figure 4.38).

Strikes or punches performed by the arm provide the maximal inertial resistance when the line of force action is along the forearm-and-hand segment (like the arm positions seen in boxing punches; see figure 4.37 again). The inertial resistance is minimal when the line of force action is perpendicular to the long axis of the arm (as during volleyball spikes). When a desired external impact force is not very large, the movement is usually performed by the distal segments of small mass. For instance, because the badminton shuttlecock is light, players perform smashes using the wrist alone (figure 4.39). Such a movement is almost never seen in tennis, where the ball is much heavier and larger forces are required. Experienced piano players use fingers alone, the wrist, the forearm-plus-hand segment, or even the entire arm to produce the desired sound.

If the joint angular displacements during a collision phase were zero (i.e., if the joint stiffness were infinite), the human body would behave as a rigid body. In this case, a larger portion of the body mass would contribute to the effect of the impact. To achieve this, experienced soccer players activate the antagonistic muscles of the leg immediately before the collision. This technique decreases the joint deformation and increases the apparent mass. The ability to attain body rigidity during collision is the key to better performance in striking and kicking.

Figure 4.39 (*a*) In badminton, wrist flexion can be sufficient for performing a smash. (*b*) In a tennis serve, the entire arm (and the whole body) is involved.

When details of the behavior of the human body during impact are not important, the body is customarily modeled as a material particle of a given mass (called the *effective mass* or *striking mass*). The dynamic response of an effective mass against a rigid surface approximates the primary response of a more complex system subjected to the same impact conditions. The effective (striking) mass is computed as the ratio

$$m_e = \frac{\int_{t_1}^{t_2} F dt}{v_i} \tag{4.53}$$

where the numerator represents the linear impulse registered during the collision (the area below the force–time curve) and v_i is the velocity of the impactor (e.g., handheld implement, fist, foot) immediately before the collision. Equation 4.53 is based on the principle of linear impulse and momentum (see Mechanics Refresher p. 53). The striking mass (m_s) can also be computed from kinematic data. The formula for conservation of linear momentum during elastic impacts is used

$$m_s v_i^{\text{bef}} + m_b v_b^{\text{bef}} = m_s v_i^{\text{aft}} + m_b v_b^{\text{aft}} \tag{4.54}$$

where subscripts i and b refer to the impactor and the colliding body, respectively, and the superscripts bef and aft designate the velocity values immediately before and immediately after the impact. As an example, let us solve for the striking mass of a golf drive (the mass of a golf ball is $m_b \approx 50$ g). The club velocity is 60 m/s before the impact and 40 m/s after the impact. The ball velocity is 80 m/s. The striking mass is thus 200 g. The striking mass is a highly variable parameter. It varies among individual athletes and among trials by the same athlete.

Equation 4.54 is based on the postulation that during the collision the momentum of the system is conserved, that is, that muscle work performed during this period can be neglected. If the contact time is large and during the contact the contacting bodies are substantially displaced, the muscles contract during this time. In this case, they do work that increases the mechanical energy of the system, the momentum of the system is not conserved, and equation 4.54 cannot be applied. To visualize the described phenomenon, consider two scenarios in which a car collides with an obstacle. In the first case, the engine is off. Hence, the linear momentum of the system is conserved. In the second case, the engine is working; thus, the momentum of the system during the collision increases. Equation 4.54 can be applied to the first situation, but it cannot be used for analyzing the second case, especially not if the time of collision is long and the obstacle as well as the car is displaced.

The concept of striking mass is comparable to the concept of apparent mass. The difference is that the apparent mass represents the inertial resistance of a kinematic chain consisting of rigid links connected by joints. Hence, only joint motion, and not link deformation, is allowed in the apparent-mass model. The concept of striking mass is less strict. Example estimates of the effective mass of the head in children and in an adult are presented in figure 4.41. In children, the effective mass of the head was approximately equal to its real mass. During falling and collision, the children relaxed their neck muscles, and therefore the trunk mass did not contribute to the impact force. In the adult, the effective mass of the head was 2.4 times larger than its real mass, due to the contribution of the rest of the body. It is not clear from this figure whether the children's behavior during falling (relaxing the neck muscles) is beneficial for preventing injuries. There is no doubt that stiffening the joints during falling increases the apparent mass. However, muscle co-contraction may be advantageous in decreasing the risk of joint dislocations or other damage. In martial arts, such as

■ ■ ■ *FROM THE LITERATURE* ■ ■ ■

Collision Phase in Kicking: Muscles Do Work, Momentum Is Not Conserved, and the Striking Mass Cannot Be Reliably Determined

Source: Tsaousidis, N., and V. Zatsiorsky. 1996. Two types of ball-effector interaction and their relative contribution to soccer kicking. *Hum. Mov. Sci.* 15: 861–76.

There are two ways of imparting momentum and kinetic energy to a ball: muscle work performed *during* the contact with the ball (throwing-like) and *before* the contact (impact-like). In throwing, only the first mechanism is manifested, but during striking, both mechanisms may contribute to the outcome. Experienced soccer players can approach the ball with a low velocity and then accelerate it in a throwing-like manner.

Soccer kicking was filmed at 4,000 frames per second. During the period of contact, the ball–foot displacement was large, 26 ± 2.3 cm. At the instant of peak deformation, the ball possessed a considerable speed of 13.4 ± 1.2 m/s, which was more than 50% of the speed at loss of contact, 24.9 ± 1.1 m/s. Thus, more than 50% of the ball's linear momentum was imparted to the ball without any contribution by the potential energy of the ball deformation. During the ball recoil (phase A3 in figure 4.40b), the foot did not decelerate in spite of the force acting on it from the decompressing ball. This is possible only because of a counterbalancing muscle force. Since the muscles are contracting during this time and also generate force, they do mechanical work, which changes the mechanical energy and momentum of the system. Hence, the linear momentum is not conserved. Determination of the striking mass is questionable in this situation.

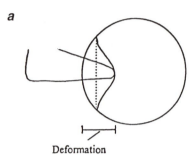

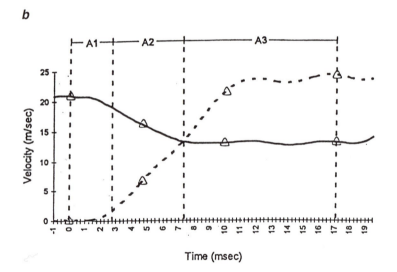

Figure 4.40 (*a*) Schematic of ball deformation during a collision phase in soccer kicking. (*b*) Speed–time curves for foot (solid line) and ball (dotted line) during the collision. The point of intersection of the two curves corresponds to the point of peak ball deformation. Period A1 is the time from the instant of initial contact with the ball to the beginning of ball displacement. A2 is the period of ball deformation during the movement. A3 is the period of ball decompression during the movement. The foot speed is almost constant during this period.

Adapted from *Human Movement Science*, 15, N. Tsaousidis and V. Zatsiorsky, Two types of ball-effector interaction and their relative contribution to soccer kicking, 861–876, 1996, with permission from Excerpta Medica Inc.

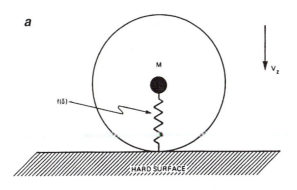

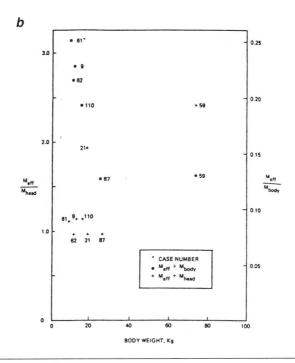

Figure 4.41 (*a*) Model of the head during an impact with a solid surface. (*b*) The effective mass of the head during impacts in children and in an adult. The data were obtained from mathematical modeling of anecdotal observations of falling accidents. The effective mass of the head was normalized with respect to its real mass (crosses, left axis) and the mass of the entire body (black dots, right axis). The numbers refer to individual subjects. Subject 59 was a 21-year-old adult. The other subjects were children, from 1 to 10 years old.

Reproduced, by permission, from D. Mohan et al., 1979, "A biomechanical analysis of head impact injuries to children," *Transactions of the ASME, Journal of Biomechanical Engineering* 101:250–260.

■ ■ ■ *FROM THE LITERATURE* ■ ■ ■

The Effective Mass of the Body During Falls

Source: Kroonenberg, A.J. van den, W.C. Hayes, and T.A. McMahon. 1995. Dynamic models for sideways falls from standing height. *J. Biomech. Eng.* 117: 309–18.

About 250,000 hip fractures occur annually in the United States. A model was developed to predict hip impact force during sideways falls from standing height (figure 4.42). The model requires knowledge of the effective mass of the body that is moving in the vertical direction prior to impact. In this study, the effective mass was defined as the ratio of the vertical impact force applied at the hip divided by the vertical acceleration of this point during the impact. A fall with a vertical trunk (the "vertical jackknife" model) resulted in more than a twofold increase in the effective mass over an impact configuration in which the trunk was inclined at 45° to the vertical. Depending on the body mass and the posture at impact, the effective mass ranged from 15.9 to 70.0 kg. The peak impact force was 2.9 kN for a woman in the 5th percentile by weight and 4.26 kN for a woman in the 95th percentile, thereby confirming the perception that "the bigger they are, the harder they fall." The majority of the predicted values for the peak impact force were below the strength of the femur of young women but above that of old women. The authors said that they could not tell whether falling while relaxed was safer than falling in a muscle-active state.

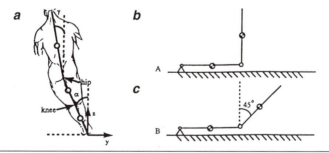

Figure 4.42   (*a*) A model used in the study of the sideways falls. (*b* and *c*) Front view of the impact configuration for two-link models: (*b*) vertical jackknife model, in which the trunk is vertical at impact, and (*c*) 45° jackknife model, in which the trunk is at a 45° angle to the vertical.

Adapted, by permission, from A.J. van der Kronenberg, W.C. Hayes, and T.A. McMahon, 1995, "Dynamic models for sideways falls from standing height," *Journal of Biomechanical Engineering* 117:309–318.

judo, instructors usually teach athletes not to relax while falling and to try to rotate the body when touching the surface. Do they know better?

4.6 SUMMARY

This chapter deals with the inertial properties of individual segments of the human body, also known as body-segment parameters, and of the body as a whole. The inertial properties are location of the center of mass (CoM), moments and products of inertia (which can be written as a tensor of inertia), and radius of gyration.

Density is the amount of mass m per unit of volume V. The center of volume is called the centroid. In a homogeneous body, the masses contained in any two equal volumes are equal. In such a body, the center of mass coincides with the centroid. The human body is nonhomogeneous. Moment of inertia is a measure of the resistance of a body to a change in angular velocity. The moment of inertia plays the same role in angular dynamics as mass plays in linear dynamics. Mass moments of inertia are expressed in $kg \cdot m^2$. The radius of gyration equals the distance from the axis at which a particle having the same mass as an entire body has the same moment of inertia. The resistance to centripetal acceleration is specified by the products of inertia. A product of inertia is zero if the axis of rotation is an axis of symmetry for the mass.

The moment of inertia about any axis equals the moment with respect to the parallel axis through the CoM (the centroidal axis) plus the product of the mass of the body and the square of distance between the axes. This statement is known as the parallel axis theorem. This theorem is commonly used for computing the moments of inertia of body segments with respect to the joint axes when the moments of inertia with respect to the centroidal axes are known.

Because an infinite number of axes can pass through the center of mass, a rigid body has an infinite number of moments of inertia. These values are, however, related in a regular manner. If the moments and products of inertia about the three coordinate axes of a body are known, the moment of inertia with respect to any axis can be computed. There always exists a reference frame such that the products of inertia about its axes vanish. Such axes are called the principal axes of inertia, and the moments of inertia about these axes are called the principal moments of inertia. The moments and products of inertia of a body with respect to three orthogonal coordinate axes can be written as a matrix called the inertia matrix, or the tensor of inertia. The diagonal terms of the inertia matrix are the moments of inertia, and the off-diagonal elements are the products of inertia. The tensor of inertia about a principal axis is a diagonal matrix. The principal axes of inertia of a body are the eigenvectors of its inertia matrix, and the principal moments of inertia are the eigenvalues. The moment of inertia of a body about any axis through a fixed center can be represented by the ellipsoid of inertia. When the orientation of the axes varies, the magnitude of the moments of inertia changes, but the ellipsoid itself remains unchanged.

Inertial properties of the entire human body can be either computed from known properties of individual body segments or determined experimentally. The data on the location of the CoM in an upright posture are presented in table 4.1. For the human body in an upright posture, the maximal moment of inertia is about the anteroposterior axis, and the minimal moment of inertia is about the longitudinal axis. The moment about the frontal axis has an intermediate value. The moments of inertia strongly depend on the body size and are approximately proportional to the height of the body to the fifth power.

This chapter also discusses the inertial properties of body parts. The individual parts move similarly to ideal rigid bodies rotating around the joint axes. The task of segmenting can be solved with only limited accuracy. When body dissection is performed relative to the bony landmarks, the individual body parts are called body segments. When the separation is performed with respect to joint axes and the parts are considered rigid, the parts are called body links. By definition, a body link is the central straight line that extends longitudinally through a body segment and terminates at both ends in axes about which the adjacent members rotate. A body link is an ideal representation of the body segment. The length of the body segments (anatomic length) and the corresponding body links (biomechanical length) may be different. For instance, the length of the upper arm link is shorter than the length of the upper arm segment. Although the principal axes of inertia of human body segments may not coincide with the principal anatomic axes, in practical applications, discrepancies are usually disregarded. Some sample data on body-segment parameters are presented in tables 4.4 and 4.5. Additional data on body-segment parameters are given in the appendices at the end of the book.

The human body is nonhomogeneous. It is a composite of biological tissues that have different density. The relative mass of the body segments is not the same in people of different height and weight. For instance, in heavier people, the relative mass of the abdomen is larger and the relative masses of the head, feet, and hands are smaller than in lighter people. The relative masses of body segments can be used only when the subject's measurements are close to the average values of the corresponding sample. Modeling inertial properties of the trunk is a particularly challenging task because it can barely be considered a single rigid body.

Several methods used for obtaining subject-specific inertial characteristics are discussed:

- Geometric modeling. With this method, the body parts are considered homogeneous solids of simple geometric shape. Density data are taken from the literature.

- Geometric scaling. For this approach, either the entire body or the body parts of different people are assumed to be geometrically similar. The simplest model of geometric scaling, known as the isometry model, assumes that human bodies are geometrically similar and differ only in size. The allometry model presumes that body proportions change systematically with body size. In both cases, certain scaling coefficients are computed. To obtain a subject-specific value, a group mean is multiplied by the scaling factor.
- Regression equations. Frequently, these equations are not based on any mechanics theory but are just curve-fitting procedures.
- Combined methods (nonlinear regression equations). These methods integrate geometric scaling with the statistical regression approach.

Voluntary control of inertial properties of body limbs and the entire body can be done by changing the body posture or its orientation with respect to the external forces and by stiffening or relaxing the joints. In particular, the moments and products of inertia of the entire body can be modulated by repositioning body segments.

The inertial resistance of the limbs (kinematic chains) to angular acceleration in joints is described by the inertia matrix of the chain. The diagonal elements of the matrix are the moments of inertia of the links with respect to the joint axes. The off-diagonal elements are the coupling inertia terms that represent inertial resistance in a joint that is due to angular acceleration in another joint. The endpoint inertia specifies a relationship between an external endpoint force and the endpoint acceleration; it is characterized by the apparent mass. Endpoint inertia can be represented by an ellipse. The apparent mass depends on the point of application and direction of the impact force and on body posture.

4.7 QUESTIONS FOR REVIEW

1. Discuss the mass moments of the first and second order.
2. Explain the application of Varignon's theorem to locating the center of mass.
3. Define homogeneous bodies.
4. Define a centroid.
5. How can a moment of inertia of a body limb be controlled?
6. What is the area moment of inertia?
7. Define a radius of gyration.

8. Discuss products of inertia.

9. Explain the meaning of the local and remote terms in the parallel axis theorem.

10. The masses of the upper, middle, and lower parts of the trunk equal m_1, m_2, and m_3, respectively. The distances of their CoMs to a selected anatomic landmark are l_1, l_2, and l_3, respectively. Write an equation to find the distance of the CoM of the entire trunk to the landmark. Hint: use Varignon's theorem.

11. The moments of inertia of the upper, middle, and lower parts of the trunk about their frontal centroidal axes are I_1, I_2, and I_3, correspondingly. Write an equation to find the moment of inertia of the entire trunk about the frontal axis through its CoM. Hint: use the parallel axis theorem.

12. The moments of inertia of the upper, middle, and lower parts of the trunk about their longitudinal centroidal axes are I_1, I_2, and I_3, correspondingly. Find the moment of inertia of the entire trunk about the longitudinal axis through its CoM. Assume trunk symmetry about the midsagittal plane.

13. To define the moment of inertia of a body with respect to any axis through the CoM, six parameters (or constants) have to be determined. Name these constants.

14. Explain the meaning of the diagonal and off-diagonal elements of the inertia matrix.

15. Discuss the principal axes of inertia and the principal moments of inertia.

16. Discuss the ellipsoid of inertia.

17. Explain the physical meaning of the eigenvectors of the inertia tensor.

18. Approximately where is the CoM of the entire human body located?

19. A gymnast performs giant swings on a high bar and then a tucked somersault dismount. Compare the frontal moments of inertia of the gymnast's body during the giant swings and in the air.

20. Explain the difference between body segments and body links. Define body links.

21. Do the principal axes of inertia of the human body segments coincide with the cardinal (principal) anatomic axes?

22. Discuss the density of the human body and its parts.

23. What is the relationship between the masses of the individual body segments and the total mass of the body?

24. Discuss the methods of geometric modeling of human body segments.

25. Discuss isometry and allometry. Explain the scaling factor.

26. Discuss the nonlinear regression equations used for computation of subject-specific values of body-segment parameters.

27. Discuss the off-diagonal elements of the inertia matrix of a kinematic chain.

28. Define endpoint inertia.

29. Suggest methods to measure (a) the mass, (b) the apparent mass, and (c) the effective mass of a leg.

30. How can the apparent mass of the leg be increased and decreased?

4.8 BIBLIOGRAPHY

Ackland, T.R., B.A. Blanksby, and J. Bloomfield. 1988. Inertial characteristics of adolescent male body segments. *J. Biomech.* 21: 319–27.

Ackland, T.R., P.W. Henson, and D.A. Bailey. 1988. The uniform density assumption: Its effect upon the estimation of body segment inertial parameters. *Int. J. Sports Biomech.* 4: 146–55.

Allum, J.H.J., and L.R. Young. 1976. The relaxed oscillation technique for the determination of the moment of inertia of limb segments. *J. Biomech.* 9: 21–25.

Andrews, J.G., and S.P. Mish. 1996. Methods for investigating the sensitivity of joint resultants to body segment parameter variations. *J. Biomech.* 29 (5): 651–54.

Arampatzis, A., J. Gao, and G.-P. Brüggemann. 1997. Influences of inertial properties on joint resultants. In *XVIth ISB Tokyo congress: Book of abstracts,* 7. Tokyo: University of Tokyo.

Aruin, A., and V. Zatsiorsky. 1988. *Biomechanics in ergonomics.* Moscow: Mashinostroenie.

Asang, E. 1978. Injury thresholds of the leg: Ten years of research on safety in skiing. In *Skiing safety II,* ed. J.M. Figueras, 103–29. Baltimore: University Park Press.

Ascenzi, A. 1993. Biomechanics and Galileo Galilei. *J. Biomech.* 26 (2): 95–100.

Baca, A. 1996. Precise determination of anthropometric dimensions by means of image processing methods for estimating human body segment parameter values. *J. Biomech.* 29 (4): 563–67.

Bahlsen, H.A., and B.M. Nigg. 1987. Estimation of impact forces using the idea of an effective mass. In *Biomechanics X-B,* ed. B. Jonsson, 837–41. Champaign, IL: Human Kinetics.

Barter, J.T. 1957. *Estimation of the mass of body segments.* Report no. WADC TR 57-260, ASTIA document no. 118222. Ohio: Wright-Patterson Air Force Base.

Behnke, A.R. 1961. Comment on the determination of whole body density and a résumé of body composition. In *Techniques for measuring body compositions,* ed. J. Brozek, 118–33. Washington, DC: National Research Council.

Behnke, A.R., B.G. Feen, and W.C. Welham. 1942. The specific gravity of healthy men. *JAMA* 118: 495–98.

Beier, G., and R. Döschl. 1974. Schwerpunktshöhe und Trägheitsmoment beim Fussgänger (Height of the center of gravity and moment of inertia in the pedestrian). *Beitr. Gerichtl. Med.* 32: 60–65.

Berg, C., and J. Rayner. 1995. The moment of inertia of bird wings and the inertial power requirement for flapping flight. *J. Exp. Biol.* 198 (Pt. 8): 1655–64.

Bergsma-Kadijk, J.A., B. Baumeister, and P. Deurenberg. 1996. Measurement of body fat in young and elderly women: Comparison between a four-compartment model and widely used reference methods. *Br. J. Nutr.* 75 (5): 649–57.

Bernstein, N.A., O.A. Salzgeber, P.P. Pavlenko, and N.A. Gurvich. 1931. *Determination of location of the centers of gravity and mass of the limbs of the living human body* (in Russian). Moscow: All-Union Institute of Experimental Medicine.

Bishop, P.J., R.W. Norman, R. Wells, D. Ranney, and B. Skleryk. 1983. Changes in the centre of mass and moment of inertia of a headform induced by a hockey helmet and face shield. *Can. J. Appl. Sport Sci.* 8 (1): 19–25.

Borelli, G.A. 1679. *De motu animalium.* Rome: Bernabò.

Bouisset, S., and E. Pertuzon. 1968. Experimental determination of the moments of inertia of limb segments. In *Biomechanics I,* ed. J. Wartenweiler, 106–9. New York: Karger.

Boyd, E. 1933. The specific gravity of the human body. *Hum. Biol.* 5: 646–72.

Braune, W., and O. Fischer. 1889. Über den Schwerpunkt des menschlichen Körpers mit Rücksicht auf die Austrustung des Deutschen Infanteristern. *Abhandlungen der Mathematische-Physische Klasse der Köninglich-Sächsischen Gesellschaft der Wissenschaften* 15:561–672. Also published in English translation as *On the center of gravity of the human body as related to the equipment of the German infantry soldiers,* trans. P.G.J. Maquet and R. Furlong, Berlin: Springer-Verlag, 1985.

Braune, W., and O. Fischer. 1892. *Bestimmung der Tragheitsmomente des menschlichen Körpers und seiner Glieder.* Leipzig: S. Hirzel. Also published in English translation as *Determination of the moments of inertia of the human body and its limbs,* trans. P. Maquet and R. Furlong, Berlin: Springer-Verlag, 1988.

Brody, H. 1985. The moment of inertia of a tennis racket. *Physics Teacher* 23: 213–17.

Brooks, C.B., and A.M. Jacobs. 1975. The gamma mass scanning technique for inertial anthropometric measurement. *Med. Sci. Sports Exerc.* 7: 290–95.

Brown, G.A., R.J. Tello, D. Rowell, et al. 1987. Determination of body segment inertia parameters. In *RESNA '87,* ed. R.D. Steele and W. Gerrey, vol. 7, 299–301. Washington, DC: Association for the Advancement of Rehabilitation Technology.

Brown, G.M. 1960. Relationship between body types and static posture in young adult women. *Res. Q.* 31: 403.

Brozek, J., and A. Keys. 1952. Body build and body composition. *Science* 116: 140–42.

Buchner, H.H., H.H. Savelberg, H.C. Schamhardt, and A. Barneveld. 1997. Inertial properties of Dutch warmblood horses. *J. Biomech.* 30 (6): 653–658.

Cappozzo, A., and N. Berme. 1990. Subject-specific segmental inertia parameter determination: A survey of current methods. In *Biomechanics of human movement: Applications in rehabilitations, sports and ergonomics,* ed. N. Berme and A. Cappozzo, 179–85. Worthington, OH: Bertec Corp.

Cavanagh, P., and R.J. Gregor. 1974. The quick-release method for estimating the moment of inertia of the shank and foot. In *Biomechanics IV,* ed. R.C. Nelson and C.A. Morehouse, 524–30. Baltimore: University Park Press.

Challis, J.H. 1996. Accuracy of human limb moment of inertia estimations and their influence on resultant joint moments. *J. Appl. Biomech.* 12 (4): 517–30.

Challis, J.H. 1999. Precision of the estimation of human limb inertial parameters. *J. Appl. Biomech.* 15: 418–28.

Challis, J.H., and D.C. Kerwin. 1992. Calculating upper limb inertial parameters. *J. Sports Sci.* 10: 275–84.

Chandler, R.F., C.E. Clauser, J.T. McConville, H.M. Reynolds, and J.W. Young. 1975. *Investigation of the inertial properties of the human body.* Technical report AMRL-TR-74-137, AD-A016-485, DOT-HS-801-430. Ohio: Aerospace Medical Research Laboratories, Wright-Patterson Air Force Base.

Chandler, R.F., C.C. Snow, and J.W. Young. 1978. Computation of mass distribution characteristics of children. In *Proceedings of the Society of the Photo-Optical Instrument Engineers,* ed. A.M. Coblentz and R.E. Herron, vol. 166, 158–61. Bellingham, WA: International Society for Optical Engineering.

Cheng, C., H. Chen, C. Chen, and C. Lee. 2000. Segment inertial properties of Chinese adults determined from magnetic resonance imaging. *Clin. Biomech.* 15 (8): 559–66.

Cheng, E.J., and S.H. Scott. 2000. Morphometry of *Macaca mulatta* forelimb. I. Shoulder and elbow muscles and segment inertial parameters. *J. Morphol.* 245 (3): 206–24.

Chesnin, K.J., M.P. Besser, L. Selby-Silverstein, W. Freedman, and R. Seliktar. 1996. A fourteen segment geometric-based multiple linear regression model for calculating segment masses. In *Twentieth annual meeting of the American Society of Biomechanics: Conference proceedings,* 255–56. Atlanta: Georgia Tech.

Clarys, J.P., and M.J. Marfell-Jones. 1986. Anatomical segmentation in humans and the prediction of segmental masses from intra-segmental anthropometry. *Hum. Biol.* 58: 771–82.

Clarys, J.P., A.D. Martin, and D.T. Drinkwater. 1984. Gross tissue weights in the human body by cadaver dissection. *Hum. Biol.* 56 (3): 459–73.

Clauser, C.E. 1963. *Moments of inertia and centers of gravity of the living human body.* Technical report AMRL-TDR-63-36. Ohio: Wright-Patterson Air Force Base.

Clauser, C.E., J.T. McConville, and J.W. Young. 1969. *Weight, volume and center of mass of segments of the human body.* Technical report no. AMRL-TDR-69-70. Ohio: Wright-Patterson Air Force Base.

Cleaveland, H.G. 1955. The determination of the center of gravity in segments of the human body. Doctoral diss., University of California, Los Angeles.

Collins, M.A., M.L. Millard-Stafford, P.B. Sparling, T.K. Snow, L.B. Rosskopf, S.A. Webb, and J. Omer. 1999. Evaluation of the BOD POD for assessing body fat in collegiate football players. *Med. Sci. Sports Exerc.* 31 (9): 1350–56.

Contini, R., R. Drillis, and M. Bluestein. 1963. Determination of body segment parameters. *Hum. Factors* 5: 493–504.

Cooper, J.M., M. Adrian, and R.B. Glassow. 1982. *Kinesiology.* St. Louis: Mosby.

Correa, S.C., U. Glitsch, M. Paris, and W. Baumann. 1995. Some aspects of two different anthropometric models and their application in motion analysis. In *International Society of Biomechanics, XVth congress, book of abstracts*, 182–83, ed. K. Häkkinen, K.L. Keskinen, P.V. Komi, and A. Mero. Jyväskylä, Finland: Jyväskylä University.

Crisco, J.J., and R.D. McGovern. 1996. Efficient calculation of mass moments of inertia for segmented homogeneous 3D objects. In *Twentieth annual meeting of the American Society of Biomechanics, conference proceedings*, 253–54. Atlanta: Georgia Tech.

Crompton, R.H., Y. Li, R.M. Alexander, W. Wang, and M.M. Gunther. 1996. Segment inertial properties of primates: New techniques for laboratory and field studies of locomotion. *Am. J. Phys. Anthropol.* 99 (4): 547–70.

Croskey, M.I., P.M. Dawson, C. Alma, I.E. Marohn, and H.E. Wright. 1922. The height of the center of gravity in man. *Am. J. Physiol.* 61: 171–85.

Cureton, T.K., and J.S. Wickens. 1935. The center of gravity in the human body in the anterior-posterior plane and its relation to posture, physical fitness and athletic abilities. *Res. Q.* 6 (Suppl. 2): 93–100.

Dainis, A. 1980. Whole body and segment center of mass determination from kinematic data. *J. Biomech.* 13: 647–51.

Dapena, J. 1978. A method to determine the angular momentum of a human body about three orthogonal axes passing its center of gravity. *J. Biomech.* 11:251–56.

Dempster, W.T. 1955. *Space requirements of the seated operator.* Technical report WADC-TR-55-159. Ohio: Wright-Patterson Air Force Base.

Dempster, W.T., W.J. Gabel, and L. Felts. 1959. The anthropometry of manual work space for the seated subject. *Am. J. Phys. Anthropol.* 17: 289–317.

Dempster, W.T., and G.R.L. Gaughran. 1967. Properties of body segments based on size and weight. *Am. J. Anat.* 120 (1): 33–54.

Diffrient, N., A.R. Tilley, and J.C. Bargadhy. 1978. *Humanscale.* Boston: MIT Press.

Dillman, C.J. 1994. Biomechanical contributions to the science of rehabilitation in sports. *Sport Sci. Rev.* 3 (2): 70–78.

Dillon, M.P., T.M. Barker, and M.D. McDonald. 1999. Modelling of body segment parameters for partial foot amputees. In *International Society of Biomechanics, XVIIth congress, book of abstracts*, ed. W. Herzog and A. Jinha, 865. Calgary, AB: University of Calgary.

Donskoy, D., and V. Zatsiorsky. 1979. *Biomechanics.* Moscow: FiS.

Dowling, J.J., and J. Durkin. 1999. Geometric solid modelling of the human forearm. In *American Society of Biomechanics, 23rd annual meeting, October 21–23, 1999, abstract book*, 88–89. Pittsburgh: American Society of Biomechanics.

Drillis, R., and R. Contini. 1966. *Body segment parameters.* Report no. PB 174 945, TR 1166.03. New York: School of Engineering and Science, New York University.

Drillis, R., R. Contini, and M. Bluestein. 1964. Body segment parameters: A survey of measurement techniques. *Artif. Limbs* 8: 44–66.

Drinkwater, D.T., A.D. Martin, W.D. Ross, and J.P. Clarys. 1984. Validation by cadaver dissection of Matiegka's equations for the anthropometric estimation of anatomical body composition in adult humans. In *Perspectives in kinanthropometry,* ed. J.A. Day, 221–27. Champaign, IL: Human Kinetics.

DuBois, J., and W.R. Santschi. 1963. *The determination of the moment of inertia of living human organism.* New York: Wiley.

Du Bois-Reymond, R. 1900. Die Grenzen der Unterstützungsflache beim Stehen. *Arch. Anat. Physiol.* 23: 562–64.

Duggar, S.C. 1963. The center of gravity of the human body. *Hum. Factors* 4: 131–48.

Durkin, J.L., and J.J. Dowling. 1998. A new technique for measuring body segment parameters using dual photon absorptiometry. In *Proceedings of NACOB '98: The third North American Congress on Biomechanics, August 14–18, 1998,* 175–76. Waterloo, ON: University of Waterloo.

Durkin, J.L., and J.J. Dowling. 1999. Predicting body segment parameters for four human populations using DPX. In *International Society of Biomechanics, XVIIth congress, book of abstracts,* ed. W. Herzog and A. Jinha, 815. Calgary, AB: University of Calgary.

Duval-Beaupere, G., and G. Robain. 1987. Visualization on full spine radiographs of the anatomical connections of the centres of the segmental body mass supported by each vertebra and measured in vivo. *Int. Orthop.* 11:261–69.

Dye, E.R. 1956. Kinematics of the human body under crash conditions. *Clin. Orthop.* 8: 305–9.

Elliott, B.C. 2000. Hitting and kicking. In *Biomechanics in sport,* ed. V.M. Zatsiorsky, 487–504. Oxford, UK: Blackwell Science.

Erdmann, W.S. 1987a. The division device for locating segmental mass centers on a picture. In *Biomechanics X-B,* ed. B. Jonsson, 1049–53. Champaign, IL: Human Kinetics.

Erdmann, W.S. 1987b. Individual moment of inertia of the lower extremity of the living young adult male. In *Biomechanics X-B,* ed. B. Jonsson, 1157–61. Champaign, IL: Human Kinetics.

Erdmann, W.S. 1995. *Badania wielkosci geometrycznych i inercyjnych tulowa mezczyzn uzyskanych metoda tomografii komputerowej.* Gdansk: Akademia Wychowania Fizycznego.

Erdmann, W.S. 1997. Geometric and inertial data of the trunk in adult males. *J. Biomech.* 30 (7): 679–88.

Erdmann, W.S., and T. Gos. 1990. Density of trunk tissues of young and medium age people. *J. Biomech.* 23: 945–47.

Finch, C.A. 1985. Estimation of body segment parameters of college age females using a mathematical model. Doctoral diss., University of Windsor.

Fischer, O. 1906. *Theoretische Grundlagen für eine Mechanik der lebenden Körper mit speziellen Anwendungen auf die Menschen.* Berlin: B.G. Teubner.

Fitzpatrick, P., C. Carello, and M.T. Turvey. 1994. Eigenvalues of the inertia tensor and exteroception by the "muscular sense." *Neuroscience* 60 (2): 551–68.

Forwood, M.R., R.J. Neal, and B.D. Wilson. 1985. Scaling segmental moments of inertia for individual subjects. *J. Biomech.* 18: 755–61.

Fox, M.G., and O.G. Young. 1954. Placement of the gravity line in anteroposterior posture. *Res. Q.* 25: 277–283.

Frohlich, C. 1979. Do springboard divers violate angular momentum conservation? *Am. J. Phys.* 47 (7): 583–92.

Fujikawa, K. 1963. The center of gravity in the parts of the human body. *Okajimas Folia Anat. Jpn.* 39: 117–25.

Gagnon, M., and D. Rodrigue. 1979. Determination of the forearm parameters by anthropometry, immersion and photography methods. *Res. Q. Exerc. Sport* 50 (2): 188–98.

Galilei, G. 1974. *Two new sciences, including centers of gravity and force of percussion.* Transl. with introduction and notes by Stillman Drake (ed). Madison: University of Wisconsin Press.

Gervais, P., and G.W. Marino. 1983. A procedure for determining angular position data relative to the principal axes of the human body. *J. Biomech.* 16: 109–13.

Gilbert, J.A., M.C. Skrzynski, and G.E. Lester. 1989. Cross-sectional moment of inertia of the distal radius from absorptiometric data. *J. Biomech.* 22 (6–7): 751–54.

Goto, K., and S. Hideki. 1956. Concerning the measure of the weight and center of gravity in the parts of human body. *Tokyo Ijishinshi* 73 (2): 45–49.

Gruber, K., H. Ruder, J. Denoth, and K. Schneider. 1998. A comparative study of impact dynamics: Wobbling mass versus rigid body models. *J. Biomech.* 31 (5): 439–44.

Gubitz, H. 1978. Zur qualitativen Bestimmung der Lage des Körperschwerpunkts. In *Biomechanische Untersuchungsmethoden im Sport,* ed. W. Marhold, W. Gutewort, and G. Hochmuth, 171–80. Karl-Marx-Stadt, East Germany.

Hale, S.A. 1990. Analysis of the swing phase dynamics and muscular effort of the above-knee amputee for varying prosthetic shank loads. *Prosthet. Orthot. Int.* 14 (3): 125–35.

Hanavan, E.P. 1964. *A mathematical model of the human body.* Report no. AMRL-TR-64-102, AD-608-463. Ohio: Aerospace Medical Research Laboratories, Wright-Patterson Air Force Base.

Harless, E. 1860. Die statische Momente der menschlichen Gliedermassen. *Abhandlungen der Mathematische-Physikalischen Klasse der Königlich-Baverischen Akademie der Wissenschaften, München* 8: 69–97.

Hatze, H. 1975. A new method for the simultaneous measurement of the moment of inertia, the damping coefficient and the location of the center of mass of a body segment in situ. *Eur. J. Appl. Physiol.* 34: 217–26.

Hatze, H. 1980. A mathematical model for the computational determination of parameter values of anthropometric segments. *J. Biomech.* 13: 833–43.

Hay, J.G. 1973. The center of gravity of the human body. In *Kinesiology III,* 20–44. Washington, DC: American Association for Health, Physical Education and Recreation.

Hay, J.G. 1974. Moment of inertia of the human body. In *Kinesiology IV,* 43–52. Washington, DC: American Association for Health, Physical Education and Recreation.

Hay, J.G., B.D. Wilson, J. Dapena, and G.G. Woodworth. 1977. A computational technique to determine the angular momentum of a human body. *J. Biomech.* 10: 269–77.

Hellebrandt, F.A., R.H. Tepper, G.L. Braun, and M.C. Elliott. 1938. The location of the cardinal anatomical orientation planes passing through the center of weight in young adult women. *Am. J. Physiol.* 121: 465–80.

Henson, P.W., T. Ackland, and R.A. Fox. 1987. Tissue density measurements using CT scanning. *Aust. Phys. Eng. Sci. Med.* 10: 162–66.

Herron, R.E., D.V. Cuzzi, and J. Hugg. 1976. *Mass distribution of the human body using biostereometrics.* Report no. AMRL-TR-75-18. Ohio: Wright-Patterson Air Force Base.

Herron, R.E., J.R. Cuzzi, D.V. Goulet, and J.E. Hugg. 1974. *Experimental determination of mechanical features of adults and children.* Report no. DOT-HS-231-2-397. Washington, DC: U.S. Department of Transportation.

Hinrichs, R.N. 1985. Regression equations to predict segmental moments of inertia from anthropometric measurements: An extension of the data from Chandler et al. (1975). *J. Biomech.* 18 (8): 621–24.

Hinrichs, R.N. 1990. Adjustments to the segment center of mass proportions of Clauser et al. (1969). *J. Biomech.* 23 (9): 949–51.

Hof, A.L. 1997. Correcting for limb inertia and compliance in fast ergometers. *J. Biomech.* 30 (3): 295–97.

Hogan, N. 1985. The mechanics of multi-joint posture and movement control. *Biol. Cybern.* 52: 315–31.

Huang, C., C.Y. Chen, and Y. Liu. 1999. The study of dynamical movement of the human center of mass measurement errors by different body segment parameters. In *International Society of Biomechanics, XVIIth congress, book of abstracts,* ed. W. Herzog and A. Jinha, 811. Calgary, AB: University of Calgary.

Huang, H.K., and F.R. Suarez. 1983. Evaluation of cross-sectional geometry and mass density distributions of humans and laboratory animals using computerized topography. *J. Biomech.* 16: 821–32.

Huang, H.K., and S.C. Wu. 1976. The evaluation of mass densities of the human body in vivo from CT scans. *Comput. Biol. Med.* 6: 337–43.

Huston, R.L., and C.E. Passerello. 1971. On dynamics of a human body model. *J. Biomech.* 4: 369–78.

Ignazi, G., A. Coblentz, H. Peneau, P. Hennion, and J. Prudent. 1972. Position du centre de gravite chez l'homme: Determination, signification fonctionelle et evolutive. *Anthropol. Appl.* 43: 72.

Ignazi, G., H. Pineau, and A. Coblentz. 1974. Le centre de gravite en relation avec les caracteres anthropologiques: Influence des facteurs internes et externes (Center of gravity in connection with anthropological characteristics: Influence of internal and external factors). *Arch. Anat. Pathol. (Paris)* 22 (4): 271–77.

Ivanitsky, M.F. 1965. *Human anatomy* (in Russian). Moscow: FiS.

Jensen, R.K. 1975. Model for body segment parameters. In *Biomechanics V-B,* ed. P. Komi, 380–86. Baltimore: University Park Press.

Jensen, R.K. 1978. Estimation of the biomechanical properties of three body types using a photogrammic method. *J. Biomech.* 11: 349–58.

Jensen, R.K. 1981. The effect of a 12-month growth period on the body moments of inertia of children. *Med. Sci. Sports Exerc.* 13 (4): 238–42.

Jensen, R.K. 1986a. Body segment mass, radius and radius of gyration proportions of children. *J. Biomech.* 19 (5): 359–68.

Jensen, R.K. 1986b. The growth of children's moment of inertia. *Med. Sci. Sports Exerc.* 18: 440–45.

Jensen, R.K. 1987. Growth of estimated segment masses between four and sixteen years. *Hum. Biol.* 59: 173–89.

Jensen, R.K. 1988. Developmental relationships between body inertia and joint torques. *Hum. Biol.* 60 (5): 693–707.

Jensen, R.K. 1989. Changes in segment inertia proportions between 4 and 20 years. *J. Biomech.* 22 (6–7): 529–36.

Jensen, R.K. 1993. Human morphology: Its role in the mechanics of movement. *J. Biomech.* 26 (Suppl. 1): 81–94.

Jensen, R.K., S. Doucet, and T. Treitz. 1996. Changes in segment mass and mass distribution during pregnancy. *J. Biomech.* 29 (2): 251–56.

Jensen, R.K., and P. Fletcher. 1993. Body segment moments of inertia of the elderly. *J. Appl. Biomech.* 9: 287–305.

Jensen, R.K., and P. Fletcher. 1994. Distribution of mass to the segments of elderly males and females. *J. Biomech.* 27 (1): 89–96.

Jensen, R.K., and G. Nassas. 1986. A mixed longitudinal description of body shape growth. In *Biostereometrics '85 SPIE,* ed. A.M. Coblentz and R.E. Herron, vol. 602. Bellingham, WA: International Society for Optical Engineering.

Jensen, R.K., and G. Nassas. 1988. Growth of segment principal moments of inertia between four and twenty years. *Med. Sci. Sports Exerc.* 20 (5): 594–604.

Jensen, R.K., T. Treitz, and S. Doucet. 1996. Prediction of human segment inertias during pregnancy. *J. Appl. Biomech.* 12 (1): 15–30.

Jensen, R.K., T. Treitz, and H. Sun. 1997. Prediction of infant segment inertias. *J. Appl. Biomech.* 13 (3): 287–99.

Johnson, R.E. 1976. Comparison of segment ratio weights and segment centers of gravity in the living human male and female. *Res. Q.* 47 (1): 105–9.

Johnson, S.H. 1994. Experimental determination of inertia ellipsoids. In *Science and golf II,* ed. A.J. Cochran and F.R. Farrally, 290–95. London: Spon.

Kaleps, I., C.E. Clauser, J.W. Young, R.F. Chandler, G.F. Zehner, and J.T. McCarville. 1984. Investigation into the mass distribution properties of the human body and its segments. *Ergonomics* 27 (12): 1225–37.

Katch, V., and E. Gold. 1976. Normative data for body segment weights, volumes and densities in cadaver and living subjects. *Res. Q.* 47: 542–47.

Kingma, I. 1998. Challenging gravity: The mechanics of lifting. Doctoral diss., Vrije Universiteit Amsterdam.

Kingma, I.D.L.M.P., H.M.K.H.G. Toussaint, and T.B.M. Bruijnen. 1996. Validation of a full body 3D dynamic linked segment model. *Hum. Mov. Sci.* 15: 833–60.

Kingma, I., H.M. Toussaint, D.A. Commissaris, M.J. Hoozemans, and M.J. Ober. 1995. Optimizing the determination of the body center of mass. *J. Biomech.* 28 (9): 1137–42.

Kingma, I., H.M. Toussaint, M.P. de Looze, and J.H. Van Dieen. 1996. Segment inertial parameter evaluation in two anthropometric models by application of a dynamic linked segment model. *J. Biomech.* 29 (5): 693–704.

Kirkpatrick, S.J. 1990. The moment of inertia of bird wings. *J. Exp. Biol.* 151: 489–94.

Klausen, R., and B. Rasmussen. 1968. On the location of the line of gravity in relation to L5 in standing. *Acta Physiol. Scand.* 72: 45–52.

Klose, H., R. Preiss, and H. Schölhorn. 1993. Experimental examination of methods to determine moment of inertia of human bodies. In *International Society of Biomechanics, XIVth congress, abstracts,* 698–99, Paris.

Kocsis, L. 1998. Investigation of the simulation of the mass moments of inertia for the trunk. In *ISBS '98, XVI International Symposium on Biomechanics in Sports, University of Konstanz, proceedings 1,* ed. H.J. Riehle and M.M. Vieten, 347–50. Konstanz, Germany: UVK-Universitätverlag Konstanz.

Kohda, E., and N. Shigematsu. 1989. Measurement of lung density by computed tomography: Implication for radiotherapy. *Keio J. Med.* 38 (4): 454–63.

Kosa, F., S. Homma, R. Itagaki, T. Ishida, and K. Haruyama. 1967. On the method of measuring moment of inertia of the human body and mechanical quantum during somersault. *Bull. Inst. Sports Sci. (Tokyo)* 13: 65–73.

Kozyrev, G.S. 1962. The center of gravity of human body in healthy people and patients (in Russian). Doctoral diss., Char'kov Medical Institute, Char'kov, Ukraine.

Krabbe, B., R. Farkas, and W. Baumann. 1997. Influence of inertia on intersegment moments of the lower extremity joints. *J. Biomech.* 30 (5): 517–9.

Kroonenberg, A.J. van den, W.C. Hayes, and T.A. McMahon. 1995. Dynamic models for sideways falls from standing height. *J. Biomech. Eng.* 117: 309–18.

Kwon, Y.-H. 1996. Effects of the method of body segment parameter estimation on airborne angular momentum. *J. Appl. Biomech.* 12: 413–30.

Lakie, M., E.G. Walsh, and G.W. Wright. 1981. Measurement of inertia of the hand and the stiffness of the forearm using resonant frequency methods with added inertia or position feedback. *J. Physiol. (Lond.)* 310: 3P–4P.

Lebiedowska, M.K., and A. Polisiakiewicz. 1997. Changes in the lower leg moment of inertia due to child's growth. *J. Biomech.* 30 (7): 723–28.

Lephart, S.A. 1984. Measuring the inertial properties of cadaver segments. *J. Biomech.* 17 (7): 537–43.

Leva, P. de. 1993. Validity and accuracy of four methods for locating center of mass of young male and female athletes. In *International Society of Biomechanics, XIVth congress,* 318–19. Paris.

Leva, P. de. 1996. Joint center longitudinal positions computed from a selected subset of Chandler's data. *J. Biomech.* 29 (9): 1231–33.

Li, Y., and P.H. Dangerfield. 1993. Inertial characteristics of children and their application to growth study. *Ann. Hum. Biol.* 20 (5): 433–54.

Lindsay, M. 1968. Moment of inertia of a human body. In *Research papers in physical education,* 46–49. Leeds, England: British Association of Sports Sciences.

Liu, W., and B.M. Nigg. 2000. A mechanical model to determine the influence of masses and mass distribution on the impact force during running. *J. Biomech.* 33 (2): 219–24.

Liu, Y.K., J.M. Laborde, and W.C. Van Buskirk. 1971. Inertial properties of a segmented cadaver trunk: Their implications in acceleration injuries. *Aerospace Med.* 42 (6): 650–57.

Liu, Y.K., and J.K. Wickstrom. 1973. Estimation of the inertial property distribution of the human torso from segmented cadaver data. In *Perspectives in biomedical engineering,* ed. R.M. Kenedi, 203–13. Baltimore: University Park Press.

Lohman, T.G., A.G. Rocke, and R. Martorelle, eds. 1988. *Anthropometric standardization reference manual.* Champaign, IL: Human Kinetics.

Magnant, D. 1986. Three dimensional measurement of the human body. In *Biostereometrics '85 SPIE,* ed. A.M. Coblentz and R.E. Herron, vol. 602, 130–35. Bellingham, WA: International Society for Optical Engineering.

Mansfield, N.J., and R. Lundstrom. 1999. The apparent mass of the human body exposed to non-orthogonal horizontal vibration. *J. Biomech.* 32 (12): 1269–78.

Martin, P.E., M. Mungiole, M.W. Marzke, and J.M. Longhill. 1989. The use of magnetic resonance imaging for measuring segment inertial properties. *J. Biomech.* 22 (4): 367–76.

Martin, R.B., and D.B. Burr. 1984. Non-invasive measurement of long bone cross-sectional moment of inertia by photon absorptiometry. *J. Biomech.* 17 (3): 195–201.

Matiegka, J. 1921. The testing of physical efficiency. *Am. J. Phys. Anthropol.* 4 (3): 223–32.

Matsui, H. 1974. *Review of our research, 1970–1973.* Nagoya, Japan: Department of Physical Education, University of Nagoya.

Matsuo, A., and T. Fukunaga. 1995. The effect of body structures on moment of inertia of whole body in Japanese female. In *International Society of Biomechanics, XVth congress, book of abstracts,* K. Häkkinen, K.L. Keskinen, P.V. Komi, and A. Mero, eds, 604–5. Jyväskylä, Finland: Jyväskylä University.

Matsuo, A., T. Fukunaga, and S. Uchino. 1990. The estimation of segment weight of human extremities from serial cross-sectional areas and densities of tissues. *Proc. Dept. Sports Sci. Univ. Tokyo* 24: 55–64.

Matsuo, A., H. Ozawa, K. Goda, and T. Fukunaga. 1995. Moment of inertia of whole body using an oscillating table in adolescent boys. *J. Biomech.* 28 (2): 219–23.

McConville, J.T., T.D. Churchill, I. Kaleps, C.E. Clauser, and J. Cuzzi. 1980. *Anthropometric relationships of body and body segment moments of inertia.* Report no. AFAMRL-TR-80-119. Ohio: Aerospace Medical Research Laboratories, Wright-Patterson Air Force Base.

McConville, J.T., and C.E. Clauser. 1976. Anthropometric assessment of the mass distribution characteristics of the living human body. In *Proceedings of 6th congress of the International Ergonomics Association,* 379–83. College Park, MD: University of Maryland.

McLean, S.P., and R.N. Hinrichs. 2000. Influence of arm position and lung volume on the center of buoyancy of competitive swimmers. *Res. Q. Exerc. Sport* 71 (2): 182–89.

Meeh, C. 1895. Volummessungen des menschlichen Körpers und seiner einzelnen Teile in den verschidenen Altersstufen. *Z. Biol.* 13: 125–47.

Milgrom, C., M. Giladi, A. Simkin, N. Rand, R. Kedem, H. Kashtan, M. Stein, and M. Gomori. 1989. The area moment of inertia of the tibia: A risk factor for stress fractures. *J. Biomech.* 22 (11–12): 1243–48.

Miller, D., and R.C. Nelson. 1973. *The biomechanics of sport: A research approach.* Philadelphia: Lea and Febiger.

Miller, D.I., and W. Morrison. 1975. Prediction of segmental parameters using the Hanavan human body model. *Med. Sci. Sports Exerc.* 7: 207–12.

Minetti, A.E., and G. Belli. 1994. A model for the estimation of visceral mass displacement in periodic movements. *J. Biomech.* 27 (1): 97–101.

Mitai, A., S. Anand, and M. Nalgirkar. 1993. Point representation versus solid representation of limb mass and its effect on the moment at a joint. *Int. J. Ind. Ergon.* 11: 67–71.

Mitchell, J.E., B.C. Elliott, and T.R. Ackland. 1993. Inertial characteristics of preadolescent female gymnasts and their effect on performance. In *International Society of Biomechanics, XIV congress, abstracts,* vol. 2, 888–89. Paris.

Mohan, D., B.M. Bowman, R.G. Snyder, and D.R. Foust. 1979. A biomechanical analysis of head impact injuries to children. *Trans. ASME J. Biomech. Eng.* 101: 250–60.

Mori, M., and T. Yamamoto. 1959. Die Massenanteile der einzelner Körperabschitte der Japaner. *Acta Anat.* 37 (4): 385–88.

Mull, R.T. 1984. Mass estimates by computed tomography. *Am. J. Roentgenol.* 143: 1101–4.

Mungiole, M., and P.E. Martin. 1990. Estimating segment inertial properties: Comparison of magnetic resonance imaging with existing methods. *J. Biomech.* 23 (10): 1039–46.

Mussa-Ivaldi, F.A., N. Hogan, and E. Bizzi. 1985. Neural, mechanical and geometric factors subserving arm posture in humans. *J. Neurosci.* 5 (10): 2732–43.

Nesbit, S.M., T.A. Hartzell, J.C. Nalevanko, R.M. Starr, M.G. White, J.R. Anderson, and J.N. Gerlacki. 1996. A discussion of iron golf club head inertia tensors and their effects on the golfer. *J. Appl. Biomech.* 12 (4): 449–69.

Nightingale, R.W., D.L. Camacho, A.J. Armstrong, J.J. Robinette, and B.S. Myers. 2000. Inertial properties and loading rates affect buckling modes and injury mechanisms in the cervical spine. *J. Biomech.* 33 (2): 191–97.

Pagano, C.C., and M.T. Turvey. 1992. Eigenvectors of the inertia tensor and perceiving the orientation of a hand-held object by dynamic touch. *Percept. Psychophys.* 52: 617–24.

Pagano, C.C., and M.T. Turvey. 1998. Eigenvectors of the inertia tensor and perceiving the orientations of limbs and objects. *J. Appl. Biomech.* 14 (4): 331–59.

Page, R.L. 1974. The position and dependence on weight and height of the centre of gravity of the young adult male. *Ergonomics* 17 (5): 603–12.

Palmer, C.E. 1928. Center of gravity of the human body during growth: 1. An improved apparatus for determining the center of gravity. *Am. J. Phys. Anthropol.* 11: 423–55.

Palmer, C.E. 1944. Studies of the center of gravity in the human body. *Child Dev.* 15 (2–3): 99–180.

Park, S.J., C.B. Kim, and S.C. Park. 1999. Anthropometric and biomechanical characteristics on body segments of Koreans. *Appl. Hum. Sci.* 18 (3): 91–99.

Parks, J.L. 1959. An electromyographic and mechanical analysis of selected abdominal exercises. Doctoral diss., University of Michigan.

Pearsall, D.J., and P.A. Costigan. 1999. The effect of segment parameter error on gait analysis results. *Gait Posture* 9 (3): 173–83.

Pearsall, D.J., L. Livingston, and J.G. Reid. 1992. Center of mass of trunk segments relative to the spine as determined by computed tomography. In *2nd North American*

Congress on Biomechanics, abstracts, ed. L. Draganich, R. Wells, and J. Bechtold, 77–78. Chicago: University of Illinois at Chigago.

Pearsall, D.J., and J.G. Reid. 1992. Line of gravity relative to upright vertebral posture. *Clin. Biomech.* 7: 80–86.

Pearsall, D.J., and J.G. Reid. 1994. The study of human body segment parameters in biomechanics: An historical review and current status report. *Sports Med.* 18 (2): 126–40.

Pearsall, D.J., J.G. Reid, and L.A. Livingston. 1996. Segmental inertial parameters of the human trunk as determined from computer tomography. *Ann. Biomed. Eng.* 24 (2): 198–210.

Pearsall, D.J., J.G. Reid, and R. Ross. 1994. Inertial properties of the human trunk of males determined from magnetic resonance imaging. *Ann. Biomed. Eng.* 22 (6): 692–706.

Peeraer, L., E. Willems, H. Stijns, A. Spaepen, and V. Stijnen. 1987. A method to determine segmental moments of inertia, segmental accelerations, and resultant joint forces using accelerometers. In *Biomechanics X-B,* ed. B. Jonsson, 1059–63. Champaign, IL: Human Kinetics.

Peyton, A.J. 1986. Determination of the moment of inertia of limb segments by a simple method. *J. Biomech.* 19 (5): 405–10.

Plagenhoef, S., F.G. Evans, and T. Abdelnour. 1983. Anatomical data for analyzing human motion. *Res. Q. Exerc. Sport* 54: 169–78.

Plagenhoef, S. 1971. *Patterns of human motion. A cinematographic analysis.* Englewood Cliffs, NJ: Prentice Hall.

Qassem, W., and M.O. Othman. 1996. Vibration effects on sitting pregnant women—subjects of various masses. *J. Biomech.* 29 (4): 493–502.

Rabuffetti, M., and G. Baroni. 1999. Validation protocol of models for centre of mass estimation. *J. Biomech.* 32 (6): 609–13.

Raitsina, L.P. 1976. Morphological peculiarities and the position of the center of gravity in female athletes (in Russian). Doctoral diss., Moscow State University.

Rames, M.H.A., A. Cosgrove, and R. Baker. 1999. Comparing methods of estimating the total body centre of mass in three dimensions in normal and pathological gaits. *Hum. Mov. Sci.* 18: 637–46.

Reid, J.G. 1984. Physical properties of the human trunk as determined by computer tomography. *Arch. Phys. Med. Rehabil.* 65: 246–50.

Reid, J.G., and R.K. Jensen. 1990. Human body segment inertia parameters: A survey and status report. *Exerc. Sport Sci. Rev.* 18: 225–41.

Reynolds, E., and R.W. Lovett. 1909. A method for determining the position of the center of gravity in its relation to certain bony landmarks in the erect position. *Am. J. Physiol.* 24: 286–93.

Reynolds, H.M. 1978. The inertial properties of the body and its segments. In *Anthropometric source book: Volume II. A handbook of anthropometric data,* 1–59. NASA reference publication 1024. Houston, TX: National Aeronautics and Space Administration.

Rodrigue, D., and M. Gagnon. 1983. The evaluation of forearm density with axial tomography. *J. Biomech.* 16: 907–13.

Rodrigue, D., and M. Gagnon. 1984. Validation of Weinbach's and Hanavan's models for computation of physical properties of the forearm. *Res. Q. Exerc. Sport* 55: 272–77.

Rork, R., and F.A. Hellebrandt. 1937. The floating ability of college women. *Res. Q.* 8: 19–28.

Sady, S., P. Freedson, V.L. Katch, and H.M. Reynolds. 1978. Anthropometric model of total body volume for males of different sizes. *Hum. Biol.* 50 (4): 529–40.

Saini, M., D.C. Kerrigan, M.A. Thirunarayan, and M. Duff-Raffaele. 1998. The vertical displacement of the center of mass during walking: A comparison of four measurement methods. *J. Biomech. Eng.* 120 (1): 133–39.

Santschi, W.R., J. DuBois, and C. Omoto. 1963. *Moments of inertia and centers of gravity of the living human body*. AMRL Tech. Doc. Report 63-36. Ohio: Wright-Patterson Air Force Base.

Sarfaty, O., and Z. Ladin. 1993. A video-based system for the estimation of the inertial properties of body segments. *J. Biomech.* 26 (8): 1011–16. See also comments in *J. Biomech.* (1995) 28 (4): 483, 485.

Schneider, K., and R.F. Zernicke. 1992. Mass, center of mass, and moment of inertia estimates for infant limb segments. *J. Biomech.* 25 (2): 145–48.

Schneider, K., R.F. Zernicke, B.D. Ulrich, T.J. Hart. 1990. Understanding movement control in infants through the analysis of limb intersegmental dynamics. *J. Mot. Behav.* 22: 493–520.

Shan, G.B., and M. Huo. 1993. A low-cost measuring method for body shape. In *International Society of Biomechanics, XIVth congress, abstracts,* vol. 2, 1230–31. Paris.

Shan, G.B., and K. Nicol. 1993. A method for obtaining anthropometrical data. In *International Society of Biomechanics, XIVth congress, abstracts,* vol. 2, 832–33. Paris.

Sheffer, D., A. Schaer, and J. Baumann. 1989. Stereophotogrammetric mass distribution parameter determination of the lower body segments for use in gait analysis. In *Biostereometrics '88 SPIE,* ed. J.E. Baumann and R.E. Herron, vol. 1030, 361–68. Bellingham, WA: International Society for Optical Engineering.

Simic, D. 1976. Odredlivanje tezista coveka pri sedenju (Determination of man's gravity center in the sitting position). *Srp. Arh. Celok. Lek.* 104 (3–4): 199–208.

Simon, S.R., J.K. Knirk, J.M. Mansour, and M.F. Koskinen. 1977. The dynamics of the center of mass during walking and its clinical applicability. *Bull. Hosp. Jt. Dis.* 38 (2): 112–16.

Sinclair, D.S. 1985. *Human growth after birth*. Oxford, UK: Oxford University Press.

Smith, A.W., and D.A. Winter. 1987. The effects of different segmental length determinations on segment inertia during planar airborne activities. In *Biomechanics X-B,* ed. B. Jonsson, 1065–69. Champaign, IL: Human Kinetics.

Smith, J.W. 1957. The forces operating at the human ankle during standing. *J. Anat.* 91: 545–64.

Sommer, H.J., III, N.R. Miller, and G.J. Pijanowski. 1982. Three-dimensional osteometric scaling and normative modeling of skeletal segments. *J. Biomech.* 15 (3): 171–80.

Southard, D. 1998. Mass and velocity: Control parameters for throwing patterns. *Res. Q. Exerc. Sport* 69 (4): 355–67.

Spivak, C.D. 1915. Methods of weighing parts of the living human body. *JAMA* 65: 1707–8.

Sprigings, E.J., D.B. Burko, L.G. Watson, and H. Laverty. 1987. An evaluation of three segmental methods used to predict the location of the total body CG for human airborne movements. *J. Hum. Mov. Stud.* 13: 57–68.

Stein, R.B., P. Zehr, M.K. Lebiedowska, D. Popovic, A. Scheiner, and H.J. Chizeck. 1996. Estimating mechanical parameters of leg segments in individuals with and without physical disabilities. *IEEE Trans. Rehabil. Eng.* 4 (3): 201–11.

Stijnen, V.A., A.J. Spaepen, and E.J. Willems. 1976. Determination of the segmental masses by film analysis and force plate. *Hermes* 10: 385–96.

Stijnen, V.V., A.J. Spaepen, and E.J. Willems. 1981. Models and methods for the determination of the center of gravity of the human body from film. In *Biomechanics VII-A,* ed. A. Morecki, K. Fidelus, K. Kedzior, and A. Wit, 558–64. Baltimore: University Park Press.

Stijnen, V.V., E.E. Willems, A.J. Spaepen, L. Peeraer, and M. Van Leemputte. 1983. A modified release method for measuring the moment of inertia of the limbs. In *Biomechanics VIII-B,* ed. H. Matsui and K. Kobayashi, 1138–43. Champaign, IL: Human Kinetics.

Sun, H., and R. Jensen. 1994. Body segment growth during infancy. *J. Biomech.* 27 (3): 265–75.

Susse, H.J., and A. Wurterle. 1967. Die Schwerpunktverschiebung im wachsenden Fetus (The shifting of the center of gravity in the growing fetus). *Zentralbl. Gynakol.* 89 (27): 980–84.

Swearingen, J.J. 1962. *Determination of centers of gravity of men.* Report no. 62-14. Oklahoma City, OK: Civil Aeromedical Research Institute, Federal Aviation Agency.

Tichonov, V.N. 1975. Distribution of body masses of a sportsman. In *Biomechanics V-B,* ed. P.V. Komi, 103–8. Champaign, IL: Human Kinetics.

Tischler, V.A., V.M. Zatsiorsky, and V.N. Seluyanov. 1981. Study of the mass-inertial characteristics of human body by the gamma-scanning method during 6-month hypokinesia (in Russian). *Kosmicheskaja Biologija i Aviakosmicheskaja Medicina (Space Biology and Aerospace Medicine)* 15 (1): 36–42.

Tsaousidis, N., and V. Zatsiorsky. 1996. Two types of ball-effector interaction and their relative contribution to soccer kicking. *Hum. Mov. Sci.* 15: 861–76.

Tsuang, Y.H., O.D. Schipplein, J.H. Trafimow, and G.B. Andersson. 1992. Influence of body segment dynamics on loads at the lumbar spine during lifting. *Ergonomics* 35 (4): 437–44.

Tsuji, T. 1997. Human arm impedance in multi-joint movements. In *Self-organization, computational maps and motor control,* ed. P. Morasso and V. Sanguinetti, 357–72. Amsterdam: Elsevier Science.

Tsuji, T., P.G. Morasso, K. Goto, and K. Ito. 1995. Human arm impedance characteristics during maintained posture. *Biol. Cybern.* 72: 475–85.

Turvey, M.T. 1998. Dynamics of effortful touch and interlimb coordination. *J. Biomech.* 31 (10): 873–82.

Vaughan, C.L., J.G. Andrews, and J.G. Hay. 1982. Selection of body segment parameters by optimization methods. *J. Biomech. Eng.* 104: 38–44.

Veeger, H.E., F.C.T. van der Helm, L.H.V. van der Woude, G.M. Pronk, and R.H. Rozendal. 1991. Inertia and muscle contraction parameters for musculoskeletal modelling of the shoulder mechanism. *J. Biomech.* 24: 615–29.

Verstraete, M.C. 1992. A technique for locating the center of mass and principal axes of the lower limb. *Med. Sci. Sports Exerc.* 24 (7): 825–31.

Vilensky, J.A. 1978. Masses, centers-of-gravity, and moments-of-inertia of the body segments of the rhesus monkey (*Macaca mulatta*). *Am. J. Phys. Anthropol.* 50 (1): 57–65.

Walker, L.B., E.H. Harris, and V.R. Pontius 1973. *Mass, volume, center of mass and mass moment of inertia of head and head and neck of the human body.* Report no. AD-762 581. Washington, DC: Department of the Navy, Office of Naval Research.

Walsh, E.G., and G.W. Wright. 1987. Inertia, resonant frequency, stiffness and kinetic energy of the human forearm. *Q. J. Exp. Physiol.* 72: 161–70.

Walton, J.S. 1970. A template for locating segmental centers of gravity. *Res. Q.* 41: 615–18.

Ward, C.H.T. 1976. Effects of mass distribution and inertia on selected mechanical and biological movement components. In *Biomechanics V-B,* 46–51. Baltimore: University Park Press.

Waterland, J.C., and G.M. Shambles. 1970. Biplane center of gravity procedures. *Percept. Mot. Skills* 30: 511–16.

Wei, C., and R.K. Jensen. 1995. The application of segment axial density profiles to a human body inertia model. *J. Biomech.* 28 (1): 103–8.

Weinbach, A.P. 1938. Contour maps, center of gravity, moment of inertia and surface area of the human body. *Hum. Biol.* 10: 356–71.

Welham, W.C., and A.R. Behnke. 1942. The specific gravity of healthy men. *JAMA* 118: 498–501.

Whitsett, C.E. 1963. *Some dynamic response characteristics of weightless man.* Report no. AMRL-TDR-63-18, AD-412 541. Ohio: Aerospace Medical Research Laboratories, Wright-Patterson Air Force Base.

Widule, C.J. 1976. Segmental moment of inertia scaling procedure. Res. Q. 47 (2): 143-47.

Willems, E., and P. Swalus. 1968. Apparatus for determining the center of gravity of the human body. In *Biomechanics I,* ed. J. Wartenweiler, 72–77. New York: Karger.

Withers, R.T., J. LaForgia, R.K. Pillans, N.J. Shipp, B.E. Chatterton, C.G. Schultz, and F. Leaney. 1998. Comparisons of two-, three-, and four-compartment models of body composition analysis in men and women. *J. Appl. Physiol.* 85 (1): 238–45.

Woodard, H.Q., and D.R. White. 1986. The composition of body tissues. Br. J. Radiol. 59 (708): 1209-18.

Yeadon, M.R. 1990. The simulation of aerial movement: II. A mathematical inertia model of the human body. *J. Biomech.* 23 (1): 67–74.

Yeadon, M.R. 1993. The biomechanics of twisting somersaults: III. Aerial twist. *J. Sports Sci.* 11: 209–18.

Yeadon, M.R., J.H. Challis, and J.A. Draper. 1992. Measurement of body segment parameters. *J. Sports Sci.* 10: 589–90.

Yeadon, M.R., J.H. Challis, and R. Ng. 1993. Personalised segmental inertia parameters. In *International Society of Biomechanics, XIVth congress, abstracts,* vol. 2, 1494–95. Paris.

Yeadon, M.R., and E.C. Mikulcik. 1996. The control of non-twisting somersaults using configuration changes. *J. Biomech.* 29 (10): 1341–48.

Yeadon, M.R., and M. Morlock. 1989. The appropriate use of regression equations for the estimation of segmental properties. *J. Biomech.* 22: 683–89.

Yokoi, T., K. Shibukawa, M. Ae, and Y. Hashihara. 1985. Body segment parameters of Japanese children. In *Biomechanics IX-B,* ed. D.A. Winter, R.W. Norman, R.P. Wells, K.C. Hayes, and A.E. Patla, 227–32. Champaign, IL: Human Kinetics.

Young, J.W., R.F. Chandler, C.C. Snow, K.M. Robinette, G.F. Zehner, and M.S. Lofberg. 1983. *Anthropometric and mass distribution characteristics of the adult female.* Technical report. Oklahoma City, OK: FAA Civil Aeromedical Institute.

Zamparo, P., G. Antonutto, C. Capelli, M.P. Francescato, M. Girardis, R. Sangoi, R.G. Soule, and D.R. Pendergast. 1996. Effects of body size, body density, gender and growth on underwater torque. *Scand. J. Med. Sci. Sports* 6 (5): 273–80.

Zatsiorsky, V.M., A.S. Aruin, and W.N. Seluyanov. 1981. Biomechanics of musculoskeletal system (in Russian). Moscow: FiS. Also published in German translation as *Biomechanik des menschlichen Bewegungsapparates.* Berlin: Sportverlag, 1984.

Zatsiorsky, V.M., L.M. Raitsin, V.N. Seluyanov, A.S. Aruin, and B.I. Prilutsky. 1993. Biomechanical characteristics of the human body. In *Biomechanics and performance in sport,* ed. W. Baumann, 71–83. Cologne: Bundesinstitut für Sportwissenschaft.

Zatsiorsky, V.M., and V.N. Seluyanov. 1979. Mass-inertia characteristics of human body segments and their relationship with anthropometric parameters (in Russian). *Voprosy Anthropologii (Problems in Anthropology)* 62: 91–103.

Zatsiorsky, V.M., and V.N. Seluyanov. 1983. The mass and inertia characteristics of the main segments of the human body. In *Biomechanics VIII-B,* ed. H. Matsui and K. Kobayashi, 1152–59. Champaign, IL: Human Kinetics.

Zatsiorsky, V.M., and V.N. Seluyanov. 1985. Estimation of the mass and inertia characteristics of the human body. In *Biomechanics IX-B,* ed. D.A. Winter, R.W. Norman, R.P. Wells, K.C. Hayes, and A.E. Patla, 233–39. Champaign, IL: Human Kinetics.

Zatsiorsky, V.M., V.N. Seluyanov, and L. Chugunova. 1990a. In vivo body segment inertial parameters determination using a gamma-scanner method. In *Biomechanics of human movement: Application in rehabilitation, sports, and ergonomics,* ed. N. Berme and A. Cappozzo, 186–202. Worthington, OH: Bertec Corp.

Zatsiorsky, V.M., V.N. Seluyanov, and L.G. Chugunova. 1990b. Methods of determining mass-inertial characteristics of human body segments. In *Contemporary problems of biomechanics,* ed. G.G. Chernyi and S.A. Regirer, 272–91. Moscow, Boca Raton: Mir, CRC Press.

Zatsiorsky, V.M., M.G. Sirota, B.I. Prilutsky, and L.M. Raitsin. 1985. Biomechanics of human body and movements after 120 days of anti-orthostatic hypokinesia (in Russian). *Kosmicheskaja Biologija i Aviakosmicheskaja Medicina (Space Biology and Aerospace Medicine)* 19 (5): 23–27.

Zhu, X.P., D.R. Checkley, D.S. Hickey, and I. Isherwood. 1986. Accuracy of area measurements made from MRI images compared with computed tomography. *J. Comput. Assist. Tomogr.* 10 (1): 96–102.

Zook, D.E. 1932. The physical growth of boys. *Am. J. Disabled Child.* 43: 1347–432.

5
CHAPTER

JOINT TORQUES AND FORCES:
THE INVERSE PROBLEM OF DYNAMICS

Beware of mathematicians and all those who make empty prophecies.
The danger already exists that the mathematicians
have made a covenant with the devil to darken the spirit
and to confine man in the bonds of Hell.

St. Augustine (354–430)

This chapter concentrates on the dynamics of human motion, specifically on the cause–effect relationships between joint torques and the time course of motion. The relationships are described by differential equations, known as *equations of motion* or *dynamic equations.* The dynamic equations are used either to predict the changes in motion caused by the given forces (*direct problem of dynamics*) or to determine the forces that produced a given motion (*inverse problem of dynamics;* figure 5.1).

In the rudimentary case of a material particle moving in the absence of friction and gravity, the equation of motion is $F = ma$, and the direct problem corresponds to finding acceleration a when mass m and force F are known. The inverse problem deals with finding force F when mass m and acceleration a are established. The adjective "inverse" refers to the fact that forces cause the change in motion, but the opposite is not true.

This chapter is concerned with the inverse problem. For human motion, the inverse problem of dynamics can be described thus: given a motion (trajectory, velocity, and acceleration) of body parts, find the joint torques and force that caused the motion. We restrict our discussion to the main ideas and the most popular methods. Fortunately, the main concepts can be grasped easily when working with relatively simple objects, such as two-link chains. In biomechanics research, the inertial characteristics of the subjects are determined by the techniques described in chapter **4**, and movement kinematics, particularly acceleration, are recorded.

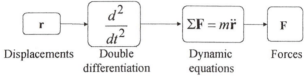

Direct problem of dynamics

Forces — Dynamic equations — Double integration — Displacements

Inverse problem of dynamics

Displacements — Double differentiation — Dynamic equations — Forces

Figure 5.1 Direct and inverse problems of dynamics.

Computation of joint torques is based on mechanics theory and experimental measurements. This chapter covers the theory but not the experimental methods. It seems that contemporary experimental methods, as seen in many studies, are accurate enough to obtain reproducible estimates of joint torques and forces in the majority of movements. Still, the reader should be aware of the main causes of potential inaccuracies:

- Displacement of the skin markers with respect to the skeleton
- Difficulties in determining the acceleration of the markers from their known positions
- Difficulties in locating the joint axes of rotation
- Inaccuracy in determining the body-segment parameters
- Inaccuracy in determining the point of application of the external forces, in particular, the GRF

A large body of contemporary research is devoted to improving the accuracy of biomechanical methods.

During impacts, such as landing in sport movements, the human body behaves in a complex fashion. During the first 20 to 30 ms after the contact, the soft parts of the segments shift relative to the bones and fluctuate in a damped manner. The basic model that assumes rigidity of all the body segments is not appropriate for studying these phenomena. Models that are more complex should be used. Such models are not described in this book.

In section **5.1** basic dynamic equations are considered. Section **5.2** is devoted to the analysis of simple planar chains. For a planar two-link chain, the closed-form dynamic equations are derived and analyzed in detail. Our attention is given mainly to the physical interpretation of the individual terms in the

The Rigid-Body Model Yields Incorrect Values of Joint Torques and Forces During Impacts

Source: Gruber, K., H. Ruder, J. Denoth, and K. Schneider. 1998. A comparative study of impact dynamics: Wobbling mass model versus rigid body models. *J. Biomech.* 31 (5): 439–44.

A drop jump from a height of 40 cm with a heel landing on a hard surface was studied. Two models were compared: the rigid-body model (basic model) and the wobbling-mass model. In the second model, the soft parts of the segments were represented as rigid "wobbling masses" that can move and rotate relative to the skeleton (figure 5.2a). The wobbling masses are coupled elastically to each skeleton part and strongly damped. A representative sample of the obtained results is presented in figure 5.2b. During the first 50 ms after the contact, the two models yield quite different results.

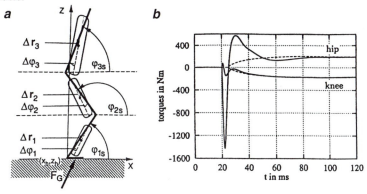

Figure 5.2 Wobbling-mass model. (a) The three-link model with wobbling masses representing soft parts of the segments. In the model, the wobbling masses are rigid bodies that are connected to the skeleton and can move and rotate about it. (b) Comparison of the torques in the knee and hip joints that result from inverse dynamics in the wobbling-mass model (dashed line) and in the basic model (solid line). The differences in the torque magnitude during the first 50 ms after the collision are evident. This is an extreme case: people usually do not land on their heels when dropping from such a large height (40 cm).

dynamic equations. Section **5.3** is concerned with the dynamics of motion in three dimensions. Subsection **5.3.1** is intended to familiarize readers with the motion of a rigid body, in particular, with the Euler equations. Then, two basic methods of obtaining the equations of motion for the multilink chains are described: the recursive Newton-Euler formulation (section **5.3.2**) and the Lagrangian formulation (section **5.3.3**). Section **5.4** discusses joint torques and forces in human movement. It focuses mainly on the origin of joint torques and forces and the effects that they produce. Throughout the chapter, a planar two-link chain serves as a convenient model for studying the inverse problem of dynamics. For this chain, the inverse dynamic problem is solved in three different ways: the elementary geometric solution, the recursive Newton-Euler formulation, and the Lagrangian formulation. Repetition and cross sections from different angles are effective pedagogical tools. In this connection, it is appropriate to quote Lewis Carroll in the *Hunting of the Snark:* "I have said it thrice: What I tell you three times is true."

5.1 BASIC DYNAMIC EQUATIONS

When solving inverse problems of dynamics, the real forces that are acting on a rigid body cannot be determined. For instance, if two forces act in opposite directions and cancel each other, we cannot know anything about their existence. The only available option is to find a resultant force and moment that caused the body to change its velocity. Various other equivalent solutions may exist.

• • • MECHANICS REFRESHER • • •

Equations of Motion of a Rigid Body

The fundamental equations of motion of a rigid body are

$$\Sigma \mathbf{F} = m\mathbf{a}_{\text{CoM}}$$

$$\Sigma \mathbf{M} = \dot{\mathbf{H}} \qquad (5.1)$$

where $\Sigma \mathbf{F}$ and $\Sigma \mathbf{M}$ are the sums of all the external forces and moments of force acting on the body, respectively; m is the body mass; $\mathbf{a}_{\text{CoM}}$ is the acceleration of the center of mass; and $\dot{\mathbf{H}}$ is the time rate of the angular momentum of the body. The moments of force are taken about the CoM. Equations 5.1 are called the *Newton-Euler equations*. $\Sigma \mathbf{F}$ and $\Sigma \mathbf{M}$ are equipollent to a resultant force vector attached at the CoM, $\mathbf{F}_R$, and a resultant couple, $\mathbf{M}_R$. The vectors $m\mathbf{a}$ and $\dot{\mathbf{H}}$ are called the *effective force* and *effective moment,* correspondingly. They are expressed in units of force and moment of force, respectively. The effective force is applied at the CoM, whereas the effective moment, like a couple, can be applied at any point on the body. Hence, equations 5.1 affirm that any system of external forces acting on a rigid body is equivalent to the system consisting of the effective force and couple (figure 5.3).

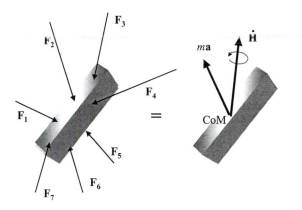

Figure 5.3   The system of external forces is equivalent to the system consisting of vector $m\mathbf{a}$ attached to the CoM and the couple of moment $\dot{\mathbf{H}}$.

The vector $-m\mathbf{a}$ is known as the *reversed effective force*. It is also frequently called the *inertial force*. Although the latter term has been criticized as inaccurate for more than 100 years, it is convenient and still in use. Analogously, the reversed effective moment is called the *inertial moment*. The equation $\mathbf{F} = m\mathbf{a}$ can also be written as $\mathbf{F} - m\mathbf{a} = 0$. Formally, the latter equation specifies a condition for equilibrium. Hence, the equations of motion that contain both the real forces and the reversed effective force are similar to the equations of static equilibrium. This statement is known as *d'Alembert's principle* (after the French scientist Jean Le Rond d'Alembert, 1717–1783). The principle is based on Newton's second law of motion and can be stated thus: the algebraic sum of external forces and the forces resisting motion is zero. The principle enables us to solve dynamic problems by using the methods of statics.

In the dynamics of rigid bodies, reducing the force system to a resultant force acting at the CoM and a couple is the most popular solution. The dynamic equations in this case are simple: one equation describes the CoM translation, while the other describes the body's rotation about the CoM (see equations 5.1). However, in biomechanics, reducing the resultant force system to the link CoM is not sufficient: it does not tell anything about the forces and moments that act at the joints and change the link motion. The terms on the left sides of equations 5.1 should be itemized; they should explicitly include the joint forces and moments.

Consider a diagram of an individual link with all the forces and moments, including the reversed effective force and couple (figure 5.4). The figure is similar to the free-body diagram of the link that describes the static balance of forces (figure 2.9), except that the reversed effective force and couple are added. Such diagrams are known as *kinetic diagrams* (other terms are also in use).

The equations of motion are analogous to the equations of static equilibrium with the reversed effective force and couple added. The equations are written in the global reference system; they specify the linear acceleration of the link's CoM and the angular acceleration of the link with respect to the environment. For the planar case, the equations are

$$\mathbf{F}_i - \mathbf{F}_{i+1} + m_i\mathbf{g} - m_i\mathbf{a}_i = 0 \tag{5.2a}$$

$$\mathbf{T}_i - \mathbf{T}_{i+1} + \mathbf{r}_i \times \mathbf{F}_i - \mathbf{r}_{i+1} \times \mathbf{F}_{i+1} - I_i\ddot{\theta}_i = 0 \tag{5.2b}$$

where $\mathbf{F}_i$ and $\mathbf{F}_{i+1}$ are the forces acting on link i at joints i and $i + 1$, respectively; $\mathbf{T}_i$ and $\mathbf{T}_{i+1}$ are the torques exerted on link i at these joints; $\mathbf{r}_i$ and $\mathbf{r}_{i+1}$ are the

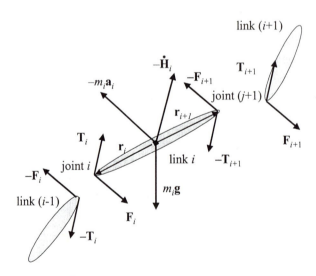

Figure 5.4 Kinetic diagram of link i. $\dot{\mathbf{H}}_i$ is the time rate of change of the angular momentum of the link; m_i is the link mass; and $\mathbf{a}_i$ is the linear acceleration of the CoM of the link. Link i has two contacts: one with a link $i + 1$ at joint $i + 1$ and one with a link $i - 1$ at joint i. Axes of a local reference frame (not shown in the figure) with an origin at the CoM of the link are aligned with the principal axes of inertia.

radii from the CoM of link i to the joint centers i and $i + 1$, correspondingly; **g** is the acceleration of gravity ($g = -9.81$ m/s²); I_i is the moment of inertia of the link about its CoM; and $\ddot{\theta}_i$ is the angular acceleration (in two dimensions; $I_i \ddot{\theta}_i = \dot{\mathbf{H}}_i$). The vector products $\mathbf{r}_i \times \mathbf{F}_i$ and $\mathbf{r}_{i+1} \times \mathbf{F}_{i+1}$ are the moments of force about the CoM of the link. The vector $-m_i \mathbf{a}_i$ is the reverse effective force (inertial force) acting at the CoM of the link, and the product $-I_i \ddot{\theta}_i$ is the reverse effective moment (inertial moment) about the link's CoM. Since the gravity vector $m_i \mathbf{g}$ is applied directly at the CoM, it does not appear in the moment equation 5.2b. Indirectly, the rotational effect of gravity is represented by a joint force. Having computed the effective force ($m_i \mathbf{a}_i$) and moment ($I_i \ddot{\theta}_i$) acting on the link, it remains to compute the joint torques ($\mathbf{T}_i$ and $\mathbf{T}_{i+1}$) and forces ($\mathbf{F}_i$ and $\mathbf{F}_{i+1}$) that caused this effective force and moment. This is the topic of the following sections.

5.2 INVERSE DYNAMICS OF SIMPLE PLANAR CHAINS

We consider the movement first of a single link and then of two- and three-link planar chains. The main goal is to discuss the physical meaning of the individual terms of the dynamic equations.

5.2.1 Single-Link Motion

Consider rotation of one link about a fixed joint axis *O-O* (figure 5.5). Gravity is neglected. If the angular velocity $\dot{\theta}$ is constant, the material particles of the link experience centripetal acceleration (see section **3.2** in *Kinematics of Human Motion*). Centripetal, or normal, acceleration is the time rate of the change in the direction of the velocity. According to Newton's second law, acceleration is always induced by a force. The force that induces the centripetal acceleration is called the *centripetal force*. The magnitude of this force equals the mass of the rotating body, m, times the centripetal (normal) acceleration of its CoM, a_n; $F = ma_n = ml_c \dot{\theta}^2$, where l_c is the distance from the axis of rotation to the CoM of the body and $\dot{\theta}$ is the angular velocity. The centripetal force acts along the radius in the direction of the axis of rotation. The centripetal force is applied to the rotating body; it causes the velocity of the individual particles of which the body is composed to change the body's direction. The centripetal force is a force of action. According to Newton's third law, an equal and opposite force of reaction acts on a body (e.g., an axle, a neighboring link) that is in contact with the first body. This force is called the *centrifugal force*. The centripetal force and the centrifugal force have the same

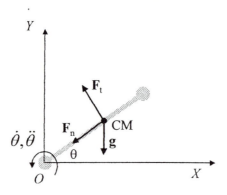

Figure 5.5   A single link rotating at an angular velocity $\dot{\theta}$ and angular accelera-tion $\ddot{\theta}$ at joint O. The distance from the axis of rotation to the CoM is l_c (not shown). $\mathbf{F}_n$ is the normal (centripetal) force that causes the change in the direction of the velocity vector, and $\mathbf{F}_t$ is a tangential force at the CoM that causes the change in the magnitude of the velocity vector.

magnitude and the same line of action but the opposite sense. They are ap-plied to different bodies.

If the body rotates nonuniformly, the velocity of individual particles changes in magnitude, and the particles of the body experience tangential acceleration in addition to centripetal acceleration. The centripetal and the tangential forces are perpendicular to each other. The fundamental equations 5.1 are still valid, and we can write the following set of equations of motion:

$$F_n = ml_c\,\dot{\theta}^2$$
$$F_t = ma_{\text{CoM}} = ml_c\,\ddot{\theta} \tag{5.3}$$
$$M = I\ddot{\theta}$$

where F_n is the normal (centripetal) force, F_t is the tangential force acting at the CoM, I is the moment of inertia of the body with respect to its CoM, M is the moment of force with respect to the CoM, and $\ddot{\theta}$ is the angular accelera-tion. Equations 5.3 are not always convenient to use.

We replace the force-couple system acting at the CoM by an equivalent force-couple system acting at the center of rotation O, as described in section **1.1.5**. At the CoM of the link, two forces are acting: (1) the centripetal force, which depends on the angular velocity, and (2) the tangential force, which depends on the angular acceleration of the link. These forces can be transmitted in the usual manner to the axis of rotation if the correspondent couples are added. The centripetal force acts along the radius; hence, the corresponding couple is zero. In full accord with the principle of transmissibility, the centripetal force can be moved along the radius. The tangential force is $F_t = ml_c\ddot{\theta}$, and its moment arm about O is l_c. The moment of the corresponding couple is $ml_c^2\ddot{\theta}$. Therefore, parallel displacement of the tangential force from the CoM to the axis of rotation should be compensated for by a couple that equals $ml_c^2\ddot{\theta}$. We obtain

$$F_n = ml_c\dot{\theta}^2 \qquad (5.4)$$

$$M_O = I_O\ddot{\theta} = (I + ml_c^2)\ddot{\theta}$$

In equations 5.4, M_O and I_O are taken with respect to the axis of rotation, O. The relationship $I_O = I + ml_c^2$ immediately follows from the parallel axis theorem. Both representations of the resultant force-couple system, at the CoM and at the joint (equations 5.3 and 5.4), are equivalent. Selection of one of them is a matter of convenience and personal preference. Note that the moment M_O causes not only the angular acceleration of the link but also the linear acceleration of its CoM.

From the foregoing discussion, it follows that the load on the joint or bearing (the *joint force,* or the *joint resultant force*) is equal to the sum of the centrifugal force and the reverse of the tangential force. If the link rotation is performed in the vertical plane, the gravity force $m\mathbf{g}$ should be included.

■ ■ ■ *FROM THE LITERATURE* ■ ■ ■

Centripetal Force Affects Head Tilt in Passengers but Not in Drivers

Source: Zikovitz, D.C., and L.R. Harris. 1999. Head tilt during driving. *Ergonomics* 42 (5): 740–46.

Drivers and passengers in cars tilt their heads when cornering. A person in a car driving around a corner is exposed to two orthogonal forces: gravity and centripetal force. The combination of these two forces is the gravito-inertial force (GIF). The direction of the GIF while cornering is tilted relative to the direction of gravity alone (figure 5.6a). The tilt angle depends on the magnitude of the centripetal force (equation 5.4). In the experiment, the centripetal forces were produced by having subjects drive a car around a number of different corners of various radii at speeds between 20 and 80 km/h. The lateral acceleration of the car and the head tilt of both the drivers and passengers were measured. The tilt angle of the GIF was determined as $\tan^{-1}$(centripetal acceleration/g), where g is the acceleration due to gravity. The visual tilt of the road, which did not depend on the speed, was also determined. In some trials, passengers rode with their eyes closed.

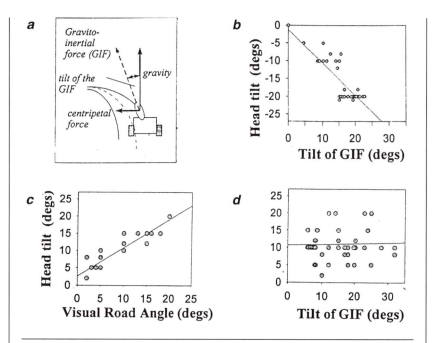

Figure 5.6 The effect of centripetal force on the head tilt of passengers and drivers. (*a*) The designations used in the study. (*b*) The head tilt of passengers (eyes closed) versus the tilt of GIF, $r^2 = .69$. (*c*) The head tilt of drivers versus the visual tilt of the road, $r^2 = .86$. (*d*) The head tilt of drivers versus the tilt of GIF, $r^2 = .01$. Compare (*b*) and (*c*).

Adapted, by permission, from C. Zikovitz and L.R. Harris, 1999, "Head tilt during driving," *Ergonomics*, 42 (5):740–746. By permission of Taylor & Francis Ltd, **http://www.tandf.co.uk/journals**

Passengers usually tilted their heads away from the center of rotation (figure 5.6*b*), while drivers always tilted their heads toward the center of rotation (figure 5.6*c*). In passengers, a strong negative correlation with the GIF tilt was observed (figure 5.6b). Contrarily, in drivers, there was a strong positive correlation between head tilt and the visual measure of road tilt (figure 5.6*c*) but none with the GIF tilt (figure 5.6*d*). The researchers concluded that drivers use a visual reference frame for the driving task. This is especially important for racing car drivers, who experience centripetal forces in excess of 4**g**, which would require a head tilt over 70°.

5.2.2 Planar Two-Link Chain

Consider a planar two-link chain similar to the chains that were analyzed in the previous chapters (figure 5.7). The links of the chain rotate at joints 1 (proximal)

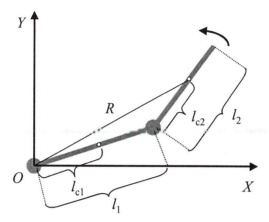

Figure 5.7 A planar two-link chain. The links are modeled as slender rods. R is the radius from joint 1 to the CoM of link 2; l_1 and l_2 are the lengths of links 1 and 2, respectively; l_{c1} and l_{c2} are the distances to the centers of mass of links 1 and 2 from joints 1 and 2, respectively. α_1, $\dot{\alpha}_1$, and $\ddot{\alpha}_1$ (not shown) are the angle, angular velocity, and angular acceleration, respectively, of joint 1; α_2, $\dot{\alpha}_2$, and $\ddot{\alpha}_2$ (not shown) are the angular position, velocity, and acceleration, respectively, of joint 2. The masses of links 1 and 2 are m_1 and m_2, correspondingly, and the moments of inertia of the links with respect to their centers of mass are I_1 and I_2. The counterclockwise direction of rotation is considered positive.

and 2 (distal). Since the chain is planar, the tensor of inertia of a link is reduced to a scalar moment of inertia. The angular velocity and acceleration are also scalars. The linear velocity and acceleration are two-component vectors. We assume that the positions, velocities, and accelerations of the links are known. The goal is to find the forces and moments that caused the observed pattern of motion.

5.2.2.1 Forces During Motion of a Two-Link Chain

As discussed previously, during rotation of a single link about a fixed axis, two forces act on the particles of the body: the centripetal force, which causes the change in the direction of the velocity vector, and the tangential force, which induces the change in the magnitude of velocity. Gravity should also be included. During movement of a two-link chain, the number of acting forces increases to six for link 2 and to nine for link 1.

The following forces are recognized at the CoM of the second link:

1. The weight of the link, $-m_2 g$, where g is the acceleration due to gravity.

2. The centripetal force due to the rotation of joint 2, $m_2 l_{c2} \dot{\alpha}_2^2$. This force is acting toward joint 2.

3. The tangential force due to the angular acceleration at joint 2, $m_2 l_{c2} \ddot{\alpha}_2$. This force is acting perpendicularly to the radius from joint 2 to the CoM of link 2.

4. The centripetal force due to the rotation at joint 1. This force is along the radius R from the CoM of link 2 to joint 1. The magnitude of the force is $m_2 R \dot{\alpha}_1^2 = m_2 (l_1^2 + l_{c2}^2 + 2 l_1 l_{c2} \cos\alpha_2)^{0.5} \dot{\alpha}_1^2$, where α_2 is the external (anatomic) angle at joint 2. With link 2 rotating at joint 2, the angle α_2 changes. Hence, radius R is not constant. Consequently, the centripetal force associated with rotation at joint 1 that is exerted on link 2 depends on the angular position at joint 2.

5. The tangential force that is due to the angular acceleration of joint 1. This force is perpendicular to R and equals $m_2 R \ddot{\alpha}_1 = m_2 (l_1^2 + l_{c2}^2 + 2 l_1 l_{c2} \cos\alpha_2)^{0.5} \ddot{\alpha}_1$.

6. The *Coriolis force,* which equals $m_2 a_{c2}^{cor}$, where a_{c2}^{cor} is the Coriolis acceleration of the CoM of link 2, $a_{c2}^{cor} = 2 l_{c2} \dot{\alpha}_1 \dot{\alpha}_2$ (see section **3.2.1.2** in *Kinematics of Human Motion*). The Coriolis force acts along the radius of the second link.

The following forces act on link 1:

1. The weight of the link, $-m_1 g$.

2. The centripetal force due to the rotation at joint 1, $m_1 l_{c1} \dot{\alpha}_1^2$. This force acts along link 1 toward the center of rotation at joint 1.

3. The tangential force due to the angular acceleration of joint 1, $m_1 l_{c1} \ddot{\alpha}_1$. This force is perpendicular to the radius from joint 1 to the CoM of link 1.

4–9. The reaction forces due to the interaction with link 2. The action forces exerted on link 2 are impressed on link 1 as reaction forces. For instance, the centripetal force associated with the rotation at joint 2 causes the particles of link 2 to change the direction of velocity vector. This force is exerted on link 2. An equal and opposite reaction force (the centrifugal force) acts on link 1. The reader is invited to trace the origin of other reaction forces acting on link 1. If all the reaction forces exerted on link 1 by link 2 are lumped together, the number of forces exerted on link 1 decreases to four.

The forces acting at the CoM of an individual link can be used for estimating the link's contribution to the ground reaction force (figure 5.8). The GRF is simply the sum of the forces at the CoMs of the links. However, most often forces at the CoMs are computed to determine the joint forces and torques.

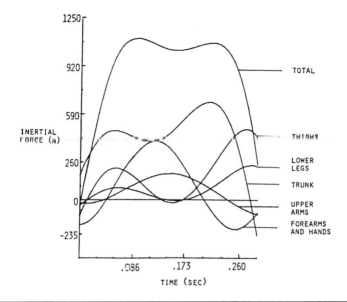

Figure 5.8 The contribution of the individual body segments to the vertical GRF during a standing vertical jump. The inertial forces were computed as the products of the mass of the segment and the vertical acceleration of the segment's CoM. The integral of a force over the studied interval is the linear impulse, which equals the change in the linear momentum of the segment (the principle of linear impulse and momentum, equation 1.58). Because the initial velocity was zero, the contribution of the individual segments to the total impulse is equal to the contribution of the segment to the velocity of the CoM of the body at the instant of takeoff. For the given subject, the contribution of the trunk amounted to 46.6%, the thighs contributed 17.3%, and the forearm and hands together contributed 10.8%. The contribution of the feet was only 1.3%.

Reprinted from D.I. Miller and D.J. East, 1976, Kinematic and kinetic correlates of vertical jumping in women. In *Biomechanics V-B*, edited by P.V. Komi (Baltimore: University Park Press), 65–72.

5.2.2.2 Dynamic Equations in the Closed Form

Representative papers: Hollerbach and Flash (1982); Sainburg et al. (1995)

Dynamic equations that contain all the variables in explicit input-output form are called *closed-form dynamic equations*. We derive the closed-form dynamic equations for a planar two-link chain using elementary geometry and algebra. This approach was selected intentionally to make the physical meaning of the terms in the equations clear. The equations are written in joint coordinates.

Refer again to figure 5.7; an immediate task is to find the joint torques T_1 and T_2 when joint positions (α_1, α_2), velocities $(\dot{\alpha}_1, \dot{\alpha}_2)$, and accelerations $(\ddot{\alpha}, \ddot{\alpha}_2)$ are given as a function of time. The following six variables influence the joint torques: gravity, centripetal and tangential acceleration due to the rotation in both joints 1 and 2, and Coriolis acceleration. Their effects on T_1 and T_2 are discussed next, using simple geometric relationships. In the equations, the symbols C and S stand for the cosine and sine, correspondingly, and their subscripts 1, 2, and 12 refer, respectively, to α_1, α_2, and $\alpha_1 + \alpha_2$.

1. Gravity. The contribution of gravity to T_1 and T_2 is

$$G_1 = -m_1 l_{c1} g C_1 - m_2 g (l_1 C_1 + l_{c2} C_{12})$$
$$G_2 = -m_2 l_{c2} g C_{12}$$

(5.5)

The geometric dimensions used in equations 5.5 are shown in figure 5.9*a*. The gravity moments depend on the chain configuration and its orientation. The moments are maximal when the chain (e.g., an arm) is fully extended in the horizontal direction.

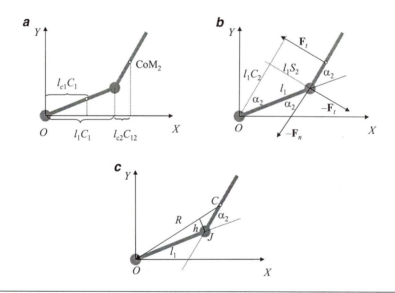

Figure 5.9   Geometric relationships used for computation of the joint torques. (*a*) Geometric dimensions used for computation of the gravity torques. (*b*) Geometric relationships for establishing the moment arms of the centrifugal force $(-\mathbf{F}_n)$ and the force of reaction to the tangential force $(-\mathbf{F}_t)$ arising from rotation of joint 2. With respect to joint 1, the moment arms are $l_1 S_2$ for the centrifugal force and $l_1 C_2$ for the tangential force. (*c*) Moment arm h about joint 2 of the centripetal force due to rotation at joint 1.

2. Centripetal acceleration due to rotation at joint 2. The centripetal force is $m_2 l_{c2} \dot{\alpha}_2^2$, and the centrifugal force acting on link 1 at joint 2 is $-m_2 l_{c2} \dot{\alpha}_2^2$. This force produces a moment about joint 1. The force acts along link 2, and its moment arm is $l_1 S_2$ (see figure 5.9b). Hence, the centrifugal effect of the rotation of link 2 on joint 1 is $-m_2 l_1 l_{c2} S_2 \dot{\alpha}_2^2$.

3. Angular acceleration at joint 2. Due to this acceleration, both the angular velocity of the link and the linear velocity of its CoM vary. The magnitude of the joint torque at joint 2 that induces acceleration $\dot{\alpha}_2$ is $\tau_{2,2} = (I_2 + m_2 l_{c2}^2) \ddot{\alpha}_2 = I_2 \ddot{\alpha}_2 + l_{c2} \cdot m_2 l_{c2} \ddot{\alpha}_2$, where $m_2 l_{c2} \ddot{\alpha}_2$ is the tangential force at the CoM of the second link ($\mathbf{F}_t$) that is due to the angular acceleration of joint 2.

When the second link accelerates, both the reaction force and moment act on the first link. They cause a torque about the first joint axis given by

$$\tau_{1,2} = -\underbrace{\left[(I_2 + m_2 l_{c2}^2) \ddot{\alpha}_2 \right.}_{\substack{\text{Torque at joint 1} \\ \text{due to the reaction} \\ \text{torque at joint 2}}} + \underbrace{\left. m_2 l_1 l_{c2} C_2 \ddot{\alpha}_2 \right]}_{\substack{\text{Torque at joint 1} \\ \text{due to the reaction} \\ \text{force } -\mathbf{F}_t \text{ at joint 2}}} =$$

$$-\left[I_2 + m_2 \left(l_{c2}^2 + l_1 l_{c2} C_2 \right) \right] \ddot{\alpha}_2 \qquad (5.6)$$

The moment arm of the reaction force $-\mathbf{F}_t$ is $l_1 C_2$ (see figure 5.9b).

4. Centripetal acceleration associated with rotation of joint 1. The line of action of the centripetal force exerted on link 1 is along link 1, and the line of action of the centripetal force acting on link 2 is along the radius R. Both lines of force action pass through joint 1, and hence these forces do not produce moments at this joint. However, the centripetal force exerted on link 2 generates a moment about the distal joint. The moment arm of this force equals the height h of triangle OJC in figure 5.9c. To find h, recall that the area A of the triangle is given by $A = 0.5Rh = 0.5 l_1 l_{c2} S_2$, where S_2 stands for $\sin\alpha_2$. Hence, $h = l_1 l_{c2} S_2 / R$, and because the centripetal force equals $m_2 R \dot{\alpha}_1^2$, the moment of this force about joint 2 is $m_2 l_1 l_{c2} S_2 \dot{\alpha}_1^2$. The moment is analogous to the centrifugal effect on joint 1 of the rotation at joint 2 ($-m_2 l_1 l_{c2} S_2 \dot{\alpha}_2^2$), which was derived previously. The inertia terms in these expressions ($m_2 l_1 l_{c2} S_2$) are the same. These inertia terms are known as the *centripetal coupling coefficients*. Hence, the centripetal effects on the joints of the two-link chain are symmetric.

5. Tangential acceleration at joint 1. The resistance to the acceleration is provided by the total moment of inertia of both links with respect to the first joint axis. The moment of inertia of link 1 is $I_1 + m_1 l_{c1}^2$, and the moment of inertia of the second link with respect to joint 1 is $I_2 + m_2 R^2$. The total moment of inertia of the chain about joint 1 is

$$I_{1,1} = I_1 + m_1 l_{c1}^2 + I_2 + m_2 (l_1^2 + l_{c2}^2 + 2 l_1 l_{c2} C_2) \qquad (5.7a)$$

and the joint torque at joint 1 expended for the angular acceleration of the entire chain is

$$\tau_{1,1} = [I_1 + m_1 l_{c1}^2 + I_2 + m_2 (l_1^2 + l_{c2}^2 + 2l_1 l_{c2} C_2)] \ddot{\alpha}_1 \qquad (5.7b)$$

The accelerated motion at joint 1 induces the force and torque at joint 2. Linear acceleration of joint 2 induced by the angular acceleration at joint 1 is $l_1 \ddot{\alpha}_1$. The force at joint 2 associated with this acceleration is $m_2 l_1 \ddot{\alpha}_1$. This force produces a moment about the CoM of link 2 (CoM$_2$) that equals $(m_2 l_1 l_{c2} C_2) \ddot{\alpha}_1$. The angular acceleration of link 2 is the same as that of link 1, $\ddot{\alpha}_1$. The moment of inertia of link 2 about joint 2 is $m_2 l_{c2}^2 + I_2$. Therefore, the joint torque at joint 2 that is due to the angular acceleration at joint 1 is

$$\tau_{2,1} = -(m_2 l_1 l_{c2} C_2 + m_2 l_{c2}^2 + I_2) \ddot{\alpha}_1 =$$
$$-[I_2 + m_2 (l_{c2}^2 + l_1 l_{c2} C_2)] \ddot{\alpha}_1 \qquad (5.8)$$

The reader is encouraged to compare equations 5.6 and 5.8. The expressions in the square brackets, called the coupling inertia or interactive inertia, are similar in both equations. Therefore, the inertial effects of the accelerated motion at one joint on the torque at another joint are symmetric. The effect of the accelerated first link's motion on the second joint is similar to the effect of the second link's motion on joint 1. This conclusion seems counterintuitive. It is much easier to imagine the effect of fast hip flexion on knee torque (the proximal-to-distal effect) than the effect of knee extension on hip torque (the distal-to-proximal effect), but in reality they are equal.

6. Coriolis effects. Coriolis acceleration of the CoM of the second link is $2l_{c2} \dot{\alpha}_1 \dot{\alpha}_2$ (see section **3.2.1.2** in *Kinematics of Human Motion*). The Coriolis force is then $\mathbf{F}_{cor} = 2m_2 l_{c2} \dot{\alpha}_1 \dot{\alpha}_2$. The force is along the second link. Its moment arm about the first joint is $l_1 S_2$, and the moment of this force about joint 1 is $2m_2 l_1 l_{c2} S_2 \dot{\alpha}_1 \dot{\alpha}_2$. The expression $2m_2 l_1 l_{c2} S_2$ (the *Coriolis coupling coefficient*), which characterizes the inertial resistance to the Coriolis force, is twice the inertial resistance to the centrifugal forces (the centripetal coupling coefficient).

The closed-form equations of motion for the two-link planar chain combine all the aforementioned terms.

$$T_1 = \underbrace{\left[I_1 + m_1 l_{c1}^2 + I_2 + m_2 \left(l_1^2 + l_{c2}^2 + 2l_1 l_{c2} C_2\right)\right] \ddot{\alpha}_1}_{\text{Moment of inertia of the chain about joint 1}} + \underbrace{\left[I_2 + m_2 \left(l_{c2}^2 + l_1 l_{c2} C_2\right)\right] \ddot{\alpha}_2}_{\substack{\text{Coupling inertia; inertial effect of angular} \\ \text{acceleration of joint 2 on joint 1}}}$$

$$\underbrace{-\left(m_2 l_1 l_{c2} S_2\right) \dot{\alpha}_2^2}_{\substack{\text{Centripetal} \\ \text{coupling} \\ \text{coefficient}}} - \underbrace{\left(2m_2 l_1 l_{c2} S_2\right) \dot{\alpha}_1 \dot{\alpha}_2}_{\substack{\text{Coriolis} \\ \text{coupling} \\ \text{coefficient}}} + \underbrace{\left[m_1 g l_{c1} C_1 + m_2 g (l_1 C_1 + l_{c2} C_{12})\right]}_{\text{Gravity term, } G_1} \qquad (5.9)$$

$$T_2 = \underbrace{\left(I_2 + m_2 l_{c2}^2\right) \ddot{\alpha}_2}_{\substack{\text{Moment of} \\ \text{inertia of link 2} \\ \text{about joint 2}}} + \underbrace{\left[I_2 + m_2 \left(l_{c2}^2 + l_1 l_{c2} C_2\right)\right] \ddot{\alpha}_1}_{\substack{\text{Coupling inertia;} \\ \text{inertial effect of angular} \\ \text{acceleration of joint 1 on joint 2}}} + \underbrace{\left(m_2 l_1 l_{c2} S_2\right) \dot{\alpha}_1^2}_{\substack{\text{Centripetal} \\ \text{coupling} \\ \text{coefficient}}} + \underbrace{m_2 g l_{c2} C_{12}}_{\text{Gravity term, } G_2}$$

Even for a simple planar two-link chain, the equations of motion are complex. The terms of equations 5.9 correspond to various forces that either resist the movement or assist it. The terms that correspond to the resisting effects are added in the equation, while those that correspond to assisting effects are subtracted. For instance, in the example under consideration (see figure 5.7 again), the joint torque is in a counterclockwise direction, while gravity generates a torque in a clockwise direction, increasing the resistance. The joint torque should counterbalance the torque due to gravity. Consequently, the gravity term is included in the right part of the equation with a positive sign. The centripetal effect is positive for T_2 and negative for T_1. The reader is encouraged to explain this difference.

5.2.2.3 Dynamic Coefficients and Interactive Effects

Representative paper: Hoy and Zernicke (1985)

Equations 5.9 can be written in a more compact form by replacing lengthy expressions with the *dynamic coefficients.*

$$T_1 = I_{1,1} \ddot{\alpha}_1 + I_{1,2} \ddot{\alpha}_2 - D \dot{\alpha}_2^2 - 2D \dot{\alpha}_1 \dot{\alpha}_2 + G_1$$
$$T_2 = I_{2,2} \ddot{\alpha}_2 + I_{2,1} \ddot{\alpha}_1 + D \dot{\alpha}_1^2 + G_2$$

(5.10)

■ ■ ■ *From the Literature* ■ ■ ■

Joint Torques and Equilibrium Trajectories During Arm Movement

Source: Latash, M.L., A.S. Aruin, and V.M. Zatsiorsky. 1999. The basis of a simple synergy: Reconstruction of joint equilibrium trajectories during unrestrained arm movements. *Hum. Mov. Sci.* 18: 3–30.

In the experiment, the subjects performed 25 fast elbow flexions and 25 fast wrist flexions (figure 5.10*a*). Equations 5.9 were used to compute the elbow and wrist torques. An example of the obtained joint torque–joint angle relationships is presented in the middle panel of figure 5.10*b*. These relationships were used to compute the joint compliant characteristics (JCC, explained in **3.6.2.1**) The instantaneous positions of the JCC were determined from the regression lines of joint angle–joint torque at a given instant (figure 5.10*b*). At a given moment, the JCC is the joint angle at which the joint torque is on average zero (the intercept of the regression line).

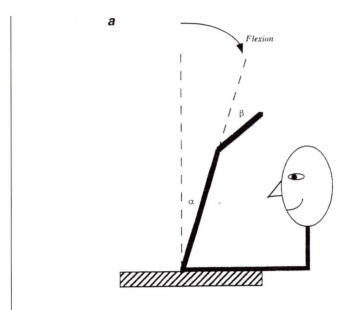

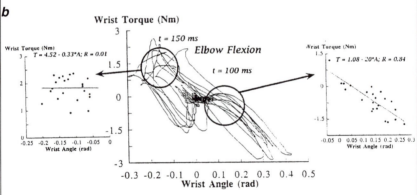

Figure 5.10 Wrist torque during elbow motion. (*a*) The experimental setup. (*b*) An example of the experimental recordings. The middle panel in (*b*) shows the torque–angle curves for the wrist joint for six elbow flexion trials (all 25 curves would make the figure too messy). The side panels show scatter plots and linear regression lines for all 25 trials within this series, measured 100 and 150 ms after the movement initiation. The circles in the middle panel show approximately when the measurements were made. In the equations, *T* is torque and *A* is angle. The JCC were successfully reconstructed at 100 ms but not at 150 ms. Compare this figure with figure 3.25 (p. 248).

Adapted by permission from Latash, M.L., A.S. Aruin, and V.M. Zatsiorsky. 1999. The basis of a simple synergy: Reconstruction of joint equilibrium trajectories during unrestrained arm movements. *Hum. Mov. Sci.* 18: 3–30.

The meaning of the dynamic coefficients is explained in the following paragraphs.

$I_{1,1}$ is the moment of inertia of the chain about joint 1. $I_{2,2}$ is the moment of inertia of the second link about joint 2. $I_{1,2}$ and $I_{2,1}$ ($I_{1,2} = I_{2,1}$) represent the coupling inertia. All the four elements $I_{i,j}$ indicate the inertial effects felt at joint i of acceleration at joint j. For instance, $I_{1,2}$ is a measure of the inertial resistance felt at joint 1 to the angular acceleration at joint 2. All the elements $I_{i,j}$ depend on the inertia of the individual links (their masses and moments of inertia) and on the chain configuration. The elements $I_{i,j}$ have the dimensionality of rotational inertia (kg · m²). The products $I_{i,j}\ddot{\alpha}_j$ represent the components of the joint torques that are due to the angular acceleration at the joints.

D designates the inertial resistance of the chain to the centrifugal and Coriolis forces. The corresponding components of the joint torques depend both on the joint angular positions and on the joint angular velocities. For a planar two-link chain, the centripetal acceleration coefficient D equals the partial derivative of the coupling inertia $I_{1,2}$ over the joint angle, $D = \partial I_{1,2}/\partial\alpha_2$, and the Coriolis coefficient is twice this value. The dimensionality of D is mass times the square of distance (unit, kg · m²). Finally, the gravity torque components, G_1 and G_2, depend on the chain position.

Note in equation 5.10 that three out of the five terms defining the torque at joint 1 depend on the movement at joint 2. Similarly, two out of the four terms for the torque at joint 2 depend on the movement at joint 1. The joint torques and forces induced by motion in other joints are called interactive forces and torques. The following simple experiment demonstrates the existence of these interactive forces and torques. Starting with an arm extended straight down at the side of the body, the elbow is flexed vigorously. If the upper arm is unsupported, the shoulder will be extended. This shoulder extension occurs despite the fact that none of the elbow flexors is also a shoulder extensor. Thus, the shoulder extension was caused by the activity of the elbow flexors.

Table 5.1 summarizes the effects of different sources on the joint torques. The centrifugal forces associated with angular motion at particular joints do not generate moments of force at these joints. Coriolis force does not cause a moment at joint 2. The reason is evident: the lines of action of these forces pass through the corresponding joint centers.

At this point, the reader is encouraged to review section **4.5.2**, where the coupling inertia terms and inertia matrix of the chain $[I^c]$ were introduced. The origin of the elements of the chain inertia matrix is now explained.

Equation 5.10 can be symbolically written as a matrix-vector equation known as the *state-space equation:*

$$\mathbf{T} = [I^c(\boldsymbol{\alpha})]\,\ddot{\boldsymbol{\alpha}} + \mathbf{V}(\boldsymbol{\alpha},\,\dot{\boldsymbol{\alpha}}) + \mathbf{G}(\boldsymbol{\alpha}) \qquad (5.11a)$$

or simply

$$\mathbf{T} = [I^c]\,\ddot{\alpha} + \mathbf{V}(\dot{\alpha}) + \mathbf{G} \tag{5.11b}$$

where **T** is the vector of joint torques, $[I^c]$ is the chain inertia matrix, $\ddot{\alpha}$ is the vector of joint angular accelerations, $\mathbf{V}(\dot{\alpha})$ is the vector of the centrifugal and Coriolis terms, and **G** is the vector of the gravity terms. Each element in $[I^c]$ and **G** is a complex function that depends on the joint configuration; each element in $\mathbf{V}(\dot{\alpha})$ depends on the joint configuration and joint angular velocities. Some dynamic coefficients may become zero at particular joint configurations.

Equation 5.11 is essentially the same as equation 5.9. In the literature, dynamic equations are presented in a detailed form, as in equation 5.9, or in a matrix-vector form, as in equation 5.11. Equations of motion can be written in joint coordinates or in any appropriate system of coordinates. Notwithstanding how different the equations look, all the dynamic equations always contain the *acceleration-related terms* (inertia times the joint angular acceleration), the *centrifugal terms* (inertia times angular velocity squared), and the *Coriolis terms* (inertia times a product of the joint angular velocities). They also in-

Table 5.1   Dynamic Coefficients in the Closed-Form Equations for a Planar Two-Link Chain

Independent variable	Dynamic coefficients	
	T_1	T_2
Angular velocity at joint 1, $\dot{\alpha}_1^2$	0	$D = m_2 l_1 l_{c2} S_2$
Angular acceleration at joint 1, $\ddot{\alpha}_1$	$I_{1,1} = m_1 l_{c1}^2 + I_1 + m_2 (l_1^2 + l_{c2}^2 + 2l_1 l_{c2} C_2) + I_2$	$I_{2,1} = m_2 l_1 l_{c2} C_2 + m_2 l_{c2}^2 + I_2$
Angular velocity at joint 2, $\dot{\alpha}_2^2$	$D = m_2 l_1 l_{c2} S_2$	0
Angular acceleration at joint 2, $\ddot{\alpha}_2$	$I_{1,2} = m_2 l_1 l_{c2} C_2 + m_2 l_{c2}^2 + I_2$	$I_{2,2} = I_2 + m_2 l_{c2}^2$
Velocity at joints 1 and 2 (Coriolis acceleration), $\dot{\alpha}_1 \dot{\alpha}_2$	$D = m_2 l_1 l_{c2} S_2$	0
Gravity	$G_1 = m_1 l_{c1} g C_1 + m_2 g(l_1 C_1 + l_{c2} C_{12})$	$G_2 = m_2 l_{c2} g C_{12}$

Input: joint movements. Output: joint torques.

clude the gravity terms. The centrifugal and Coriolis terms are the *velocity-related terms*. Some of the velocity-related dynamic coefficients are zero. For instance, the centrifugal force does not generate torque at the focal joint (the joint that generates the motion); however, it produces a torque in another joint. The general rule is that the forces and moments acting on a given link in a multilink chain are affected by the motions of the other links of the chain.

5.2.3 Three-Link Planar Chains

Representative papers: Plagenhoef (1971), T.C. Huang, Roberts, and Youm (1982)

In a three-link chain (figure 5.12), the following 10 forces are impressed on

■ ■ ■ *From the Literature* ■ ■ ■

State-Space Equations of Pedaling

Source: Fregly, B.J., and F.E. Zajac. 1996. A state-space analysis of mechanical energy generation, absorption, and transfer during pedaling. *J. Biomech.* 29 (1): 81–90.

In this study, the dynamic equations were used to determine the contribution of different forces to the angular acceleration of the modeled segments and to compute mechanical power. A planar 3-DoF model of pedaling on a stationary bicycle ergometer was developed (figure 5.11). The closed-form state-space equations were written in the following form:

$$\mathbf{M}(\boldsymbol{\theta})\ddot{\boldsymbol{\theta}} = \mathbf{T}(\boldsymbol{\theta}) + \mathbf{V}(\boldsymbol{\theta}, \dot{\boldsymbol{\theta}}) + \mathbf{G}(\boldsymbol{\theta}) + \mathbf{F} \qquad (5.12)$$

where

- $\boldsymbol{\theta}$ is the vector of the crank angle θ_1 and the two foot-segment angles θ_2 and θ_3, $\boldsymbol{\theta} = [\theta_1 \ \theta_2 \ \theta_3]^T$;
- $\mathbf{M}(\boldsymbol{\theta})$ is the 3 × 3 inertia matrix of the chain;
- $\mathbf{T}(\boldsymbol{\theta})$ is the 3 × 1 column matrix due to the joint torques (note that the angles θ are *not* the joint angles);
- $\mathbf{V}(\boldsymbol{\theta}, \dot{\boldsymbol{\theta}})$ is the 3 × 1 column matrix due to angular velocity of the segments (i.e., the centripetal and Coriolis forces);
- $\mathbf{G}(\boldsymbol{\theta})$ is the 3 × 1 column matrix due to gravity forces; and
- $\mathbf{F}$ is 3 × 1 column matrix due to friction forces (pedal force).

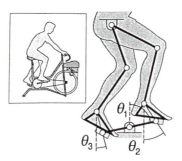

Figure 5.11　Model of pedaling. The crank angle θ_1 and the two foot-segment angles θ_2 and θ_3 with respect to an inertial reference frame are the three generalized coordinates. All joints are assumed to be frictionless and revolute, and both hips are assumed to remain stationary. The total load that the cyclist experiences at the crank due to all ergometer components is modeled by an effective inertia and an effective friction.

Reprinted from *Journal of Biomechanics*, 29 (1), B.J. Fregly and F.E. Zajac, A state-space analysis of mechanical energy generation, absorption, and transfer of energy during pedaling, 81–90, 1996, with permission from Elsevier Science.

Because the inertia matrix always has an inverse (see **4.5.2**), equation 5.12 was written as

$$\ddot{\theta} = \underbrace{\mathbf{M}^{-1}(\boldsymbol{\theta})\mathbf{T}(\boldsymbol{\theta})}_{\ddot{\theta}\atop \text{Joint torques}} + \underbrace{\mathbf{M}^{-1}(\boldsymbol{\theta})\mathbf{V}(\boldsymbol{\theta},\dot{\boldsymbol{\theta}})}_{\ddot{\theta}\atop \text{Velocity}} + \underbrace{\mathbf{M}^{-1}(\boldsymbol{\theta})\mathbf{G}(\boldsymbol{\theta})}_{\ddot{\theta}\atop \text{Gravity}} + \underbrace{\mathbf{M}^{-1}(\boldsymbol{\theta})\mathbf{F}}_{\ddot{\theta}\atop \text{Friction}} \tag{5.13}$$

where individual terms in the right side of the equation represent the contribution of the joint torques, velocity, gravity, and friction into the angular acceleration of the crank and the two foot links. Equation 5.13 allows determining how any torque or force in the model contributes to the angular acceleration of any modeled segment (e.g., the crank).

the third link: (1) gravity, $m_3\mathbf{g}$; (2–4) the centripetal forces, which are proportional to the squared angular velocity at joints 1, 2, and 3 and equal $m_3 R_{13} \dot{\alpha}_1^2$, $m_3 R_{23} \dot{\alpha}_2^2$, and $m_3 l_{c3} \dot{\alpha}_3^2$, correspondingly; (5–7) forces associated with the angular acceleration at joints 1, 2, and 3, which equal $m_3 R_{13} \ddot{\alpha}_1$, $m_3 R_{23} \ddot{\alpha}_2$, and $m_3 l_{c3} \ddot{\alpha}_3$; and (8–10) the Coriolis forces, which are proportional to the products of the joint angular velocities $\dot{\alpha}_1 \dot{\alpha}_2$, $\dot{\alpha}_1 \dot{\alpha}_3$, and $\dot{\alpha}_2 \dot{\alpha}_3$.

Eight inertial forces act on the third link, five on the second link, and two on the first link, for a total of 15 inertial forces. According to Newton's third law, for every force exerted on a link, the link exerts a reaction force on the

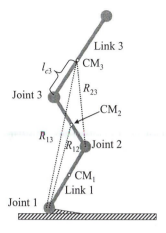

Figure 5.12   A planar three-link chain. R_{12} is the radius from joint 1 to the CoM of link 2; R_{13} is the radius from joint 1 to the CoM of link 3, and R_{23} is the radius from joint 2 to the CoM of link 3. l_{c3} is the distance from joint 3 to the CoM of link 3. The feet are not included in the model.

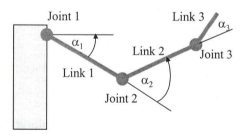

Figure 5.13   A three-link planar model of the human arm.

adjacent link. The forces are "propagated" along the chain. In general, movement of every link affects every other link. Therefore, the closed-form dynamic equations for a planar three-link chain are rather complex. Figure 5.13 shows a model of a human arm that includes the shoulder, elbow, and wrist joints. The state-space equation of the chain in joint coordinates is

$$\begin{bmatrix} T_1 \\ T_2 \\ T_3 \end{bmatrix} = \begin{bmatrix} I_{1,1} & I_{1,2} & I_{1,3} \\ I_{2,1} & I_{2,2} & I_{2,3} \\ I_{3,1} & I_{3,2} & I_{3,3} \end{bmatrix} \begin{bmatrix} \ddot{\alpha}_1 \\ \ddot{\alpha}_2 \\ \ddot{\alpha}_3 \end{bmatrix} + \begin{bmatrix} v(\alpha,\dot{\alpha})_1 \\ v(\alpha,\dot{\alpha})_2 \\ v(\alpha,\dot{\alpha})_3 \end{bmatrix} + \begin{bmatrix} G(\alpha)_1 \\ G(\alpha)_2 \\ G(\alpha)_3 \end{bmatrix} \qquad (5.14)$$

Because inertia matrices are symmetric ($I_{1,2} = I_{2,1}$, $I_{1,3} = I_{3,1}$, and $I_{2,3} = I_{3,2}$), only six elements of the matrix must be determined. These elements are the three moments of inertia, $I_{1,1}$, $I_{2,2}$, and $I_{3,3}$,

$$I_{1,1} = I_1 + m_1 l_{c1}^2 + I_2 + m_2(l_1^2 + l_{c2}^2 + 2l_1 l_{c2} C_2) + I_3 +$$
$$m_3[l_1^2 + l_2^2 + l_{c3}^2 + 2l_1 l_2 C_2 + 2l_1 l_{c3} C_{12} + 2l_2 l_{c3} C_3]$$
$$I_{2,2} = I_2 + m_2 l_{c2}^2 + I_3 + m_3(l_2^2 + l_{c3}^2 + 2l_2 l_{c3} C_3) \qquad (5.15)$$
$$I_{3,3} = I_3 + m_3 l_{c3}^2$$

and three coupling inertia terms, $I_{1,2}$ (which accounts for the effect on the torque at joint 1 of angular acceleration at joint 2), $I_{1,3}$, and $I_{2,3}$,

$$I_{1,2} = m_2(l_{c2}^2 + l_1 l_{c2} C_2) + I_2 +$$
$$m_3[l_2^2 + l_{c3}^2 + l_1 l_2 C_2 + l_1 l_{c3} C_{12} + 2l_2 l_{c3} C_3] + I_3$$
$$I_{1,3} = m_3[l_{c3}^2 + l_1 l_{c3} C_{23} + l_2 l_{c3} C_3] + I_3 \qquad (5.16)$$
$$I_{2,3} = m_3(l_{c3}^2 + l_2 l_{c3} C_3) + I_3$$

The centrifugal and Coriolis terms are

$$v(\alpha, \dot{\alpha})_1 = -[(m_2 l_1 l_{c2} + m_3 l_1 l_2)S_2 + m_3 l_1 l_{c3} S_{23}] (2\dot{\alpha}_1 \dot{\alpha}_2 + \dot{\alpha}_2^2) -$$
$$[m_3 l_1 l_{c3} S_{23} + m_3 l_2 l_{c3} S_3] (2\dot{\alpha}_1 \dot{\alpha}_3 + 2\dot{\alpha}_2 \dot{\alpha}_3 + \dot{\alpha}_3^2)$$
$$v(\alpha, \dot{\alpha})_2 = -[(m_2 l_1 l_{c2} + m_3 l_1 l_2)S_2 + m_3 l_1 l_{c3} S_{23}] \dot{\alpha}_1^2 - \qquad (5.17)$$
$$m_3 l_2 l_{c3} S_3 (2\dot{\alpha}_1 \dot{\alpha}_3 + 2\dot{\alpha}_2 \dot{\alpha}_3 + \dot{\alpha}_3^2)$$
$$v(\alpha, \dot{\alpha})_3 = (m_3 l_1 l_{c3} S_{23} + m_3 l_2 l_{c3} S_3) \dot{\alpha}_1^2 + m_3 l_2 l_{c3} S_3 (2\dot{\alpha}_1 \dot{\alpha}_2 + \dot{\alpha}_2^2)$$

The gravity terms are

$$G(\alpha)_1 = m_1 l_{c1} g C_1 + m_2 g(l_1 C_1 + l_{c2} C_{12}) + m_3 g(l_1 C_1 + l_2 C_{12} + l_{c3} C_{123})$$
$$G(\alpha)_2 = m_2 g l_{c2} C_{12} + m_3 g(l_2 C_{12} + l_{c3} C_{123}) \qquad (5.18)$$
$$G(\alpha)_3 = m_3 g l_{c3} C_{123}$$

For planar chains with more than three links, the closed-form equations include many terms. These equations are derived with special computer programs. In three dimensions, the problem becomes even more perplexing.

5.3 MOVEMENT IN THREE DIMENSIONS

Most human movements occur in three dimensions. When analysis is limited to one plane, for instance, when walking is analyzed only in the sagittal plane, the deviation from the sagittal plane and rotation of the links about their long axes are neglected. Whether such a simplification is acceptable or not depends on the situation at hand. For instance, a two-dimensional analysis may be insufficient for documenting gait asymmetry in some patients.

In three dimensions, the dynamic equations 5.1, which include an equation describing a linear motion and an equation describing a rotational motion, remain valid. The linear dynamic equations are much simpler than the rotational ones. The resultant of the external forces and the linear acceleration $\mathbf{a}_{CoM}$ are related by a scalar m and thus are always parallel. In contrast, the resultant couple and the rate of change of angular momentum are related by an inertia tensor $[I]$ and therefore may be not parallel. A further complication arises because the inertia tensor changes when the body rotates. To grasp this idea, consider rotation at the shoulder joint around the mediolateral axis (in anatomic nomenclature, this rotation corresponds to shoulder flexion). Compare two movements starting at different joint positions: with the arm at the side of the body or with the arm abducted (moved away from the sagittal plane) 90°. When the arm is abducted, the moment of inertia of the arm about the mediolateral axis is much smaller. Hence, any arm abduction changes the moment of inertia of the limb with respect to the "flexion-extension" axis. In general, in three-dimensional space, the moments and products of inertia of body segments are not constant but change as the segments rotate. Kinematics of motion in space are also complex. Finite rotations are not commutative, and angular velocity is not integrable: the final position of the body cannot be found by integrating the angular velocity vector. Therefore, dynamic solutions in space are rather sophisticated. To derive the dynamic equations for a multilink chain, it is first necessary to develop the dynamic equations for a single rigid body.

5.3.1 Motion of a Rigid Body in Three Dimensions

The goal of this section is to derive the equations of motion of a rigid body in three dimensions. When a rigid body spins around a fixed axis, the equation of motion is very simple: $M = I\ddot{\theta}$. If only a point is fixed, the axes of rotation through the point are not constant. The dynamic equations of rotation about a fixed point are much more complicated. Rotation about a fixed point is of wider interest than might appear at first sight. Any movement of a rigid body can be represented as a translation of the body's CoM and rotation around the CoM. Rotation of a rigid body around its CoM is the same as if the CoM were fixed.

If one point in the body is held fixed, it is often convenient to choose this point as the origin of a reference frame. For instance, to analyze hip motion in three dimensions, the center of the hip joint can be selected as an origin of the frame. If the motion is not constrained, the CoM is usually chosen as the origin. We derive the dynamic equations for motion of a rigid body with a fixed point in two steps. First, we take a time derivative of an angular momentum vector. We return to basic kinematics and derive an equation for the time derivative of a vector (any vector, not just the vector of angular momentum) with respect to a rotating frame. Then, we derive the dynamic equation for a rigid body with a fixed point.

5.3.1.1 Time Derivative of a Vector With Respect to a Rotating Frame

The following equation of the time derivative of a vector with respect to a rotating frame was provided without proof in *Kinematics of Human Motion* (p. 205)

$$(\dot{\mathbf{P}})_G = (\dot{\mathbf{P}})_L + \dot{\boldsymbol{\theta}} \times \mathbf{P} \tag{5.19}$$

where $(\dot{\mathbf{P}})_G$ and $(\dot{\mathbf{P}})_L$ are the time derivative of vector $\mathbf{P}$ as seen from the global and local reference frames, respectively, and $\dot{\boldsymbol{\theta}}$ is the angular velocity vector of the local frame measured from the fixed axes. We are going to prove this equation. Consider an instant when the global frame $O\text{-}XYZ$ and the local frame $o\text{-}xyz$ coincide. To prove equation 5.19, we first resolve $\mathbf{P}$ into its rectangular components,

$$\mathbf{P} = P_x \mathbf{i} + P_y \mathbf{j} + P_z \mathbf{k} \tag{5.20}$$

and then take time derivatives of it in both the local and global frames. In the local frame, vector $\mathbf{P}$ does not change its orientation with respect to the coordinate axes (the unit vectors $\mathbf{i}$, $\mathbf{j}$, and $\mathbf{k}$ are fixed) and the time rate of change of $\mathbf{P}$ is simply

$$\dot{\mathbf{P}}_L = \dot{P}_x \mathbf{i} + \dot{P}_y \mathbf{j} + \dot{P}_z \mathbf{k} \tag{5.21}$$

In the global frame, the unit vectors $\mathbf{i}$, $\mathbf{j}$, and $\mathbf{k}$ are rotating. Therefore, the time rate of change of $\mathbf{P}$ in the global frame is

$$\dot{\mathbf{P}}_G = \dot{P}_x \mathbf{i} + \dot{P}_y \mathbf{j} + \dot{P}_z \mathbf{k} + P_X \frac{d\mathbf{i}}{dt} + P_Y \frac{d\mathbf{j}}{dt} + P_Z \frac{d\mathbf{k}}{dt} \tag{5.22}$$

The first three terms in equation 5.22 are equal to $\dot{\mathbf{P}}_L$, and the last three represent the velocity of the tip of vector $\mathbf{P}$. Because the magnitude of the unit vectors $\mathbf{i}$, $\mathbf{j}$, and $\mathbf{k}$ is constant by definition ($i, j, k = 1$), the time derivatives $d\mathbf{i}/dt$, $d\mathbf{j}/dt$, and $d\mathbf{k}/dt$ represent the change in only the direction of vector $\mathbf{P}$. This change is due to the rotation of local frame $o\text{-}xyz$ with respect to the global frame $O\text{-}XYZ$. The last three terms in equation 5.22 can be written as the cross product

$$P_x \frac{d\mathbf{i}}{dt} + P_y \frac{d\mathbf{j}}{dt} + P_z \frac{d\mathbf{k}}{dt} = \dot{\boldsymbol{\theta}} \times \mathbf{P} \tag{5.23}$$

Hence, equation 5.22 is essentially the same as equation 5.19. These equations state that the rate of change of a vector with respect to a frame with coordinate axes of fixed orientation can be represented as a sum of (1) the rate of change of the vector with respect to a rotating local frame and (2) the vector product $\boldsymbol{\theta} \times \mathbf{P}$ associated with the rotation of the local frame.

5.3.1.2 Motion of a Rigid Body About the Center of Mass

Consider a rigid body rotating about its CoM (figure 5.14). The linear momentum of a particle i is $\mathbf{L}_i = m_i \mathbf{v}_i$ (equation 1.55), and the angular momentum with respect to the CoM is

$$\mathbf{H}_i = \mathbf{r}_i \times m_i \mathbf{v}_i \qquad (5.24)$$

where the vectors $\mathbf{r}_i$ and $\mathbf{v}_i$ represent the position and velocity of the particle in a reference frame with an origin at the CoM and m_i is the mass of the particle. The orientation of the reference frame does not change.

Using the equality $\mathbf{v}_i = \boldsymbol{\theta} \times \mathbf{r}_i$ we obtain

$$\mathbf{H}_i = \mathbf{r}_i \times (\boldsymbol{\theta} \times \mathbf{r}_i) m_i \qquad (5.25)$$

where $\dot{\theta}$ is the angular velocity of body rotation at the instant considered. For the entire body, the angular momentum about the CoM is

$$\mathbf{H} = \int_m \left[\mathbf{r} \times (\dot{\boldsymbol{\theta}} \times \mathbf{r}) \right] dm \qquad (5.26)$$

Equation 5.26 yields the angular momentum of the body relative to the centroidal axes of fixed orientation.

In general, when the body rotates, its inertia tensor changes, and hence the moments and products of inertia may vary and should be determined as a function of time. It would be beneficial to use a system of coordinates in which the moments and products of inertia maintain the same values during the motion.

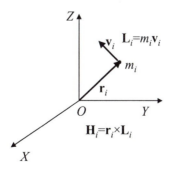

Figure 5.14 Linear and angular momenta of a material particle.

Reference frames that are rigidly attached to the body satisfy this requirement. The inertia matrix of the body with respect to such a frame remains constant irrespective of whatever motion the body executes. In other frames, the inertia matrix transforms according to equation 4.21, and there is no reason to expect it to remain constant. Hence, there is an advantage to writing dynamic equations in a frame rigidly attached to the body. To derive the dynamic equations, we differentiate the angular momentum **H** over time.

Since **H** is a vector, equation 5.19 is valid. Replacing **P** with **H** and $\dot{\mathbf{P}}$ with $\dot{\mathbf{H}}$, we obtain

$$(\dot{\mathbf{H}})_G = (\dot{\mathbf{H}})_L + \dot{\boldsymbol{\theta}} \times \mathbf{H} \tag{5.27}$$

where **H** is the angular momentum of the body in its rotation about the CoM, $(\dot{\mathbf{H}})_G$ is the rate of change of the angular momentum with respect to the frame of fixed orientation, $(\dot{\mathbf{H}})_L$ is the rate of change of **H** with respect to the rotating frame attached to the body, and $\dot{\boldsymbol{\theta}}$ is the angular velocity. Let H_x, H_y, and H_z be the instantaneous values of the projections of **H** onto the x, y, and z axes. Then

$$(\dot{\mathbf{H}})_L = \dot{H}_x \mathbf{i} + \dot{H}_y \mathbf{j} + \dot{H}_z \mathbf{k} \tag{5.28}$$

where the unit vectors **i**, **j**, and **k** rotate with the body. The projections of $(\dot{\mathbf{H}})_L$ onto the axes of the rotating frame are

$$\dot{H}_x = I_{xx} \ddot{\theta}_x + I_{xy} \ddot{\theta}_y + I_{xz} \ddot{\theta}_z$$
$$\dot{H}_y = I_{yx} \ddot{\theta}_x + I_{yy} \ddot{\theta}_y + I_{yz} \ddot{\theta}_z \tag{5.29}$$
$$\dot{H}_z = I_{zx} \ddot{\theta}_x + I_{zy} \ddot{\theta}_y + I_{zz} \ddot{\theta}_z$$

If the coordinate axes of the body-attached reference frame are chosen to be the principal axes of inertia, the products of inertia in equation 5.29 vanish, and the equation assumes a simple form:

$$\dot{H}_x = I_{xx} \ddot{\theta}_x$$
$$\dot{H}_y = I_{yy} \ddot{\theta}_y \tag{5.30}$$
$$\dot{H}_z = I_{zz} \ddot{\theta}_z$$

The cross product $\dot{\boldsymbol{\theta}} \times \mathbf{H}$ equals

$$\dot{\boldsymbol{\theta}} \times \mathbf{H} = \begin{vmatrix} \mathbf{i} & \mathbf{j} & \mathbf{k} \\ \dot{\theta}_x & \dot{\theta}_y & \dot{\theta}_z \\ H_x & H_y & H_z \end{vmatrix} = \left(H_z \dot{\theta}_y - H_y \dot{\theta}_z \right) \mathbf{i} + \left(H_x \dot{\theta}_z - H_z \dot{\theta}_x \right) \mathbf{j} + \left(H_y \dot{\theta}_x - H_x \dot{\theta}_y \right) \mathbf{k} \tag{5.31}$$

where H_x, H_y, and H_z can be obtained from

$$H_x = I_{xx}\dot{\theta}_x + I_{xy}\dot{\theta}_y + I_{xz}\dot{\theta}_z$$
$$H_y = I_{yx}\dot{\theta}_x + I_{yy}\dot{\theta}_y + I_{yz}\dot{\theta}_z \qquad (5.32)$$
$$H_z = I_{zx}\dot{\theta}_x + I_{zy}\dot{\theta}_y + I_{zz}\dot{\theta}_z$$

According to equation 5.1, $(\dot{\mathbf{H}})_G - \Sigma\mathbf{M}$. Substitution of $\Sigma\mathbf{M}$ for $(\dot{\mathbf{H}})_G$ in equation 5.27 yields

$$\Sigma\mathbf{M} = (\dot{\mathbf{H}})_L + \dot{\boldsymbol{\theta}} \times \dot{\mathbf{H}} \qquad (5.33)$$

where $\Sigma\mathbf{M}$ is the sum of the moments of force with respect to the CoM of the body. Because the tensor of inertia $[I]$ remains constant in the body frame, equation 5.33 can be written as

$$\mathbf{M}_R = [I]\ddot{\boldsymbol{\theta}} + \dot{\boldsymbol{\theta}} \times ([I]\dot{\boldsymbol{\theta}}) \qquad (5.34)$$

where $\mathbf{M}_R$ is the moment resultant ($\mathbf{M}_R = \Sigma\mathbf{M}$) and $[I]$ is the inertia matrix of the body in a frame whose origin is located at the CoM. The term $\dot{\boldsymbol{\theta}} \times ([I]\dot{\boldsymbol{\theta}})$ is known as the gyroscopic term. The gyroscopic term results from the changes in the inertia tensor during the body rotation. Equation 5.34 can be written in the component form. If the axes of the body-fixed frame are chosen along the principal axes of inertia, the equations of motion are

$$\Sigma M_x = I_x\ddot{\theta}_x - (I_y - I_z)\,\dot{\theta}_y\dot{\theta}_z$$
$$\Sigma M_y = I_y\ddot{\theta}_y - (I_z - I_x)\dot{\theta}_z\dot{\theta}_x \qquad (5.35)$$
$$\Sigma M_z = I_z\ddot{\theta}_z - (I_x - I_y)\dot{\theta}_x\dot{\theta}_y$$

Comparing equation 5.35 with equations 5.30 and 5.34, we see that the products $I\ddot{\theta}$ represent the rate of change of the angular momentum in the frame attached to the body and the terms containing the products of angular velocities represent the gyroscopic terms. In equations 5.33 and 5.34, the angular velocity vector $\dot{\boldsymbol{\theta}}$ is considered in the frame of fixed orientation. However, in equation 5.35 $\dot{\theta}_x$, $\dot{\theta}_y$, and $\dot{\theta}_z$ are the components of $\dot{\boldsymbol{\theta}}$ along the rotated axes x, y, and z. The symbols $\ddot{\theta}_x$, $\ddot{\theta}_y$, and $\ddot{\theta}_z$ denote the time derivatives of the magnitudes of $\dot{\theta}_x$, $\dot{\theta}_y$, and $\dot{\theta}_z$. Equations 5.33 through 5.35 are known as the *Euler dynamic equations* (after Leonhard Euler, 1707–1783). They are the dynamic equations expressed in the frame attached to the body and having its origin at the CoM.

It should be realized that the Euler equations do not contain the angular velocity of the body in the body-fixed (rotating) reference frame. In such a frame, all particles of the body retain their position, and the vectors from the origin to the particles do not change their orientation (nor their length). There-fore, the angular velocity of the body in the body-fixed frame is always zero. In contrast, the vector of angular velocity $\boldsymbol{\theta}$ in the frame of fixed orientation can always be determined. When the vector $\boldsymbol{\theta}$ is obtained, its components in any reference frame, including the rotating frame, can be computed. This pro-cedure is used in the Euler equations. (Rotation of a body about the axes at-tached to it can be visualized by introducing two reference frames. One frame changes its orientation in a steplike fashion and is at rest at any given instant of time. The second frame rotates with the body continuously.)

The Euler dynamic equations are broadly used to analyze the motion of the rigid body about its mass center. The dynamic equations can also be written in reference frames that are not fixed to the body links; for instance, they can be written with respect to the mediolateral, anteroposterior, and vertical axes. Al-though these equations may look similar to equation 5.34, they are fundamen-tally different. The matrix of inertia in equation 5.34 is constant, whereas the inertia elements in an unattached reference frame are time dependent.

The inertia tensor in the body-fixed principal axes of inertia is diagonal and remains constant. However, the principal axes change their orientation with time. Therefore, to solve dynamic equations in three dimensions, repeated co-ordinate transformations are required. During computations, repetitive trans-formations from the absolute system of coordinate into the movable system and vice versa are performed. Such coordinate transformations are the price that is paid for the simplicity and convenience of the equations in movable axes. We do not have much choice: equations in the absolute reference frame are even more complex and require continuous recomputation of the tensors of inertia.

In the following two sections, **5.3.2** and **5.3.3**, we return to the main object of our concern, the motion of multilink chains.

5.3.2 Recursive Newton-Euler Formulation

The closed-form dynamic equations are useful for understanding the investi-gated motion. With these equations, the contribution of individual effects (in-ertial forces, centrifugal forces, etc.) to the joint torques can be computed. However, when the number of links in the chain increases, the complexity of the equations becomes prohibitive. In this case, a recursive procedure is usu-ally used. The computations are performed in individual steps (iterations), for one link at a time.

5.3.2.1 The Newton-Euler Equations

The Newton-Euler equations for the individual links in three dimensions are

$$\mathbf{F}_i - \mathbf{F}_{i+1} + m_i\mathbf{g} - m_i\mathbf{a}_i = 0$$

$$\mathbf{T}_i - \mathbf{T}_{i+1} + \mathbf{r}_i \times \mathbf{F}_i - \mathbf{r}_{i+1} \times \mathbf{F}_{i+1} - [I_i]\ddot{\boldsymbol{\theta}}_i - \dot{\boldsymbol{\theta}}_i \times ([I_i]\dot{\boldsymbol{\theta}}_i) = 0$$

(5.36)

where the designations are similar to those introduced in figure 5.4. $\mathbf{F}_i$ and $\mathbf{F}_{i+1}$ are the forces acting on link i at joints i and $i + 1$, respectively; $\mathbf{T}_i$ and $\mathbf{T}_{i+1}$ are the torques exerted on link i at these joints; $\mathbf{r}_i$ and $\mathbf{r}_{i+1}$ are the radii from the CoM of link i to the joint centers i and $i + 1$, respectively; and $\mathbf{g}$ is the acceleration of gravity. Also, $[I_i]$ is the matrix of inertia, and $\dot{\boldsymbol{\theta}}_i$ and $\ddot{\boldsymbol{\theta}}_i$ are the vectors of the link angular velocity and acceleration, respectively. Equation 5.36 can be also written as

$$\underbrace{\mathbf{F}_i - \mathbf{F}_{i+1} + m_i\mathbf{g}}_{\text{Effective force}} = m_i\mathbf{a}_i$$

$$\underbrace{\mathbf{T}_i - \mathbf{T}_{i+1} + \mathbf{r}_i \times \mathbf{F}_i - \mathbf{r}_{i+1} \times \mathbf{F}_{i+1}}_{\text{Effective couple}} = \underbrace{[I_i]\ddot{\boldsymbol{\theta}}_i + \underbrace{\dot{\boldsymbol{\theta}}_i \times \left([I_i]\dot{\boldsymbol{\theta}}_i\right)}_{\text{Gyroscopic term}}}_{\dot{\mathbf{H}}_i, \text{ the rate of change of} \atop \text{angular momentum}}$$

(5.37)

When the inverse solution is sought, the right-hand sides of equation 5.37 serve as the input to the equations.

5.3.2.2 The Recursive Procedure

Representative papers: Zatsiorsky and Aleshinsky (1976); Aleshinsky and Zatsiorsky (1978)

To illustrate the method, let as consider a planar multilink chain. The recursive computation usually starts from a distal link and goes inward to the proximal links. For an unloaded distal link i, the equations are

$$\mathbf{F}_{i-1,i} - m_i\mathbf{g} - m_i\mathbf{a}_i = 0 \qquad (5.38a)$$

$$\mathbf{T}_{i-1,i} + \mathbf{r}_i \times \mathbf{F}_{i-1,i} - I_i\ddot{\theta}_i = 0 \qquad (5.38b)$$

where $\mathbf{F}_{i-1,i}$ is the force exerted on link i by link $i - 1$ and $\mathbf{T}_{i-1,i}$ is the joint torque. First, provided that acceleration of the CoM of the distal link $\mathbf{a}_i$ is known, the force $\mathbf{F}_{i-1,i}$ is determined. Then, the known force $\mathbf{F}_{i-1,i}$ is substituted in equation 5.38b and the joint torque $\mathbf{T}_{i-1,i}$ acting on link i is computed. To do that, the angular acceleration of link i $\ddot{\theta}_i$ and its angular velocity $\dot{\theta}_i$, should be known. The force $\mathbf{F}_{i-1,i}$ arises from an interaction of link i with link $i - 1$. According to Newton's third law of motion, an equal and opposite force $-\mathbf{F}_{i-1,i}$ acts on link $i - 1$. A dynamic equation for the force exerted on link $i - 1$ by link $i - 2$ is

$$\mathbf{F}_{i-2,i-1} = -\mathbf{F}_{i-1,i} - m_{i-1}\mathbf{g} + m_{i-1}\mathbf{a}_{i-1} \qquad (5.39)$$

and an equation for the joint torque acting between the links $i-1$ and $i-2$ is

$$\mathbf{T}_{i-2,i-1} = -\mathbf{T}_{i-1,i} - \mathbf{r}_i \times \mathbf{F}_{i-1,i} + \mathbf{r}_{i-1} \times \mathbf{F}_{i-2,i-1} + I_{i-1}\ddot{\theta}_{i-1} \tag{5.40}$$

where $\mathbf{r}_i$ and $\mathbf{r}_{i-1}$ are the radii from the CoM of link $i-1$ to the joint centers i and $i-1$.

The forces $\mathbf{F}_{k,k+1}$ and $-\mathbf{F}_{k,k+1}$, acting according to Newton's third law between the adjacent segments k and $k+1$, are known as the *coupling forces*. The torques $\mathbf{T}_{k,k+1}$ and $-\mathbf{T}_{k,k+1}$ are called the coupling torques. The adjective "coupling" in this context refers to the method of obtaining these quantities. It indicates that the forces and torques were determined from Newton's third law.

The coupling forces and moments for all the joints are computed recursively by solving the dynamic equations one by one from the distal link to the proximal. Each force and torque appears twice in the equations: once as an unknown in the equation for the distal link and the second time in the equation for the proximal link (with an opposite sign). The torque at a given joint cannot be calculated until the joint reaction force at this joint has first been computed. To compute the force and torque at a proximal joint, the torque and force acting at the distal joint must first be determined.

When a coupling force and torque acting on the trunk are computed, they are saved until all the other forces and torques that act on the trunk, except those of the last link, are determined. All these saved forces and moments are used to determine the coupling force and torque in the last joint of the analysis. For instance, during the support period in running, the coupling forces and torques are determined for two shoulder joints, the neck–trunk articulation, and one hip joint. Finally, this information is used to compute the coupling force and torque in the hip joint of the supporting leg.

For a chain with n links, $2n$ equations must be solved. For instance, for a two-link chain, two sets of equations with two equations per set—one for a force and one for a moment—must be solved. If the number of joints exceeds the number of links, there are more unknowns than equations, and a unique solution cannot be obtained. This happens, for instance, during double-support periods in walking. Forces exerted on the ground by each leg cannot be computed by the methods of inverse dynamics alone.

The Newton-Euler equations contain the coupling forces and moments; the equations by themselves do not tell whether the moments are due to the muscle-tendon torques or to the anatomic constraints at the joints. The equations are written in an externally fixed reference frame. They do not immediately allow partitioning the estimated forces and moments into the components associated with individual kinematic variables, such as the centrifugal components or Coriolis components. However, the recursive Newton-Euler formulation provides a convenient computational tool for a numerical solution of the inverse problem of dynamics: both the joint torques and joint forces can be computed.

■ ■ ■ **FROM THE LITERATURE** ■ ■ ■

Inverse Dynamics in Three Dimensions

Source: Aleshinsky, S.Y., and V.M. Zatsiorsky. 1978. Human locomotion in space analyzed biomechanically through a multi-link chain model. *J. Biomech.* 11 (3): 101–8.

The goal of the study was to determine joint moments and forces in all the major joints during casual walking, race walking, and sprint running. A 15-link model of the human body was used (see figure 1.29 in *Kinematics of Human Motion*). The coordinates of the markers fixed to the subjects were determined by optical methods, and the inverse problem of dynamics was solved in three dimensions. For individual body links, the Euler dynamic equations (equations 5.33 to 5.35) were used. To solve the inverse problem for the entire chain, the recursive Newton-Euler formulation was employed. An example of the outcome of the study is presented in figure 5.15.

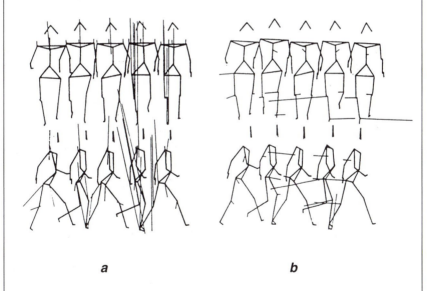

a   b

Figure 5.15 Vectors of (*a*) the joint forces and (*b*) the joint moments in walking: front and side views.

Adapted from *Journal of Biomechanics*, 11, S. Yu. Aleshinsky and V.M. Zatsiorsky, Human locomotion in space analyzed biochemically through a multi-link chain moment, 101–108, 1978, with permission from Elsevier Science.

5.3.2.3 External Contact Forces

Representative paper: Hof (1992)

If external contact forces (besides the GRF) act on the body links, the forces and couples acting at the contact should be included in the equations. The external contact is treated as an additional joint.

Ground reaction force represents the inertial and gravitational forces transmitted through the feet to the ground. For single-support periods, the GRF can be computed from the equations of inverse dynamics. To do that, the supporting object is treated as link $n + 1$ added to an n-link chain, and the ground reaction force and moments are computed as the coupling force and moment in the "joint" between links n and $n + 1$. Although the GRF can be estimated from the equations of inverse dynamics, the direct measurement of GRF with force plates is recommended. The computed GRF values can be compared with the GRF recorded experimentally. The comparison allows estimating and improving the accuracy of the inverse dynamics solution.

When a recursive procedure is used, the error accumulates from joint to joint. Because the computations start from the most distal free segment, the error of computation is the largest for the joint torques of the supporting leg, which are computed at the last stage of the procedure. Especially large errors are associated with the inclusion of the trunk segment, with its large mass and doubtful rigidity. To avoid this impediment, the joint torques during single-support periods can be computed starting from the supporting leg and using the GRF as an input variable to the dynamic equations. In this case, however, the opportunity for checking the accuracy of the inverse solution by comparing the computed and measured values of the GRF is lost.

During double-support periods, individual GRFs exerted by each leg cannot be determined by the methods of inverse dynamics. The task is indeterminate (the so-called *closed-loop problem*). However, if the GRF acting on one leg is measured, the force generated by the second leg can be determined from inverse dynamics. When several external forces are acting on a human body, the inverse problem of dynamics can be solved only if all the forces except one are directly measured (figure 5.16).

In dynamics, contrary to statics, knowledge of the external forces acting on the body in itself does not allow determining the joint torques and joint reaction forces. Equation 2.16, ($\mathbf{T} = \mathbf{J}^T \overline{\mathbf{F}}$), cannot be immediately applied. Velocity and acceleration of body parts must be known, and their effects should be included in the calculations. Consider the situation in which a force and moment $\overline{\mathbf{F}}$ are exerted on the environment, for instance, lifting a barbell. ($\overline{\mathbf{F}}$ is a six-component vector.) The joint torques can be partitioned into the three components that resist (1) the inertial resistance of the body parts, (2) the weight of

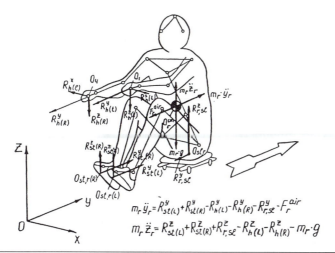

$$m_r \ddot{y}_r = R^y_{st(L)} + R^y_{st(R)} - R^y_{h(L)} - R^y_{h(R)} - R^y_{r,se} - F^{air}_r$$

$$m_r \ddot{z}_r = R^z_{st(L)} + R^z_{st(R)} + R^z_{r,se} - R^z_{h(L)} - R^z_{h(R)} - m_r \cdot g$$

Figure 5.16 External forces acting on a rower's body in a single scull in the first half of the stroke phase. The analysis is limited to the sagittal plane. Seven forces are acting: (1 and 2) the reaction forces of the boat's stretchers, $R_{st(L)}$ and $R_{st(R)}$; (3 and 4) the forces of the oar handle reaction, $R_{h(L)}$ and $R_{h(R)}$; (5) the reaction force of the sliding seat, $R_{r,sl}$; (6) air resistance, R_{air}; and (7) gravity force, $m_r g$. To compute the joint torques and forces, at least five external forces, in addition to the weight of the athlete and the kinematics of the athlete's movement, must be measured. O indicates the support points of the 15-link model of the rower's body; ⊕ indicates the rower's CoM.

Reprinted, by permission, from V.M. Zatsiorsky and N. Yakunin, 1991, "Mechanics and biomechanics of rowing," *International Journal of Sport Biomechanics* 7(3):229–281.

■ ■ ■ *FROM THE LITERATURE* ■ ■ ■

Solving the Inverse Problem by Using the GRF and Kinematic Data

Source: Hof, A.L. 1992. An explicit expression for the moment in multibody systems. *J. Biomech.* 25 (10): 1209–11.

Equations were developed for computing the joint forces and torques by using the GRF as an input variable. This approach has certain advantages when the goal is to compute the joint forces and torques of the supporting leg.

Consider the case of a moving subject who has one foot on a force plate (figure 5.17). The ground reaction force **F**$_r$ is measured. Positions and accelerations are determined by other means (e.g., from videotape). It is assumed that the foot does not exert a free moment on the force plate. The biomechanical model is a linked chain of n rigid bodies, the stance foot being link 1. To compute the joint force $\mathbf{F}_{k,k+1}$ and moment $\mathbf{M}_{k,k+1}$ between the adjacent links k and $k + 1$, a subset of links from 1 to k is considered. The force $\mathbf{F}_{k,k+1}$ and moment $\mathbf{M}_{k,k+1}$ are now considered "external" to this subset of links. The joint force $\mathbf{F}_{k,k+1}$ equals

$$\mathbf{F}_{k,k+1} = -\mathbf{F}_r - \sum_{i=1}^{k} m_i \mathbf{g} + \sum_{i=1}^{k} m_i \mathbf{a}_i \qquad (5.41)$$

where subscript i refers to an individual body link ($i = 1, \ldots , k$). Joint torque $\mathbf{M}_{k,k+1}$ is equal to

$$\mathbf{M}_{k,k+1} = -\underbrace{\left(\mathbf{r}_r - \mathbf{r}_{k,k+1}\right) \times \mathbf{F}_r}_{\text{Moment of the GRF}} - \underbrace{\sum_{i=1}^{k}\left[\left(\mathbf{r}_r - \mathbf{r}_{k,k+1}\right) \times m_i \mathbf{g}\right]}_{\text{Moment of the weights of the links}} +$$

$$\underbrace{\sum_{i=1}^{k}\left[\left(\mathbf{r}_i - \mathbf{r}_{k,k+1}\right) \times m_i \mathbf{a}_i\right]}_{\substack{\text{Moments of the effective forces} \\ \text{acting at the CoM of the links}}} + \underbrace{\sum_{i=1}^{k}\frac{d}{dt}\left(I_i \dot{\theta}_i\right)}_{\substack{\text{Moments due to} \\ \text{rotational acceleration}}} \qquad (5.42)$$

where $\mathbf{r}_r$, $\mathbf{r}_{k,k+1}$, and $\mathbf{r}_i$ are the position vectors from the origin of an inertial system of coordinates to the point of force application on the force platform, to the joint between the links k and $k + 1$, and to the CoM of link i, respectively. The point of force application can be computed because the free moment is zero. If the last two terms of the equation were omitted, the first two would represent the quasi-static solution (explained later in the text).

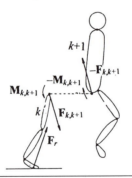

Figure 5.17 Model of a subject walking over a force plate. The n-link chain is divided into two subsets, one containing the links from 1 to k and the other from $k + 1$ to n.

Reprinted from *Journal of Biomechanics*, 25 (10), A.L. Hof, An explicit expression for the moment in multibody system, 1209–1211, 1992, with permission from Elsevier Science.

the body parts, and (3) external force. Symbolically, this statement can be written as

$$\mathbf{T} = \boldsymbol{\tau}_{\text{inertia}} + \boldsymbol{\tau}_{\text{gravity}} + \mathbf{J}^T \overline{\mathbf{F}} \tag{5.43}$$

where $\mathbf{T}$ is the vector of the joint torques, $\boldsymbol{\tau}_{\text{inertia}}$ is the vector of the torque components associated with the inertial resistance (the acceleration-related terms and the velocity-related terms of the dynamic equations, in particular, the centrifugal and Coriolis terms), $\boldsymbol{\tau}_{\text{gravity}}$ is the gravity term, and $\mathbf{J}^T \overline{\mathbf{F}}$ is the vector of the torque components that resist the external force $\overline{\mathbf{F}}$. For very slow movements, if the joint torques due to the external forces and the body weight are much larger than the inertial forces, the joint torques and forces can be approximately estimated using the methods of statics. This technique is called a *quasi-static solution*. In the quasi-static technique, the inertial forces ($\boldsymbol{\tau}_{\text{inertia}}$) are neglected. When external contact forces are impressed on the body, the dynamic equations are commonly written in compact form as

$$\mathbf{T} = \left[I^c(\alpha)\right]\ddot{\boldsymbol{\alpha}} + \mathbf{V}(\alpha,\dot{\alpha}) + \mathbf{G}(\alpha) + \overline{\mathbf{F}} \tag{5.44}$$

where $\overline{\mathbf{F}}$ stands for the vector of the external contact forces and couples and the other symbols are the same as in equation 5.11.

During isokinetic dynamometer testing, the moment measured by the machine may differ from the active joint torque due to the weight of the distal segment. For such testing, a correction for gravity is recommended. In general,

---

••• *MECHANICS REFRESHER* •••

Kinetic and Potential Energy

Energy is the capacity to do work. The *kinetic energy* of a material particle of mass m moving with velocity $\mathbf{v}$ is a positive scalar: $K = (m\mathbf{v} \cdot \mathbf{v})/2 = mv^2/2$.

A force that depends only on the location of the particle is called the *conservative force*. The conservative force does not depend on the velocity, acceleration, or path followed by the particle in its movement from one location to another. The *potential energy, Q,* of the particle equals the work done by the conservative forces to move the particle from location B to location A, the standard location of the particle. The potential energy of the particle depends only on the location of the particle with respect to A.

The total energy of a particle is the sum of its kinetic and potential energies: $E = K + Q$.

> ### ▪ ▪ ▪ *FROM THE LITERATURE* ▪ ▪ ▪
>
> ## Decomposition of Force
> ## Applied to the Pedal in Cycling
>
> Source: Kautz, S.A., and M.L. Hull. 1993. A theoretical basis for interpreting the force applied to the pedal in cycling. *J. Biomech.* 26 (2): 155–65.
>
> In cycling, muscular forces accelerate the legs, while the foot–pedal connection constrains the movement so that the foot follows a circular path about the crank axis. Consequently, pedal forces not only reflect muscular activity (joint torques) but also depend on weight and inertia forces. In this study, the pedal force was decomposed into a component due to joint torques ("muscular component," F_m) and a "nonmuscular component," F_n, due to gravitational and inertial effects (figure 5.18). Component F_m equals the pedal force that would be measured if the pedal were fixed in the observed configuration in the absence of gravity. F_n is the force that would be measured if the leg were externally driven by a motor attached to the crank, as long as the kinematics were identical and there were no joint torques. However, because the leg is not externally driven, the "nonmuscular" component is mostly the indirect result of muscular efforts.
>
>   In the experiment, a cyclist pedaled at 70, 90, and 110 rpm against a constant workload of 167 J per revolution (corresponding to powers of 194, 250, and 306 W, respectively). To solve the inverse problem of dynamics, the Newton-Euler equations were used. The nonmuscular component increased in magnitude as the cadence increased from 70 to 110 rpm. Even at the slowest pedaling rate of 70 rpm, the magnitude of the nonmuscular component was substantial.

force exerted on the environment by a performer is due to (1) active joint torques, (2) inertial forces, and (3) weight of the body parts.

5.3.3 Lagrangian Formulation

The dynamic equations can be derived using different techniques. All methods generate equivalent sets of equations, but some equations may be better suited for a particular analysis. In the *Lagrangian formulation* (named after Joseph Louis Lagrange, 1736–1813), the equations of motion are based on the kinetic and potential energies of the entire body. With this approach, the dynamic equations can be obtained in the closed form, allowing detailed analysis of the

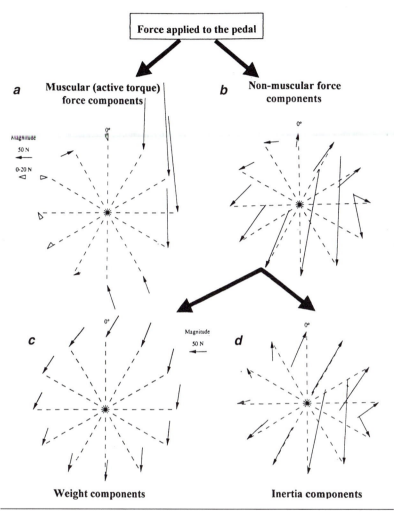

Figure 5.18 Decomposition of the force applied to the pedal. (*a*) Force due to the joint torques, F_m. (*b*) F_n, the pedal force due to gravity and inertia, nonmuscular force components. (*c*) Force component due to the weight of the limb. At some crank angles, large horizontal forces due to the weight component were observed, although the gravity force is directed vertically downward. To explain this observation, consider the entire leg to be one body that is suspended at the hip and supported by the pedal. Because the hip joint shares the applied weight, the weight subcomponent at the pedal is a complicated function of the leg configuration and not simply a downward vertical force equal to the weight of the leg. (*d*) The inertial subcomponent of the force applied to the pedal.

Adapted from *Journal of Biomechanics*, 26 (2), S.A. Kautz and M.L. Hull, A theoretical basis for interpreting the force applied to the pedal in cycling, 155–163, 1993, with permission from Elsevier Science.

motion. The number of equations is equal to the number of degrees of freedom. The forces of constraint in the joints do not appear in the equations; therefore, in the Lagrangian method, the joint reaction forces are not determined. The method does not constitute a new theory. It is based entirely on Newton's laws of motion. The results of a Lagrangian analysis and a Newtonian analysis are the same for any mechanical system. However, technically, the method is an alternative to the Newton-Euler approach. This section is limited to a very brief discussion of the main ideas of the Lagrangian formulation for a specific case of kinematic chains.

A scalar function L called the *Lagrangian* is defined as the difference between the kinetic energy K and potential energy Q of the system:

$$L = K - Q \tag{5.45}$$

Since energy is a scalar quantity, the Lagrangian is invariant to coordinate transformation, including the transformation to *generalized coordinates*. In the biomechanics of human motion, the Lagrangian function is usually computed in joint coordinates. For a kinematic chain, the equations of motion in the Lagrangian formulation are

$$T_i = \frac{d}{dt}\left(\frac{\partial L}{\partial \dot{\alpha}_i}\right) - \frac{\partial L}{\partial \alpha_i} \tag{5.46a}$$

or

$$T_i = \frac{d}{dt}\left(\frac{\partial K}{\partial \dot{\alpha}_i}\right) - \frac{\partial K}{\partial \alpha_i} + \frac{\partial Q}{\partial \alpha_i} \tag{5.46b}$$

where α_i is the coordinates in which the kinetic and potential energy are expressed (for a kinematic chain, these coordinates are the joint angles), $\dot{\alpha}_i$ is the corresponding velocity, and T_i is the corresponding generalized force (if the coordinates are joint angles, the T_i terms are the torques acting about the joint axes). For instance, in the absence of gravity, the potential energy Q of a material particle is zero, and the kinetic energy K is $mv^2/2$. The partial derivative

$$\frac{\partial K}{\partial v} = \frac{\partial}{\partial v}\left(\frac{mv^2}{2}\right) = mv \tag{5.47}$$

is simply the linear momentum of the particle. Further application of equation 5.47 yields the equation "Generalized force = ma," which is simply Newton's second law. However, the Newton-Euler approach places the emphasis on the forces (vectors) acting on a body, whereas the Lagrangian method deals only with the scalar quantities that are associated with the body. Equations of motion 5.46 are known as the Lagrangian equations of the second type.

To solve the Lagrangian equations, the potential and kinetic energy of the links must be known. The (gravitational) potential energy of link i equals the amount of work required to raise the CoM of the link from a reference plane to its present position, $Q_i = m_i g h_i$, where h_i is the height of the CoM of the link above the reference level. The total potential energy of the body Q equals the sum of the potential energies stored in the individual links. The potential energy is a function of chain configuration. The most complicated term in the Lagrangian function is the total kinetic energy of the body.

5.3.3.1 Kinetic Energy of Individual Links

Representative papers: Fenn (1930); Ralston and Lukin (1969)

Consider an ith link of a kinematic chain moving in three-dimensional space. The linear velocity of the CoM of the link is $\mathbf{v}_i$, and its angular velocity about the CoM is $\dot{\theta}_i$. The kinetic energy K_i of the link is then given by

$$K_i = \frac{1}{2} m_i \mathbf{v}_i^T \mathbf{v}_i + \frac{1}{2} \dot{\theta}_i^T [I_i] \dot{\theta}_i \qquad (5.48)$$

where m_i is the mass of the link and $[I_i]$ is the (3×3) inertia tensor. All the variables in equation 5.48 are expressed in the global reference frame of fixed orientation. The first term on the right side of the equation is called the *translational kinetic energy,* and the second term represents the *rotational kinetic energy* of the link. The expression for the translational kinetic energy is analogous to the familiar expression for the kinetic energy of a material particle, $K = mv^2/2$. The rotational kinetic energy can also be written as the scalar product of the vectors,

$$K_{i\,(\text{rot})} = \frac{1}{2} \dot{\theta}_i \mathbf{H}_i \qquad (5.49)$$

where $\mathbf{H}_i$ is the angular momentum of the link with respect to its CoM. If the link rotates about a fixed point, $\mathbf{H}_i$ can be computed with respect to the center of rotation, in which case equation 5.49 yields the total kinetic energy of the link. In matrix form, the rotational kinetic energy of a single link is

$$K_{\text{rot}} = \frac{1}{2} \begin{bmatrix} \dot{\theta}_x \\ \dot{\theta}_y \\ \dot{\theta}_z \end{bmatrix}^T \begin{bmatrix} I_{xx} & I_{xy} & I_{xz} \\ I_{yx} & I_{yy} & I_{yz} \\ I_{zx} & I_{zy} & I_{zz} \end{bmatrix} \begin{bmatrix} \dot{\theta}_x \\ \dot{\theta}_y \\ \dot{\theta}_z \end{bmatrix} \qquad (5.50)$$

where the origin of the coordinate frame is at the CoM. The multiplication yields

$$K_{\text{rot}} = \frac{1}{2} I_{xx} \dot{\theta}_x^2 + \frac{1}{2} I_{yy} \dot{\theta}_y^2 + \frac{1}{2} I_{zz} \dot{\theta}_z^2 +$$
$$I_{xy} \dot{\theta}_x \dot{\theta}_y + I_{xz} \dot{\theta}_x \dot{\theta}_z + I_{yz} \dot{\theta}_y \dot{\theta}_z \tag{5.51a}$$

or

$$K_{\text{rot}} = \frac{1}{2} \sum_i \sum_j I_{ij} \dot{\theta}_i \dot{\theta}_j \tag{5.51b}$$

If the vector of angular velocity is along one of the coordinate axes, equation 5.50 reduces to

$$K_{\text{rot}} = \frac{1}{2} I \dot{\theta}^2 \tag{5.52}$$

where I is the moment of inertia about the axis of rotation and $\dot{\theta}$ is the angular velocity (scalar).

5.3.3.2 Kinetic Energy of the Entire Body

Since energy is additive, the kinetic energy of the human body is the sum of the kinetic energies of the individual links:

$$K = \Sigma K_i \tag{5.53}$$

To derive the equations of motion, we have to express the kinetic energy as a function of joint angles and joint angular velocities. To do that, we use the Jacobian method discussed in section **3.1.2.2.2** of *Kinematics of Human Motion*. If we regard each link in the same way as we regarded the end effector, the linear velocity of the CoM of link i can be written as

$$\mathbf{v}_i = \mathbf{J}_{Li} \dot{\boldsymbol{\alpha}} \tag{5.54}$$

where $\mathbf{J}_{Li}$ is the positional Jacobian that maps the instantaneous joint velocities to the linear velocity of the CoM of link i, and $\dot{\boldsymbol{\alpha}}$ is the vector of joint angular velocities. The motion of link i depends only on joints 1 through i. For instance, motion of the second link does not depend on the third joint. To account for that, the column vectors in the Jacobian are set to zero for $j > i$. The translational kinetic energy of link i is then

$$K_{i\,(\text{tran})} = \frac{1}{2} \mathbf{v}_i^T m_i \mathbf{v}_i = \frac{1}{2} \left(\mathbf{J}_{Li} \dot{\boldsymbol{\alpha}} \right)^T m_i \left(\mathbf{J}_{Li} \dot{\boldsymbol{\alpha}} \right) =$$
$$\frac{1}{2} \dot{\boldsymbol{\alpha}}^T \left(\mathbf{J}_{Li}^T m_i \mathbf{J}_{Li} \right) \dot{\boldsymbol{\alpha}} \tag{5.55}$$

In a similar way, the rotational kinetic energy of the ith link can be expressed as

$$K_{i \, (\text{rot})} = \frac{1}{2} \dot{\boldsymbol{\alpha}}^T \left(\mathbf{J}_{Ai}^T [I]_i \mathbf{J}_{Ai} \right) \dot{\boldsymbol{\alpha}} \tag{5.56}$$

where $\mathbf{J}_{Ai}$ is the orientation Jacobian of link i that maps the instantaneous joint rates to the instantaneous angular velocity of the link. Substituting equations 5.55 and 5.56 into equations 5.48 and 5.53 yields

$$K = \frac{1}{2} \dot{\boldsymbol{\alpha}}^T \left[\sum_{i=1}^{n} \left(\mathbf{J}_{Li}^T m_i \mathbf{J}_{Li} + \mathbf{J}_{Ai}^T [I]_i \mathbf{J}_{Ai} \right) \right] \dot{\boldsymbol{\alpha}} = \frac{1}{2} \dot{\boldsymbol{\alpha}}^T [I^c] \dot{\boldsymbol{\alpha}} \tag{5.57}$$

where K is the total kinetic energy of the chain and $[I^c]$ is the inertia matrix of the chain, which is analogous to the inertia matrix introduced previously for planar chains (equation 5.11). The difference is that in three dimensions, the inertia matrix of the chain is a matrix of matrices; the elements of the matrix are the (3×3) inertia matrices of the individual links. The inertia tensor of the chain has properties similar to those of individual inertia tensors. It is a symmetric matrix, and it always has an inverse. Because equation 5.57 is quadratic, the kinetic energy is always positive, unless the system is at rest.

Equation 5.57 can be written in a scalar form:

$$K = \frac{1}{2} \sum_{i=1}^{n} \sum_{i=1}^{n} I_{ij} \dot{\alpha}_i \dot{\alpha}_j \tag{5.58}$$

To formulate the Lagrangian dynamic equations, the derivatives of the potential and kinetic energies of the chain with respect to joint angular velocities, joint angles, and time must be determined. The equations allow computing the torques at joints.

5.3.3.3 Deriving the Lagrangian Equations

Consider an n-link chain driven by joint torque actuators. The links are unloaded; that is, external forces and moments acting on the links are zero (with the exception of the GRF). Our immediate goal is to obtain explicit expressions for the members of equation 5.46b, $\dfrac{d}{dt}\left(\dfrac{\partial K}{\partial \dot{\alpha}_i}\right)$, $\dfrac{\partial K}{\partial \alpha_i}$, and $\dfrac{\partial Q}{\partial \alpha_i}$.

Differentiating equation 5.58, we obtain

$$\frac{d}{dt}\left(\frac{\partial K}{\partial \dot{\alpha}_i}\right) = \frac{d}{dt}\left(\sum_{j=1}^{n} I_{ij} \dot{\alpha}_j\right) = \sum_{j=1}^{n} I_{ij} \ddot{\alpha}_j + \sum_{j=1}^{n} \frac{dI_{ij}}{dt} \dot{\alpha}_j \tag{5.59}$$

Because the elements of the inertia matrix of the chain I_{ij} depend on the joint angles, the time derivative of I_{ij} can be presented as

$$\frac{dI_{ij}}{dt} = \sum_{k=1}^{n} \frac{\partial I_{ij}}{\partial \alpha_k} \frac{d\alpha_k}{dt} = \sum_{k=1}^{n} \frac{\partial I_{ij}}{\partial \alpha_k} \dot{\alpha}_k \tag{5.60}$$

Therefore, equation 5.59 can now be rewritten as

$$\frac{d}{dt}\left(\frac{\partial K}{\partial \dot{\alpha}_i}\right) = \sum_{j=1}^{n} I_{ij}\ddot{\alpha}_j + \sum_{j=1}^{n}\sum_{k=1}^{n} \frac{\partial I_{ij}}{\partial \alpha_k} \dot{\alpha}_j \dot{\alpha}_k \tag{5.61}$$

The partial derivative of the kinetic energy with respect to the joint angles is

$$\frac{\partial K}{\partial \alpha_i} = \frac{\partial}{\partial \alpha_i}\left(\frac{1}{2}\sum_{j=1}^{n}\sum_{k=1}^{n} I_{jk}\dot{\alpha}_j\dot{\alpha}_k\right) = \frac{1}{2}\sum_{j=1}^{n}\sum_{k=1}^{n} \frac{\partial I_{jk}}{\partial \alpha_i}\dot{\alpha}_j\dot{\alpha}_k \tag{5.62}$$

Substituting equations 5.61 and 5.62 into 5.42 yields

$$\sum_{j=1}^{n} I_{ij}\ddot{\alpha}_j + \sum_{j=1}^{n}\sum_{k=1}^{n} D_{ijk}\dot{\alpha}_j\dot{\alpha}_k + G_i = T_i \tag{5.63}$$

where $G_i = \partial Q/\partial \alpha_i$ is the partial derivative of the potential energy with respect to joint angle and

$$D_{ijk} = \frac{\partial I_{ij}}{\partial \alpha_k} + \frac{1}{2}\frac{\partial I_{jk}}{\partial \alpha_i} \tag{5.64}$$

The first term in equation 5.63 represents the components of the joint torques that are associated with the angular accelerations at the joints. The interactive inertia torques in this term arise from the off-diagonal elements I_{ij} ($i \neq j$) of the tensor of inertia of the chain. The second term in the equation accounts for the centrifugal and Coriolis effects, and the third term is due to gravity. Equation 5.63 is a closed-form equation. The equation is similar to the closed-form equations derived previously for simple planar chains (equations 5.11).

5.3.3.4 Lagrangian Equations for a Two-Link Planar Chain

To illustrate the method, we derive the Lagrangian equations for a two-link planar chain (see figure 5.7 again). Due to the simple nature of the object, many of the possible difficulties do not arise.

Kinetic energy of the chain is

$$K = \frac{1}{2}\left(m_1 v_1^2 + I_1\dot{\theta}_1^2 + m_2 v_2^2 + I_2\dot{\theta}_2^2\right) \tag{5.65}$$

where all the variables are in the global reference frame (segment coordinates). We have to express the kinetic energy in the joint space using the joint angles and their time derivatives as the input variables.

The positional Jacobians $\mathbf{J}_{L1}$ and $\mathbf{J}_{L2}$ for the first and the second links, respectively, are

$$\mathbf{J}_{L1} = \begin{bmatrix} -l_{c1}\sin\alpha_1 & 0 \\ l_{c1}\cos\alpha_1 & 0 \end{bmatrix} \qquad (5.66a)$$

and

$$\mathbf{J}_{L2} = \begin{bmatrix} -l_1\sin\alpha_1 - l_{c2}\sin(\alpha_1+\alpha_2) & -l_{c2}\sin(\alpha_1+\alpha_2) \\ l_1\cos\alpha_1 + l_{c2}\cos(\alpha_1+\alpha_2) & l_{c2}\cos(\alpha_1+\alpha_2) \end{bmatrix} \qquad (5.66b)$$

The orientation Jacobians are (1×2) vectors, $\mathbf{J}_{A1} = [1\ 0]$ and $\mathbf{J}_{A2} = [1\ 1]$. Substituting equations 5.66a and 5.66b into the inertia tensor of the chain,

$$[I^c] = \sum_{i=1}^{n}\left(\mathbf{J}_{Li}^T m_i \mathbf{J}_{Li} + \mathbf{J}_{Ai}^T [I]_i \mathbf{J}_{Ai}\right) \qquad (5.67)$$

after some transformations yields

$$[I^c] = \begin{bmatrix} m_1 l_{c1}^2 + I_1 + m_2\left(l_1^2 + l_{c2}^2 + 2l_1 l_{c2} C_2\right) + I_2 & I_2 + m_2\left(l_{c2}^2 + l_1 l_{c2} C_2\right) \\ I_2 + m_2\left(l_{c2}^2 + l_1 l_{c2} C_2\right) & m_2 l_{c2}^2 + I_2 \end{bmatrix} \qquad (5.68)$$

As expected, the elements of the inertia matrix of the chain are the same as the corresponding dynamic coefficients in equation 5.9 and table 5.1. The elements of the matrix have the following physical meaning:

$$[I^c] = \begin{bmatrix} \text{Moment of inertia of the chain about joint 1} & \text{Coupling inertia; inertial resistance at joint 1 associated with acceleration at joint 2} \\ \text{Coupling inertia; inertial resistance at joint 2 associated with acceleration at joint 1} & \text{Moment of inertia of link 2 about joint 2} \end{bmatrix}$$

These elements are the coefficients I_{ij} of the first term of equation 5.63. To obtain the coefficients D_{ijk} of the second term, which represents the centrifugal and Coriolis effects, the elements of matrix $[I^c]$ should be substituted into equation 5.64. The obtained dynamic coefficients and their associated velocities are presented in table 5.2.

The coefficients are the same as in equation 5.10 and table 5.1. It is left to the reader to determine the gravity coefficients and to check that they are identical to the coefficients obtained previously in **5.2.2.1** with another method. In

Table 5.2 Centrifugal and Coriolis Dynamic Coefficients

Torque at joint	Velocity at joint(s)		
	1	2	1 and 2
1	$D_{111} = 0$	$D_{122} = -m_2 l_1 l_{c2} S_2$	$D_{112} + D_{121} = -2m_2 l_1 l_{c2} S_2$
2	$D_{211} = m_2 l_1 l_{c2} S_2$	$D_{222} = 0$	$D_{212} + D_{221} = 0$

summary, the Lagrangian formulation allows us to obtain the dynamic equations in the closed form.

5.4 JOINT TORQUES AND JOINT FORCES IN HUMAN MOTION

This section is concerned with the joint torques and forces that act in human movements. Subsection **5.4.1** deals with interpreting joint torques; subsection **5.4.2** addresses joint forces; and subsection **5.4.3** discusses interactive joint torques and forces.

5.4.1 Interpreting Joint Torques

This section discusses the effects of the joint torques on joint motion, passive torques, and the scaling of the torques with the speed of motion.

5.4.1.1 Joint Torques and Joint Motion

Joint torques *do not* represent joint motion nor joint acceleration. Joint torques can be in opposite direction of the joint motion (figure 5.19) or acceleration. What joint torques do is *contribute* to the angular acceleration at the joints.

Joint angular acceleration also depends on the joint torques and the joint forces acting on the link in other joints (see equation 5.37). The right-hand part of the equation, $[I_i]\ddot{\boldsymbol{\theta}}_i + \dot{\boldsymbol{\theta}}_i \times ([I_i]\dot{\boldsymbol{\theta}}_i)$, represents the effective (inertial) moment acting on the link. Thus, a joint torque can be presented as the sum

$$\mathbf{T}_i = -\mathbf{T}_{i+1} \underbrace{- \mathbf{r}_i \times \mathbf{F}_i + \mathbf{r}_{i+1} \times \mathbf{F}_{i+1}}_{\substack{\text{Moment of the joint forces} \\ \text{about the center of mass}}} + \underbrace{[I_i]\ddot{\boldsymbol{\theta}}_i + \dot{\boldsymbol{\theta}}_i \times ([I_i]\dot{\boldsymbol{\theta}}_i)}_{\substack{\text{Inertial (effective)} \\ \text{moment}}} \tag{5.69}$$

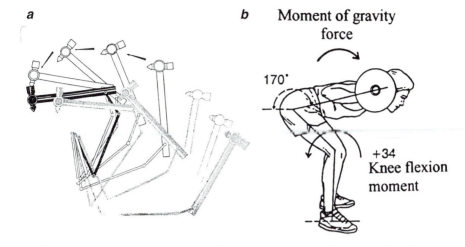

Figure 5.19 Joint torques and joint movement may be in opposite directions. (*a*) Hammering. The arrows mark the portion of the movement where the movement (backward) is in the direction opposite to the joint torques (not shown). The hammer and the arm decelerate during this time. (*b*) Squatting with a barbell. In this particular body position, the weight of the system located above the knees generates a clockwise moment of force above the knee joints. To prevent body rotation and falling, the knee joint muscles generate a torque in flexion. The direction of the knee joint torque (flexion) is opposite to the direction of the knee movement (extension).

Figure 5.19*a* reprinted from Bernstein, N.A. 1947. *On construction of movement.* Moscow: Medgiz. (Original recording was done in 1923.)

Figure 5.19*b* adapted, by permission, from V.M. Zatsiorsky, 1995, *Science and Practice of Strength Training* (Champaign, IL: Human Kinetics), 141.

■ ■ ■ *FROM THE LITERATURE* ■ ■ ■

Eccentric Joint Moment Induces Positive Acceleration

Source: Kepple, T.M., K.L. Siegel, and S.J. Stanhope. 1997. Relative contribution of the lower extremity joint moments to forward progression and support during gait. *Gait Posture* 6 (1): 1–8.

Subjects were examined while walking with their arms crossed against their chests. The joint moments were computed by solving the inverse problem of dynamics. Then the acceleration of the head-arms-trunk (HAT) segment produced by individual joint moments was computed using equation 5.13 (figure 5.20).

During the interval immediately following midstance, the ankle plantar flexors of the support leg act eccentrically. Eccentric activity is characterized by energy absorption (discussed in chapter **6**), which might be assumed to produce a negative acceleration of the HAT segment. However, it was found that this eccentric plantar flexion generated a forward acceleration of the HAT segment.

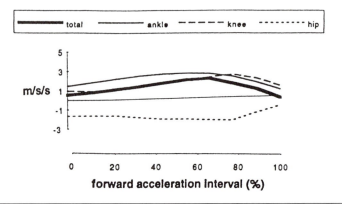

Figure 5.20 A representative sample of the contribution of individual joint moments of the support leg to the forward acceleration of the head, arms, and trunk. The data are for the forward acceleration interval, the second half of the support period starting at midstance. The accelerations produced by the moments at the ankle, knee, and hip and the total acceleration are shown. The ankle and knee joint moments are the primary contributors to forward acceleration.

Adapted from *Gait and Posture*, 6 (1), T.M. Kepple, K.L. Siegel, S.J. Stanhope, Relative contribution of the lower extremity joint moments to forward progression and support during gait, 1–8, 1997, with permission from Elsevier Science.

The forward acceleration from eccentric plantar flexion arises because the acceleration produced by a joint moment is independent of the sense of the joint angular velocity (with the exception of the small contribution through the Coriolis term).

Equation 5.69 does not immediately depict the cause–effect relationships (the joint torque $\mathbf{T}_i$ is not due to the joint force $\mathbf{F}_i$). The equation portrays only the inverse solution. Since we derive the joint torque from kinematics data, $\ddot{\boldsymbol{\theta}}_i$ and $\dot{\boldsymbol{\theta}}_i$, we have to account in the dynamic equation for all the forces and moments that act on the link.

As already mentioned, the torques obtained from the inverse solution are called the coupling torques. The coupling torques are obviously the net joint torques that were defined in chapter **2**. This means that the equations of inverse dynamics do not tell us whether a torque is active (due to active contraction of muscles crossing the joint) or passive (supported by the skeleton and other connective tissues). The passive torques arise near the limits of joint motion, where bones, ligaments, and other passive structures resist the motion. In joints with one or two degrees of freedom, when the torque vector is not along the anatomically permitted axis of rotation, passive torques contribute to the total joint torque. For instance, the knee torque components about the longitudinal axis of the shank and the varus/valgus components are mainly passive. To convert the coupling torque into active joint torque, equation 2.9 should be used.

■ ■ ■ *FROM THE LITERATURE* ■ ■ ■

Effects of Different Foot-and-Pedal Platforms on Knee-Joint Load Components

Source: Ruby, P., and M.L. Hull. 1993. Response of intersegmental knee loads to foot/pedal platform degrees of freedom in cycling. *J. Biomech.* 26 (11): 1327–40.

Overuse knee injuries account for approximately 25% of all reported cycling injuries and trouble cyclists of all levels. Researchers tested the hypothesis that foot-and-pedal platforms that allow relative motion between the foot and pedal decrease the knee-joint torques and forces. Four platforms were designed to interface with a dynamometer, which attached the cyclist's cleat to the pedal:

1. A fixed platform that acted like a standard step-in cycling pedal

2. A toe-in/toe-out platform that allowed the foot to adduct and abduct $\pm15°$

3. A platform that allowed $\pm10°$ of inversion/eversion from the neutral foot position

4. A platform that permitted ±7 mm of medial/lateral translation

Compared with the fixed platform, the inversion/eversion platform significantly reduced the varus/valgus moments, and the abduction/adduction platform significantly reduced both the axial and varus/valgus knee moments (figure 5.21). The medial/lateral translation platform did not cause significant differences in knee-joint torque or force.

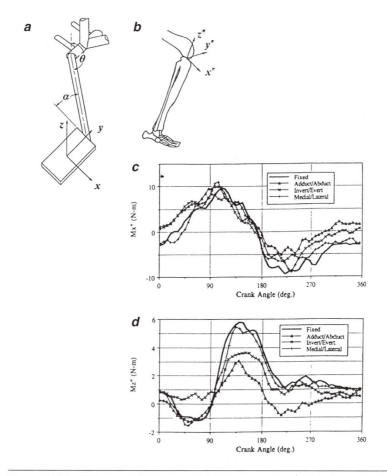

Figure 5.21 Effects of different pedal designs on the axial and varus/valgus components of knee-joint moments. (*a* and *b*) Local coordinate systems at the pedal and the knee joint. (*c* and *d*) Knee moments for four foot-and-pedal interfaces. Shown are the (*c*) varus/valgus and (*d*) axial moments exerted by the tibia on the femur. Knee moments produced with the foot fixed provide the basis for comparison. When the foot was allowed to move, both the abduction/adduction and inversion/eversion platforms caused changes in the knee moments. The abduction/adduction platform reduced the magnitude of the valgus knee moments ($-M_{x''}$) and produced near-zero axial moment ($M_{z''}$). Allowing the foot to invert and evert reduced the magnitude of both the varus and valgus knee moments ($M_{x''}$) as well as the extremes of $M_{z''}$. The medial/lateral platform caused little change in the knee moments.

Adapted from *Journal of Biomechanics*, 26 (11), P. Ruby and M.L. Hull, Response of intersegmental knee loads to foot/pedal platform degrees of freedom in cycling, 1327–1340, 1993, with permission from Elsevier Science.

5.4.1.2 Passive Mechanical Resistance at Joints

Representative paper: Yoon and Mansour (1982)

During movement, the anatomic structures surrounding the joint provide resistance to the joint motion. In the majority of joints, the passive resistance in the middle range of motion is small. For instance, the contribution of passive resistance to the net ankle-joint torque in the sagittal plane during walking is less than 6% (Siegler, Moskowitz, and Freedman 1984). However, in the phalangeal articulations of the fingers, passive resistance is large, up to 50% of the total resistance. Passive resistance is also large in trunk bending, extension, and twisting. Passive resistance is increased in some people with certain disorders (figure 5.22).

The passive joint resistance depends on the joint angular position. The resistance increases as the joint approaches the limits of its range of motion. The existence of passive resistance that depends on joint velocity (damping) was not confirmed in the majority of studies on the topic. However, in swollen joints, the damping may be noticeable due to the motion of liquids around the joint. The friction between articular surfaces is remarkably low and in healthy subjects can be neglected (friction coefficients in the joints are around 0.001 to 0.005).

5.4.1.3 Scaling Joint Torques With Movement Speed

Representative paper: Hollerbach and Flash (1982)

When movement speed changes, so do the joint torques. The torque components are proportional to the joint angular accelerations and the product of joint angular velocities (with the exception of gravity terms, which do not depend on the rate of motion). Consider a planar motion described by the time function of joint angles $\theta(t)$. If this movement is performed faster or more slowly, the corresponding time function is $\theta(kt)$ where a coefficient k is the scaling factor. When the movement speeds up, $k > 1$; when the movement slows down, $k < 1$. The angular velocities change as $\dot{\theta}(kt) = k\dot{\theta}(t)$, and the angular accelerations are scaled as $\ddot{\theta}(kt) = k^2\ddot{\theta}(t)$. Thus, when the speed of performance is scaled by a factor k, the new torques are related to the old by a scaling factor k^2. When the speed doubles, the torques quadruple.

5.4.2 Interpreting Joint Forces

We consider first the origin of joint forces and then their effects on the movement of kinematic chains. If not mentioned otherwise, we analyze the basic model, a kinematic chain with hinge joints powered by torque actuators.

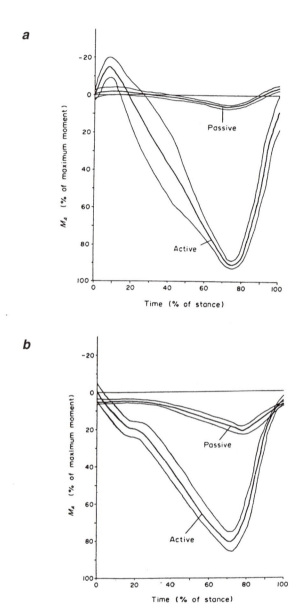

Figure 5.22 Passive and active components of the ankle-joint torque during the stance phase of level walking in (*a*) a normal adult and (*b*) a hemiparetic patient. The passive and active components are shown as percentage of maximal total joint torque.

Reprinted from *Journal of Biomechanics*, 17 (9), S. Siegler, G.D. Moskowitz, and W. Freedman, Passive and active component of the internal moment developed about the ankle joint during human ambulation, 647–652, 1984, with permission from Elsevier Science.

5.4.2.1 Origin of Joint Forces

Joint force is the resultant force exerted by one link on another at a joint (figure 5.24). The sign selection is arbitrary; the positive direction can be either from a proximal link to a distal one or from the distal link to the proximal one. The joint force is an external force with respect to an adjacent link and thus should be included in the dynamic equations and free-body diagrams. The joint force accounts for the inertial effect of the link movement both in rotation and in translation. It accounts also for gravity force impressed at the joint. In multilink chains, joint forces are transmitted from one joint to another. They reflect the inertial and gravitational effects of the movement of the entire chain.

■ ■ ■ *FROM THE LITERATURE* ■ ■ ■

Passive Joint Torques Are Influenced by Neighboring Joints

Source: Riener, R., and T. Edrich. 1999. Identification of passive elastic joint moments in the lower extremities. *J. Biomech.* 32 (5): 539–44.

Passive elastic joint moments can be measured when no reflexes occur and the subjects do not voluntarily contract the muscles. This assumption is true for slow joint speeds. If two-joint muscles are present, the passive elastic moment at a joint is influenced by the angular position of the adjacent joint. The authors measured passive joint torque–joint angle relationships at the ankle, knee, and hip at different positions of the neighboring joints. The adjacent joint angles were fixed with an adjustable orthosis or a backrest (figure 5.23).

The passive moments at the joints depended on both the joint angle and the angular position of the adjacent joint or joints. Passive resistance in the middle range of hip motion was low. The data were approximated by double-exponential functions. For instance, for the hip joint, the empirical equation was

$$M_H = \exp(1.4655 - 0.0034\phi_K - 0.0750\phi_H) - \exp(1.3403 - 0.022\phi_K + 0.0305\phi_H) + 8.072 \qquad (5.70)$$

where ϕ_A, ϕ_K, and ϕ_H are the ankle, knee, and hip joint angles, respectively, in degrees. The authors concluded that ignoring passive elastic moments could lead to a considerable over- or underestimation of active joint torques.

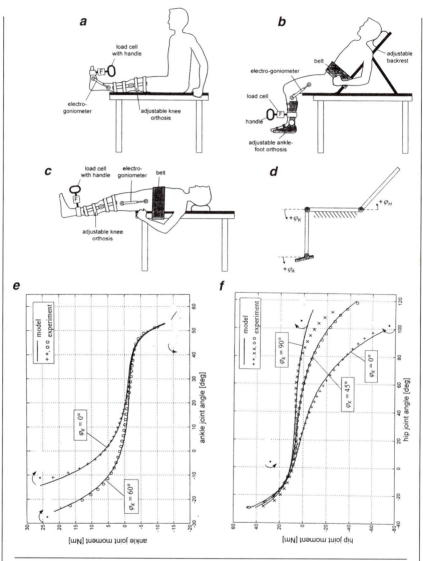

Figure 5.23 Effects of joint angular positions on passive mechanical resistance at joints. Experimental setups for measuring the passive joint moments in the (*a*) ankle, (*b*) knee, and (*c*) hip joints. (*d*) Geometric definitions of joint angles. (*e*) Ankle joint moment versus ankle angle at different knee angles. (*f*) Hip joint moment versus hip angle at different knee angles. Averaged experimental results and the fitted curve from the model (continuous line) are shown.

Adapted from *Journal of Biomechanics*, 32 (5), R. Reiner and T. Edrich, Identification of passive elastic joint moments in the lower extremities, 539–544, 1999, with permission from Elsevier Science.

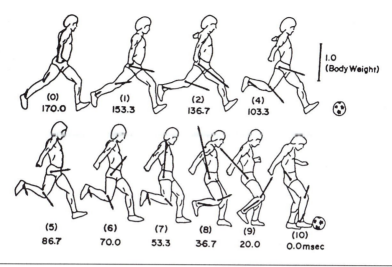

Figure 5.24 An example of joint resultant forces during a fast kick. The arrows depict the force vectors at the ankle, knee, and hip joints. During 15 fast kicks, average maximum joint resultant forces were as follows: ankle joint = 0.46 × body weight, knee joint = 0.68 × body weight, and hip joint = 1.34 × body weight.

Reprinted, by permission, from R.F. Zernicke and E.M. Roberts, 1976, Human lower extremity kinetic relationships during systematic variations in resultant limb velocity. In *Biomechanics V-B*, edited by P. Komi (Baltimore: University Park Press), 20–25.

■ ■ ■ *FROM THE LITERATURE* ■ ■ ■

Joint Forces and Torques in Baseball Pitching: Implications for Injury Mechanisms

Source: Fleisig, G.S., J.R. Andrews, C.J. Dillman, and R.F. Escamilla. 1995. Kinetics of baseball pitching with implications about injury mechanisms. *Am. J. Sports Med.* 23 (2): 233–39.

By solving the inverse problem of dynamics, shoulder and elbow torques and joint resultant forces were determined for 26 highly skilled pitchers (figure 5.25) selected from 264 pitchers tested. Each athlete performed 10 fastball pitches. The three fastest pitches thrown by each pitcher into the strike zone were analyzed. A pitcher was considered a highly skilled athlete if his fastest three pitches thrown into the strike zone during testing averaged at least 84 mph (37 m/s).

Two critical instants with the greatest risk of injury were identified: (1) shortly before the arm reached maximal external rotation, when 67 N · m of shoulder internal rotation torque and 64 N · m of elbow pronation torque were generated (figure 5.25c), and (2) shortly after ball release, when 1,090 N of compression force, 400 N of posterior force, and 97 N · m of horizontal abduction torque were produced at the shoulder (figure 5.25d).

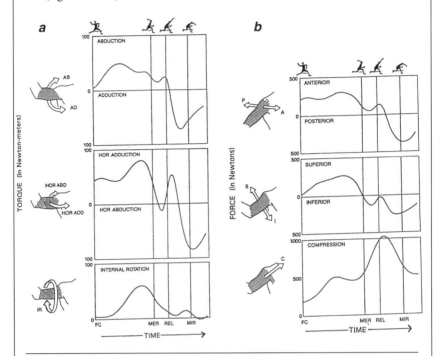

Figure 5.25 Shoulder torques and joint resultant forces in highly skilled pitchers. (a) Torques applied to the arm at the shoulder in abduction (AB) and adduction (AD), horizontal abduction (HOR ABD) and horizontal adduction (HOR ADD), and internal rotation (IR). The instants of foot contact (FC), maximal external rotation (MER), ball release (REL), and maximum internal rotation (MIR) are shown. (b) Joint resultant forces applied to the arm at the shoulder in anterior and posterior (AP), superior and inferior (SI), and compression directions. Note a large compression force after the ball release. (c) The first critical instant. At this instant, the arm was externally rotated 165°, and the elbow was flexed 95°. Among the loads generated at this time were 67 N · m of internal rotation torque and 310 N of anterior force at the shoulder and 64 N · m of varus torque at the elbow. (d) The second critical instant. At this instant, the arm was externally rotated 64°, and the elbow was flexed 25°. Among the loads generated at this time was 1,090 N of compressive force at the shoulder.

Reprinted, by permission, from G.S. Fleisig et al., 1995, "Kinetics of baseball pitching with implications about injury mechanisms," *The American Journal of Sports Medicine*, 23 (2):233–239.

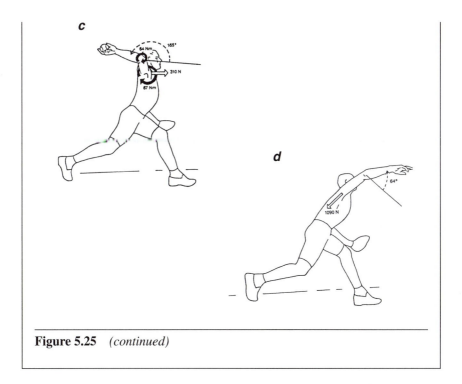

Figure 5.25 *(continued)*

5.4.2.2 Joint Forces and Bone-on-Bone Forces

It should be reiterated that until now we have been considering a model in which joint torques either were produced by torque actuators or were resisted by the skeleton. The joint forces in this model account for only the inertial forces and gravity. They do not represent the internal contact forces (bone-on-bone forces, ligament-on-bone forces, etc.) that act in the joints. The contact forces may be much larger than the joint forces determined by the methods of inverse dynamics (figure 5.26). The bone-on-bone (contact) forces depend on the level of muscle activity. The bone-on-bone forces and the joint forces may also act in different directions. Consider, as an example, an arm at rest, hanging straight down at the side of the body. The elbow joint force in this posture equals the weight of the forearm and the hand. Hence, the *joint* force acts on the upper arm downward. If the muscles are completely relaxed, the bone-on-bone force is zero. However, if the muscles crossing the elbow joint are co-contracted, the bone-on-bone force acting on the humerus is directed upward. The magnitude of this force equals the difference between the total force exerted by the muscles and the forearm-plus-hand weight. We defer detailed analysis of bone-on-bone forces to the ensuing volume of the series.

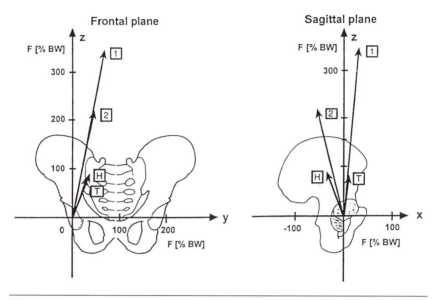

Figure 5.26 Bone-on-bone forces acting on the human acetabulum during walking as a percentage of body weight. The data are from direct force measurements with an instrumented femoral prosthesis. The squares mark force data at defined events: H = heel strike, 1 = first maximum of force, 2 = second maximum of force, T = toe-off. Note that the bone-on-bone force acting at the hip joint during walking can exceed three body weights.

Reprinted, by permission, from H. Witte, F. Eckstein, and S. Rechnagel, 1997, "A calculation of the forces acting on the human acetabulum during walking," *Acta Anatomica* 160:269–280.

■ ■ ■ *FROM THE LITERATURE* ■ ■ ■

Direct Measurement of Femoral Forces

Source: Bassey, E.J., J.J. Littlewood, and S.J.G. Taylor. 1997. Relations between compressive axial forces in an instrumented massive femoral implant, ground reaction forces, and integrated electromyographs from vastus lateralis during various "osteogenic" exercises. *J. Biomech.* 30 (3): 213–23.

The subject of this study suffered from cancer of the bone, which was treated 30 months previously by surgical replacement of the upper half of the femur with an endoprosthesis instrumented to measure axial strain (figure 5.27a). The subject performed several exercises (slow jumping, jogging in place, etc.). The implant forces were registered on the strain gauges, and the signals were transmitted telemetrically. The implant forces were 250% to 400% of the body weight. The values were about twice the magnitude of the ground reaction forces and significantly correlated with them.

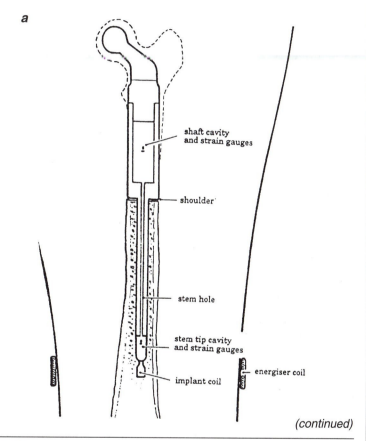

a

shaft cavity
and strain gauges

shoulder

stem hole

stem tip cavity
and strain gauges

energiser coil

implant coil

(continued)

Figure 5.27   Forces acting on the femoral head. (*a*) A schematic of the instrumented endoprosthesis. (*b*) The ground reaction force and the associated implant force during slow jumping. The correlation coefficients were statistically significant for takeoff (squares), $r = .87$, and for landing (circles), $r = .88$. Filled symbols represent values while wearing 0.4-kg training shoes, and open symbols represent values while barefoot. All data were included in the regression analysis. The implant forces were much larger than the corresponding ground reaction forces.

Reprinted from *Journal of Biomechanics*, 30 (3), E.J. Bassey, J.J. Littlewood, and S.J.G. Taylor, Relations between compressive axial forces in an instrumental massive femoral implant, ground reaction forces, and integrated electromyographs from vastus lateralis during various 'osteogenic' exercises, 213–223, 1997, with permission from Elsevier Science.

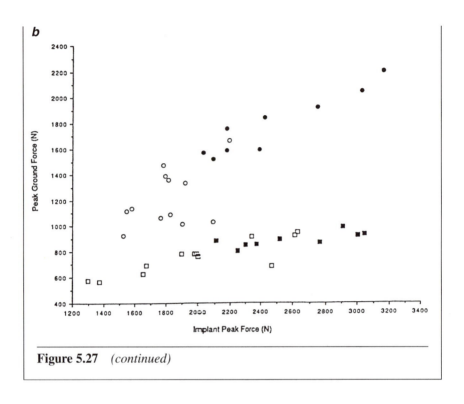

Figure 5.27   *(continued)*

5.4.2.3 Effects of Joint Forces: Whiplike Movement

Representative paper: C.A. Putnam (1993)

Joint forces induce both translation and rotation of the body parts on which they are exerted. For instance, while a person rises from a chair, the head moves upward due to the vertical force exerted at the atlantooccipital articulation, and during the swing phase in walking, the knee joint force contributes to the knee extension (shank rotation). While the role of joint forces in linear acceleration of the body parts is easily understood, their role as a cause of joint angular acceleration is less apparent. The mechanics is straightforward, however: if a joint force vector does not pass through the CoM of the link, it generates a moment of force about the CoM and hence an angular acceleration. People with prostheses above the knee cannot generate knee joint moment. In spite of this, they can extend the knee during the swing phase of walking. To comprehend this, consider a shank-plus-foot segment moving in translation at a horizontal velocity v (figure 5.28a). Gravity is neglected. If the knee joint force $-F$ is exerted on the shank in the direction opposite to v (figure 5.28b), the dynamic equations are

$$-F = ma_{\text{CoM}}$$
$$M = (-F)(-r) = I\ddot{\theta}$$

(5.71)

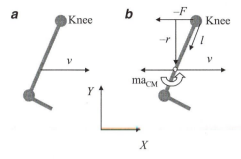

Figure 5.28 The knee joint force causes shank rotation and can increase the speed of the foot.

where $-r$ is the distance from the line of force action to the CoM of the shank-plus-foot system. The values for force F and distance r are negative to agree with the directions of the coordinate axes. According to the dynamic equations, the force applied at the knee causes the shank to rotate.

We use the example in figure 5.28 to explain a phenomenon that frequently occurs in human motion: the increase in endpoint velocity caused by negative acceleration (braking, deceleration) of the proximal joint. In figure 5.28*b*, the acceleration a of a point at location l along the shank-plus-foot system is made up of two parts: $-a = -F/m$, which is the same at all points, and $l\ddot{\theta} = (-F)(-r)l/I$, which varies from point to point. If l is sufficiently large, the endpoint of the chain accelerates in the positive direction and its velocity exceeds v. This performance is known as the *whiplike movement*. It is broadly used in throwing activities; during the delivery phase, acceleration and subsequent deceleration of body segments occur in the proximal-to-distal direction (figure 5.29). The proximal-to-distal progression of limb motion in throwing and striking was known to practitioners well before scientific investigation of these activities began. The following verse by T. Kinkaid (1687, cited in Herring and Chapman 1992) referring to a golf stroke attests to that.

All motions with the strongest joynts performe
Lett the weaker second and perfect the same
The stronger joynt its motion first must end
Before the nixt to move in the least intend.

5.4.3 Interactive Torques and Forces

There exist branched chains (e.g., the trunk). In such chains, movement of one link (e.g., the right forearm) may not affect the torques in another joint

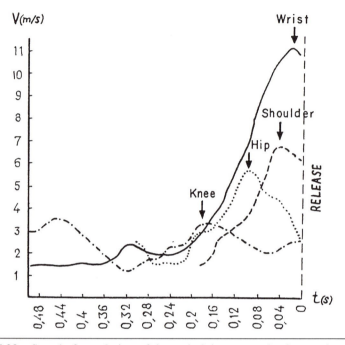

Figure 5.29 Speed of translation of the main joint centers in shot putting (right side of the body). Arrows indicate the maximal speeds and show that deceleration begins at the knee and proceeds progressively upward to the hip, shoulder, and wrist.

Reprinted, by permission, from V.M. Zatsiorsky, Y.E. Lanka, and A.A. Shalmanov, 1981, "Biomechanical analysis of shot putting technique," *Exercise and Sports Science Reviews* 9:353–389. Copyright © 1981 by Lippincott Williams & Wilkens.

■ ■ ■ *From the Literature* ■ ■ ■

Temporal Torque Sequence and Torque Reversal in Throws

Source: Herring, R.M., and A.E. Chapman. 1992. Effects of changes in segmental values and timing of both torque and torque reversal in simulated throws. *J. Biomech.* 25 (10): 1173–84.

The goal of the study was to determine the temporal sequence of joint torques that give the maximal throwing distance. An overarm throw in the sagittal plane was simulated using a three-link model representing the upper arm, forearm, and hand plus ball (figure 5.30*a*). Torque inputs to each joint were turned on at systematically varied times. The direct

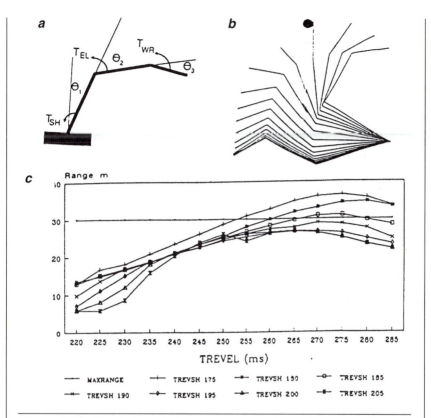

Figure 5.30 Simulation of planar throws. (*a*) The model of the upper limb oriented at the starting angles used in simulations. The view is of the vertically oriented sagittal plane for a throw to the left. θ_1, θ_2, and θ_3 represent the shoulder, elbow, and wrist angles, respectively, and T_{SH}, T_{EL}, and T_{WR} represent torques applied about the shoulder, elbow, and wrist, respectively. (*b*) Stick figures of the later stages of the simulated throw (at 20-ms time intervals). The ball is shown on the stick figure at the instant of release. The ball release occurred close to peak angular velocity of the wrist. (*c*) The relationship between the maximal throwing distance (ordinate) and time of torque reversal at the elbow (TREVEL, abscissa) for various times of torque reversal at the shoulder (TREVSH). For example, TREVSH 175 indicates shoulder torque reversal at 175 ms after the start of the throw. The horizontal line represents the best throwing range without torque reversal. The best throwing distance was achieved with torque reversals at the shoulder at 175 ms and at the elbow at 275 ms. Hence, the optimal delay for the torque reversals between these two joints was 100 ms.

Adapted from *Journal of Biomechanics*, 25 (10), R.M. Herring and A.E. Chapman, Effects of changes in segmental values and timing on both torque and torque reversal in simulated throws, 1173–1184, 1992, with permission from Elsevier Science.

problem of dynamics was solved (figure 5.30*b*). The benefit of arresting proximal segment motion was investigated by varying both the magnitude and timing of torque reversal at the two most proximal joints. To estimate the timing, the value for the most proximal joint was subtracted from that of its distal neighbor. For example, the time of initiation of torque at joint 2 (elbow) was subtracted from that of joint 1 (shoulder). A positive value for this comparison indicated a proximal-to-distal sequence, while a negative value indicated a distal-to-proximal sequence.

The best throws proved to be ones in which joint torques were turned on in a proximal-to-distal temporal sequence. The torque reversal led to increased output if performed late in the throw in a proximal-to-distal sequence (figure 5.30*c*).

(e.g., the left elbow joint). Contrarily, in serial multilink chains, movement of any link influences the torques in all joints of the chain. This section addresses the interaction effects between the links.

5.4.3.1 Partitioning the Joint Torques

Representative papers: Schneider et al. (1989); Sainburg et al. (1995); Sainburg, Ghez, and Kalakanis (1999)

Individual terms of the closed-form dynamic equations can be clustered in separate groups. We previously clustered the terms for angular accelerations in individual joints, joint angular velocities, and acceleration due to gravity. Other combinations are also possible. Consider again equation 5.9 for a planar two-link chain. For simplicity, let us consider only the torque T_2 at the distal joint 2 and neglect the gravity term. The torque can be partitioned in two components: τ_1, which accounts for the inertia of the second link, and τ_2, arising from the mechanical interaction between the links. The torque component τ_2 is an *interactive torque*.

$$T_2 = \tau_1 + \tau_2 = \underbrace{\left(I_2 + m_2 l_{c2}^2\right)\ddot{\alpha}_2}_{\substack{\text{Torque at joint 2} \\ \text{to overcome inertia} \\ \text{of link 2}}} + \underbrace{\left[I_2 + m_2\left(l_{c2}^2 + l_1 l_{c2} C_2\right)\right]\ddot{\alpha}_1 + \left(m_2 l_1 l_{c2} S_2\right)\dot{\alpha}_1^2}_{\substack{\text{Interactive torque; the torque due to movement} \\ \text{in joints other than joint 2}}} \quad (5.72)$$

Similar partitioning can be performed for the torques at other joints. The joint torques can be partitioned in many different ways. For instance, it is possible to compute the component of the joint torque at the proximal joint that is due only to the angular acceleration of the proximal link, and so on. In the literature, different decompositions are used. Unfortunately, the terminology is not standardized, and such expressions as "net joint torque" or "inertial joint torque" may have different connotations in different studies (see table 5.3). It is always advisable to look at the dynamic equations in a research paper.

Table 5.3   Decomposition of the Joint Torques

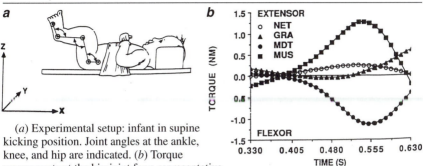

a (a) Experimental setup: infant in supine kicking position. Joint angles at the ankle, knee, and hip are indicated. (b) Torque components at the hip joint for a representative leg kick in a 3-month old baby. The positive values of the torques correspond to extension. The torques are shown for the period of time during which the net torque was positive to slow the leg flexion and begin hip extension.

The following decomposition of the joint torques has been performed:

Name in the paper	Abbreviation in the figure	Physical meaning	Name in the book and/or mathematical expression
Net torque	NET	Joint torque associated with the acceleration at the joint	A special name is not given. The equation is $T_j(\ddot{\alpha}_j) = I_j\ddot{\alpha}_j$, where $T_j(\ddot{\alpha}_j)$ is the torque, I_j is the chain inertia felt at the joint, and $\ddot{\alpha}_j$ is the joint angular acceleration.
Motion-dependent torque	MDT	Torque component at a joint associated with acceleration and velocity at other joints	Interactive torque
Muscle torque	MUS	Joint torque computed from the equations of inverse dynamics. As an example, see the expressions for T_1 and T_2 in equation 5.9.	The active joint torques produced by muscle-tendon complexes
Gravitational torque	GRA	Torque due to gravity	Same as in the paper

Journal of Motor Behavior, 22, 4, 493–520. Reprinted with permission of the Helen Dwight Reid Educational Foundation. Published by Heldref Publications, 1319 Eighteenth St., NW, Washington, DC 20036-1802. Copyright © 1990.

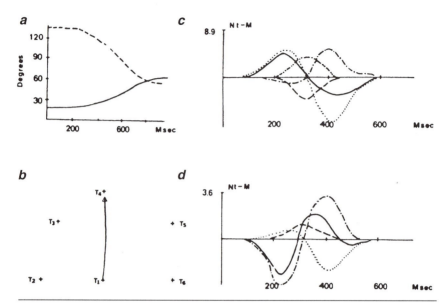

Figure 5.31 Plots of (*a*) joint angles, (*b*) the hand path, and components of joint interaction torques at (*c*) the shoulder and (*d*) the elbow during an arm extension. The arm was modeled as a planar two-link chain; equation 5.9 was used.

Reprinted, by permission, from *Biological Cybernetics*, Dynamic interaction between limb segments during planar arm movement, J.M. Hollerbach and T. Flash, 44, 1982. Copyright © of Springer-Verlag.

5.4.3.2 Contribution of Interactive Torques and Forces

Representative papers: Dounskaia, Swinnen et al. (1998); Latash, Aruin, and Zatsiorsky (1999)

In fast movements, the contribution of the interactive torques and forces to a joint torque can be quite large. Figure 5.31 illustrates the interactive contribution to the shoulder and elbow joint torques during arm extension in the horizontal plane. In the experiment from which these data are drawn, the hand started close to the shoulder and moved forward along a straight line. At the shoulder joint, the Coriolis and centripetal torques canceled each other out, and the torque associated with the angular acceleration at the elbow joint (the transferred "inertial" torque) has made the most substantial contribution. At the elbow, the centripetal torque (associated with the angular velocity at the shoulder) contributed about 30% to the joint torque. The contribution of the various interactive sources to the joint torques depended on the initial and final position of the arm.

The relative contribution of various interactive torques and forces to a joint torque depends on the chain configuration. Consider a two-link chain at two configurations, completely extended and flexed 90° (figure 5.32). The chain models a leg during the swing phase of walking. The upper (proximal) joint represents the hip, and the lower joint represents the knee. When the hip joint rotates at a given angular velocity, the centripetal force acts at the CoM of the

Movement of the proximal link

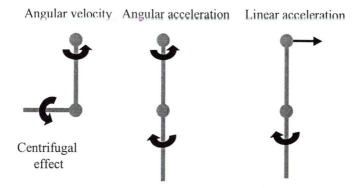

Angular velocity   Angular acceleration   Linear acceleration

Centrifugal
effect

Maximal effects on the rotation of the distal segment

Figure 5.32 Effects of angular velocity, angular acceleration, and linear acceleration of the proximal segment on the rotation of the distal segment at different joint configurations.

■ ■ ■ *From the Literature* ■ ■ ■

Sequential Motion of Body Segments in Kicking

Source: Putnam, C.A. 1993. Sequential motions of body segments in striking and throwing skills: Descriptions and explanations. *J. Biomech.* 26 (Suppl. 1): 125–35.

The author of this paper suggested the following explanation of the mechanics of kicking motion. The sequential motion of the thigh and lower leg in kicking starts with the forward rotation of the thigh. In the beginning of the kick, the lower leg experiences a backward acceleration, and the knee flexes. This acceleration is caused mainly by the interacting moment resulting from the forward acceleration of the thigh. The interaction effect is large because the knee is close to full extension (figure 5.33*a*).

During the next phase, the knee extends. The forward acceleration of the lower leg starts while the thigh is accelerating forward and continues while the proximal segment is accelerating backward. The lower leg con-

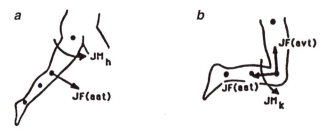

Figure 5.33 The sequence of link interaction during kicking. (*a*) At the beginning of the kick, the thigh accelerates forward. The knee joint force acting on the lower leg induces linear acceleration of the lower leg forward and angular acceleration of the lower leg in the direction of knee flexion. The angular acceleration effect is large because the leg is close to full extension. JM_h is the hip moment, and $JF(aat)$ is the component of the knee joint force associated with the angular acceleration of the thigh. (*b*) The knee torque and the interactive moment due to the hip angular velocity are responsible for accelerating the lower leg in the direction of knee extension.

Adapted from *Journal of Biomechanics*, 26, C.A. Putnam, Sequential motions of body segments in striking and throwing skills: Descriptions and explanations, 125–135, 1993, with permission from Elsevier Science.

tinues to accelerate forward just before impact. The forward acceleration of the lower leg is caused mainly by the knee extensor moment and by the interactive effect associated with the thigh angular velocity. The latter effect is especially large when the thigh is rotating rapidly and the knee angle is close to 90° (figure 5.33*b*). Both the joint moment and the interactive moment are approximately of equal magnitude throughout the kick. The forward rotation of the thigh is slowed primarily by the interactive moments resulting from the angular velocity and acceleration of the lower leg. The hip musculature does not play a major role in slowing the thigh. In simple words, the thigh decelerates because the shank accelerates.

distal link. If the leg is fully extended, the force increases the load on the knee but does not induce knee rotation. When the leg is flexed, the centripetal force tends to rotate the distal segment with respect to the proximal. In general, the angular velocity–dependent interaction is maximal when the adjacent segments are at right angles to each other. This conclusion follows from equation 5.9 and table 5.1: the dynamic coefficient $D = m_2 l_1 l_{c2} S_2$ is maximal when $S_2 = 1$. By the same token, the angular acceleration–dependent interaction is maximal when the segments are collinear: the inertia coupling coefficients $I_{1,2} = I_{2,1}$ are maximal when $C_2 = 1$ (see table 5.1 again). Finally, the linear acceleration of the proximal link induces the maximal turning effect when the acceleration vector is at a right angle to the longitudinal axis of the distal segment.

In multilink chains, the interactive effects are complex. The methods of mechanics provide a tool to compute them (at least in theory). For example, the effect of the acceleration of a pitcher's left knee on the joint torque at the right wrist can be computed. However, in human movement, there are hundreds of such effects. The effects are too numerous to comprehend in their entirety. The motor control system manages somehow to master all these interactive effects. Our movements are smooth and accurate. How does the CNS manage to control all the interactive effects? The answer is simple: nobody knows.

5.5 SUMMARY

The chapter deals with the following problem: given a motion of body parts, find the joint torques that caused the motion. This problem is known as the inverse problem of dynamics. The relationship between the forces acting on a body and the manner in which the body velocity varies with time is described by the equations of motion (or dynamic equations).

The basic equations of motion of a rigid body, called the Newton-Euler equations, are

$$\Sigma \mathbf{F} = m\mathbf{a}_{\text{CoM}}$$
$$\Sigma \mathbf{M} = \dot{\mathbf{H}} \tag{5.1}$$

where $\Sigma \mathbf{F}$ and $\Sigma \mathbf{M}$ are the sums of all the external forces and moments of force acting on the body, respectively; m is the body mass; $\mathbf{a}_{\text{CoM}}$ is the acceleration of the CoM; and $\dot{\mathbf{H}}$ is the time rate of the angular momentum of the body. The moments of force are computed about the CoM. Equations 5.1 do not allow the forces and the torques acting at the joints to be computed.

For individual links of a planar kinematic chain, the dynamic equations are

$$\mathbf{F}_i - \mathbf{F}_{i+1} + m_i\mathbf{g} - m_i\mathbf{a}_i = 0$$
$$\mathbf{T}_i - \mathbf{T}_{i+1} + \mathbf{r}_i \times \mathbf{F}_i - \mathbf{r}_{i+1} \times \mathbf{F}_{i+1} - I_i\ddot{\theta}_i = 0 \tag{5.2}$$

where $\mathbf{F}_i$ and $\mathbf{F}_{i+1}$ are the joint forces acting on link i at joints i and $i + 1$, respectively; $\mathbf{T}_i$ and $\mathbf{T}_{i+1}$ are the joint torques exerted on link i at these joints; $\mathbf{r}_i$ and $\mathbf{r}_{i+1}$ are the radii from the CoM of link i to the joint centers i and $i + 1$, correspondingly; and $\mathbf{g}$ is the acceleration of gravity.

To clarify the main aspects of the inverse problem of dynamics, several simple chains are analyzed. Two forces are acting on a body in rotation: the centripetal force, which causes the change in the direction of the velocity vector, and the tangential force, which induces the change in the magnitude of velocity. During movement of a two-link chain, the number of acting forces increases.

The forces are associated with the angular velocity at the joints, joint angular acceleration, Coriolis acceleration, and gravity. In kinematic chains, movement of any joint in the chain affects the joint torques and forces at all the other joints (interactive effects). The coefficients of the dynamic equations that relate the kinematic parameters, such as angular velocity and acceleration, with the components of the joint torques are called the dynamic coefficients. The effects of accelerated motion at one joint on the torque at another joint are characterized by the coupling inertia (interactive inertia). The coupling inertia is symmetric: the torque induced at joint i by the accelerated motion at joint j at a given magnitude equals the torque at j due to the angular acceleration at i at the similar magnitude. The dynamic coefficients that relate the squared angular velocity values with the torque values are called the centripetal coupling coefficients. Like the coupling inertia, the centripetal coupling coefficients are equal. For a two-link chain, the Coriolis coupling coefficient is twice the inertial resistance to the centrifugal forces.

The dynamic equations that include the angular positions, velocities, and accelerations of all the joints of the chain as inputs and the joint torques as outputs are called the closed-form dynamic equations. The closed-form equations always contain acceleration-related terms (inertia times joint angular acceleration), centrifugal terms (inertia times angular velocity squared), Coriolis terms (inertia times the product of the joint angular velocities), and gravity terms. The centrifugal and Coriolis terms are the velocity-related terms. If written in the matrix-vector form, the equations contain (1) the product of the inertia matrix and the vector of joint angular accelerations; (2) a vector of the centripetal and Coriolis terms, and (3) a vector of gravity terms. The inertia matrix of the chain is symmetric.

In three dimensions, the matrices of inertia of body segments change as the segments rotate, and the dynamic equations are rather complex. To simplify the equations, they are customarily written in a frame rigidly attached to the body. In this frame, the tensor of inertia is constant, and the dynamic equation, called the Euler dynamic equation, is

$$\left(\frac{d\mathbf{H}}{dt}\right)_{space} = \left(\frac{d\mathbf{H}}{dt}\right)_{body} + \dot{\boldsymbol{\theta}} \times \mathbf{H} = \mathbf{M} \tag{5.27}$$

where $\mathbf{H}$ is the angular momentum, $\dot{\boldsymbol{\theta}}$ is the vector of joint angular velocity, and $\mathbf{M}$ is the moment resultant. The term $\dot{\boldsymbol{\theta}} \times ([I]\dot{\boldsymbol{\theta}}) = \dot{\boldsymbol{\theta}} \times \mathbf{H}$ is called the gyroscopic term. It accounts for the change in the tensor of inertia due to the body rotation in three dimensions.

Two main methods of solving the inverse problem of dynamics are the recursive Newton-Euler method and the Lagrangian method. In the recursive Newton-Euler method, the dynamic equations are solved consecutively, link by link, starting from the distal link and working proximally. Each force and torque appears twice in the equations: once as an unknown in the equation for the distal link and the second time as input in the equation for the proximal link (with the opposite sign). The method does not immediately allow partitioning the estimated forces and moments into the components associated with individual kinematic variables, such as the centrifugal components or Coriolis components.

In the Lagrangian formulation, the equations of motion are based on the kinetic and potential energy of the entire body. The dynamic equations are obtained in the closed form, allowing detailed analysis of the motion. The number of equations is equal to the number of degrees of freedom. The forces of constraint in the joints do not appear in the equations; therefore, the joint forces are not determined in the Lagrangian method.

The origin and effects of joint torques and forces are discussed. The joint torques contribute to the angular acceleration at the joints, but they do not directly represent the joint motion or angular acceleration. Joint angular acceleration also depends on the joint torques and joint forces that act on the link in other joints. The equations of inverse dynamics do not distinguish between active torques (due to muscle activity) and passive torques (supported by the skeleton and other connective tissues). This information must be obtained from other sources. When the speed of performance is scaled by a factor k, the torques are scaled by a factor k^2. When the speed doubles, the torques quadruple.

Joint forces can contribute substantially to the angular acceleration of body segments. The relative contribution of various reactive torques and moments to the acceleration of body segments depends on the body configuration.

5.6 QUESTIONS FOR REVIEW

1. Define the inverse problem of dynamics.
2. Write the Newton-Euler equations for a link of a kinematic chain. Use two equipollent force systems: (a) the resultant force acting at the CoM and (b) the forces and moments acting at the joints.
3. A rigid body rotates about a fixed axis. Discuss the acceleration experienced by the particles of the body.
4. A single link is rotating at a joint. Discuss the origin of the joint reaction force.
5. Angular motion is performed at the joints of a two-link planar chain. Discuss the forces acting at the centers of mass of the individual links.
6. Describe the dynamic coefficients. In particular, discuss the coupling inertia coefficients, centripetal coupling coefficients, and the Coriolis coupling coefficients.
7. Name the individual terms in the closed-form dynamic equations.

8. Explain why some dynamic coefficients are equal to zero. Give an example.

9. Explain why the inertia matrix of a chain is symmetric.

10. Write a dynamic equation of a chain in matrix-vector form.

11. What is the origin of the interactive joint torques and forces?

12. Explain the difficulties associated with solving the inverse problem of dynamics for movement in three dimensions.

13. Please complete this statement: "In three dimensions, the moments and products of inertia of a body maintain the same values with respect to the following system of coordinates . . . "

14. Discuss the Euler dynamic equations.

15. Define the gyroscopic term in the dynamic equations.

16. Discuss the recursive Newton-Euler method of solving the inverse problem of dynamics.

17. Explain the main ideas of the Lagrangian method.

18. Define rotational and translational kinetic energies.

19. Compare the Newton-Euler and Lagrangian methods of obtaining the dynamic equations for multilink chains.

20. How do joint torques affect human movement?

21. How do joint forces affect human movement?

22. What are coupling forces and torques?

23. Discuss the whiplike movement. Give examples.

24. Discuss interactive joint torques and forces and their role in human movement.

5.7 BIBLIOGRAPHY

Ae, M., M. Miyashita, K. Shibukawa, T. Yokoi, and Y. Hashihara. 1985. Body segment contributions during the support phase while running at different velocities. In *Biomechanics IX-B,* ed. D.A. Winter, R.W. Norman, R.P. Wells, K.C. Hayes, and A.E. Patla, 343–49. Champaign, IL: Human Kinetics.

Aleshinsky, S.Y., and V.M. Zatsiorsky. 1978. Human locomotion in space analyzed biomechanically through a multi-link chain model. *J. Biomech.* 11 (3): 101–8.

Allard, P., A. Cappozzo, A. Lundberg, and C.L. Vaughan, eds. 1997. *Three-dimensional analysis of human locomotion.* Chichester, NY: Wiley.

Amis, A.A., D. Dowson, and V. Wright. 1980. Analysis of elbow forces due to high-speed forearm movement. *J. Biomech.* 13:825–31.

An, K.-N., K.R. Kaufman, and E.Y.S. Chao. 1995. Estimation of muscle and joint forces. In *Three-dimensional analysis of human movement,* ed. P. Allard, I.A.P. Stokes, and J.-P. Blanchi, 201–14. Champaign: IL: Human Kinetics.

Andrews, J.G. 1995. Euler's and Lagrange's equations for linked rigid-body models of three-dimensional human motion. In *Three-dimensional analysis of human movement,* ed. P. Allard, I.A.F. Stokes, and J.-P. Blanchi, 145–75. Champaign, IL: Human Kinetics.

Andrews, J.G., and S.P. Mish. 1996. Methods for investigating the sensitivity of joint resultants to body segment parameter variations. *J. Biomech.* 29 (5): 651–54.

Andriacchi, T.P., R.N. Natarajan, and D.E. Hurwitz. 1997. Musculoskeletal dynamics, locomotion and clinical applications. In *Basic orthopedic biomechanics,* 2nd ed., ed. V.C. Mow and W.C. Hayes, 37–68. Philadelphia: Lippincott-Raven.

Anglin, C., and U.P. Wyss. 2000. Arm motion and load analysis of sit-to-stand, stand-to-sit, cane walking and lifting. *Clin. Biomech.* 15 (6): 441–48.

Arampatzis, A., J. Gao, and G.-P. Brüggemann. 1997. Influences of inertial properties on joint resultants. In *XVIth ISB Tokyo Congress: Book of abstracts,* 7. Tokyo: University of Tokyo.

Atkeson, C.G. 1989. Learning arm kinematics and dynamics. *Annu. Rev. Neurosci.* 12: 157–83.

Bahrami, F., R. Riener, P. Jabedar-Maralani, and G. Schmidt. 2000. Biomechanical analysis of sit-to-stand transfer in healthy and paraplegic subjects. *Clin. Biomech.* 15 (2): 123–33.

Ball, K.A., and M.R. Pierrynowski. 1998. Pliant surface modelling: Use of invasive and external observations to determine skeletal movement. In *Proceedings of NACOB '98: The third North American Congress on Biomechanics, August 14–18, 1998,* 565–66. Waterloo, ON: University of Waterloo.

Barbenel, J.C. 1983. The application of optimization methods for the calculation of joint and muscle forces. *Eng. Med.* 12 (1): 29–43.

Bassey, E.J., J.J. Littlewood, and S.J.G. Taylor. 1997. Relations between compressive axial forces in an instrumented massive femoral implant, ground reaction forces, and integrated electromyographs from vastus lateralis during various "osteogenic" exercises. *J. Biomech.* 30 (3): 213–23.

Bauby, C.E., and A.D. Kuo. 2000. Active control of lateral balance in human walking. *J. Biomech.* 33 (11): 1433–40.

Bay, B.K., A.J. Hamel, S.A. Olson, and N.A. Sharkey. 1970. Statically equivalent load and support conditions produce different joint contact pressures and periacetabular strains. *J. Biomech.* 30 (2): 193–96.

Bennett, D.J. 1993. Torques generated at the human elbow joint in response to constant position errors imposed during voluntary movements. *Exp. Brain Res.* 95: 488–98.

Bergmann, G., G. Deuretzbacher, M. Heller, F. Graichen, A. Rohlmann, J. Strauss, and G.N. Duda. 2001. Hip contact forces and gait patterns from routine activities. *J. Biomech.* 34: 859–71.

Bergmann, G., F. Graichen, and A. Rohlmann. 1999. Hip joint forces in sheep. *J. Biomech.* 32: 769–77.

Bergmann, G., F. Graichen, and A. Rohlmann. 1993. Hip joint loading during walking and running, measured in two patients. *J. Biomech.* 26: 969–90.

Bergmann, G., F. Graichen, and A. Rohlmann. 1995. Is staircase walking a risk for the fixation of hip implants? *J. Biomech.* 28: 535–53.

Bergmann, G., F. Graichen, A. Rohlmann, and H. Linke. 1997. Hip joint forces during load carrying. *Clin. Orthop.* (335): 190–201.

Bergmann, G., H. Kniggendorf, F. Graichen, and A. Rohlmann. 1995. Influence of shoes and heel strike on the loading of the hip joint. *J. Biomech.* 28 (7): 817–28.

Bergmann, G.A., and A. Rohlmann. 1996. Die Belastung des Hüftenlelenkes—Ein Ueberblick. *Med. Orth. Tech.* 116: 143–50.

Bergmann, G., J. Siraky, A. Rohlmann, and R. Koelbel. 1984. A comparison of hip joint forces in sheep, dog and man. *J. Biomech.* 17: 907–21.

Bernstein, N.A. 1947. *On construction of movement.* Moscow: Medgiz.

Bogert, A.J. van den, and H.C. Schamhardt. 1993. Multibody modelling and simulation of animal locomotion. *Acta Anat.* 146 (2–3): 95–102.

Bizzi, E., and F.A. Mussa-Ivaldi. 1989. Geometrical and mechanical issues in movement planning and control. In *Foundations of cognitive science,* ed. M.I. Posner. Cambridge, MA: MIT Press.

Bobbert, M.F., and J.P. van Zandwijk. 1999. Dynamics of force and muscle stimulation in human vertical jumping. *Med. Sci. Sports Exerc.* 31 (2): 303–10.

Bock, O. 1994. Scaling of joint torque during planar arm movements. *Exp. Brain Res.* 101: 346–52.

Bogert, A.J. van den. 1994. Analysis and simulation of mechanical loads on the human musculoskeletal system: A methodological overview. *Exerc. Sport Sci. Rev.* 22: 23–55.

Bogert, A.J. van den, L. Read, and B.M. Nigg. 1996. A method for inverse dynamic analysis using accelerometry. *J. Biomech.* 29 (7): 949–54.

Bogert, A.J. van den, L. Read, and B.M. Nigg. 1999. An analysis of hip joint loading during walking, running, and skiing. *Med. Sci. Sports Exerc.* 31 (1): 131–42.

Boyd, T.F., R.R. Neptune, and M.L. Hull. 1997. Pedal and knee loads using a multi-degree-of-freedom pedal platform in cycling. *J. Biomech.* 30 (5): 505–11.

Bresler, B., and J.P. Frankel. 1950. The forces and moments in the leg during level walking. *J. Appl. Mech. Trans. Am. Soc. Mech. Eng.* 72: 27.

Burdett, R.G. 1982. Forces predicted at the ankle during running. *Med. Sci. Sports Exerc.* 14: 308–16.

Cairns, M.A., R.G. Burdett, J.C. Pisciotta, and S.R. Simon. 1986. A biomechanical analysis of racewalking gait. *Med. Sci. Sports Exerc.* 18 (4): 446–53.

Callaghan, J.P., A.E. Patla, and S.M. McGill. 1999. Low back three-dimensional joint forces, kinematics, and kinetics during walking. *Clin. Biomech.* 14 (3): 203–16.

Cappozzo, A. 1983a. Force actions in the human trunk during running. *J. Sports Med. Phys. Fitness* 23 (1): 14–22.

Cappozzo, A. 1983b. The forces and couples in the human trunk during level walking. *J. Biomech.* 16 (4): 265–77.

Cappozzo, A., F. Fugura, M. Marchetti, and A. Pedotti. 1976. The interplay of muscular and external forces in human ambulation. *J. Biomech.* 9 (1): 35–43.

Cappozzo, A., F. Gazzani, and N. Berme. 1990. Internal resultant loads. In *Biomechanics of human movement: Applications in rehabilitation, sports and ergonomics,* ed.

N. Berme and A. Cappozzo, 284–88. Worthington, OH: Bertec Corp.

Cavanagh, P.R., and R.J. Gregor. 1975. Knee joint torque during the swing phase of normal treadmill walking. *J. Biomech.* 8: 337–44.

Celigueta, J. 1996. Multibody simulation of the human body in sports. In *Proceedings of the XIVth International Symposium on Biomechanics in Sports,* ed. J.M.C.S. Abrantes, 81–94. Lisbon: Edições FMH.

Chadwick, E.K., and A.C. Nicol. 2000. Elbow and wrist joint contact forces during occupational pick and place activities. *J. Biomech.* 33 (5): 591–600.

Challis, J.H. 1996. Accuracy of human limb moment of inertia estimations and their influence on resultant joint moments. *J. Appl. Biomech.* 12 (4): 517–30.

Challis, J.H., and D.G. Kerwin. 1996. Quantification of the uncertainties in resultant joint moments computed in a dynamic activity. *J. Sports Sci.* 14 (3): 219–31.

Chapman, A.E., R. Lonergan, and G.E. Caldwell. 1984. Kinetic sources of lower-limb angular displacement in the recovery phase of sprinting. *Med. Sci. Sports Exerc.* 16 (4): 382–88.

Chester, V.L., and R.K. Jensen. 1998. Changes in segment inertia and hip angular impulses during the swing phase of the first three months of independent walking. In *Proceedings of NACOB '98: The third North American Congress on Biomechanics, August 14–18, 1998,* 567–68. Waterloo, ON: University of Waterloo.

Chèze, L., and J. Dimnet. 1995. Modeling human body motions by the techniques known to robotics. In *Three-dimensional analysis of human movement,* ed. P. Allard, I.A.F. Stokes, and J.-P. Blanchi, 177–200. Champaign, IL: Human Kinetics.

Cholewicki, J., S.M. McGill, and R.W. Norman. 1995. Comparison of muscle forces and joint load from an optimization and EMG assisted lumbar spine model: Towards development of a hybrid approach. *J. Biomech.* 28 (3): 321–31.

Chou, L.-S., and L.F. Draganich. 1997. Stepping over an obstacle increases the motions and moments of the joints of the trailing limb in young adults. *J. Biomech.* 30 (4): 331–37.

Chow, J.W. 1999. Knee joint forces during isokinetic knee extensions: A case study. *Clin. Biomech.* 14 (5): 329–38.

Clayton, H.M., J.L. Lanovaz, H.C. Schamhardt, M.A. Willemen, and G.R. Colborne. 1998. Net joint moments and powers in the equine forelimb during the stance phase of the trot. *Equine Vet. J.* 30 (5): 384–89.

Clayton, H.M., H.C. Schamhardt, M.A. Willemen, J.L. Lanovaz, and G.R. Colborne. 2000. Net joint moments and joint powers in horses with superficial digital flexor tendinitis. *Am. J. Vet. Res.* 61 (2): 197–201.

Colborne, G.R., J.L. Lanovaz, E.J. Sprigings, H.C. Schamhardt, and H.M. Clayton. 1997. Joint moments and power in equine gait: A preliminary study. *Equine Vet. J.* 23: 33–36.

Colborne, G.R., J.L. Lanovaz, E.J. Sprigings, H.C. Schamhardt, and H.M. Clayton. 1998. Forelimb joint moments and power during the walking stance phase of horses. *Am. J. Vet. Res.* 59 (5): 609–14.

Cole, G.K., B.M. Nigg, G.H. Fick, and M.M. Morlock. 1995. Internal loading of the foot and ankle during impact in running. *J. Appl. Biomech.* 11: 25–46.

Collins, J.J. 1990. Joint mechanics: Modelling of the lower limb. Doctoral diss., University of Oxford, England.

Cooke, J.D. 1979. Dependence of human arm movements on limb mechanical properties. *Brain Res.* 165 (2): 366–69.

Cooper, R.A., M.L. Boninger, S.D. Shimada, and B.M. Lawrence. 1999. Glenohumeral joint kinematics and kinetics for three coordinate system representations during wheelchair propulsion. *Am. J. Phys. Med. Rehabil.* 78 (5): 435–46.

Costigan, P.A., U.P. Wyss, and K.J. Deluzio. 1998. Force and moment at the normal knee during stair climbing. In *Proceedings of NACOB '98: The third North American Congress on Biomechanics, August 14–18, 1998,* 35–36. Waterloo, ON: University of Waterloo.

Cristofolini, L., M. Vicenonti, A. Toni, and A. Giunti. 1995. Influence of thigh muscles on the axial strains in a proximal femur during early stance in gait. *J. Biomech.* 28 (5): 617–24.

Crowninshield, R.D., and R.A. Brand. 1981. The prediction of forces in joint structures: Distribution of intersegmental resultants. *Exerc. Sport Sci. Rev.* 9: 159–81.

Crowninshield, R.D., R.C. Johnston, J.G. Andrews, and R.A. Brand. 1978. A biomechanical investigation of the human hip. *J. Biomech.* 11: 75–85.

Czerniecki, J.M., and A. Gitter. 1992. Insights into amputee running: A muscle work analysis. *Am. J. Phys. Med. Rehabil.* 71 (4): 209–18.

Czerniecki, J.M., A. Gitter, and C. Munro. 1991. Joint moment and muscle power output characteristics of below knee amputees during running: The influence of energy storing prosthetic feet. *J. Biomech.* 24 (1): 63–75.

Darling, W.G., and K.J. Cole. 1990. Muscle activation patterns and kinetics of human index finger movements. *J. Neurophysiol.* 63 (5): 1098–108.

Davy, D.T., and M.L. Audu. 1987. A dynamic optimization technique for predicting muscle forces in the swing phase of gait. *J. Biomech.* 20 (2): 187–201.

Davy, D.T., G.M. Kotzar, R.H. Brown, K.G. Heiple, V.M. Goldberg, J. Heiple, J. Berilla, and A.H. Burstein. 1988. Telemetric force measurements across the hip after total arthroplasty. *J. Bone Joint Surg.* 70A: 45–50.

de Boer, R.W., J. Cabri, W. Vaes, J.P. Clarijs, A.P. Hollander, G. de Groot, and G.J. van Ingen Schenau. 1987. Moments of force, power, and muscle coordination in speed-skating. *Int. J. Sports Med.* 8 (6): 371–78.

de Looze, M.P., A.R. Stassen, A.M. Markslag, M.J. Borst, M.M. Wooning, and H.M Toussaint. 1995. Mechanical loading on the low back in three methods of refuse collecting. Ergonomics 38 (10): 1993-2006.

De Lorenzo, D.S., and M.L. Hull. 1999. Quantification of structural loading during off-road cycling. *J. Biomech. Eng.* 121 (4): 399–405.

Devita, P., and W.A. Skelly. 1992. Effect of landing stiffness on joint kinetics and energetics in the lower extremity. *Med. Sci. Sports Exerc.* 24 (1): 108–15.

Devita, P., and J. Stribling. 1991. Lower extremity joint kinetics and energetics during backward running. *Med. Sci. Sports Exerc.* 23 (5): 602–10.

Dillman, C.J. 1994. Biomechanical contributions to the science of rehabilitation in sports. *Sport Sci. Rev.* 3 (2): 70–78.

Dixon, S.J. 1998. The influence of heel lift manipulation on Achilles tendon loading in running. *J. Appl. Biomech.* 14 (4): 374–89.

Domen, K., M.L. Latash, and V.M. Zatsiorsky. 1999. Reconstruction of equilibrium trajectories during whole-body movements. *Biol. Cybern.* 80 (3): 195–204.

Donkers, M.J., K.N. An, E.Y. Chao, and B.F. Morrey. 1993. Hand position affects elbow joint load during push-up exercises. *J. Biomech.* 26 (6): 625–32.

Doorenbosch, C.A., T.G. Welter, and G.J. van Ingen Schenau. 1997. Intermuscular coordination during fast contact control leg tasks in man. *Brain Res.* 751 (2): 239–46.

Dounskaia, N.V., S.P. Swinnen, C.B. Walter, A.J. Spaepen, and S.M. Verschueren. 1998. Hierarchical control of different elbow-wrist coordination patterns. *Exp. Brain Res.* 121 (3): 239–54.

Duda, G.N., E. Schneider, and E.Y. Chao. 1997. Internal forces and moments in the femur during walking. *J. Biomech.* 30 (9): 933–41.

Durey, A., and R. Journeaux. 1995. Application of three-dimensional analysis in sports. In *Three-dimensional analysis of human movement,* ed. P. Allard, I.A.F. Stokes, and J.-P. Blanchi, 327–47. Champaign, IL: Human Kinetics.

Edrich, T., R. Riener, and J. Quintern. 2000. Analysis of passive elastic joint moments in paraplegics. *IEEE Trans. Biomed. Eng.* 47 (8): 1058–65.

Elftman, H. 1939. Forces and energy changes in the leg during walking. *Am. J. Physiol.* 125: 339–56.

Eng, J.J., and D.A. Winter. 1995. Kinetic analysis of the lower limbs during walking: What information can be gained from a three-dimensional model? *J. Biomech.* 28 (6): 753–58.

Eng, J.J., D.A. Winter, and A.E. Patla. 1997. Intralimb dynamics simplify reactive control strategies during locomotion. *J. Biomech.* 30 (6): 581–88.

Engin, A.E. 1979. Passive resistive torques about the long axes of the major human joints. *Aviat. Space Environ. Med.* 50: 1052–57.

English, T.A., and M. Kilvington. 1978. A direct telemetric method for measuring hip load. In *Orthopaedic engineering,* ed. J.D. Harris, and K. Copeland, 198–201. London: Biological Engineering Society.

Ericson, M.O., Å. Bratt, R. Nisell, G. Németh, and J. Ekholm. 1986. Load moments about the hip and knee joints during ergometer cycling. *Scand. J. Rehabil. Med.* 18: 165–72.

Feltner, M.E. 1989. Three-dimensional interactions in a two-segment kinetic chain: II. Application to the throwing arm in baseball pitching. *Int. J. Sport Biomech.* 5: 420–50.

Feltner, M.E., and J. Dapena. 1989. Three-dimensional interactions in a two-segment kinetic chain: I. General model. *Int. J. Sport Biomech.* 5: 403–19.

Feltner, M.E., F.J. Fraschetti, and R.J. Crisp. 1999. Upper extremity augmentation of lower extremity kinetics during countermovement vertical jumps. *J. Sports Sci.* 17 (6): 449–66.

Fenn, W.O. 1930. Frictional and kinetic factors in the work of sprint running. *Am. J. Physiol.* 92:583–611.

Fleisig, G.S., J.R. Andrews, C.J. Dillman, and R.F. Escamilla. 1995. Kinetics of baseball pitching with implications about injury mechanisms. *Am. J. Sports Med.* 23 (2): 233–39.

Fleisig, G.S., S.W. Barrentine, N. Zheng, R.F. Escamilla, and J.R. Andrews. 1999. Kinematic and kinetic comparison of baseball pitching among various levels of development. *J. Biomech.* 32 (12): 1371–75.

Fowler, N.K., and A.C. Nicol. 1999. Measurement of external three-dimensional interphalangeal loads applied during activities of daily living. *Clin. Biomech.* 14 (9): 646–52.

Fowler, N.K., and A.C. Nicol. 2000. Interphalangeal joint and tendon forces: Normal model and biomechanical consequences of surgical reconstruction. *J. Biomech.* 33 (9): 1055–62.

Frankel, V.H. 1986. Biomechanics of the hip joint. *Instr. Course Lect.* 35: 3–9.

Fregly, B.J., and F.E. Zajac. 1996. A state-space analysis of mechanical energy generation, absorption, and transfer during pedaling. *J. Biomech.* 29 (1): 81–90.

Fregly, B.J., F.E. Zajac, and C.A. Dairaghi. 1996. Crank inertial load has little effect on steady-state pedaling coordination. *J. Biomech.* 29 (12): 1559–67.

Fujie, H., G.A. Livesay, M. Fujita, and S.L. Woo. 1996. Forces and moments in six-DOF at the human knee joint: Mathematical description for control. *J. Biomech.* 29 (12): 1577–85.

Fukashiro, S., and P.V. Komi. 1987. Joint moment and mechanical power flow of the lower limb during vertical jump. *Int. J. Sports Med.* 8 (Suppl. 1): 15–21.

Fukashiro, S., P.V. Komi, M. Jarvinen, and M. Miyashita. 1993. Comparison between the directly measured Achilles tendon force and the tendon force calculated from the ankle moment during vertical jumps. *Clin. Biomech.* 8: 25–30.

Gagnon, M., G. Robertson, and R. Norman. 1987. Kinetics. In *Standardizing biomechanical testing in sport,* ed. D.A. Dainty and R.W. Norman, 21–57. Champaign, IL: Human Kinetics.

Galea, V., and R.W. Norman. 1985. Bone-on-bone forces at the ankle joint during a rapid dynamic movement. In *Biomechanics IX-A,* ed. D. Winter, R.W. Norman, R.P. Wells, K.C. Hayes, and A.E. Patla, 71–76. Champaign, IL: Human Kinetics.

Geesaman, D.L., and T.S. Buchanan. 1998. An improved rigid body spring model solving for moments at the wrist joint. In *Proceedings of NACOB '98: The third North American Congress on Biomechanics, August 14–18, 1998,* 529–30. Waterloo, ON: University of Waterloo.

Gerritsen, K.G., A.J. van den Bogert, and B.M. Nigg. 1995. Direct dynamics simulation of the impact phase in heel-toe running. *J. Biomech.* 28 (6): 661–68.

Gerritsen, K.G.M., W. Nachbauer, and A.J. van den Bogert. 1996. Computer stimulation of landing movement in downhill skiing: Anterior cruciate ligament injuries. *J. Biomech.* 29 (7): 845–54.

Ghez, C., and R. Sainburg. 1995. Proprioceptive control of interjoint coordination. *Can. J. Physiol. Pharmacol.* 73 (2): 273–84.

Gignoux, P., L. Chèze, J.P. Carret, and J. Dimnet. 1994. Hip joint loading computation of a walking patient during stance phase. *Clin. Mater.* 15 (4): 247–52.

Gitter, A., J. Czerniecki, and M. Meinders. 1997. Effect of prosthetic mass on swing phase work during above-knee amputee ambulation. *Am. J. Phys. Med. Rehabil.* 76 (2): 114–21.

Giurintano, D.J., A.M. Hollister, W.L. Buford, D.E. Thompson, and L.M. Myers. 1995. A virtual five-link model of the thumb. *Med. Eng. Phys.* 17 (4): 297–303.

Glitsch, U., and W. Baumann. 1997. The three-dimensional determination of internal loads in the lower extremity. *J. Biomech.* 30 (11–12): 1123–31.

Gonzalez, H., and M.L. Hull. 1989. Multivariable optimization of cycling biomechanics. *J. Biomech.* 22 (11–12): 1151–61.

Gottlieb, G.L., C.H. Chen, and D.M. Corcos. 1995. Relations between joint torque, motion, and electromyographic patterns at the human elbow. *Exp. Brain Res.* 103 (1): 164–67.

Gottlieb, G.L., Q. Song, D.A. Hong, and D.M. Corcos. 1996. Coordinating two degrees of freedom during human arm movement: Load and speed invariance of relative joint torques. *J. Neurophysiol.* 76 (5): 3196–206.

Granata, K.P., and W.S. Marras. 1995. An EMG-assisted model of trunk loading during free-dynamic lifting. *J. Biomech.* 28 (11): 1309–17.

Gregor, R.J., P.R. Cavanagh, and M. Lafortune. 1985. Knee flexor moments during propulsion in cycling: A creative solution to Lombard's paradox. *J. Biomech.* 18 (5): 307–16.

Gregor, R.J., P.V. Komi, R.C. Browning, and M. Jarvinen. 1991. A comparison of the triceps surae and residual muscle moments at the ankle during cycling. *J. Biomech.* 24: 287–98.

Gribble, P.L., and D.J. Ostry. 1999. Compensation for interaction torques during single- and multijoint limb movement. *J. Neurophysiol.* 82 (5): 2310–26.

Gruber, K., H. Ruder, J. Denoth, and K. Schneider. 1998. A comparative study of impact dynamics: Wobbling mass model versus rigid body models. *J. Biomech.* 31 (5): 439–44.

Gu, M.J., A.B. Schultz, N.T. Shepard, and N.B. Alexander. 1996. Postural control in young and elderly adults when stance is perturbed: Dynamics. *J. Biomech.* 29 (3): 319–29.

Hale, S.A. 1990. Analysis of the swing phase dynamics and muscular effort of the above-knee amputee for varying prosthetic shank loads. *Prosthet. Orthot. Int.* 14 (3): 125–35.

Happee, R. 1994. Inverse dynamic optimization including muscular dynamics. *J. Biomech.* 27: 953–60.

Happee, R., and F.C.T. Van der Helm. 1995. The control of shoulder muscles during goal directed movements, an inverse dynamic analysis. *J. Biomech.* 28 (10): 1179–91.

Harding, D.C., K.D. Brandt, and B.M. Hillberry. 1993. Finger joint force minimization in pianists using optimization techniques. *J. Biomech.* 26 (12): 1403–12.

Harrison, R.N., A. Lees, P.J.J. McCullagh, and W.R. Rowe. 1987. Bioengineering analysis of muscle and joint forces in the human leg during running. *Biomechanics X-B,* ed. B. Johnson, 855–61. Champaign, IL: Human Kinetics.

Hasler, E.M., and W. Herzog. 1998. Quantification of in vivo patellofemoral contact forces before and after ACL transection. *J. Biomech.* 31 (1): 37–44.

Hatze, H. 1977. A complete set of control equations for the human musculoskeletal system. *J. Biomech.* 10: 799–805.

Hatze, H. 2000. The inverse dynamics problem of neuromuscular control. *Biol. Cybern.* 82 (2): 133–41.

Hawkins, D., and M. Smeulders. 1999. Relationship between knee joint torque, velocity, and muscle activation: Considerations for musculoskeletal modeling. *J. Appl. Biomech.* 15 (3): 141–57.

Heise, G.D., and A. Cornwell. 1997. Relative contributions to the net joint moment for a planar multijoint throwing skill: Early and late in practice. *Res. Q. Exerc. Sport* 68 (2): 116–24.

Herman, R., R. Wirta, S. Bampton, and F.R. Finley. 1976. Human solutions for locomotion: I. Single limb analysis. In *Neural control of locomotion*, ed. R.M. Herman, S.G. Grillner, P.S. Stein, and D.G. Stuart, 13–49. New York: Plenum Press.

Herring, R.M., and A.E. Chapman. 1992. Effects of changes in segmental values and timing of both torque and torque reversal in simulated throws. *J. Biomech.* 25 (10): 1173–84.

Herzog, W. 1988. The relation between the resultant moments at a joint and the moments measured by an isokinetic dynamometer. *J. Biomech.* 21 (1): 5–12.

Herzog, W. 1998. Torque: Misuse of a misused term. *J. Manipulative Physiol. Ther.* 21 (1): 57–59.

Herzog, W. 1999. Commentary: Torque misuse revisited [letter]. *J. Manipulative Physiol. Ther.* 22 (5): 347–48.

Hilding, M.B., H. Lanshammar, and L. Ryd. 1995. A relationship between dynamic and static assessments of knee joint load: Gait analysis and radiography before and after knee replacements in 45 patients. *Acta Orthop. Scand.* 66 (4): 317–20.

Hof, A.L. 1992. An explicit expression for the moment in multibody systems. *J. Biomech.* 25 (10): 1209–11.

Hof, A.L., and J.W. van den Berg. 1977. Linearity between the weighted sum of the EMG of the human triceps surae and the total torque. *J. Biomech.* 10: 529–39.

Hof, A.L., C.N. Pronk, and J.A. van Best. 1987. Comparison between EMG to force processing and kinetic analysis for the calf muscle moment in walking and stepping. *J. Biomech.* 20 (2): 167–78.

Holden, J.P., and S.J. Stanhope. 1998. The effect of variation in the knee center location estimates on net knee moments. *Gait Posture* 7: 1–6.

Hollerbach, J.M., and T. Flash. 1982. Dynamic interaction between limb segments during planar arm movements. *Biol. Cybern.* 44: 67–77.

Hooper, D.M., V.K. Goel, K.M. Bolte, and M.H. Pope. 1996. Predictions of muscle and L3-L4 and L4-L5 disc loads during freestyle asymmetric lifting. In *Twentieth annual meeting of the American Society of Biomechanics, conference proceedings*, 17–18. Atlanta: Georgia Tech.

Hore, J., S. Watts, and D. Tweed. 1999. Prediction and compensation by an internal model for back forces during finger opening in an overarm throw. *J. Neurophysiol.* 82 (3): 1187–97.

Houdijk, H., J.J. de Koning, G. de Groot, M.F. Bobbert, and G.J. van Ingen Schenau. 2000. Push-off mechanics in speed skating with conventional skates and klapskates. *Med. Sci. Sports Exerc.* 32 (3): 635–41.

Hoy, M.G., W.C. Whiting, and R.F. Zernicke. 1982. Stride kinematics and knee joint kinetics of child amputee gait. *Arch. Phys. Med. Rehabil.* 63 (2): 74–82.

Hoy, M.G., and R.F. Zernicke. 1985. Modulation of limb dynamics in the swing phase of locomotion. *J. Biomech.* 18 (1): 49–60.

Hoy, M.G., and R.F. Zernicke. 1986. The role of intersegmental dynamics during rapid limb oscillations. *J. Biomech.* 19 (10): 867–77.

Hoy, M.G., R.F. Zernicke, and J.L. Smith. 1985. Contrasting roles of inertial and muscle moments at knee and ankle during paw-shake response. *J. Neurophysiol.* 54 (5): 1282–94.

Huang, S.C. 1995. Modeling human body motion with application in crash victim simulation. *J. Appl. Biomech.* 11: 322–36.

Huang, T.C., E.M. Roberts, and Y. Youm. 1982. Biomechanics of kicking. In *Human body dynamics: Impact, occupational and athletic aspects,* ed. D.N. Ghista, 409–43. Oxford, UK: Clarendon Press.

Hughes, R.E. 2000. Effect of optimization criterion on spinal force estimates during asymmetric lifting. *J. Biomech.* 33 (2): 225–29.

Hull, M.L., and H. Gonzalez. 1988. Bivariate optimization of pedalling rate and crank arm length in cycling. *J. Biomech.* 21 (10): 839–49.

Hurwitz, D.E., K.C. Foucher, D.R. Sumner, T.P. Andriacchi, A.G. Rosenberg, and J.O. Galante. 1998. Hip motion and moments during gait relate directly to proximal femoral bone mineral density in patients with hip osteoarthritis. *J. Biomech.* 31 (10): 919–25.

Hurwitz, D.E., D.R. Sumner, T.P. Andriacchi, and D.A. Sugar. 1998. Dynamic knee loads during gait predict proximal tibial bone distribution. *J. Biomech.* 31 (5): 423–30.

Huston, R.L., and C.E. Passerello. 1982. The mechanics of human body motion. In *Human body dynamics: Impact, occupational and athletic aspects,* ed. D.N. Ghista, 203–47. Oxford, UK: Clarendon Press.

Hutchins, E.L., R.V. Gonzalez, and R.E. Barr. 1993. Comparison of experimental and analytical torque-angle relationship of the human elbow joint complex. *Biomed. Sci. Instrum.* 29: 17–24.

Iossifidou, A.N., and V. Baltzopoulos. 1998. Inertial effects on the assessment of performance in isokinetic dynamometry. *Int. J. Sports Med.* 19 (8): 567–73.

Jacobs, R., and G.J. van Ingen Schenau. 1992. Intermuscular coordination in a sprint push-off. *J. Biomech.* 25 (9): 953–65.

Jäger, M., and A. Luttmann. 1989. Biomechanical analysis and assessment of lumbar stress during load lifting using a dynamic 19-segment model. *Ergonomics* 32: 93–112.

Jansen, M.O., A.J. van den Bogert, D.J. Riemersma, and H.C. Schamhardt. 1993. In vivo tendon forces in the forelimb of ponies at the walk, validated by ground reaction force measurements. *Acta Anat.* 146 (2–3): 162–67.

Jensen, R.K. 1988. Developmental relationships between body inertia and joint torques. *Hum. Biol.* 60 (5): 693–707.

Kaneko, Y., and F. Sato. 1999. Comparison of transformation of joint torque to ground reaction force under different running velocities. In *International Society of Biomechanics, XVIIth congress, book of abstracts,* ed. W. Herzog and A. Jinha, 505. Calgary, AB: University of Calgary.

Karlsson, D., and B. Peterson. 1992. Towards a model for force prediction in the human shoulder. *J. Biomech.* 25: 189–99.

Kaufman, K.R., K.-N. An, and E.Y.S. Chao. 1995. A comparison of intersegmental joint dynamics to isokinetic dynamometer measurements. *J. Biomech.* 28 (10): 1243–56.

Kaufman, K.R., K.-N. An, W.J. Litchy, B.F. Morrey, and E.Y.S. Chao. 1991. Dynamic joint forces during knee isokinetic exercise. *Am. J. Sports Med.* 19: 305–16.

Kautz, S.A., and M.L. Hull. 1993. A theoretical basis for interpreting the force applied to the pedal in cycling. *J. Biomech.* 26 (2): 155–65.

Kautz, S.A., R.R. Neptune, and F.E. Zajac. 2000. General coordination principles elucidated by forward dynamics: Minimum fatigue does not explain muscle excitation in dynamic tasks [comment]. *Motor Control* 4 (1): 75–80; discussion 97–116.

Kearney, R.E., and I.W. Hunter. 1990. System identification of human joint dynamics. *Crit. Rev. Biomed. Eng.* 18 (1): 55–87.

Kellis, E., and V. Baltzopoulos. 1999. The effects of the antagonist muscle force on intersegmental loading during isokinetic efforts of the knee extensors. *J. Biomech.* 32 (1): 19–25.

Kepple, T.M., K.L. Siegel, and S.J. Stanhope. 1997. Relative contribution of the lower extremity joint moments to forward progression and support during gait. *Gait Posture* 6 (1): 1–8.

Ketchum, L.D., P.W. Brand, D. Thompson, and G.S. Pocock. 1978. The determination of moments for extension of the wrist generated by the muscles of the forearm. *J. Hand Surg.* 3: 205–10.

Kingma, I. 1998. Challenging gravity: The mechanics of lifting. Doctoral diss., Vrije Universiteit Amsterdam.

Kingma, I.D.L.M.P., H.M.K.H.G. Toussaint, and T.B.M. Bruijnen. 1996. Validation of a full body 3D dynamic linked segment model. *Hum. Mov. Sci.* 15: 833–60.

Ko, H. 1996. Animating human locomotion with inverse dynamics. *IEEE Comput. Graphics Appl.* 16 (2): 50–56.

Komistek, R.D., J.B. Stiehl, D.A. Dennis, R.D. Paxson, and R.W. Soutas-Little. 1998. Mathematical model of the lower extremity joint reaction forces using Kane's method of dynamics. *J. Biomech.* 31 (2): 185–90.

Koolstra, J.H., T.M. van Euden, W.A. Weijs, and M. Naeije. 1988. A three-dimensional mathematical model of the human masticatory system predicting maximum possible bite forces. *J. Biomech.* 21: 563–76.

Koopman, B., H.J. Grootenboer, and H.J. de Jongh. 1995. An inverse dynamic model for the analysis, reconstruction and prediction of bipedal walking. *J. Biomech.* 28 (11): 1369–76.

Kotzar, G.M., D.T. Davy, V.M. Goldberg, K.G. Heiple, J. Berilla, K.G. Heiple Jr., R.H. Brown, and A.H. Burstein. 1991. Telemeterized in vivo hip joint force data: A report on two patients after total hip surgery. *J. Orthop. Res.* 9 (5): 621–33.

Krabbe, B., R. Farkas, and W. Baumann. 1997. Influence of inertia on intersegment moments of the lower extremity joints. *J. Biomech.* 30 (5): 517–19.

Krebs, D.E., C.E. Robbins, L. Lavine, and R.W. Mann. 1998. Hip biomechanics during gait. *J. Orthop. Sports Phys. Ther.* 28 (1): 51–59.

Kromodihardjo, S., and A. Mital. 1986. Kinetic analysis of manual lifting activities: 1. Development of a three-dimensional computer model. *Int. J. Ind. Ergon.* 1: 77–90.

Kuo, A.D. 1998. A least-squares estimation approach to improving the precision of inverse dynamics computations. *J. Biomech. Eng.* 120 (1): 148–59.

Kyröläinen, H., A. Belli, and P.V. Komi. 1999. Lower limb mechanics with increasing running speed. In *International Society of Biomechanics, XVIIth congress, book of abstracts,* ed. W. Herzog and A. Jinha, 262. Calgary, AB: University of Calgary.

Labossière, R., D.J. Ostry, and A.G. Feldman. 1996. The control of multi-muscle systems: Human jaw and hyoid movements. *Biol. Cybern.* 74: 373–84.

Lan, N., and P.E. Crago. 1992. A noninvasive technique for in vivo measurement of joint torques of biarticular muscles. *J. Biomech.* 25: 1075–79.

Latash, M.L., A.S. Aruin, and V.M. Zatsiorsky. 1999. The basis of a simple synergy: Reconstruction of joint equilibrium trajectories during unrestrained arm movements. *Hum. Mov. Sci.* 18: 3–30.

Lee, R.Y., and J. Munn. 2000. Passive moment about the hip in straight leg raising. *Clin. Biomech.* 15 (5): 330–34.

Li, Z.M., V.M. Zatsiorsky, and M.L. Latash. 2000. Contribution of the extrinsic and intrinsic hand muscles to the moments in finger joints. *Clin. Biomech.* 15 (3): 203–11.

Looze, M.P. de, I. Kingma, J.B.J. Bussmann, and H.M. Toussaint. 1992. Validation of a dynamic linked segment model to calculate joint moments in lifting. *Clin. Biomech.* 7: 161–69.

Looze, M.P. de, H.M. Toussaint, J.H. van Dieen, and H.C.G. Kemper. 1993. Joint moments and muscle activity in the lower extremities and lower back in lifting and lowering tasks. *J. Biomech.* 26: 1067–76.

Lu, T.-W., J.J. O'Connor, S.J.G. Taylor, and P.S. Walker. 1998. Validation of a lower limb model with in vivo femoral forces telemetered from two subjects. *J. Biomech.* 31 (1): 63–69.

Luhtanen, P., and P. Komi. 1978. Segmental contribution to forces in vertical jump. *Eur. J. Appl. Physiol.* 38: 181–88.

Marras, W., and C. Sommerich. 1991. A three-dimensional motion model of loads on the lumbar spine: I. Model structure. *Hum. Factors* 33: 123–37.

Marsh, A.P., P.E. Martin, and D.J. Sanderson. 2000. Is a joint moment-based cost function associated with preferred cycling cadence? *J. Biomech.* 33 (2): 173–80.

Marshall, R.N., R.K. Jensen, and G.A. Wood. 1985. A general Newtonian simulation of an *n*-segment open chain model. *J. Biomech.* 18 (5): 359–67.

Marshall, R.N., G.A. Wood, and L.S. Jennings. 1989. Performance objectives in human movement: A review and application to the stance phase of normal walking. *Hum. Mov. Sci.* 8: 571–94.

Martin, P.E., and P.R. Cavanagh. 1990. Segment interactions within the swing leg during unloaded and loaded running. *J. Biomech.* 23 (6): 529–36.

McCaw, S.T., and P. Devita. 1995. Errors in alignment of center of pressure and foot coordinates affect predicted lower extremity torques. *J. Biomech.* 28 (8): 985–88.

McGill, S.M., and R.W. Norman. 1986. Partitioning of the L4-L5 dynamic moment into disc, ligamentous, and muscular components during lifting. *Spine* 11: 666–76.

McGinnis, P.M., and L.A. Bergman. 1986. An inverse dynamic analysis of the pole vault. *Int. J. Sport Biomech.* 2: 186–201.

McLaughlin, T.M., and N.R. Miller. 1980. Techniques for the evaluation of loads on the forearm prior to impact in tennis stroke. *J. Mech. Des.* 102: 701–10.

Mikosz, R.P., T.P. Andriacchi, and G.B.J. Andersson. 1988. Model analysis of factors influencing the prediction of muscle forces at the knee. *J. Orthop. Res.* 6: 205–14.

Miller, D.I., and D.J. East. 1976. Kinematic and kinetic correlates of vertical jumping in women. In *Biomechanics V-B,* ed. P.V. Komi, 65–72. Baltimore: University Park Press.

Morrison, J.B. 1970. The mechanics of the knee joint in relation to normal walking. *J. Biomech.* 3: 51–61.

Murray, S., R. Warren, J. Otis, M. Kroll, and T. Wickiewicz. 1984. Torque-velocity relationships of the knee extensor and flexor muscles in individuals sustaining injuries of the anterior cruciate ligaments. *Am. J. Sports Med.* 12: 436–40.

Neptune, R.R., and M.L. Hull. 1998. Evaluation of performance criteria for simulation of submaximal steady-state cycling using a forward dynamic model. *J. Biomech. Eng.* 120 (3): 334–41.

Neptune, R.R., and M.L. Hull. 1999. A theoretical analysis of preferred pedaling rate selection in endurance cycling. *J. Biomech.* 32 (4): 409–15.

Nisell, R., G. Nemeth, and H. Ohlsen. 1986. Joint forces in extension of the knee. *Acta Orthop. Scand.* 57: 41–46.

Novacheck, T.F. 1998. The biomechanics of running. *Gait Posture* 7 (1): 77–95.

Olney, S.J., and D.A. Winter. 1985. Prediction of knee and ankle moments of force in walking from EMG and kinematic data. *J. Biomech.* 18: 9–20.

Pandy, M.G., and N. Berme. 1990. A general recursive technique for formulating mathematical models of human gait: Theory and application. In *Biomechanics of human movement: Applications in rehabilitation, sports and ergonomics,* ed. N. Berme and A. Cappozzo, 324–33. Worthington, OH: Bertec Corp.

Pandy, M.G., and K.B. Shelburne. 1997. Dependence of cruciate-ligament loading on muscle forces and external load. *J. Biomech.* 30 (10): 1015–24.

Patla, A.E., and S.D. Prentice. 1995. The role of active forces and intersegmental dynamics in the control of limb trajectory over obstacles during locomotion in humans. *Exp. Brain Res.* 106 (3): 499–504.

Paul, J.P. 1965. Bioengineering studies of the forces transmitted by joints. In *Biomechanics and related engineering topics,* ed. R.M. Kenedi, 369–80. Oxford: Pergamon Press.

Paul, J.P. 1978. Torques produce torsions. *J. Biomech.* 11: 87.

Paul, J.P. 1990. Joint loads. In *Biomechanics of human movement: Applications in rehabilitation, sports and ergonomics,* ed. N. Berme and A. Cappozzo, 304–14. Worthington, OH: Bertec Corp.

Pedersen, D.R., R.A. Brand, and D.T. Davy. 1997. Pelvic muscle and acetabular contact forces during gait. *J. Biomech.* 30 (9): 959–65.

Pedotti, A., V.V. Krishnan, and L. Stark. 1978. Optimization of muscle-force sequencing in human locomotion. *Math. Biosci.* 38: 57–76.

Phillips, S.J., E.M. Roberts, and T.C. Huang. 1983. Quantification of intersegmental reactions during rapid swing motion. *J. Biomech.* 16 (6): 411–18.

Piazza, S.J., and S.L. Delp. 1996. The influence of muscles on knee flexion during the swing phase of gait. *J. Biomech.* 29 (6): 723–33.

Plagenhoef, S. 1971. *Patterns of human motion: A cinematographic analysis.* Englewood Cliffs, NJ: Prentice Hall.

Popovic, D., M.N. Oguztöreli, and R.B. Stein. 1995. Optimal control for an above-knee prosthesis with two degrees of freedom. *J. Biomech.* 28 (1): 89–98.

Praagman, M., M. Stokdijk, H.E. Veeger, and B. Visser. 2000. Predicting mechanical load of the glenohumeral joint, using net joint moments. *Clin. Biomech.* 15 (5): 315–21.

Prilutsky, B.I. 2000. Eccentric muscle action in sport and exercise. In *Biomechanics in sport,* ed. V.M. Zatsiorsky, 56–86. Oxford, UK: Blackwell Science.

Prilutsky, B.I., R.J. Gregor, and M.M. Ryan. 1998. Coordination of two-joint rectus femoris and hamstrings during the swing phase of human walking and running. *Exp. Brain Res.* 120 (4): 479–86.

Prilutsky, B.I., W. Herzog, and T.L. Allinger. 1997. Forces of individual cat ankle extensor muscles during locomotion predicted using static optimization. *J. Biomech.* 30 (10): 1025–33.

Prilutsky, B.I., T. Isaka, A.M. Albrecht, and R.J. Gregor. 1998. Is coordination of two-joint leg muscles during load lifting consistent with the strategy of minimum fatigue? *J. Biomech.* 31 (11): 1025–34.

Procter, P., and J.P. Paul. 1982. Ankle joint biomechanics. *J. Biomech.* 15 (9): 627–34.

Putnam, C.A. 1991. A segment interaction analysis of proximal-to-distal sequential segment motion patterns. *Med. Sci. Sports Exerc.* 23 (1): 130–44.

Putnam, C.A. 1993. Sequential motions of body segments in striking and throwing skills: Descriptions and explanations. *J. Biomech.* 26 (Suppl. 1): 125–35.

Ralston, H.J., and L. Lukin. 1969. Energy levels of human body segments during level walking. *Ergonomics* 12: 399–46.

Redfield, R., and M.L. Hull. 1986a. On the relation between joint moments and pedalling rates at constant power in bicycling. *J. Biomech.* 19: 317–29.

Redfield, R., and M.L. Hull. 1986b. Prediction of pedal forces in bicycling using optimization methods. *J. Biomech.* 19 (7): 523–40.

Reilly, D.T. 1988. Dynamic loading of normal joints. *Rheum. Dis. Clin. N. Am.* 14 (3): 497–502.

Reinschmidt, C., A.J. van den Bogert, B.M. Nigg, and N. Murphy. 1997. Effect of skin movement on the analysis of skeletal knee joint motion during running. *J. Biomech.* 30 (7): 729–32.

Reinschmidt, C., and B.M. Nigg. 1995. Influence of heel height on ankle joint moments in running. *Med. Sci. Sports Exerc.* 27 (3): 410–16.

Reisman, M., R.G. Burdett, S.R. Simon, and C. Norkin. 1985. Elbow moment and forces at the hands during swing-through axillary crutch gait. *Phys. Ther.* 65 (5): 601–5.

Reynolds, H.M., M.C. Beal, and R.C. Hallgren. 1993. Quantifying passive resistance to motion in the straight-leg-raising test on asymptomatic subjects. *J. Am. Osteopath. Assoc.* 93 (9): 913–14, 917–20.

Riener, R., and T. Edrich. 1999. Identification of passive elastic joint moments in the lower extremities. *J. Biomech.* 32 (5): 539–44.

Riley, P.O., and D.C. Kerrigan. 1998. Torque action of two-joint muscles in the swing period of stiff-legged gait: A forward dynamic model analysis. *J. Biomech.* 31 (9): 835–40.

Rodgers, M.M., S. Tummarakota, and J. Lieh. 1998. Three-dimensional dynamic analysis of wheelchair propulsion. *J. Appl. Biomech.* 14 (1): 80–92.

Rohlmann, A., G. Bergmann, and F. Graichen. 1997. Loads on an internal spinal fixation device during walking. *J. Biomech.* 30 (1): 41–47.

Rohrie, H., R. Scholten, C. Sigolotto, W. Sollbach, and H. Kellner. 1984. Joint forces in the human pelvis–leg skeleton during walking. *J. Biomech.* 17 (6): 409–24.

Ruby, P., and M.L. Hull. 1993. Response of intersegmental knee loads to foot/pedal platform degrees of freedom in cycling. *J. Biomech.* 26 (11): 1327–40.

Ruby, P., M.L. Hull, and D. Hawkins. 1992. Three-dimensional knee joint loading during seated cycling. *J. Biomech.* 25 (1): 41–53.

Ruby, P., M.L. Hull, K.A. Kirby, and D.W. Jenkins. 1992. The effect of lower-limb anatomy on knee loads during seated cycling. *J. Biomech.* 25 (10): 1195–207.

Runge, C.F., C.L. Shupert, F.B. Horak, and F.E. Zajac. 1999. Ankle and hip postural strategies defined by joint torques. *Gait Posture* 10 (2): 161–70.

Runge, C.F., F.E. Zajac, J.H. Allum, D.W. Risher, A.E. Bryson Jr., and F. Honegger. 1995. Estimating net joint torques from kinesiological data using optimal linear system theory. *IEEE Trans. Biomed. Eng.* 42 (12): 1158–64.

Sainburg, R.L., C. Ghez, and D. Kalakanis. 1999. Intersegmental dynamics are controlled by sequential anticipatory, error correction, and postural mechanisms. *J. Neurophysiol.* 81 (3): 1045–56.

Sainburg, R.L., M.F. Ghilardi, H. Poizner, and C. Ghez. 1995. Control of limb dynamics in normal subjects and patients without proprioception. *J. Neurophysiol.* 73 (2): 820–35.

Sainburg, R.L., and D. Kalakanis. 2000. Differences in control of limb dynamics during dominant and nondominant arm reaching. *J. Neurophysiol.* 83 (5): 2661–75.

Sainburg, R.L., H. Poizner, and C. Ghez. 1993. Loss of proprioception produces deficits in interjoint coordination. *J. Neurophysiol.* 70 (5): 2136–47.

Schneider, K., R.F. Zernicke, R.A. Schmidt, and T.J. Hart. 1989. Changes in limb dynamics during the practice of rapid arm movements. *J. Biomech.* 22 (8–9): 805–17.

Schneider, K., R.F. Zernicke, B.D. Ulrich, J.L. Jensen, and E. Thelen. 1990. Understanding movement control in infants through the analysis of limb intersegmental dynamics. *J. Mot. Behav.* 22: 493–520.

Schwameder, H., R. Roithner, E. Muller, W. Niessen, and C. Raschner. 1999. Knee joint forces during downhill walking with hiking poles. *J. Sports Sci.* 17 (12): 969–78.

Scuindt, F., W.P. Cooney, R.L. Linscheid, K.N. An, and E.Y.S. Chao. 1995. Force and pressure transmission through the normal wrist. *J. Biomech.* 28 (5): 587–601.

Scultz, A.B., N.B. Alexander, and J.A. Ashton-Miller. 1992. Biomechanical analysis of rising from a chair. *J. Biomech.* 25 (12): 1383–91.

Seireg, A., and R.J. Arvikar. 1973. A mathematical model for evaluation of forces in lower extremities of the musculo-skeletal system. *J. Biomech.* 6: 313–26.

Seireg, A., and R.J. Arvikar. 1975. The prediction of muscular load sharing and joint forces in the lower extremities during walking. *J. Biomech.* 8: 89–102.

Sepulveda, F., D.M. Wells, and C.L. Vaughan. 1993. A neural network representation of electromyography and joint dynamics in human gait. *J. Biomech.* 26: 101–10.

Shelburne, K.B., and M.G. Pandy. 1977. A musculoskeletal model of the knee for evaluating ligament forces during isometric contractions. *J. Biomech.* 30 (2): 163–76.

Siegler, S., and W. Liu. 1997. Inverse dynamics in human locomotion. In *Three-dimensional analysis of human locomotion,* ed. P. Allard, A. Cappozzo, A. Lundberg, and C.L. Vaughan, 191–210. Chichester, NY: Wiley.

Siegler, S., G.D. Moskowitz, and W. Freedman. 1984. Passive and active components of the internal moment developed about the ankle joint during human ambulation. *J. Biomech.* 17: 647–52.

Siegler, S., R. Seliktar, and W. Hyman. 1982. Simulation of human gait with the aid of a simple mechanical model. *J. Biomech.* 15: 415–25.

Silva, M.P.T., J.A.C. Ambrósio, and M.S. Pereira. 1997. Biomechanical model with joint resistance for impact simulation. *Multibody Syst. Dyn.* 1: 65–84.

Simonsen, E.B., P. Dyhre-Poulsen, M. Voigt, P. Aagaard, G. Sjogaard, and F. Bojsen-Moller. 1995. Bone-on-bone forces during loaded and unloaded walking. *Acta Anat. (Basel)* 152 (2): 133–42.

Smith, J.W. 1957. The forces operating at the human ankle during standing. *J. Anat.* 91: 545–64.

Spoor, C.W., J.L. van Leeuwen, F.H.J. de Windt, and A. Huson. 1989. A model study of muscle forces and joint-force direction in normal and dysplasic neonatal hips. *J. Biomech.* 22: 873–84.

Stokes, I.A.F., W.C. Hutton, and J.R.R. Scott. 1979. Forces acting in the metatarsals during normal walking. *J. Anat.* 129: 579–90.

Stone, C., and M.L. Hull. 1995. The effect of rider weight on rider-induced loads during common cycling situations. *J. Biomech.* 28 (4): 365–75.

Stuart, M.J., D.A. Meglan, G.E. Lutz, E.S. Growney, and K.N. An. 1996. Comparison of intersegmental tibiofemoral joint forces and muscle activity during various closed kinetic chain exercises. *Am. J. Sports Med.* 24 (6): 792–99.

Suzuki, M., Y. Yamazaki, and K. Matsunami. 2000. Simplified dynamics model of planar two-joint arm movements. *J. Biomech.* 33 (8): 925–31.

Taylor, S.J., P.S. Walker, and R. Woledge. 1998. Direct measurement of femoral and knee forces using a telemeterised distal femoral displacement. *J. Biomech.* 32 (Suppl. 1): 25.

Theeuwen, M., C.C. Gielen, and B.M. van Bolhuis. 1996. Estimating the contribution of muscles to joint torque based on motor-unit activity. *J. Biomech.* 29 (7): 881–89.

Thelen, E., and D.M. Fisher. 1983. The organization of spontaneous leg movements in newborn infants. *J. Mot. Behav.* 15: 353–77.

Thorpe, S.K., Y. Li, R.H. Crompton, and R.M. Alexander. 1998. Stresses in human leg muscles in running and jumping determined by force plate analysis and from published magnetic resonance images. *J. Exp. Biol.* 201 (Pt. 1): 63–70.

Tolbert, J.R., W.F. Blair, J.G. Andrews, and R.D. Crowninshield. 1985. The kinetics of normal and prosthetic wrists. *J. Biomech.* 18: 887–97.

Toussaint, H.M., A.F. de Winter, Y. de Haas, M.P. de Looze, J.H. Van Dieen, and I. Kingma. 1995. Flexion relaxation during lifting: Implications for torque production by muscle activity and tissue strain at the lumbo-sacral joint. *J. Biomech.* 28 (2): 199–210.

Tsuang, Y.H., O.D. Schipplein, J.H. Trafimow, and G.B. Andersson. 1992. Influence of body segment dynamics on loads at the lumbar spine during lifting. *Ergonomics* 35 (4): 437–44.

Vaughan, C.L. 1984. Biomechanics of running gait. *Crit. Rev. Biomed. Eng.* 12 (1): 1–48.

Vaughan, C. 1996. Are joint torques the Holy Grail of human gait analysis? *Hum. Mov. Sci.* 15: 423–43.

Vaughan, C.L., B.L. Davis, and J.C. O'Connor. 1992. *Dynamics of human gait.* Champaign, IL: Human Kinetics.

Virji-Babul, N., and J.D. Cooke. 1995. Influence of joint interactional effects on the coordination of planar two-joint arm movements. *Exp. Brain Res.* 103 (3): 451–59.

Virji-Babul, N., J.D. Cooke, and S.H. Brown. 1994. Effects of gravitational forces on single joint arm movements in humans. *Exp. Brain Res.* 99 (2): 338–46.

Voronov, A.V., E.K. Lavrovsky, and V.M. Zatsiorsky. 1995. Modelling of rational variants of the speed-skating technique. *J. Sports Sci.* 13 (2): 153–70.

Vrahas, M.S., R.A. Brand, T.D. Brown, and J.G. Andrews. 1990. Contribution of passive tissues to the intersegmental moments at the hip. *J. Biomech.* 23 (4): 357–62.

Weiss, P.L., R.E. Kearney, and I.W. Hunter. 1986a. Position dependence of ankle joint dynamics: I. Passive mechanics. *J. Biomech.* 19: 727–35.

Weiss, P.L., R.E. Kearney, and I.W. Hunter. 1986b. Position dependence of ankle joint dynamics: II. Active mechanics. *J. Biomech.* 19: 737–51.

Werner, F.W., K.-N. An, A.K. Palmer, and E.Y.S. Chao. 1991. Force analysis. In *Biomechanics of the wrist joint,* ed. K.-N. An, R.A. Begrer, and W.P.I. Cooney, 77–98. New York: Springer-Verlag.

Whittle, M.W. 1995. Musculoskeletal applications of three-dimensional analysis. In *Three-dimensional analysis of human movement,* ed. P. Allard, I.A.F. Stokes, and J.-P. Blanchi, 295–309. Champaign, IL: Human Kinetics.

Whittlesey, S.N., and J. Hamill. 1996. An alternative model of the lower extremity during locomotion. *J. Appl. Biomech.* 12: 269–79.

Wilk, K.E., R.F. Escamilla, G.S. Fleisig, S.W. Barrentine, J.R. Andrews, and M.L. Boyd. 1996. A comparison of tibiofemoral joint forces and electromyographic activity during open and closed kinetic chain exercises. *Am. J. Sports Med.* 24 (4): 518–27.

Williams, K.R. 2000. The dynamics of running. In *Biomechanics in sport,* ed. V.M. Zatsiorsky, 161–83. Oxford, UK: Blackwell Science.

Winter, D.A. 1983. Moments of force and mechanical power in jogging. *J. Biomech.* 16 (1): 91–97.

Winter, D.A. 1984. Biomechanics of human movement with application to the study of human locomotion. *Crit. Rev. Biomed. Eng.* 9: 287–314.

Winter, D.A., and D.G.E. Robertson. 1978. Joint torque and energy pattern in normal gait. *Biol. Cybern.* 29: 137–42.

Winters, J.M. 1995. Concepts in neuromuscular modeling. In *Three-dimensional analysis of human movement,* ed. P. Allard, I.A.F. Stokes, and J.-P. Blanchi, 257–93. Champaign, IL: Human Kinetics.

Wisleder, D., R.F. Zernicke, and J.L. Smith. 1990. Speed-related changes in hindlimb intersegmental dynamics during the swing phase of cat locomotion. *Exp. Brain Res.* 79 (3): 651–60.

Witte, H., F. Eckstein, and S. Recknagel. 1997. A calculation of the forces acting on the human acetabulum during walking: Based on in vivo force measurements, kinematic analysis and morphometry. *Acta Anat.* 160 (4): 269–80.

Wolchok, J.C., M.L. Hull, and S.M. Howell. 1998. The effect of intersegmental knee moments on patellofemoral contact mechanics in cycling. *J. Biomech.* 31 (8): 677–83.

Wu, G., and Z. Ladin. 1996. Limitations of quasi-static estimation of human joint loading during locomotion. *Med. Biol. Eng. Comput.* 34 (6): 472–76.

Wychowanski, M. 1989. Flexion torque in ankle joint measuring set. *J. Biomech.* 22 (10): 1102–7.

Yack, H.J. 1984. Techniques for clinical assessment of human movement. *Phys. Ther.* 64 (12): 1821–30.

Yamaguchi, G.T., and F.E. Zajac. 1989. A planar model of the knee joint to characterize the knee extensor mechanism. *J. Biomech.* 22: 1–10.

Yang, J.F., D.A. Winter, and R.P. Wells. 1990a. Postural dynamics in the standing human. *Biol. Cybern.* 62 (4): 309–20.

Yang, J.F., D.A. Winter, and R.P. Wells. 1990b. Postural dynamics of walking in humans. *Biol. Cybern.* 62 (4): 321–30.

Yoganandan, N., and F.A. Pintar. 1997. Inertial loading of the human cervical spine. *J. Biomech. Eng.* 119 (3): 237–40.

Yoon, Y.S., and J.M. Mansour. 1982. The passive elastic moment at the hip. *J. Biomech.* 15: 905–10.

Young, R.P., and R.G. Marteniuk. 1995. Changes in inter-joint relationships of muscle moments and powers accompanying the acquisition of a multi-articular kicking task. *J. Biomech.* 28 (8): 701–14.

Young, R.P., and R.G. Marteniuk. 1998. Stereotypic muscle-torque patterns are systematically adopted during acquisition of a multi-articular kicking task. *J. Biomech.* 31: 806–19.

Zajac, F.E., and M.E. Gordon. 1989. Determining muscle force and action in multiarticular movement. *Exerc. Sport Sci. Rev.* 17: 187–230.

Zarrugh, M.Y. 1981. Kinematic prediction of intersegment loads and power at the joints of the leg in walking. *J. Biomech.* 14: 713–25.

Zatsiorsky, V.M. 1995. *Science and practice of strength training.* Champaign, IL: Human Kinetics.

Zatsiorsky, V.M., and S.Y. Aleshinsky. 1976. Simulation of human locomotion in space. In *Biomechanics V-B,* ed. P.V. Komi, 387–94. Baltimore: University Park Press.

Zatsiorsky, V.M., Y.E. Lanka, and A.A. Shalmanov. 1981. Biomechanical analysis of shot putting technique. *Exerc. Sport Sci. Rev.* 9: 353–89.

Zatsiorsky, V.M., and N. Yakunin. 1991. Mechanics and biomechanics of rowing: A review. *Int. J. Sport Biomech.* 7 (3): 229–81.

Zernicke, R.F., and E.M. Roberts. 1976. Human lower extremity kinetic relationships during systematic variations in resultant limb velocity. In *Biomechanics V-B,* ed. P.V. Komi, 20–25. Champaign, IL: Human Kinetics.

Zernicke, R.F., and E.M. Roberts. 1978. Lower extremity forces and torques during systematic variation of non–weight bearing motion. *Med. Sci. Sports.* 10 (1): 21–26.

Zheng, N., G.S. Fleisig, R.F. Escamilla, and S.W. Barrentine. 1998. An analytical model of the knee for estimation of internal forces during exercise. *J. Biomech.* 31 (10): 963–67.

Zikovitz, D.C., and L.R. Harris. 1999. Head tilt during driving. *Ergonomics* 42 (5): 740–46.

6

CHAPTER

MECHANICAL WORK AND ENERGY IN HUMAN MOVEMENT

The Duke of Marlborough
Had twenty thousand men.
He marched them up a hill
And marched them down again.

What was the total amount of work performed?

A child's education starts with well-established knowledge (e.g., $2 \times 2 = 4$) and fairy tales. The fairy tales always end happily. When a student enters a college, the same principle pertains: the student studies well-established knowledge. Fairy tales for students also exist: they are called "problems" in the textbooks. Textbook problems contain all the necessary information and are always solvable. Science is different. Many problems cannot be solved because the necessary information is not available; some problems are not solvable at all. Still, it is important to understand the problem and the difficulties associated with its solution. This chapter deals with an unsolved problem of biomechanics: the problem of determining the mechanical energy expended for motion.

The human body spends chemically bound energy to move. Through diverse metabolic processes, the energy is transformed into mechanical work and heat. Examination of mechanical work and energy in human movements is an important field of biomechanics research. This research potentially benefits from understanding a basic law of nature: the law of conservation of energy. Similar opportunities do not exist in other areas of biomechanics, where fundamental laws have not yet been discovered.

Though mechanical work in human movement was first estimated as early as 1836 (English translation in Weber and Weber 1992), the determination of mechanical power and work in human movement is still a challenging task for biomechanists. There are two reasons for this: the complexity of the studied object (a human body) and the limited abilities of contemporary experimental methods. Existing methods do not provide sufficient information for computing mechanical work in diverse human movements in an unambiguous way. This computation is possible only for some movements. Establishing a relationship between mechanical work and the metabolic energy expended for it is even more complex. In contemporary science, this issue remains largely unsolved.

The problem of work and energy in human and animal movement is an interdisciplinary one; it is studied in biophysics, biochemistry, physiology, motor control, and biomechanics. This chapter addresses only the biomechanical aspects of the problem.

Even from a purely mechanical standpoint, the task of determining the mechanical work in human movement is not trivial. The human body is a multilink system powered by force actuators (muscles). Forces generated by the individual muscles are usually unknown. The muscles spend energy not only for work production but also for force generation. The muscles also spend energy when they resist their forcible stretching. In addition, the muscle-tendon complexes and ligaments can store elastic potential energy that can later be recoiled. The proportions of stored and recoiled energy are generally unknown. All the mentioned phenomena make determining the mechanical energy expenditure in human movement a perplexing task. Some of the problems will be overcome in the future by new experimental methods (e.g., methods that allow individual muscle forces in an intact human body to be determined). However, equally important is the understanding of the problem itself. In the contemporary literature, various techniques for determining "mechanical work" in human movement are used. Often, these techniques yield dissimilar results when applied to the same movement. The reasons for these discrepancies should be understood. Some techniques measure not work (as defined in classical mechanics) but worklike quantities that differ from the characteristics used in classical mechanics.

The goal of the chapter is to discuss the biomechanical background of the problem rather than the mathematical techniques or biological complexities in solving it. The main model on which the discussion is based is similar to the basic model used in previous chapters, the model with rigid links and hinge joints served by torque actuators. However, much more attention than before is paid to the muscle actuators, in particular, muscles crossing either one or two joints. The majority of examples and formulas are limited to planar movements. For three-dimensional models, mathematical description would be much more complicated without changing the conclusions. "Purely" biological aspects of the problem—such as negative muscle work, storage and recoil of elastic energy, and so on—will be covered in a subsequent volume of the series.

When people move, they do work on the environment; they also do work on their own bodies, changing the kinetic and potential energy of the body

••• MECHANICS REFRESHER •••

Work of a Force

Consider a particle on which a force **F** acts (figure 6.1). The particle undergoes an infinitesimal displacement *d***r**. The *work* (or *elementary work*) done by **F** in this displacement is defined as the scalar product of the vectors **F** and *d***r**

$$dW = \mathbf{F} \cdot d\mathbf{r} \tag{6.1a}$$

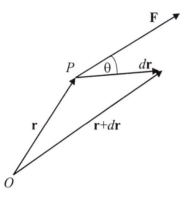

Figure 6.1   The work of force **F** corresponding to the displacement *d***r**.

The work can also be written as

$$dW = Fdr\cos\theta \tag{6.1b}$$

where *F* and *dr* are the magnitudes of the force and displacement and θ is the angle formed by the vectors **F** and *d***r**. The expression $F\cos\theta$ represents the component of the force in the direction of displacement. Thus, the work of a force equals the product of the displacement of its point of action and the force component in the direction of the displacement. The work is positive if the angle θ is acute, and it is negative when the angle is obtuse.

The work of a force in a succession of small displacements equals the work done by the force in the resultant displacement. The work during a finite displacement from P_1 to P_2 is obtained by integrating equation 6.1 along the path traveled by the point of force application

$$W\Big|_{P_1}^{P_2} = \int_{P_1}^{P_2} \mathbf{F} d\mathbf{r} \tag{6.2}$$

The work of a couple that produces a free moment **M** is

$$dW = Md\theta \qquad (6.3)$$

where M is the magnitude of the moment and $d\theta$ is the small angle through which the body rotates. The work is positive when M and $d\theta$ are in the same direction.

Power is the amount of work done per unit of time:

$$P = \frac{dW}{dt} = \frac{\mathbf{F} \cdot d\mathbf{r}}{dt} = \mathbf{F} \cdot \mathbf{v} \qquad (6.4)$$

where **v** is the velocity of the particle.

The SI unit of work is the *joule* (J), which is obtained by multiplying the unit of force (newton, N) by the unit of length (meter, m), J = N · m. The unit of power is the watt (W), which is 1 J/s or 1 (N · m)/s.

---

segments. The work or power done on external objects is called *external work* or external power and the work or power done on the body links (to change their *total mechanical energy*) is named *internal work* or internal power. For instance, in cycling, the power of the force applied to the pedals is external power, and the power to move the legs themselves is internal power. In routine functional testing with a bicycle ergometer, when the mechanical power is determined as the product of the moment of force applied to the pedals and the angular velocity of the pedaling, external power is registered, whereas in treadmill tests, internal power is measured.

This chapter concentrates mainly on the determination of internal work and power. Unless mentioned otherwise, the work done by the performer on the environment and the work done by the environment on the performer are not addressed. The chapter starts with a discussion of the concept of work and power in section **6.1**. Although the main definitions are simple and straightforward, their applications are not obvious. In biomechanics, mechanical work is computed either as the work of a force (section **6.1.1**) or as the work done on the body (section **6.1.2**). In section **6.2**, we begin to apply the idea of work to human movement. We look at the work done at a single joint and then at the work done on a single body segment. In section **6.3**, work in multilink kinematic chains is discussed. The energy balance equations are introduced. Then we examine mechanical energy expenditure and energy compensation. Several models of energy expenditure are discussed. In section **6.4**, we consider how the energy for human movement can be minimized. First, we discuss the methods for quantifying energy expenditure, then we examine how energy efficiency is obtained in human movement.

■ ■ ■ *FROM THE LITERATURE* ■ ■ ■

External Power in Human Movement

Power in Cycling

Sources: Broker, J.P., C.R. Kyle, and E.R. Burke. 1999. Racing cyclist power requirements in the 4000-m individual and team pursuits. *Med. Sci. Sports Exerc.* 31 (11): 1677–85.

Bassett, D.R., Jr., C.R. Kyle, L. Passfield, J.P. Broker, and E.R. Burke. 1999. Comparing cycling world hour records, 1967–1996: Modeling with empirical data. *Med. Sci. Sports Exerc.* 31 (11): 1665–76.

The power of the force applied to the pedal was measured in the members of the U.S. cycling pursuit team. The power at 60 km/h (37 miles/h) averaged 607 W in lead position (100%), 430 W in second position (70.8%), 389 W in third position (64.1%), and 389 W in fourth position (64.0%). Thus, a team member requires about 75% of the energy necessary for a cyclist riding alone at the same speed: $(100 + 70.8 + 64.1 + 64.0)/4 \approx 75\%$. External power averaged about 520 W when the 4,000-m individual pursuit record of 4 min 11.114 s was set. To break the world hour record, an average power of 440 W will be required (at sea level).

Power Against Air Resistance

Sources: Margaria, R. 1976. *Biomechanics and energetics of muscular exercise.* Oxford: Clarendon Press.

Whitt, F.R., and D.G. Wilson. 1974. *Bicycling science.* Cambridge, MA: MIT Press.

Zatsiorsky, V.M., S.Y. Aleshinsky, and N.A. Yakunin. 1982. *Biomechanical basis of endurance* (in Russian). Moscow: FiS. Also published in German translation as *Biomechanische Grundlagen der Ausdauer.* Berlin: Sportverlag, 1987.

Air resistance is proportional to the velocity squared, $F_{air} = kv^2$, where v is the velocity of the performer with respect to the air (see equation 1.77). Hence, the work done against the air resistance is also proportional to the squared values of the velocity with respect to the air ($W_{air} = F_{air}d = kv^2d$, where d stands for the displacement). The power is proportional to the velocity cubed, $P_{air} = F_{air}v = kv^3$. Figure 6.2 illustrates the experimentally determined dependence of the power on the speed against air resistance.

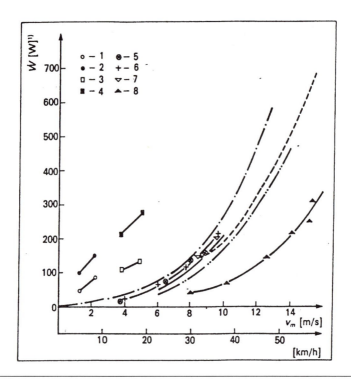

Figure 6.2 Dependence of power against air resistance on speed in several activities. (1) Treadmill walking, 10 m/s head wind; (2) treadmill walking, 14.1 m/s head wind; (3) treadmill running, the head wind 10 m/s; (4) treadmill running, 14.1 m/s head wind (Pugh 1971); (5) running (Luhtanen and Komi 1978); (6) running, 70-kg male subject (Margaria 1968); (7) running (Cavagna, Komarek, and Mazzoleni 1971); (8) speed skating (di Prampero et al. 1976); —— cycling, the racing position (Pugh 1971); – · – · – cycling, the racing position (Sharp 1898); – – – cycling, the racing position (Nonweiler 1956); – ·· – ·· – cycling, the racing position (Whitt and Wilson 1974).

Reprinted from V. Zatsiorsky, S. Aleshinsky, and N. Yakunin, 1982, *Biomechanical Basis of Endurance* (Moscow: FiS Publishers).

The power of the force expended against air resistance in running can be estimated from the empirical equation

$$P = 0.00035v^3 \tag{6.5}$$

where P is power in kilograms of force times meters per second (1 kg · m/s = 9.8 W) per kilogram of body mass, v is the velocity of the performer with respect to the air in the direction of running in meters per second (Margaria 1976). Equation 6.5 does not take into account such factors as the body size of the runner, the runner's attire, shielding, and altitude; it provides only an estimate of the power. In such sports as cycling and skating, where the athlete's speed exceeds 10 m/s, the power expended in overcoming the air resistance is large. For instance, in cycling at a speed of 10.2 m/s, the power expended in overcoming the air resistance amounts to 80% of the power of the forces exerted by the athlete on the pedals (Whitt and Wilson 1974).

The external power expended in overcoming air resistance is the power of the forces that act between the human body and the air, not the power of the force that is actively exerted by the performer. If a bicyclist suddenly stops pedaling, his or her body still delivers power to the environment for a certain period although the athlete is doing nothing but sitting on the bicycle. Air resistance is overcome at the expense of the kinetic energy of the body.

Work Done When Throwing Implements

Source: Bartonietz, K. 2000. Javelin throwing: An approach to performance development. In *The IOC encyclopaedia of sports medicine. Vol. 9, Biomechanics in sport: Performance enhancement and injury prevention,* ed. V.M. Zatsiorsky, 401–34. Oxford, UK: Blackwell Science.

A top javelin thrower was filmed during a competition and in training. For a 800-g javelin thrown with a velocity of 30 m/s, the ratio between the kinetic and potential energy of the implement at release was 30:1. During throws of a 7.26-kg shot with both arms, the ratio was 3.7:1. The work done on the javelin was 372 J (360 J for the change in kinetic energy plus 12 J for the change in potential energy). The work done on the shot was 662 J (522 J + 140 J). The duration of the delivery phase was shorter in javelin throwing (0.12 s) than in shot throwing (0.35 s). The average power of the javelin throws (3.10 kW) was significantly larger than that of the shot throws (1.90 kW). It was concluded that overhead shot throwing requires less power than javelin throwing and hence produces different training effects.

6.1 THE CONCEPT OF WORK

This section is intended to familiarize readers with some basic ideas from classical mechanics. The concept of mechanical work is based on the classical definition of the work of a force (equation 6.1). Section **6.1.1** discusses some consequences of this definition. An immediate consequence is that in order to calculate the work of a force, the force under consideration must be explicitly defined. Mechanical work cannot be determined until the force under consideration is specified. For instance, attempts to determine "mechanical work in walking" are confusing if the forces of interest are not identified. Various methods can be used to determine the work of different forces.

Experimental limitations, in particular the impossibility of directly measuring the involved forces, inspired another way of analyzing the mechanical work of human movement. Instead of the work of a force, the work done on the body segments is computed. The relationship between these two approaches is discussed in section **6.1.2**.

6.1.1 Work of a Force

Since power is just a time rate of doing work, many conclusions about work are also valid for power. For brevity's sake, instead of the expression "work and power," the word "work" or "power" is used. Hence, workless forces are also powerless forces.

6.1.1.1 Workless Forces

The definition of the work of a force (equation 6.1) implies that not all forces do work. A force does work only when its point of application undergoes a displacement in the direction of the force.

If the force is perpendicular to the displacement, no work is done. In particular, centripetal forces do no work. When the CoM of a body moves horizontally, the gravity force does no work. If a body is displaced both horizontally and vertically, the work done by the gravity force equals the weight multiplied by the vertical displacement of the CoM, whatever the horizontal displacement is. In speed skating, the athletes push off at a right angle to the gliding direction. Hence, the push-off force does no work (but the friction force between the blade and ice does). A similar situation occurs during a takeoff in ski jumping (figure 6.3). Note that when the force and displacement are orthogonal, the virtual coefficient (equation 1.41) is zero.

When the forces are applied at motionless points, the work is zero. No work is done during static efforts. For instance, when one holds a barbell stationary at arm's length, the task is tiring, but the work is zero. In mechanics, the expression "static work" is a misnomer. If there is no slippage between the foot and the supporting surface and the surface does not deform, the contact forces do no work. A jumper performing a takeoff or people walking upstairs exert contact forces on the ground, but the reaction forces, though large, do no work on the performer. The internal forces do. The simple mass–spring system presented in figure 6.4 illustrates this assertion. As an aside, during rolling contact

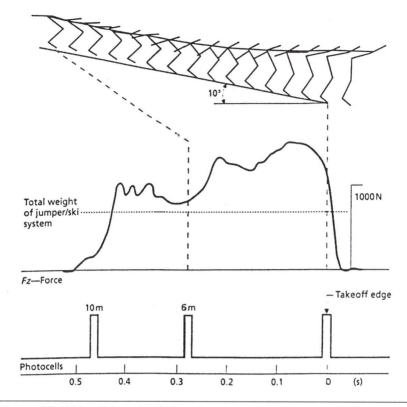

Figure 6.3 Takeoff force in ski jumping. The takeoffs in ski jumping are short in duration—from 0.25 to 0.30 s—during which the athletes cover a distance around 7 m (23 ft). Nonetheless, because the takeoff force is normal to the supporting surface, it does no work on the jumper-and-ski system. At the contact between the skis and the supporting surface, the work is done only by the shear force.

Reprinted, by permission, from P.V. Komi and M. Virmavirta, 2000, Determinants of successful ski-jumping performance. In *Biomechanics in Sport: Performance Enhancement and Injury Prevention. The IOC Encyclopedia of Sports Medicine*, edited by V.M. Zatsiorsky (Oxford, UK: Blackwell Science), 10:349–362.

without slipping (like seen in cycling or roller skating), the point of contact at any instant does not move; thus the work of the friction force acting on the rolling body is zero.

The displacement $d\mathbf{r}$ in equation 6.1a is the displacement of a particle or body on which a force $\mathbf{F}$ acts. If solely the point of force application—not the body itself—moves, the work is zero. For instance, if during a takeoff, the point of application of the GRF is displaced but the foot is not (no slipping occurs), the work of the GRF is zero.

Figure 6.4 A mass–spring system that models a takeoff. Internal force **F** does work but the external force GRF does not.

No work is done on a rigid body by the forces acting between the particles of the body. Two adjacent particles of a rigid body exert two equal and opposite forces, **F** and –**F**, on each other. If the body moves in translation, all the particles undergo equal displacement. If the body rotates, the displacement of the individual particles is different, but the components of the displacement along the line of force action ($r \cos\theta$, see equation 6.1*b*) are equal. Otherwise, the particles would not remain at the same distance. Therefore, these forces do work that is equal in magnitude and opposite in sign. Their sum is zero: the *positive work* of one force cancels the *negative work* of the other.

6.1.1.2 Invariance of Work

Work of a force is invariant; it does not depend on the system of coordinates. The same is valid for power. Obviously, the work is invariant when the reference frames differ by translation. Consider the case when the reference systems differ by rotation. The work of force **F** computed in the local reference system L is

$$dW_L = \mathbf{F}_L d\mathbf{r}_L \tag{6.6}$$

In the global system of coordinates, the work is

$$dW_G = \mathbf{F}_G d\mathbf{r}_G \tag{6.7}$$

The differentials of the coordinates satisfy the same transformation equation as the coordinates themselves, so

$$d\mathbf{r}_G = [R]d\mathbf{r}_L \tag{6.8}$$

where $[R]$ is a rotation matrix. Also, $\mathbf{F}_G = [R]\mathbf{F}_L$. Because matrices are involved, we rewrite equation 6.7 in matrix form:

$$dW_G = [F_G]^T[dr_G] \qquad (6.9)$$

The substitution yields

$$dW_G = [F_G]^T[dr_G] = [F_L]^T[R]^T[R][dr_L] = [F_L]^T[dr_L] \qquad (6.10)$$

It is thus seen that dW does not change under rotational transformation; in other words, the work is invariant.

••• MECHANICS REFRESHER •••

Conservative Forces

If the work of a force does not depend on the path followed by its point of application, the force is said to be a conservative force. Conservative forces depend only on the position of their point of application. For any conservative force $\mathbf{F}$, the following relationship is valid:

$$\mathbf{F} = F_x\mathbf{i} + F_y\mathbf{j} + F_z\mathbf{k} = -\left(\frac{\partial Q}{\partial X}\mathbf{i} + \frac{\partial Q}{\partial Y}\mathbf{j} + \frac{\partial Q}{\partial Z}\mathbf{k}\right) = -\nabla_v \qquad (6.11)$$

where $-\nabla_v$ is the gradient of the scalar potential function (potential energy) of the force $\mathbf{F}$. When the work of a conservative force $\mathbf{F}$ is positive, the potential energy decreases. In contrast, the potential energy increases when the work of $\mathbf{F}$ is negative. A system in which work is done only by conservative forces is called a *conservative system.*

Gravity force (weight) and elastic forces are conservative forces. Friction and hydrodynamic resistance are not conservative forces. The work done by these forces depends on the path: the longer the path, the larger the work.

6.1.1.3 The Law of Conservation of Energy and Conservative Systems

The *law of conservation of energy,* also known as the *first law of thermodynamics,* states that energy cannot be created or destroyed. Energy can only change its form (mechanical energy, thermal energy, radiation energy, etc.). The law is one of the main laws of nature. The impossibility of perpetual motion follows from this law. Change in the energy of a system equals the inflow or outflow of energy:

$$\left\{ \begin{array}{c} \text{Energy input to the system} \\ \text{(mechanical work, heat transfer,} \\ \text{chemical energy, etc.)} \end{array} \right\} = \left\{ \begin{array}{c} \text{Change in energy of a system} \\ \text{(kinetic energy, gravitational} \\ \text{potential energy, heat, etc.)} \end{array} \right\}$$

In a conservative system, the sum of the kinetic and potential energies remains constant amid all changes. An ideal pendulum serves as an example of a conservative system: the total mechanical energy of a swinging pendulum remains constant (figure 6.5). In such a system, the changes in kinetic energy and potential energy mirror each other, and the maximal values of the kinetic and potential energy are equal. In human movement, potential energy consists of gravitational and elastic energy.

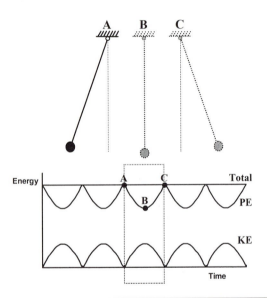

Figure 6.5 Energy changes during one swing of an ideal pendulum. The pendulum is released at point A and moves to point C. Kinetic energy of the pendulum is maximal at the bottom point B; the potential energy is minimal at this point. The *gravitational potential energy* is maximal at the highest positions of the pendulum, A and C. The sum of the kinetic and potential energy is constant. The maximal values of potential energy (measured with respect to point B) and kinetic energy are equal. The changes in the kinetic and potential energy mirror each other.

6.1.1.4 Energy Exchange During Positive and Negative Work of a Force

When two rigid bodies *A* and *B* interact, the forces of action and reaction produce equal amounts of work of opposite signs, positive and negative. The energy of the body on which the positive work is done, say, body *A*, increases,

■ ■ ■ *FROM THE LITERATURE* ■ ■ ■

Conservative Versus Nonconservative Models of Paraplegic Crutch Ambulation

Source: Rovick, J.S., and D.S. Childress. 1988. Pendular model of paraplegic swing-through crutch ambulation. *J. Rehabil. Res. Dev.* 25 (4): 1–16.

Crutch ambulation is energy demanding. It is not clear what necessitates the high metabolic energy input: the large amount of mechanical work performed or the poor physiological efficiency of the task. Swing-through crutch gait has two major phases separated by two transitional phases. The major phases are (1) the body-swing phase, in which the weight is supported on the crutches and the feet are off the ground, and (2) the body-stance phase, in which the body weight is entirely on the feet and the crutches are off the ground. In the transitional phases, both the feet and the crutches are on the ground and support is being transferred from one to the other. The body-swing phase is presumably a phase of large metabolic activity.

The authors modeled the body-swing movement by a three-link planar pendular model (figure 6.6*a*). The links were defined as (1) a rigid body composed of the arms and crutches, with a single pivot at the crutch tips and a single pivot at the estimated center of rotation of the glenohumeral joints; (2) a second rigid body composed of the head, neck, and trunk, sharing a common pivot with link 1 at the shoulders and having a second pivot at the estimated center of rotation of the hips; and (3) a third rigid body composed of the legs, feet, and orthoses, sharing a common pivot with link 2 at the hips.

The model was solved as both a conservative system with freely pivoting joints at the crutch tips, shoulders, and hips (figure 6.6*b*) and as a nonconservative force system (figure 6.6*c*). The difference between the conservative model and the real data was large (figure 6.6*b*). In the next phase of the modeling, the restriction to analysis as a fully conservative system was removed. Because no torque exists at the pivot of the crutch tips and paraplegic patients are unable to produce muscular torques at the hips, mechanical work can be done only at the shoulder joints. The nonconservative model corresponded well with observed kinematic data (figure 6.6*c*). It seems that the crutch users do not move in a purely pendular fashion. Mechanical work must be performed at the shoulder joints to ambulate.

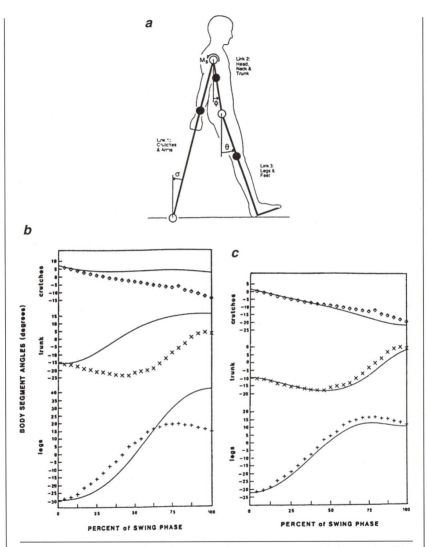

Figure 6.6 Crutch ambulation. (*a*) Schematic diagram of the model. The links are the crutches plus arms, the trunk, and the legs. (*b*) Solution of the conservative model. All pivots are frictionless and unrestricted. No mechanical work is performed at the shoulder joints. (*c*) Solution to the nonconservative model. Mechanical work is performed at the shoulder joints. Continuous curves are the modeled solutions, and points are the measured data. Plots begin at toe-off and end at heel strike.

Adapted from J.S. Rovick and D.S. Childress, 1988, "Pendular model of paraplegic swing-through crutch ambulation," *Journal of Rehabilitation Research and Development* 25 (4):1–6.

while the energy of the second body *B* decreases by the same amount. The total amount of energy in the *A* + *B* system does not change.

The force that does the positive or negative work should be explicitly described and understood. Consider a performer who is lifting and slowly lowering a heavy object. When the displacement is upward, the force of gravity and the displacement are in opposite directions. Therefore, the work of the gravity force is negative, and the potential energy of the body increases. The work of the force that is exerted on the load by the performer is positive since the force and displacement are in the same direction (the performer does the work on the object against the gravity force). In contrast, when the object is displaced downward, the work done by the weight of the object is positive. The potential energy of the object decreases. The work of the force exerted by the performer in this case is negative (the weight of the object does the work on the performer).

• • • *Mechanics Refresher* • • •

Work Done by Several Forces on a Rigid Body

The total work done by several forces on a rigid body equals the work of the resultant force and couple, $dW = \mathbf{F} \cdot d\mathbf{r} + \mathbf{M} \cdot d\boldsymbol{\theta}$, where $d\mathbf{r}$ is the infinitesimal displacement of the CoM of the body. It is also equal to the change in the body's kinetic energy (the *work–energy principle for a rigid body*).

6.1.2 Work Done on a Body

The total *work done on a body* by all the forces and couples is manifested in a change of the body's energy. This section discusses the conceptual differences between the work of a force and the work done on a body. To determine the work of a force, both the force and the displacement of its point of application must be known. This knowledge is not always available in biomechanics. Often, only the motion of body segments is known. This knowledge is sufficient to compute the mechanical energy of the segments and the work done on the segment but not the work of the actual forces.

Consider a rigid body upon which conservative and nonconservative forces are exerted. The portion of the work done by the conservative forces equals the difference in the body's potential energy at the end and at the beginning of the period studied. This work does not change the total energy of the body: when the potential energy of the body increases (decreases), the kinetic energy decreases (increases) by equal amount. On the contrary, the work of the nonconservative forces changes the total mechanical energy of the body. There-fore, the work–energy principle can be written as

$$\underbrace{K_{t_1} + Q_{t_1}}_{\substack{\text{Energy of the body} \\ \text{at the instant } t_1}} + W_{nc} = \underbrace{K_{t_2} + Q_{t_2}}_{\substack{\text{Energy of the body} \\ \text{at the instant } t_2}}$$

(6.12)

where K and Q are the kinetic and potential energy, correspondingly, and W_{nc} is the work done by the nonconservative forces.

• • • MECHANICS REFRESHER • • •

König's Theorem

König's theorem states that the total kinetic energy of a system of material particles equals the sum of the kinetic energy of the CoM (assuming that the entire mass is concentrated at the CoM) and the kinetic energy of the particles in their motion relative to the CoM.

The kinetic energy of a rigid body equals the sum of the energy associ-ated with the movement of the CoM (translational kinetic energy) and the energy associated with the body's rotation around the CoM.

For a rigid body in planar motion in the absence of elastic forces, equation 6.12 can be written as

$$W_{nc} = E_f - E_i \left(mgh_f + \frac{mv_f^2}{2} + \frac{I\dot{\theta}_f^2}{2} \right) - \left(mgh_i + \frac{mv_i^2}{2} + \frac{I\dot{\theta}_i^2}{2} \right)$$

(6.13)

where E stands for the total mechanical energy of the body (the sum of the potential and kinetic energy), m stands for mass, h is vertical location of the body's CoM, I is moment of inertia, v is the linear velocity of the body's CoM, $\dot{\theta}$ is the angular velocity, g is acceleration due to gravity, and subscripts f and i refer to the final and initial values, correspondingly. Equation 6.13 can be rewritten as

$$\underbrace{W_{nc} + mg(h_i - h_f)}_{\substack{\text{Total work done on the body} \\ \text{by all the nonconservative} \\ \text{and conservative forces}}} = \underbrace{\left(\frac{mv_f^2}{2} + \frac{I\dot{\theta}_f^2}{2} \right) - \left(\frac{mv_i^2}{2} + \frac{I\dot{\theta}_i^2}{2} \right)}_{\substack{\text{Change in the kinetic energy} \\ \text{of the body}}}$$

(6.14)

Equation 6.14 represents the work–energy principle for a rigid body in planar motion when elastic forces are neglected. The gravity term can be in-

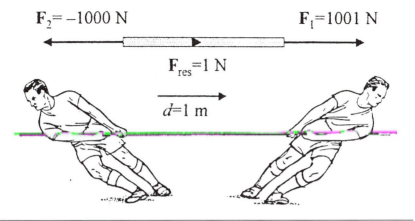

Figure 6.7   Forces in a tug-of-war. See explanation in the text.

cluded either in the energy-change side (as in equation 6.13) or in the work side of the equation (as in equation 6.14), but not in both. According to the work–energy principle, the body's initial kinetic energy plus the work done by all the forces exerted on the body is equal to the body's final kinetic energy. From equations 6.12 to 6.14, only the work done on the body (the net work) can be computed. This work is equal to the work of a resultant force and couple.

The work of the resultant (the work done on the body) can sharply differ from the work of individual forces exerted on the body. This happens when the forces act in opposite directions and their action is partly canceled. Consider the simple example in figure 6.7, a tug-of-war. Two forces, F_1 and $-F_2$ ($|F_1| > |F_2|$), act on the body in opposite directions along the same line.

The work of three forces can be considered: F_1, F_2, and their sum F_{res}. Assume that $F_1 = 1{,}001$ N, $F_2 = -1{,}000$ N, $F_{res} = 1$ N, and displacement d is 1 m. Correspondingly, the magnitudes of the work are $W_1 = 1{,}001$ N · m, $W_2 = -1{,}000$ N · m, and $W_{res} = 1$ N · m. The work W_{res}, as compared with W_1 and W_2, is negligibly small. The positive mechanical work W_1 was almost canceled by the negative work W_2. The computed values represent the work of different forces. W_1 and W_2 measure the work of actual forces F_1 and F_2. W_{res} is the work of the resultant (imaginary) force done on the body (a rope). If only this imaginary force were applied to a rigid body, its movement would be the same as under the common action of the actual forces. However, the performed work would be different. In particular, the work of actual forces can exceed many times the net work (as in the tug-of-war example). If only the work done on the body (the rope) is measured, the work of the actual forces remains unknown.

6.2 WORK AT A JOINT AND WORK ON A BODY SEGMENT

Representative papers: van Ingen Schenau and Cavanagh (1990); Zatsiorsky and Gregor (2000)

We now proceed to consider the work performed in human movement. This section discusses the work performed at a single joint (section **6.2.2**) and the work done on a single body segment (section **6.2.3**). Section **6.2.1** specifies the topic of the discourse.

••• PHYSIOLOGY REFRESHER •••

Muscle Action

The term *muscle action* refers to the production of muscle force. Depending on the change in the muscle length, muscle actions are classified as *isometric,* without change in the muscle length; *concentric,* when muscle shortens; or *eccentric,* when the muscle lengthens while producing a force. During an eccentric muscle action, the muscle tension is overpowered by an external force.

6.2.1 From Muscle Power to Joint Power

In a lifeless world, static forces are usually exerted without expending energy. The weight of a book acts on a desk without energy expenditure. In contrast, muscles require energy to generate force. They also require energy to do work. (In an inanimate world, *some* forces also require energy. For example, to activate an electromagnet and generate force, electric energy must be spent.) An active skeletal muscle spends more energy than it does at rest, regardless of whether the muscle acts isometrically (the work is zero), concentrically (the muscle does positive work), or eccentrically (an external force does work on the muscle). This fact considerably changes the determination of mechanical work and energy in human movement. In classical mechanics, energy represents the capacity to do work. In biomechanics, energy represents the capacity to do work and exert muscle forces. We concentrate, however, only on a purely mechanical approach, that is, on the relationship between mechanical work and mechanical energy expenditure.

Mechanical energy from individual muscles is the primary source of human movements. If we were able to directly measure the forces and velocities of shortening (or lengthening) of the skeletal muscle–tendon complexes in a body, we would be able to determine the *muscle power.* However, we currently cannot do this, except for several selected muscles and under unique conditions. Mechanical power in human movements is, therefore, *estimated* indirectly by measuring the external forces acting on the body and the movement performed. This approach does not take into account the energy expended

1. to overcome antagonistic activity of the muscles,

2. to displace the muscles relative to the skeleton,

3. against elastic and nonelastic internal forces (e.g., ligament extension and friction, displacement of internal organs relative to the spine), and

4. to activate muscles and to maintain isometric force.

Only the fraction of the muscle power generated in joints and expended to overcome external resistance and to change the mechanical energy of body segments is computed. It is known as the *joint power.* Joint power is the power of the joint torques. Indirectly, joint power is also manifested as the power of joint forces (in other joints). In general, joint power is not expected to equal muscle power. To calculate the joint power, a system of actual muscle forces is replaced by a resultant system of joint torques and forces.

Ideally, mechanical work and power in human movement should be measured as muscle work and power. However, the limited experimental capabilities of contemporary biomechanics do not allow that. This analysis is customarily performed at the basic level of determining the work and power of the joint torques. The work of other forces, for instance, the work done on body

---

■ ■ ■ *FROM THE LITERATURE* ■ ■ ■

Muscle Power Versus Joint Power

Source: Neptune, R.R., and A.J. van den Bogert. 1998. Standard mechanical energy analyses do not correlate with muscle work in cycling. *J. Biomech.* 31 (3): 239–45.

The authors computed the work of muscle forces during cycling by solving the direct problem of dynamics and compared it with the experimentally determined work of joint torques and work done on the body links. The work of joint torques and the work done on the body segments underestimated the mechanical energy expenditure of the muscles, with differences ranging from 5% to 40%. The differences were attributed mainly to the inability of the joint torques to account for co-contractions of antagonistic muscles.

segments, is compared with this benchmark. The fact that the joint torques are resultants of the moments of force produced by individual muscles is disregarded in this approach.

6.2.2 The Work at a Joint

Both the joint torque and joint force do the work at the joint. The power and work of the joint torque are called the joint power and *joint work*.

6.2.2.1 Joint Power and Joint Work

In three dimensions, joint power is computed as the scalar product of the vectors of joint torque ($\mathbf{T}$) and relative angular velocity ($\dot{\boldsymbol{\alpha}}$) at a joint, both in the global reference system:

$$P \text{ (watts)} = \mathbf{T} \cdot (\dot{\boldsymbol{\theta}}_d - \dot{\boldsymbol{\theta}}_p) = \mathbf{T} \cdot \dot{\boldsymbol{\alpha}} \tag{6.15}$$

where $\dot{\boldsymbol{\theta}}$ is the vector of the angular velocity of the body link rotation and subscripts d and p refer to the distal and proximal links adjacent to the joint, correspondingly (figure 6.8). In two dimensions, instead of vector equation 6.13, a scalar equation can be used:

$$P\text{(watts)} = T \cdot (\dot{\theta}_d - \dot{\theta}_p) = T \cdot \dot{\alpha} \tag{6.16}$$

The joint work during the period from t_1 to t_2 is computed as the time integral of the joint power:

$$W\Big|_{t_1}^{t_2} = \int_{t_1}^{t_2} T\dot{\theta}_d \, dt - \int_{t_1}^{t_2} T\dot{\theta}_p \, dt = \int_{t_1}^{t_2} T\left(\dot{\theta}_d - \dot{\theta}_p\right) dt = \int_{t_1}^{t_2} T\dot{\alpha} \, dt \tag{6.17}$$

The larger the joint torque and the larger the difference between the angular velocities of the segments (the joint angular velocity), the larger the joint power.

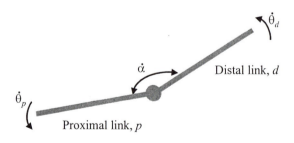

Figure 6.8 A two-link system with a hinge joint.

The joint power is positive when the joint torque and the joint angular velocity are in the same direction ($T\dot{\alpha} > 0$ or $-T \cdot -\dot{\alpha} > 0$). The joint power is negative when the joint torque and joint angular velocity are in opposite directions ($-T\dot{\alpha} < 0$ or $T \cdot -\dot{\alpha} < 0$).

Consider a case in which only one-joint muscles cross the joint. If the joint power is positive, the muscles produce positive work and the mechanical energy of the system increases (the energy goes from the muscles to the body segments). When the joint power is negative, the total mechanical energy of the adjacent segments decreases (the energy goes from the body segments to the muscle-tendon complexes), and the energy is used to forcibly stretch the muscles and tendons. This energy in part is dissipated as heat and in part is stored as the potential energy of elastic deformation. The stored energy is recoiled in subsequent phases that immediately follow the eccentric muscle action. During these periods, the work is performed in part by the conservative forces. The amount of stored elastic energy cannot exceed the amount of negative work done at the joint. Hence, the negative work at the joint provides an upper limit for stored potential energy.

The expression "joint torque" is just an abbreviated designation for the two equal and opposite moments of force acting at the joint on the adjacent segments. From equations 6.15 to 6.17, it is also possible to compute the power and work of each of the moments that changes the energy of the adjacent body segments:

$$P_d = T\dot{\theta}_d, \qquad P_p = -T\dot{\theta}_p$$
$$W_d = \int T\dot{\theta}_d dt, \quad W_p = -\int T\dot{\theta}_p dt \tag{6.18}$$

■ ■ ■ *From the Literature* ■ ■ ■

Joint Power in Hemiplegic Cerebral Palsy Gait

Source: Olney, S.J., H.E. MacPhail, D.M. Hedden, and W.F. Boyce. 1990. Work and power in hemiplegic cerebral palsy gait. *Phys. Ther.* 70 (7): 431–38.

The power generated at the joints of the affected limb during walking was studied in 10 children with spastic hemiplegia secondary to cerebral palsy. The results revealed that the ankle plantar flexors generated much less power in the patients than in normal subjects (figure 6.9). Primary, but partial, compensations are provided by the hip extensors during the stance phase and by the hip flexors at pull-off.

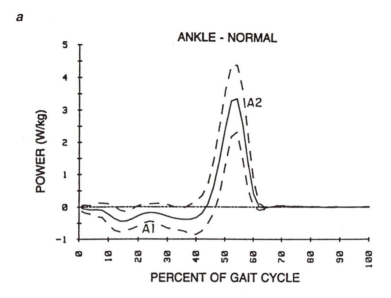

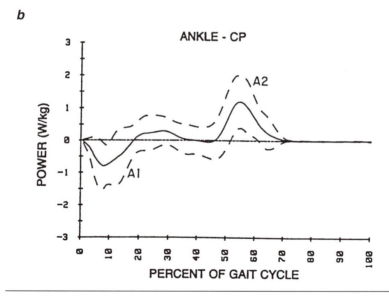

Figure 6.9 Average power patterns of the ankle for (*a*) healthy subjects and (*b*) subjects with cerebral palsy (means and 1 SD). A1 = negative power of ankle plantar flexors; A2 = positive power at push-off.

Adapted, by permission, from Olney/S. et al., Work and power in hemiplegic cerebral palsy gait, *Physical Therapy*, 1990, 70, 434, with permission of the American Physical Therapy Association.

■ ■ ■ *FROM THE LITERATURE* ■ ■ ■

Can an Ankle Joint Be Replaced by a Torsion Spring?

Source: Czerniecki, J.M., A. Gitter, and C. Munro. 1991. Joint moment and muscle power output characteristics of below knee amputees during running: The influence of energy storing prosthetic feet. *J. Biomech.* 24 (1): 63–75.

Joint power outputs were determined in five normal and five below-knee amputee subjects running at 2.8 m/s. The amputees were studied sequentially on three different prosthetic feet: the SACH (solid ankle, cushion heel) foot and two energy-storing feet. In four of the amputees, the energy-storing prosthetic feet generated two to three times more energy than the SACH foot. One of the amputees, however, exhibited increased abnormalities with the energy-storing prosthetic foot. The amount of energy restored relative to the amount of energy absorbed was greater with the energy-storing feet than the SACH foot. In healthy subjects, the ankle joint generates two to three times more energy than it absorbs. Therefore, a torsion spring at the ankle is not able to replace a functional ankle joint.

If the links are rotating in the same direction ($\dot{\theta}_d$ and $\dot{\theta}_p$ are of the same sign) the power (for brevity's sake, instead of the expression "work and power," the word "work" or "power" is used) delivered by a joint torque to the adjacent links is positive for one link and negative for the other; the joint torque increases the energy of one adjacent link and decreases the energy of the second link. If the adjacent links are rotating in opposite directions, the power of the joint torque acting on both has the same sign, positive or negative for both the links.

6.2.2.2 Energy Transfer Between Adjacent Segments

Representative papers: Robertson and Winter (1980); Aleshinsky (1986e)

Joint torques can transfer energy between body segments. The following example illustrates the *energy transfer*. From the position shown in figure 6.10*a*, the subject performs a fast hip extension, "unjacking" the body. Initially, the legs possess a certain amount of kinetic energy; the kinetic energy of the upper part of the body is zero. At an intermediate position (figure 6.10*b*), the performer activates the hip flexors, producing a flexion torque at the hip joints. Due to this torque, the legs are checked in their relative movement with respect

Figure 6.10 Energy transfer in human movement. The kinetic energy of the legs is transferred to the upper part of the body. See explanation in the text.

to the trunk, and the trunk starts rotating. The kinetic energy of the legs is transferred to the upper part of the body, and thereafter the entire body rotates. The legs lost the kinetic energy, and the upper part of the body acquired the energy. The total energy of the system did not change; it was only redistributed between the body segments. Such a transfer of energy among body segments is a common event in human movement.

Consider again the two-link chain with a hinge joint that we discussed previously (figure 6.8). The joint is served by single-joint muscles. Apart from a trivial case (when only one segment is rotating and hence the second segment neither gives off nor acquires energy), two distinct situations are possible: the adjacent segments can rotate either in opposite directions or in the same direction. When the adjacent links rotate in opposite directions, energy is not transferred between the links. When the links rotate in the same direction, transfer of energy occurs.

When the adjacent segments rotate in the same direction with *equal* angular velocities ($\dot{\theta}_d = \dot{\theta}_p$), the joint angular velocity is zero. Hence, the joint power is also zero. The absolute values of the power of the moments of force acting on the adjacent segments are equal, $\left|T\dot{\theta}_d\right| = \left|-T\dot{\theta}_p\right|$. If $T \neq 0$, one segment loses energy, and the second gains energy in the same amount. A "pure" energy transfer from one segment to the adjacent segment occurs. Energy is transferred between the segments while the total energy of the human body is conserved. The muscles that cross the joint act isometrically and do no work. They behave like nonextensible struts. This action of the muscles is known as their *tendon action* (or, more accurately, *ligamentous action*). This situation is presented in figure 6.10*b* and *c*.

When the adjacent segments rotate in the same direction with *unequal* angular velocities ($\dot{\theta}_d$ is not equal to $\dot{\theta}_p$), the joint angular velocity is $\dot{\alpha} = \dot{\theta}_d - \dot{\theta}_p$. The power flow to and from the individual segments is $P_d = T\dot{\theta}_d$ and $P_p = -T\dot{\theta}_p$. One segment gives energy to another segment; the second segment acquires the energy. Because $\dot{\theta}_d \neq \dot{\theta}_p$, the energy gain of one segment and the energy loss of another segment are dissimilar. The difference equals the joint power, $P = T \cdot (\dot{\theta}_d - \dot{\theta}_p)$ or $P = T \cdot \dot{\alpha}$ (equation 6.16). Two flows of energy exist: (a) from segment to segment and (b) between the segment or segments and the joint actuators (the muscles and tendons). The joint actuators transfer the energy from one segment to another, and in addition, they either provide mechanical energy to the system (when the joint power is positive) or absorb the energy (when the joint power is negative). The entire situation is summarized in table 6.1 and figure 6.11.

Table 6.1 Mechanical Energy Generation, Absorption, and Transfer

Segment rotation	Muscle action	Angular velocities	Muscle function	Amount and direction of power flow		Joint torque T (muscles)
				Segment 1	Segment 2	
In opposite directions, joint angle decreasing	Concentric	1 / 2	Mechanical energy generation	Receives energy generated by T at the rate $T\dot{\theta}_1$	Receives energy generated by T at the rate $T\dot{\theta}_1$	Generates energy at the rate $T(\dot{\theta}_1 + \dot{\theta}_2)$
In opposite directions, joint angle increasing	Eccentric	1 / 2	Mechanical energy absorption	Gives off energy at the rate $-T\dot{\theta}_1$	Gives off energy at the rate $-T\dot{\theta}_2$	Absorbs energy at the rate $-T(\dot{\theta}_1 + \dot{\theta}_2)$
In same direction, joint angle decreasing ($\dot{\theta}_1 > \dot{\theta}_2$)	Concentric	1 / 2	Mechanical energy generation and transfer	Receives energy at the rate $T\dot{\theta}_1$	Gives off energy at the rate $-T\dot{\theta}_2$	Generates energy at the rate $T(\dot{\theta}_1 - \dot{\theta}_2)$ and transfers the energy from segment 2 to segment 1 at the rate $T\dot{\theta}_2$

(continued)

Table 6.1 *(continued)*

Segment rotation	Muscle action	Angular velocities	Muscle function	Amount and direction of power flow		
				Segment 1	Segment 2	Joint torque T (muscles)
In same direction, joint angle increasing ($\dot\theta_1 < \dot\theta_2$)	Eccentric	1 ／ 2 ↙	Mechanical energy absorption and transfer	Receives energy at the rate $T\dot\theta_1$	Gives off energy at the rate $-T\dot\theta_2$	Absorbs energy at the rate $-T(\dot\theta_1 - \dot\theta_2)$ and transfers the energy from segment 2 to segment 1 at the rate $-T\dot\theta_1$*
In same direction, joint angle constant ($\dot\theta_1 = \dot\theta_2$)	Isometric	1 ／ 2 ↙	Mechanical energy transfer	Receives energy at the rate $T\dot\theta_1$	Gives off energy at the rate $-T\dot\theta_2$	Transfers the energy from segment 2 to segment 1 at the rate $T\dot\theta_2$

The joint torque T is a flexion torque; the angular velocities of the segments 1 and 2 are $\dot\theta_1$ and $\dot\theta_2$, respectively. A thicker arrow signifies a larger velocity.

*Segment 2 gives off energy at the rate $-T\dot\theta_2$, which is partly absorbed by the joint structures at the rate $-T(\dot\theta_2 - \dot\theta_1)$ and partly transported to segment 1 at the rate $-T\dot\theta_1$. Hence, $-T\dot\theta_2 = -T(\dot\theta_2 - \dot\theta_1) + (-T\dot\theta_1)$.

Based on the idea suggested by Robertson and Winter (1980).

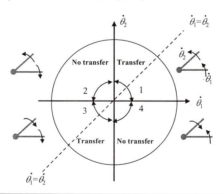

Figure 6.11   Regions of energy transfer for a two-link system. The muscles transfer energy when the angular velocities of the links are in the same directions, in quadrants 1 and 3. There is no energy transfer in quadrants 2 and 4, where the links are rotating in opposite directions. Link i acquires energy when $T\dot\theta_i > 0$, and it gives off energy when $T\dot\theta_i < 0$.

6.2.2.3 Energy Transfer Between Nonadjacent Segments

Representative papers: Elftman (1940); Morrison (1970); Prilutsky and Zatsiorsky (1994)

When muscles cross two joints, energy transfer may occur between nonadjacent segments. This happens when the segments to which a two-joint muscle is attached rotate in the same direction. In this case, the joint power in one of the joints is negative, while the power in the second joint is positive. To determine the energy production, absorption, and transfer, joint power in both joints must be known. At a certain combination of angular velocities, the length of a two-joint muscle may remain constant. In this case, the muscle acts isometrically and does no work. This behavior is known as the *tendon action of two-joint muscles*. In this particular case, the amount of positive power generated by the muscle in one joint equals the amount of negative power absorbed by the same muscle in the second joint. As an example, consider a takeoff in a standing vertical jump. During the takeoff, the monoarticular hip extensor muscles contract and the hip extension occurs. If the two-joint rectus femoris acts isometrically, it transfers the mechanical energy generated by the hip extensors from the hip to the knee joint. Consequently, the knee joint torque does work, albeit the knee joint muscles do not do any work. The tendon action of two-joint muscles has been known since the 19th century (Cleland 1867).

■ ■ ■ *From the Literature* ■ ■ ■

Large Joint Power at the Ankle During Takeoffs Is Due to Energy Transfer From Proximal Joints

Source: Gregoire, L., H.E. Veeger, P.A. Huijing, and G.J. van Ingen Schenau. 1984. Role of mono- and biarticular muscles in explosive movements. *Int. J. Sports Med.* 5 (6): 301–5.

Subjects performed standing vertical jumps. In the second part of the push-off, a high power output of 3,000 to 4,000 W was observed in the ankle joints during plantar flexion. Such high power is attributed to a sequential energy flow from hip to knee and ankle joints. Power delivered by the extensors of the hip and knee joints was transported distally via the biarticular muscles (gastrocnemius) to the ankle joints. This power transport is important in the execution of explosive movements.

To estimate the amount of energy transferred by two-joint muscles between joints, $P_j(t)$, let us consider the equality

$$P_j(t) = P_j^T(t) - \Sigma_j P^M(t) \qquad (6.19)$$

where $P_j^T(t)$ is the joint power at joint j (the power of the joint torque) and $\Sigma_j P^M(t)$ is the sum of the powers developed by all the muscles serving the jth joint. If all the muscles crossing the joint are one-joint muscles, the difference $P_j(t)$ equals zero. However, if there are two-joint muscles among the muscles serving the jth joint, the difference $P_j(t)$ is not necessarily equal to zero. Hence, the difference $P_j(t)$ represents the rate at which the energy is transferred to or from a given joint through two-joint muscles. Table 6.2 explains the five main ways that two-joint muscles transfer energy.

Table 6.2   Mechanical Energy Transfer by Two-Joint Muscles

	Joint power, $P_j^T(t)$	Summed muscle power, $\Sigma_j P^M(t)$	Difference $P_j(t)$	Energy transfer Rate	Energy transfer Direction	Muscles of the jth joint generate or absorb power at a rate of
1	> 0	≥ 0	> 0	$\lvert P_j(t)\rvert$	To joint j from adjacent joint(s)	$\lvert P_j^T(t)\rvert - \lvert P_j(t)\rvert$, generated
2	≥ 0	> 0	< 0	$-\lvert P_j(t)\rvert$	From joint j to adjacent joint(s)	$\lvert P_j^T(t)\rvert - \lvert P_j(t)\rvert$, generated
3	≤ 0	< 0	> 0	$\lvert P_j(t)\rvert$	From joint j to adjacent joint(s)	$-[\lvert P_j^T(t)\rvert + \lvert P_j(t)\rvert]$, absorbed
4	< 0	≤ 0	< 0	$-\lvert P_j(t)\rvert$	From joint j to adjacent joint(s)	$-[\lvert P_j^T(t)\rvert - \lvert P_j(t)\rvert]$, absorbed
5	≥ 0 < 0	≥ 0 < 0	$= 0$	(a) $\lvert P_j(t)\rvert = 0$ (b) $\lvert P_j(t)\rvert$ and $-\lvert P_j(t)\rvert$	(a) No energy transfer or (b) to joint j from one adjacent joint and from joint j to the other adjacent joint	$\lvert P_j^T(t)\rvert$, generated, or $-\lvert P_j^T(t)\rvert$, absorbed

Adapted by permission from Prilutsky, B.I., and V.M. Zatsiorsky. 1994. Tendon action of two-joint muscles: Transfer of mechanical energy between joints during jumping, landing, and running. *J. Biomech.* 27 (1): 25-34.

In summary, joint torques can generate, absorb, and transfer mechanical energy.

6.2.2.4 Work of a Joint Force

When a joint center—which is a point of application of the joint force—moves, the joint force does work on the adjacent links. The power P_j of the joint force $\mathbf{F}_j$ equals

$$P_j = \mathbf{F}_j \mathbf{v}_j \tag{6.20}$$

where $\mathbf{v}_j$ is the velocity of the joint center. The power is positive if the angle between the vectors of the joint force and velocity is acute. Otherwise, the power is negative.

■ ■ ■ *From the Literature* ■ ■ ■

The Tendon Action of Two-Joint Muscles

Source: Prilutsky, B.I., and V.M. Zatsiorsky. 1994. Tendon action of two-joint muscles: Transfer of mechanical energy between joints during jumping, landing, and running. *J. Biomech.* 27 (1): 25–34.

Subjects were filmed while performing various jumping and running tasks. The ground reaction forces were registered. The joint torques were computed by solving the inverse problem of dynamics. A two-dimensional musculoskeletal model of a leg consisting of four links (foot, shank, thigh, and pelvis) was developed. The model included eight muscles (tibialis anterior, soleus, gastrocnemius, hamstrings, vastus femoris, rectus femoris, iliacus, and gluteus maximus). To estimate the muscle forces, some simplifying assumptions were made. In particular, it was assumed that none of the one-joint antagonists were active. Under these assumptions, the transfer of mechanical energy by two-joint muscles between joints of the lower extremities was estimated. The computations were based on equation 6.19. Figure 6.12*a* gives the experimental results. During a squat vertical jump, the joint power at the hip is smaller than the power developed by the muscles serving the hip joint. Hence, the difference $P_j(t)$ is negative. This means that the power is transferred *from* the hip joint *to* other joints. The situation corresponds to case 2 (see row 2) in table 6.2. For the knee joint, the difference $P_j(t)$ is positive; hence, the power is transferred *to* the knee (case 1 in table 6.2, see row 1). In general, during the push-off phases, power is transferred from the proximal joints to the distal, while during the shock-absorbing phases, power flows from the distal to the proximal joints (figure 6.12*b*).

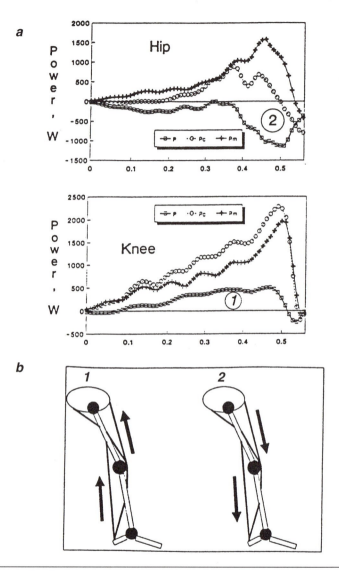

Figure 6.12 Transfer of energy by two-joint muscles. (*a*) Power developed by the joint torque (P_c), total power developed by the muscles serving the joint (P_m), and the difference between them (P) during a squat vertical jump. The numbers in circles denote the variant of energy transfer by two-joint muscles (table 6.2). (*b*) Transfer of mechanical energy by the two-joint muscles of the leg during running: (1) shock-absorbing phase and (2) push-off phase.

Adapted from *Journal of Biomechanics*, 27 (1), B.I. Prilutsky and V.M. Zatsiorsky, Tendon action of two-joint muscles: Transfer of mechanical energy between joints during jumping, landing and running, 25–34, 1994, with permission from Elsevier Science.

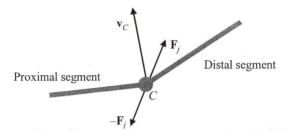

Figure 6.13 Joint resultant forces do work of equal magnitude and opposite sign on the adjacent segments. The power acquired by the distal segment is $P_d = \mathbf{F}_j \mathbf{v}_C$, and the power lost by the proximal segment is $P_p = -\mathbf{F}_j \mathbf{v}_C$.

Like "joint torque," the expression "joint force" designates two forces—action and reaction—acting between the two adjacent segments at a joint. Consider two body links with a permanent contact at a point C (figure 6.13). The links exert two equal and opposite forces $\mathbf{F}_j$ and $-\mathbf{F}_j$ on each other. While small displacements of the adjacent links are different, the displacement of the point of contact C is the same. Otherwise, either the links impinge on each other or the contact is broken. Therefore, the power of $\mathbf{F}_j$ is equal in magnitude and opposite in sign to the power of $-\mathbf{F}_j$, and their sum is zero. As the result, the mechanical energy of one of the adjacent segments increases, and the mechanical energy of the second segment decreases by an equal amount, but the total mechanical energy of the entire system does not change. The larger the

■ ■ ■ **FROM THE LITERATURE** ■ ■ ■

Power Flow Due to Joint Forces

Source: Martin, P.E., and P.R. Cavanagh. 1990. Segment interactions within the swing leg during unloaded and loaded running. *J. Biomech.* 23 (6): 529–36.

Male distance runners were filmed as they performed treadmill running (3.35 m/s) under different conditions. During the swing phase of the running cycle, the energy flow into and out of the leg segments was dominated by energy transfers between adjacent segments attributable to the joint reaction forces, which redistributed mechanical energy within the system. These contributions were considerably greater than those of the joint torques, which primarily reflected the generation and dissipation of mechanical energy by the muscles (figure 6.14).

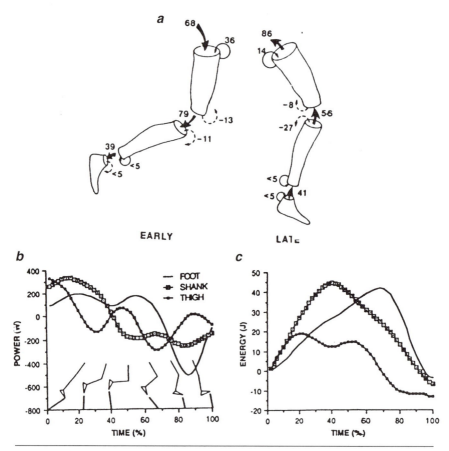

Figure 6.14 (*a*) Contribution of the reaction forces (bold arrows) and joint torques (small arrows) to segment energy fluctuations (joules) during the swing phase. All data are expressed in joules. The energy transfer due to the joint reaction forces dominates the energy flux of the leg during the swing phase. The transfer of mechanical energy occurs distally during the early portion of the swing and proximally in the later stages of the swing. (*b*) The transfer of power to and from the segments. (*c*) The mechanical energy of the segments.

Reprinted from *Journal of Biomechanics*, 23 (6), P.E. Martin and P.R. Cavanagh, Segment interactions within the swing leg during unloaded and loaded running, 529–536, 1990, with permission from Elsevier Science.

joint force and velocity of the joint translation, the greater is the power flow, that is, the amount of energy transmitted per unit of time from one body segment to another. Joint forces redistribute mechanical energy among body segments without changing the total mechanical energy of the whole human body.

In summary, joint forces neither generate nor absorb mechanical energy; they only transfer it between adjacent segments.

6.2.3 Power and Work on a Body Segment

Representative papers: Aleshinsky (1986a); Zatsiorsky and Gregor (2000)

To distinguish the work done *by* the individual forces and moments from the work done *on* the body segments, the following terminology is used. Forces and moments (joint torques) acting on the system are called the *sources of mechanical energy* (sources for short). The kinetic energy of a link associated with the horizontal, vertical, and rotational motion and the link potential energy are called the *fractions of mechanical energy* (or merely fractions), E_i ($i =$ 1, 2, 3, 4). The equations expressing the relationship between the power of the sources acting on a body link and the time rate of the mechanical energy are called the *power equations*.

6.2.3.1 Computation of Power and Work

During movement, the mechanical energy of the body segments does not stay constant. For instance, during walking at a constant speed, the energy of the body parts and its individual fractions increase and decrease (figure 6.16), and during the delivery phase in javelin throwing, the energy of the javelin increases while the energy of the trunk and the legs decreases (figure 6.17).

■ ■ ■ *From the Literature* ■ ■ ■

The Power Equation for a Single Body Segment

Source: Koning, J.J. de, and G.J. van Ingen Schenau. 1994. On the estimation of mechanical power in endurance sports. *Sport Sci. Rev.* 3 (2): 34–54.

The following power equation for the shank was developed:

$$M_{a,2}\omega_s + \mathbf{F}_{a,2} \cdot \mathbf{v}_3 + \mathbf{F}_{air,2} \cdot \mathbf{v}_4 + \mathbf{F}_{k,1} \cdot \mathbf{v}_5 + M_{k,1}\omega_s = \frac{dE_s}{dt} \qquad (6.21)$$

where symbol M designates the joint moment (torque); ω designates the segment angular velocity; and the subscripts a, k, and s stand for the ankle, knee, and shank, correspondingly (see figure 6.15).

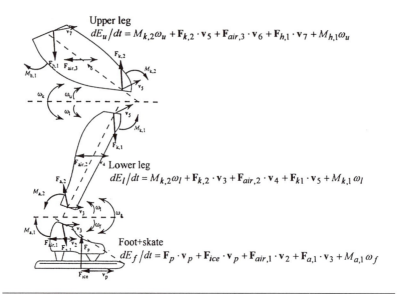

Upper leg

$$dE_u/dt = M_{k,2}\omega_u + \mathbf{F}_{k,2} \cdot \mathbf{v}_5 + \mathbf{F}_{air,3} \cdot \mathbf{v}_6 + \mathbf{F}_{h,1} \cdot \mathbf{v}_7 + M_{h,1}\omega_u$$

Lower leg

$$dE_l/dt = M_{k,2}\omega_l + \mathbf{F}_{k,2} \cdot \mathbf{v}_3 + \mathbf{F}_{air,2} \cdot \mathbf{v}_4 + \mathbf{F}_{kl} \cdot \mathbf{v}_5 + M_{k,1}\omega_l$$

Foot+skate

$$dE_f/dt = \mathbf{F}_p \cdot \mathbf{v}_p + \mathbf{F}_{ice} \cdot \mathbf{v}_p + \mathbf{F}_{air,1} \cdot \mathbf{v}_2 + \mathbf{F}_{a,1} \cdot \mathbf{v}_3 + M_{a,1}\omega_f$$

Figure 6.15 A model of a skater's leg. The sources of power for each link are the joint torques and forces.

Reprinted by permission from Koning, J.J. de, and G.J. van Ingen Schenau. 1994. On the estimation of mechanical power in endurance sports. *Sport Sci. Rev.* 3 (2): 34–54.

The time rate of change of the energy E of the segment is the measure of power P of the resultant force (A) and couple (B) acting on the segment:

$$P = \frac{dE}{dt} = \frac{d}{dt}\left(mgh + \frac{mv^2}{2} + \frac{I\dot{\theta}^2}{2}\right) = \underbrace{\frac{d}{dt}\left(mgh + \frac{mv^2}{2}\right)}_{A} + \underbrace{\frac{d}{dt}\left(\frac{I\dot{\theta}^2}{2}\right)}_{B} \quad (6.22)$$

If the power is positive, the mechanical energy of the segment increases. When the power is negative, the mechanical energy of the segment decreases: the energy either is transferred to other objects (body segments, the environment) or is transformed into another form (elastic energy or thermal energy).

In a planar case, translational kinetic energy can be written as a sum of the kinetic energies due to the components of the movement in the horizontal and vertical directions:

$$\frac{mv^2}{2} = \frac{mv_x^2}{2} + \frac{mv_y^2}{2} \quad (6.23)$$

Therefore, the mechanical energy of the body equals the sum of its potential energy and the fractions of the kinetic energy associated with the movement of the CoM in the horizontal and the vertical directions and with the body rota-

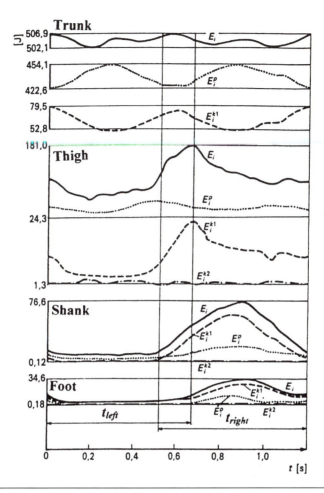

Figure 6.16 The time course of the fractions of the mechanical energy of body links during walking at 4.2 km/h. E_i is the total mechanical energy of the link, E_i^p is the potential energy, E_i^{k1} is the translational kinetic energy, and E_i^{k2} is the rotational kinetic energy, t_{left} and t_{right} are the stance periods.

Reprinted from V. Zatsiorsky, S. Aleshinsky, and N. Yakunin, 1982, *Biomechanical Basis of Endurance* (Moscow: FiS Publishers).

tion about its CoM. On the whole, the time rate of the mechanical energy of the body segment can be presented as the sum of the four fractions

$$P = \frac{dE}{dt} = \frac{d}{dt}(mgh) + \frac{d}{dt}\left(\frac{mv_x^2}{2}\right) + \frac{d}{dt}\left(\frac{mv_y^2}{2}\right) + \frac{d}{dt}\left(\frac{I\dot\theta^2}{2}\right) \qquad (6.24)$$

where $h = Y$ and $d(mgh)/dt = mg\dot{h} = mg\dot{Y}$.

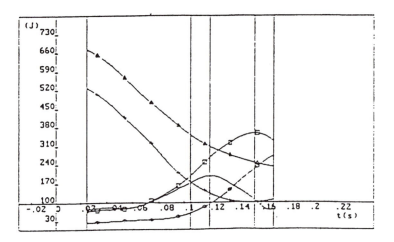

Figure 6.17 Kinetic energy of the body segments during a delivery phase in javelin throwing. The vertical lines mark five characteristic instants: the left foot contact (beginning of the delivery phase), maximal arched position of the body, maximal kinetic energy of the upper arm, maximal kinetic energy of the forearm, and javelin release. Triangles indicate the trunk; crosses, the lower legs; solid line, the upper arm; squares, the forearm; and circles, the javelin. The kinetic energy was computed from equation 5.48.

Reprinted, by permission, from E. Navarro et al., 1995, A kinetic energy model of human body applied to 3-D javelin throwing. In *XVth Congress of the International Society of Biomechanics, Book of Abstracts*, edited by K. Häkkinen et al. (Juväskylä, Finland), 668–669.

The relationship between the power of the joint moments and forces (sources) acting on a body segment and the time rate of the fractions of the mechanical energy is

$$\mathbf{F}_p \mathbf{v}_p - \mathbf{F}_d \mathbf{v}_d + T_p \dot{\theta} - T_d \dot{\theta} = \frac{d}{dt}(mgh) + \frac{d}{dt}\left(\frac{mv_x^2}{2}\right) + \frac{d}{dt}\left(\frac{mv_y^2}{2}\right) + \frac{d}{dt}\left(\frac{I\dot{\theta}^2}{2}\right) \quad (6.25)$$

where the subscripts p and d refer to the proximal and distal ends of the link, respectively. Equation 6.25 is known as the *energy balance equation* for a single link. Note that all sources of energy (joint forces and moments) are independent; each of them can either increase or decrease the energy of the link. For instance, a joint force can do positive work on a segment while the joint torque does negative work. The reader is invited to think of an example of such a situation. In general, computation of the power flow into and out of the segment is a straightforward operation that can always be performed.

To compute the work done on a body link over an interval from t_1 to t_2, equation 6.13 is used. Writing this equation in a slightly different form yields

$$W_{nc}\Big|_{t_1}^{t_2} = \underbrace{mg\left(h_f - h_i\right)}_{\substack{\text{Gravitational} \\ \text{potential energy} \\ \text{change}}} + \underbrace{\left\{\left(\frac{mv_f^2}{2} + \frac{I\dot{\theta}_f^2}{2}\right) - \left(\frac{mv_i^2}{2} + \frac{I\dot{\theta}_i^2}{2}\right)\right\}}_{\substack{\text{Change of the kinetic translational} \\ \text{and rotational energy}}} = \Delta E_p + \Delta E_K \tag{6.26}$$

where subscripts f and i refer to the final (at t_2) and initial (at t_1) values, correspondingly. Hence, the work done on the segment is represented as a sum of the changes in the potential energy (ΔE_p) and kinetic energy (ΔE_K) of the link, respectively. An example of a recording of such changes is presented in figure 6.18. To compute the work done on the segment during the period from t_1 to t_2, all the energy gains ($\Delta E > 0$) and losses ($\Delta E < 0$) during this period should be summed up.

Equation 6.26 works fine when the energy of the segment either only decreases or only increases. However, it is not practical when a rhythmic movement is performed. Consider level walking at a constant speed. Although energy is expended for the movement, the mechanical energy of the body does not change. Therefore, according to equation 6.26, the mechanical work *done on* the body segment for an arbitrary number of steps is zero. We discuss this fact later in the text.

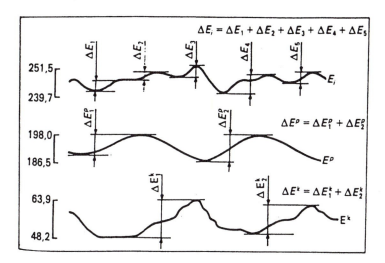

Figure 6.18 Kinetic, potential, and total mechanical energy of the upper part of the trunk during one stride of walking by a male subject (height 1.89 m, body weight 81.8 kg, speed 2.27 m/s). Only positive gains in energy are marked. To compute the work done on the segment, both the positive and negative changes in energy (gains and losses) should be added.

Adapted from V. Zatsiorsky, S. Aleshinsky, and N. Yakunin, 1982, *Biomechanical Basis of Endurance* (Moscow: FiS Publishers).

■ ■ ■ FROM THE LITERATURE ■ ■ ■

Work in Level Walking

Source: Webb, P., W.H. Saris, P.F. Schoffelen, G.J. van Ingen Schenau, and F. Ten Hoor. 1988. The work of walking: A calorimetric study. *Med. Sci. Sports Exerc.* 20 (4): 331–37.

Experiments were designed to test the traditional assumption that during level walking all of the metabolic energy from oxidation of fuel appears as heat, and no work on the environment is done. While wearing a suit calorimeter in a respiration chamber, five women and five men walked for 70 to 90 min on a level treadmill at 2.5, 4.6, and 6.7 km/h and pedaled a cycle ergometer for 70 to 90 min against 53-W and 92-W loads. They also walked with a weighted backpack and against a horizontal load. During cycling, metabolic energy expenditure matched heat loss plus the power measured by the ergometer. During walking, however, energy from fuel exceeded that which appeared as heat, meaning that work was done. Vertical and horizontal loads increased the fuel cost and heat loss of walking but did not alter the power output. The authors concluded that work is done during level walking to compress the heel of the shoe and to bend the sole.

6.2.3.2 Conservation of Energy in One-Link Motion: Pendulum-Like Motion

Representative papers: Fenn (1930a, 1930b); Zatsiorsky, Werner, and Kaimin (1994); Whittlesey, van Emmerik, and Hamill (2000)

The term "energy transformation" designates the conversion of mechanical energy (1) from translational kinetic into rotational kinetic form and back or (2) from kinetic into potential form and back. The first type of energy transformation is realized in a whiplike motion and the second in a pendular motion. Both transformations are examples of the *conservation of energy*.

The dynamics of whiplike motion were discussed in **5.4.2.3**. We consider it again briefly from the standpoint of energy transformation. Consider a joint force **F** exerted on a body link. The power delivered by the force is **Fv**, where **v** is the velocity of the joint center. Velocity **v** can be represented as the sum $\mathbf{v} = \mathbf{v}_{CM} + \mathbf{v}_{C/CM}$, where $\mathbf{v}_{CM}$ is the velocity of the center of mass and $\mathbf{v}_{C/CM}$ is the velocity of the joint center C relative to the center of mass. Hence, the power of force **F** is

$$\mathbf{Fv} = \mathbf{Fv}_{CM} + \mathbf{Fv}_{C/CM} \qquad (6.27)$$

The first term in the right side of the equation represents the time rate of the translational kinetic energy and the second term represents the time rate of the rotational kinetic energy. For a given **F** and **v**, an increase in one of these fractions means a decrease in the second fraction, and vice versa. Hence, the energy change associated with the power of the joint force is conserved.

Energy transformation from potential energy into kinetic energy and back is also quite common in human motion. The changes in the total energy of the segment shown in figure 6.18 were smaller than the sum of the changes in the kinetic and potential energy. This phenomenon habitually occurs in cyclic movements, such as walking and running. It indicates the conservation of energy. When a segment moves downward, a part of its gravitational potential energy is used to accelerate the segment and is transformed into kinetic energy. When the segment moves up, the kinetic energy of the segment can be used to lift the segment (to do work against gravity).

The pendular motion of body segments has long attracted the curiosity of researchers (English translation in Weber and Weber 1992), and its presence in human movement is sometimes taken for granted. However in each individual case, the existence of pendular motion must be tested biomechanically. As mentioned in **6.1.1.3**, for any ideal pendulum,

1. the total mechanical energy (the sum of the kinetic and potential energy) is constant over an oscillation cycle;
2. the changes in kinetic and potential energy of the body are exactly out of phase; and
3. the motion is performed due to the transformation of the potential energy into kinetic energy, and vice versa.

Consider from this standpoint leg movements during the swing phases of walking (figure 6.19). Visually, the leg movements resemble the movements of a compound pendulum. The legs are periodically raised and lowered under the influence of gravity. Therefore, the idea of pendulum-like leg movements during walking seems very natural. However, the peak values of leg kinetic energy during the swing phase greatly exceed the peak values of leg potential energy. In addition, the total mechanical energy of the leg during a stride is not constant. This is especially evident for the foot. During the support period, when the foot is on the ground, both its potential and kinetic energy fractions are at a minimum: The foot assumes the lowest possible position and does not translate. Thus, the leg during walking can barely be considered a free pendulum (a conservative system). In contrast, the total energy of the trunk at a certain speed fluctuates only slightly, and the changes in the trunk potential and kinetic energy are out of phase (figure 6.20). Hence, the trunk rather than the legs moves in such a way that energy is conserved.

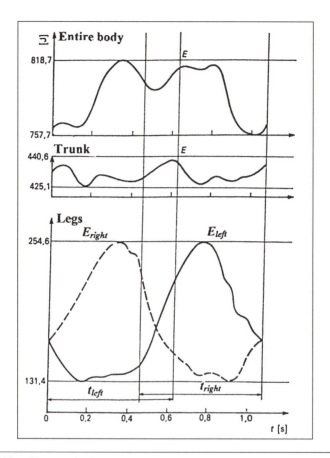

Figure 6.19 Changes in the total mechanical energy of the legs, trunk, and entire body in one cycle of walking at 1.2 m/s (4.3 km/h).

Reprinted from V. Zatsiorsky, S. Aleshinsky, and N. Yakunin, 1982, *Biomechanical Basis of Endurance* (Moscow: FiS Publishers).

The sum $(\Delta E_K + \Delta E_P)$ is known as the *quasi-mechanical work*. It represents the imaginary work that would be done on the segment by the resultant force and couple if there were no transformation between the gravitational potential energy and kinetic energy. Quasi-mechanical work is a pure theoretical construct. The difference between the sum of the gains of the kinetic and gravitational potential energy $(\Delta E_K + \Delta E_P)$ and the gain of the total energy of the segment, ΔE, can be used for estimating the magnitude of the conserved energy. It should be done with care: a positive difference $D = [(\Delta E_K + \Delta E_P) - \Delta E]$ is a necessary but not sufficient condition for the energy to transform from the potential into the kinetic form and vice versa. In other words, if the energy transformation takes place, then $D > 0$. However, if $D > 0$, we cannot claim

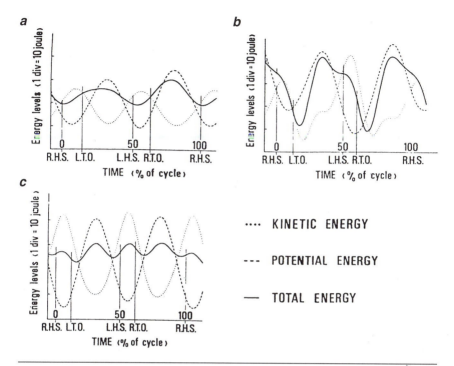

Figure 6.20   Energy levels of the upper part of the body during walking at (*a*) 2.9 km/h, (*b*) 8 km/h, and (*c*) 5 km/h. At 5 km/h, the kinetic and potential energies change synchronously in opposite directions. Therefore, the total energy remains almost constant. At 2.9 km/h and 8 km/h, such perfectly antiphase changes of the kinetic and potential energy do not occur; thus, the total energy of the trunk fluctuates substantially. R.H.S. and L.H.S. refer to the right and left heel strike, respectively, while R.T.O. and L.T.O. refer to the right and left takeoffs.

Reprinted, by permission, from A. Cappozzo et al., 1978, Movements and mechanical energy changes of the upper part of the human body during walking. In *Biomechanics VI-A*, edited by E. Asmussen and K. Jørgensen (Baltimore: University Park Press), 272–279.

that this is due to the energy transformation. For instance, if a person lowers and lifts a barbell, it is quite possible that $D > 0$; however, one cannot claim that the weight lifting was performed at the expense of the kinetic energy that the barbell had during its downward movement.

The *energy conservation coefficient* (CC) is defined as the ratio of D to the quasi-mechanical work:

$$CC = \frac{\left(\Delta E_K + \Delta E_P\right) - \Delta E}{\left(\Delta E_K + \Delta E_P\right)} \times 100\% \tag{6.28}$$

The coefficient is intended to characterize the proportion of the mechanical energy conserved in a given movement for the work of the resultant force and couple reduced to the CoM of the segment.

One of the ways to estimate the possibility of energy conservation during locomotion is to compare the maximal values of the kinetic and potential energy of a link during one cycle. If the values are different, the energy must be supplied by external sources. The kinetic energy increases as a quadratic function of the speed, while the changes in the potential energy with speed are less obvious. When the speed of a body link doubles, its kinetic energy quadruples, but potential energy of the link in a cycle does not fluctuate that much. We do not lift the legs much higher when we walk fast. Therefore, the opportunity for energy conservation due to the transformation of kinetic energy into potential energy and vice versa depends on the speed of ambulation. For a material particle, the kinetic energy is $mv^2/2$, and its potential energy is mgh. When maximal values of the kinetic and potential energy are equal, $v^2 = 2gh$, where v is the maximal velocity of the particle, and h is its maximal vertical displacement. Hence, the maximal energy conservation is possible when $v = \sqrt{2gh} \approx 1.4\sqrt{gh}$. The ratio

$$\mathrm{Fr} = \frac{v}{\sqrt{gh}} \tag{6.29}$$

is known as the *Froude number* (after English naval engineer William Froude, 1810–1879). In the biomechanics of terrestrial locomotion, the Froude number is used mainly for scaling speed to body size. The number can be seen as a dimensionless speed. In this case, h is any appropriate linear dimension of the body, such as height or leg length. When two persons of different size, for instance, a child and an adult, move at the same Froude number, the same proportion between the maximal values of the kinetic and potential energy is preserved.

6.2.3.3 Negative Change in the Body's Mechanical Energy Versus the Negative Work of Joint Torques

The previous discussion was based on the presumption that human body segments do not deform. In such a model, the mechanical energy of the body decreases due to the negative work of the joint torques and external forces, such as air resistance. In reality, body segments are not rigid. Part of the mechanical energy of the body—in some cases, the largest part—is expended for the deformation of the body segments. For instance, during falling accidents, the mechanical energy of the body decreases without any work being produced by the joint moments. In general, when the mechanical energy of the body decreases, a portion of the energy is dissipated without negative work of joint moments, through the "passive" forces that are developed without energy spending by the joint-moment sources. Hence, the negative change of the body's mechanical energy is usually larger than the negative work of joint moments. The ratio

(Negative work of joint moments)/(Reduction in total mechanical energy)

called the *index of softness of landing,* represents the proportion of negative work done on the body by the joint moments. When the index equals 1.0, the landing is ideally soft (all the energy is dissipated by the muscles and tendons involved in generating the joint moments). When the index equals zero, the mechanical energy of the body is dissipated passively in the bones, articular cartilage, and other tissues without negative work of joint moments.

■ ■ ■ *FROM THE LITERATURE* ■ ■ ■

Softness of Landing

Source: Zatsiorsky, V.M., and B.I. Prilutsky. 1987. Soft and stiff landing. In *Biomechanics X-B,* ed. B. Johnson, 739–44. Champaign, IL: Human Kinetics.

The aim of the study was to measure the softness of different landings and to compare the index of softness at landing (ISL) with the GRF. The bare-foot subject (age 25 years, height 182 cm, weight 80 kg) performed drop jumps on a force platform from heights of 25 and 50 cm. Upon landing, the subject had to stand motionless for some time. The subject was asked to land in different trials with different degrees of softness, from very softly to extremely stiffly. The basic model of the human body was used. The values of the total mechanical energy of the body immediately before the ground contact and at the lowest point during the landing were determined. The negative work of the joint moments during landing was computed and used to calculate the ISL. The slope of the initial peak of the vertical GRF was determined (figure 6.21*a*) and compared with the ISL (figure 6.21*b*). The values of ISL ranged from the extremes of 99.5% to 25.0% for the dropping height of 25 cm and from 98.7% to 75.6% for the height of 50 cm. Hence, in some landings, up to 75% of the mechanical energy of the body was dissipated passively, without negative work of joint moments. There was a strong correlation between the softness of landing determined by ISL and the slope of the initial peak of the GRF registered by the force platform.

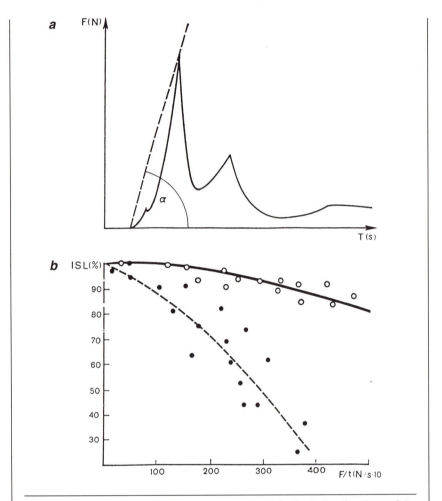

Figure 6.21 Softness of landing. (*a*) The method of calculating the slope of the initial peak of the vertical GRF (a schematic). (*b*) The relationship between the index of softness at landing (ISL) and the slope of the dynamogram at landing.

Reprinted by permission from Zatsiorsky, V.M., and B.I. Prilutsky. 1987. Soft and stiff landing. In *Biomechanics X-B,* ed. B. Johnson, 740, 742. Champaign, IL: Human Kinetics.

When a passive muscle is forcibly extended, the external force does the work on the muscle. The energy "flows" to the muscle. The energy is partly dissipated as heat in the muscle tissue and partly stored as potential energy of elastic deformation. When the muscle is active, there are two flows of energy: to and from the muscle. As with the passive muscle, external force does positive work on the muscle, and the energy goes to the muscle. In addition, the muscle spends energy to provide resistance against the external force. It is said that the muscle does *negative work.* The negative muscle work corresponds to periods of eccentric muscle action. The energy flow resembles that during cycling downhill: gravity force produces work on the cyclist–bicycle system, and the cyclist spends energy for braking. In the mechanics of inanimate (passive) objects, the expression "negative work" means that the work is done on the body, and hence the body does not expend energy for the work. In contrast, active muscles spend energy to do negative work. When active muscles absorb energy from an external source while being forcibly stretched, they also expend energy to resist the extension. It is a pity that the same term, "negative work," is used to designate these different quantities: the work done *on* the muscle by the external force and the work done *by* the muscle to resist the extension.

From equation 6.26, it follows that the flow to a body segment of energy associated with joint torques and forces equals the time rate of the mechanical energy of the segment. This equality is valid unless the segment is not deformed. For instance, the equation is not valid for the foot during foot strike and push-off in running because of the large foot deformation (figure 6.22).

If the deformation is negligible, determining the power flow to and from individual body segments is an unambiguous operation. Contemporary experimental methods can collect the information that is necessary for performing the computations. However, computing total power produced by several joint torques is a controversial procedure that cannot be performed correctly without additional information. Unfortunately, the information necessary for exact computation of total work and power is usually not available.

6.3 ENERGY, WORK, AND POWER IN MULTILINK KINEMATIC CHAINS

Representative papers: Aleshinsky (1986b); van Ingen Schenau (1998)

When a motor task involves several body segments, two main problems arise: (1) what is the total amount of work (or power) done *on* the entire system (this work equals the change in the system's energy) and (2) what is the total amount of work done *by* the sources? The solution to the first problem is based on classical mechanics and is not complicated by conceptual difficulties. It is the second problem where difficulties emerge and for which nonclassical empirical measures are suggested.

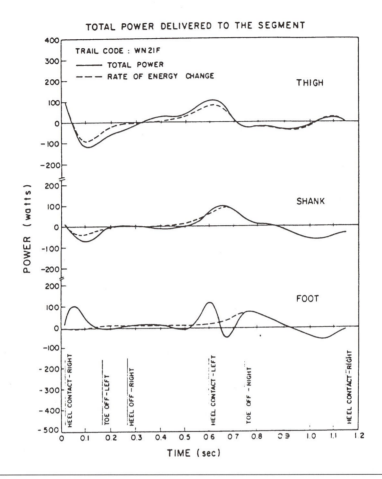

Figure 6.22 Power supplied to the lower-limb segments by joint forces and torques and the rates of change in mechanical energy during one cycle of walking. With the exception of the foot after heel contact, there is good correspondence between the two measures.

Reprinted from *Journal of Biomechanics*, 13 (10), D.G.E. Robertson and D.A. Winter, Mechanical energy generation absorption and transfer amongst segments during walking, 845–854, 1980, with permission from Elsevier Science.

6.3.1 Work Done on the System: Energy Balance Equations

Representative paper: van Ingen Schenau and Cavanagh (1990)

Energy is additive, and the total energy of the system equals the sum of the energies of the individual body segments. Since both work (or power) and kinetic energy are scalars, the work–energy equation for the human body can be obtained by summing up the work–energy equations for the individual links. Hence, the system's final kinetic energy equals the system's initial kinetic energy plus the work done by all the forces acting on the system.

$$\Sigma K_1 + \Sigma W_{1-2} = \Sigma K_2 \tag{6.30}$$

where ΣK_1 and ΣK_2 are the total kinetic energy of the system at the beginning and at the end of the studied period, respectively, and ΣW_{1-2} is the total work done on the system during the studied period. The sum ΣW_{1-2} represents the work of all the forces and moments, both external and internal, that act on the body segments. The internal forces are joint forces. They occur in pairs of equal and opposite forces. In each pair, the points of application of the joint forces move through the same distance. Therefore, the total work of the joint forces is zero. In contrast, the adjacent body segments possess different angular velocities and rotate through different angular distances. Therefore, the joint torques modulate the total energy of the body. External forces and moments are not necessary for that. For instance, in airborne movements, an athlete can increase the kinetic energy of the body by assuming a tucked position during somersaulting. In the air, the sum of the potential energy and translational kinetic energy of the body is constant, and only the kinetic rotational energy changes. The angular momentum of the body H is conserved ($H = I\dot{\theta}$ = a constant, where I is the moment of inertia and $\dot{\theta}$ is the angular velocity of the body), but the rotational kinetic energy ($I\dot{\theta}^2/2 = H\dot{\theta}/2$) is not. When a diver assumes a tucked position, the moment of inertia of the body decreases, the angular velocity correspondingly increases from $\dot{\theta}$ to $\dot{\theta}'$, but H stays the same. The rotational kinetic energy of the body in the tucked position increases and equals $H\dot{\theta}'/2$. Hence, the increase in the body's energy is solely due to internal forces (joint torques).

If work of the external forces is neglected (which can be done, for instance, for walking on a firm surface without skid or air resistance), the expression for the total energy change ΔE during the time interval from t_1 to t_2 is

$$\Delta E \Big|_{t_1}^{t_2} = \int_{t_1}^{t_2}\left[\sum_{i=1}^{N-1} T_{i,i+1}\left(\dot{\theta}_{i+1} - \dot{\theta}_i\right)\right]dt = \int_{t_1}^{t_2}\left[\sum_{i=1}^{N-1} T_{i,i+1}\dot{\alpha}_{i,i+1}\right]dt = \int_{t_1}^{t_2}\sum_{i=1}^{N-1} P_{i,i+1}dt \tag{6.31}$$

where i and $i + 1$ refer to adjacent body links. Thus, the energy change of the entire system equals the time integral of the sum of joint power in individual joints. In accordance with the work–energy principle (equation 6.30), it is also equal to the total work done on the system.

If external forces, such as air resistance and friction, are large, the energy balance equation that relates the power inflow from the joint torques ($P_{i,i+1}$) with the time rate of the mechanical energy of the system is

$$\sum P_{i,i+1} = \frac{d}{dt}\sum E_i - \mathbf{F}_e\mathbf{v}_e \tag{6.32}$$

where the dot product $\mathbf{F}_e \mathbf{v}_e$ represents the time rate of energy losses to the environment. Equation 6.32 is valid when external forces impede the movement. If external forces do positive work on the body—as in springboard diving, for instance—the sign of the product $\mathbf{F}_e \mathbf{v}_e$ should be reversed. Equation 6.32 does not include the power loss or gain associated with the external free moments that may act on the body.

According to equation 6.30, if the total energy of the human body is not changed, the work done on the body equals zero. For instance, in the absence of energy losses to the environment, the mechanical work done on the body in walking and running for an arbitrary number of steps with constant average speed is zero. The work is also zero for all movements beginning and ending at rest at the same vertical location (running a marathon, mountaineering on Everest, etc.). Because mechanical energy must be expended to execute such an activity, some authors, puzzled by this phenomenon, coined it the *zero-*

■ ■ ■ FROM THE LITERATURE ■ ■ ■

Power Production During Speed Skating

Source: Koning, J.J. de, and G.J. van Ingen Schenau. 1994. On the estimation of mechanical power in endurance sports. *Sport Sci. Rev.* 3 (2): 34–54.

The authors studied elite speed skaters. Speed skating is a sport in which the main portion of power is expended in overcoming air resistance and ice friction. The vertical displacement of the trunk in this sport is negligible. The power involved in acceleration and deceleration of the legs is low. The fluctuations in horizontal velocity are unessential. The values of negative joint power are small and observed only at the ankle joint for a short duration. Neglecting all these factors, the authors computed the mean generated power as

$$P_{\text{tot}} = f \int_0^T \sum_j P_j dt \tag{6.33}$$

where f is stroke frequency, P_j is the average power at joint j per stroke (j designates the hip, knee, and ankle joints of the support leg), T is the stroke time (figure 6.23). The power output calculated according to equation 6.33 was 3.69 ± 0.56 W/kg. The power loss determined from independent measurements of the air resistance and ice friction was 3.60 ± 0.51 W/kg. Hence, the two methods employed to compute power yielded similar results.

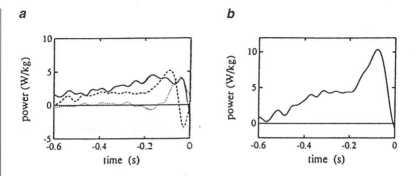

Figure 6.23 Joint power during speed skating. (*a*) Joint power at the hip (solid line), knee (dashed line) and ankle (dotted line), calculated with inverse dynamics. (*b*) Total power calculated by summation of the joint power in the three joints (equation 6.33). Note that at the end of the takeoff, the power at the ankle joint is negative. Summation according to equation 6.33 means that the values of the (negative) power at the ankle were subtracted from the positive power at the hip and knee. The maximal negative power at the ankle was around −3 W/kg; at this instant, the summed joint power at the hip and knee was about 8 W/kg. Hence, according to equation 6.33, the instantaneous value of the total power was 5 W/kg. If absolute values, rather than the real values, of the power were summed, the answer would be 11 W/kg. If the negative power were completely neglected, the total power would be 8 W/kg. Fortunately, for this particular activity, different answers do not substantially influence the obtained estimates of the average power output (3.69 ± 0.56 W/kg). Researchers studying other movements may be not so lucky.

Reprinted, by permission, from J.J. de Koning and G.J. van Ingen Schenau, 1994, "On the estimation of mechanical power in endurance sports," *Sport Science Review* 3(2):34–54.

work paradox. In reality, there is nothing paradoxical in the "zero-work paradox." The origin of the misunderstanding is in incorrect terminology. The work done *on the body* was mistaken for the work *of forces.* In the preceding examples, the mechanical work done *on* the body (by all the forces, including gravity force) is zero, but the work done *by* the individual forces exerted by the performer is not (because these forces are not conservative).

Consider the following example. A human with left arm raised is lifting the right arm and simultaneously lowering the left one. The joint power values at all joints of the right arm are positive, and the joint power values at all joints of the left arm are negative. In addition, the joint power values for the right and left arms are equal in magnitude, $|P_{left}| = |-P_{right}|$ (we call such movements *antisymmetric*). For antisymmetric movements, the total power equals zero (this is correct because the energy of the entire body stays constant), but the power supplied by individual sources (joint torques of each arm) is clearly not equal to zero.

In this example, different forces do work on the different bodies (the right and left arms) that are regarded as one system. The expression "total power" in this case has two meanings: (1) the total energy flow to the system (the rate of change in the system's energy), which equals zero, and (2) the rate at which the work is done by the performer. Assume that the same activity is performed by two people, each of them raising and lowering one arm. Formally, we can consider the performers as one system and compute the total power supplied to the system (two arms). It will be zero. We can also compute the power generated by each performer. It will differ from zero. It seems evident that the total amount of power produced by the two performers should also differ from zero: energy expended by one performer for generating positive power cannot be decreased or cancelled by actions of the second performer.

The following section discusses the problem of determining the total power and work of several joint torques.

6.3.2 Work and Power of Sources of Mechanical Energy

Total work done on the human body by all the applied forces can always be computed. The computation is in strict accordance with classical mechanics. However, the computed values of work and power, although correct, are often useless. There is not much useful information in the fact that the total work done on a runner's body during a marathon is zero. Likewise, during the antisymmetric movement, each source of energy generates power and hence expends energy. The total power is zero, but the energy expenditure is not. The main topic of interest is usually not the total power and work done on the body but the mechanical energy expended by the sources.

Determination of the total mechanical energy expended by the joint torques for moving a multilink system can be done unambiguously only for some movements. For an arbitrary movement, the problem is not solvable without additional information or assumptions. To estimate the mechanical energy expended for motion, instead of the work, some "worklike" measures have been suggested.

6.3.2.1 Introductory Examples

Representative paper: Zatsiorsky and Gregor (2000)

The goal of this section is to explain by examples the difficulties encountered in computing the total work done by the sources in a multilink motion. Consider a slow, horizontal arm extension with a load in the hand (figure 6.24). The mass of the body parts and the work to change the kinetic energy of the

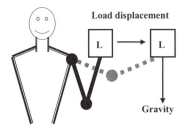

Figure 6.24 A slow, horizontal arm extension with a load in the hand. The work done on the load is zero, but the work of joint torques is not.

Reprinted, by permission, from V.M. Zatsiorsky and R.J. Gregor, 2000, Mechanical power and work in human movement. In *Energetics of Human Activity*, edited by W.S. Sparrow (Champaign, IL: Human Kinetics), 196.

load are neglected. During the movement, represented by the broken line, the muscles of the shoulder joint perform positive (concentric) work: they generate an abduction moment and elevate the arm. The elbow joint extends while producing a flexion moment against the weight of the load. Therefore, the flexors of the elbow produce negative (eccentric) work. The work of the force exerted by the hand on the load is zero. The direction of the gravity force is at a right angle to the direction of the load displacement, and hence, the potential energy of the load does not change. The total work done *on* the system (arm plus load) is zero. What is the total amount of work done *by* the subject? Is it zero? The problem is whether the negative work at the elbow joint cancels the positive work at the shoulder joint (or whether the positive work at the shoulder compensates for the negative work at the elbow, which is the same question). The correct answer depends on information that was not provided in the preceding text. To explain, we consider several additional examples.

Consider a three-link kinematic chain with two frictionless joints (figure 6.25). The total mechanical energy of the system does not change. Therefore, the net power (the time rate of mechanical energy of the system) is zero. What is the magnitude of the total power of the sources (the time rate of the mechanical energy generated by all the sources combined)? Several answers are possible. They depend on the assumptions made (or the model studied).

1. Two one-joint muscles serve the joints; production of negative power requires energy. If we add the real values of power in the joints, as we did previously, their sum is zero ($P_{tot} = \Sigma P = 0$). This answer correctly portrays the time rate of change in the mechanical energy of the system (the work done on the system): the mechanical energy of the system does not change. However, the rate of the energy generated by the sources is obviously not zero: energy is expended. In this example, the negative power absorbed in joint 2 does not

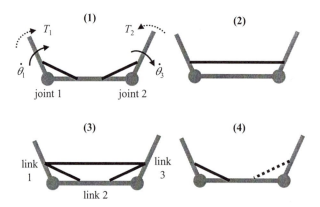

Figure 6.25 Three-link systems that move in the same manner but are served by different muscles. In all four cases, links 1 and 3 rotate with the same angular velocities ($\dot{\theta}_1 = \dot{\theta}_3$), and similar joint torques are acting ($T_1 = |-T_2|$). The joint powers are P_1 and $-P_2$ ($P_1 = |-P_2|$). See explanation in the text.

compensate for the power production at joint 1. Since two different sources generate the power, summation of the real values of P_1 and P_2 does not provide an estimate of the total power generated at the joints. If we instead compute the total power of the sources as a sum of the absolute values of the joint power, the total power in this case equals $2P_1$:

$$P_{tot} = |P_1| + |-P_2| = 2P_1 \qquad (6.34)$$

Equation 6.34 is intended to characterize the total mechanical power produced by two independent sources of mechanical energy (single-joint muscles).

Negative work by the muscles requires less metabolic energy from the organism than positive work. To account for this fact, a correction factor k ($0 \leq k \leq 1$) can be used. In this case, the equation becomes

$$P_{tot} = |P_1| + k|-P_2| \qquad (6.35)$$

2. One two-joint muscle serves both joints. Total produced power equals zero:

$$P_{tot} = P_1 + (-P_2) = 0 \qquad (6.36)$$

Negative power from decelerating link 3 is used to increase the mechanical energy of accelerating link 1. In the example under consideration, the length of the muscle is kept constant, and the muscle does not produce mechanical power. The muscle only transfers the power from one link to another.

Using the terminology introduced in **2.1.2**, the joint torques produced by two-joint muscles are equivalent torques, while those produced by one-joint muscles are actual torques. Hence, equation 6.34 describes the total power supplied by the actual joint torques, while equation 6.36 represents the situation when the torques are equivalent.

3. Two one-joint muscles and one two-joint muscle serve the two joints under consideration, and all three muscles develop power in joints 1 and 2 of the same magnitude. In addition, production of negative power requires energy. In this example, the power in each joint (P_j, $j = 1, 2$) can be presented as a sum of the powers produced by the one joint muscle ($P_j^{(1)}$) and the two joint muscle ($P_j^{(2)}$). Similar to case 2, the two-joint muscle acts isometrically and only transfers the power from one link to another. Because the muscle does not generate power,

$$P^{(2)} = P_1^{(2)} + (-P_2^{(2)}) = 0 \tag{6.37}$$

Since $P_1^{(1)} = P_1/2$, $P_2^{(1)} = P_2/2$, and $|P_1| = |-P_2|$, the total power from the two one-joint muscles equals P_1, similar to case 1:

$$\left|P_1^{(1)}\right| + \left|-P_2^{(1)}\right| = \left|\frac{P_1}{2}\right| + \left|\frac{P_2}{2}\right| = P_1 \tag{6.38}$$

4. Two one-joint muscles serve the joints; the negative work does not require metabolic energy. This may occur when the resistance to an external force is provided by friction or the deformation of passive tissues. The total power of the joint torques in this case equals P_1.

$$P_{\text{tot}} = P_1 + (-P_2) = P_1 + 0 = P_1 \tag{6.39}$$

In summary, knowledge of the power values at individual joints in itself does not allow the unequivocal determination of the total power developed in all the joints. Additional information about the sources of energy is needed to arrive at the total power generated by the joint torques.

6.3.2.2 Mechanical Energy Expenditure

In the preceding sections and earlier in this chapter, we considered several examples in which some sources did positive work on a system while others did negative work. The examples were the tug-of-war (figure 6.7), the horizontal arm extension (figure 6.24), and a three-link chain (figure 6.25). In all these cases, the work done on the system sharply differed from the mechanical energy expended by the sources and was not a very informative measure. For instance, the total work was equal to zero when the individual sources produced work and expended the energy. The total amount of mechanical energy expended by the sources is often a more practical quantity. Recall that the total work is computed as a sum of the *real* values of the work done by the individual sources. To compute the mechanical energy expended by the sources, summation of the *absolute* values was suggested.

The *mechanical energy expenditure* (MEE) of several sources is defined as an integral over time of the absolute values of power produced by the sources of energy. Thus by definition,

$$\text{MEE} = \int_{t_1}^{t_2} \sum_s |P_s| \, dt \tag{6.40a}$$

The MEE can also be written as the sum of the absolute values of work done to accelerate and decelerate the body segments

$$\text{MEE} = \underbrace{\int_{t_1}^{t_2} \sum_s (+P_s) \, dt}_{\substack{\text{Work for} \\ \text{acceleration}}} + \underbrace{\int_{t_1}^{t_2} \sum_s |-P_s| \, dt}_{\substack{\text{Work for} \\ \text{deceleration}}} \tag{6.40b}$$

MEE is a "worklike" measure used in some biomechanics and robotics research. It was suggested in an attempt to find a parameter that could be useful in determining the amount of energy expended for motion. If the negative work of the sources requires a fixed proportion of energy k ($0 \leq k \leq 1$) expended by the sources for an equal amount of positive work, the total MEE can be written as

$$\text{MEE} = \underbrace{\int_{t_1}^{t_2} \sum_s (+P_s) \, dt}_{\substack{\text{Work for} \\ \text{acceleration}}} + k \underbrace{\int_{t_1}^{t_2} \sum_s |-P_s| \, dt}_{\substack{\text{Work for} \\ \text{deceleration}}} \tag{6.40c}$$

At present, however, the value of k is not known for the majority of movements. It is also quite possible that k varies during a movement. In the ensuing discussion, the MEE is defined as in equation 6.40a. Note that the MEE and work are different measures. They are equal only in some situations, which are elucidated later in the text. Also note that the MEE does not account for the energy expended for static efforts (isometric muscle actions).

6.3.2.3 Energy Compensation

Representative papers: Aleshinsky (1986a); Zatsiorsky and Gregor (2000)

To scrutinize MEE further, we introduce the concepts intercompensation of sources and compensation of sources during time (recuperation).

6.3.2.3.1 Intercompensation of Sources

The sources are called *intercompensated* if the energy expended by one source can be compensated by the energy absorbed by the second source. For instance, a two-joint muscle can simultaneously produce negative power in one joint and positive in the second. Hence, when two joints are served by one two-joint muscle, the power from one joint can be used in the neighboring joint, and the sources of energy (the joint torques) are intercompensated. In contrast, if the joints are crossed only by single-joint muscles, the negative power from a joint cannot be used in another joint. In this case, the joint torques are *nonintercompensated sources* of energy.

When a source is intercompensated, the total power produced by the source equals the algebraic sum of the powers developed by it in individual joints:

$$P_{tot} = \sum_s P_s \qquad (6.41)$$

where P_s is the power of the individual sources of energy (forces, moments). Equation 6.41 explains why the total power of intercompensated sources in the absence of friction can be zero when a movement is performed, as in the horizontal arm extension in figure 6.24. As already mentioned, the joint torques in the joints served by two-joint muscles are the intercompensated sources.

In engineering, an example of intercompensated sources is a robot with electrical motors as torque actuators. The motors convert electrical energy into mechanical energy and mechanical energy into electrical energy. When an external force does positive work at a joint (so that the work of the joint actuator is negative), the mechanical energy is converted into electrical current and used in another joint. This arrangement allows energy to be saved.

When the sources are not intercompensated, the total power that they produce equals the sum of the absolute values of the power that each produces:

$$P_{tot} = \sum_s |P_s| \qquad (6.42)$$

The joint moments generated by one-joint muscles are not intercompensated sources of mechanical energy. If the muscles, rather than the joint torques, are considered the sources of energy, the sources are not intercompensated. As a rule, energy expended by one muscle cannot return to the system through the simultaneous absorption of energy by another muscle.

Like MEE, equation 6.42 is valid only for the specific case of nonintercompensated sources of energy when the positive and negative work requires the same amount of energy from the system. The equation yields the total amount of power generated or absorbed by several nonintercompensated sources of energy. Note that the total power *produced* (*source power*) may differ sharply from the time rate of the mechanical energy of the system upon which the sources act (*net power*). Consider again an antisymmetric movement: the energy of the system does not change, but the performer expends energy for motion. The net power is zero, but the source power is not.

6.3.2.3.2 Compensation During Time (Recuperation)

When muscles develop force eccentrically (in the direction opposite the motion), the external force does positive work on the muscle. Hence, the energy from the external source goes to the muscle. The energy can be stored temporarily as the elastic energy of muscle and tendon deformation and then recoiled to the system later in time. This feature is called *energy recuperation,* or *energy compensation during time.* For a source of energy compensated during time, the MEE equals the absolute value of the integral over the time of the power developed by the source:

$$\text{MEE} = \left| \int_{t_1}^{t_2} P_s \, dt \right| \tag{6.43}$$

■ ■ ■ FROM THE LITERATURE ■ ■ ■

Intercompensation Influences Mechanical Energy Expenditure During Locomotion

Source: Prilutsky, B.I., L.N. Petrova, and L.M. Raitsin. 1996. Comparison of mechanical energy expenditure of joint moments and muscle forces during human locomotion. *J. Biomech.* 29 (4): 405–16.

The mechanical energy expenditures (MEE) of two human lower-extremity models with different sources of mechanical energy—(1) one- and two-joint muscles and (2) joint moments—were compared theoretically. The joint moments were assumed to be the nonintercompensated sources. Hence, the joint moment model is similar to a model with only one-joint muscles. It was shown that during the same motion, the model with two-joint muscles (intercompensated sources) could spend less mechanical energy than the model without two-joint muscles (with nonintercompensated sources). Economy of mechanical energy is possible if (1) a two-joint muscle produces positive power at one joint and negative power at another joint, (2) moments produced by that muscle at each of the two joints have the same direction as the net joint moments at these joints, and (3) muscles crossing these two joints from the opposite side do not produce force.

The existence of these three conditions during human locomotion was checked experimentally. In particular, it was shown that during some phases of walking and running, the powers at adjacent joints have opposite signs. Hence, condition 1 is satisfied (figure 6.26).

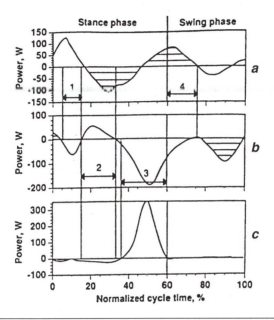

Figure 6.26 Powers of the joint torques at the (*a*) ankle, (*b*) knee, and (*c*) hip joints in the sagittal plane during one cycle of walking at 1.82 m/s. Unshaded and shaded areas represent powers produced by extensor and flexor joint torques, respectively. The arrows and numbers indicate phases when the powers at adjacent joints have opposite signs.

Reprinted from *Journal of Biomechanics*, 29 (4), L.N. Petrova and L.M. Raitsin, Comparison of mechanical energy expenditure of joint moments and muscle forces during human locomotion, 405–415, 1996, with permission from Elsevier Science.

It was concluded that the MEE of different sources of mechanical energy—intercompensated versus nonintercompensated—was different during certain periods of the gait cycle. If one-joint muscles were not active during the phases when powers at the adjacent joints have opposite signs, the economy would amount to 47% of the total MEE of the joint torques. However, the difference in the magnitude of the MEE during the swing phase in walking was relatively small.

Energy compensation over time occurs when the sources of energy include conservative elements. Consider an example. Assume that a muscle during the eccentric phase behaves like a spring; it does not expend energy to resist the forcible stretching and can store elastic energy. During a forcible extension of the muscle-tendon complex by an external force, the force does 10 J of work on the complex. This energy is absorbed by the complex. During a recoil phase, the positive work is 15 J. Assuming that the entire 10 J are recuperated, the energy expended by the muscle for motion is MEE = |–10 + 15| = 5 J. The 5 J are delivered by metabolic processes.

Contrary to the previous example, if the muscle does not recuperate the mechanical energy and expends energy to resist the forcible extension (performs negative work), the total amount of energy expended for the motion is MEE = |–10| + |15| = 25 J. All 25 J are liberated in metabolic processes. In general, for a nonrecuperative source,

$$\text{MEE} = \int_{t_1}^{t_2} |P_s| \, dt \tag{6.44}$$

Note the different location of the vertical bars between equations 6.43 and 6.44. In the literature, MEE computed by equation 6.44 is sometimes called the absolute work.

Recuperation of energy is used in technical systems to decrease energy expenditure. For instance, in underground transportation, a substantial proportion of the energy is expended to accelerate the train after stops. To save energy, the stations are usually located higher than the portions of the tracks between them (figure 6.27). Gravity helps accelerate the train departing from a station. The rather abstract terminology introduced in this section (intercompensation, recuperation) is motivated by the necessity of neutral terms that can be used with equal ease for both biological and technical systems.

In summary, equation 6.42, $P_{tot} = \sum_s |P_s|$, signifies the prohibition of energy exchange between different sources of energy, and equation 6.44, $\text{MEE} = \int_{t_1}^{t_2} |P_s| \, dt$, signifies the prohibition of energy recuperation. Hence, when the absolute val-

Station A   Station B

Figure 6.27   Recuperation of energy in a technical system. The hill-and-valley profile of the railroad helps decrease the energy expenditure. Gravity assists in accelerating a train when it departs from station A. It also aids in decelerating the train when the train approaches station B. The energy for train acceleration after the stops is stored as potential gravitational energy.

ues, rather than the real values, of power are summed, the operation corresponds to an implicit assumption that the sources of energy are not intercompensated. When the absolute values of power are integrated over time, the operation corresponds to the assumption that the sources of energy are not compensated during time. Conversely, if the sources of energy are intercompensated, the total power generated by the sources is $P_{tot} = \Sigma P_s$. In a like manner, when a source is compensated during time, its mechanical energy expenditure is expressed as $\text{MEE} = \int P_s dt$.

6.3.3 Main Models

In contemporary biomechanics, three main methods are employed to measure the mechanical energy expended or work done in human motion. In the first method, known as the *source approach,* the joint power is determined. The second technique, coined the *fraction approach,* relies on determining the change of mechanical energy of individual body segments. In the third approach, the *center of mass model,* the energy associated with the movement of the CoM of the body is determined.

Mechanical energy of the human body and the change in it (the total work done *on* the body by all the applied forces and moments) can be determined from the kinematic data and mass-inertia characteristics of the body segments. To determine the mechanical energy expended for motion, additional information about the sources of energy, their intercompensation and recuperation, is necessary. The four main situations that occur correspond to the possible answers to these two questions: (1) Are two-joint muscles included in the model (is there intercompensation)? (2) Do muscles and tendons store and recoil elastic energy (is there recuperation)? In addition, to apply the methods, a decision about the cost of negative work must be made: whether the negative work of joint torques requires metabolic energy from the system (is active) or not (is passive). Unfortunately, the precise proportion of intercompensated energy (transferred from joint to joint) and recuperated energy (stored in elastic form and later returned to the system) in human movement is commonly not known. Also, the cost of the negative work is known only for some activities. Consequently, studies are limited to models that do not represent reality exactly. However, these models are still useful in some applications as an intermediate step to the conclusive solution of the problem.

6.3.3.1 Energy Expenditure of Joint Torques (Source Approach)

Representative papers: Elftman (1939); Aleshinsky (1986a, 1986b)

The source approach is based on the classical definition of work as the integral of the product of joint moment and angular velocity over time.

■ ■ ■ *FROM THE LITERATURE* ■ ■ ■

Changes in Joint Power As Infants Learn to Reach

Source: Zernicke, R.F., and K. Schneider. 1993. Biomechanics and developmental neuromuscular control. *Child Dev.* 64: 982–1004.

During the first year of life, human infants undergo dramatic changes in motor control. A group of infants was studied weekly from 3 weeks to 30 weeks of age and every 2 weeks thereafter until 52 weeks of age as they gained dexterity in a standard task of reaching for a toy. For all reaching trials, the infants were seated in an infant seat with the torso secured to the backrest with a snug but comfortable broad band. Three-dimensional biomechanical analysis was performed. Representative examples of the joint power profiles for one of the infants are shown in figure 6.28. At 12 weeks (figure 6.28*b*), the shoulder and elbow power profiles were quite erratic. Later, a certain pattern emerged. At 52 weeks (figure 6.28*c*), the coordinated reaches had a triphasic relationship between elbow and shoulder power: (1) power generation at both elbow and shoulder, then (2) elbow absorption and power shoulder generation, followed by (3) power absorption at both the elbow and shoulder joints as the infant finally grasped the toy.

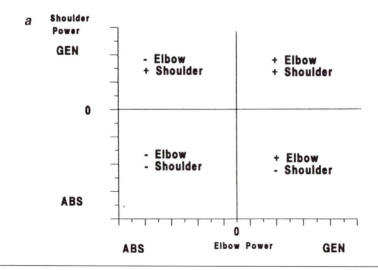

Figure 6.28 Joint power in reaching by infants at different ages. ABS = energy absorbed; GEN = energy generated. See explanation in the text.

Reprinted, by permission, from R.F. Zernicke and K. Schneider, 1993, "Biomechanics and developmental neuromuscular control," *Child Development* 64:982–1004.

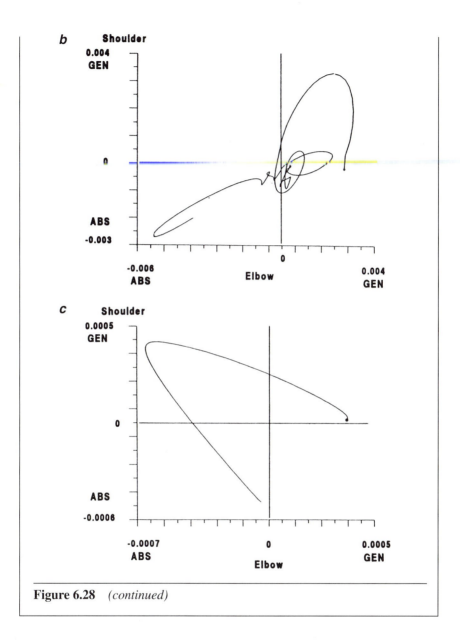

Figure 6.28   *(continued)*

A basic model of this approach has the following characteristics:

1. The sources of energy are not intercompensated. In other words, only one-joint muscles are permitted.

2. The sources of energy are not recuperative. When the sources absorb energy, the energy cannot return into the system for later use. In other words,

storage and recoil of elastic deformation energy is not permitted in the model.

3. The costs of the positive and negative power or work are equal. The model does not account for the energy expended for isometric actions.

With these assumptions, the total power produced by several joint torques equals the sum of the absolute values of the power that each develops (equation 6.42). The MEE of a single source equals the integral over time of the absolute value of the developed power (equation 6.44). Therefore, the total MEE of the several sources of energy is

$$\left. \text{MEE} \right|_{t_1}^{t_2} = \int_{t_1}^{t_2} \sum_{i=1}^{N-1} \left| P_{i,i+1} \right| dt \tag{6.45}$$

Equation 6.45 is similar to equation 6.40a. The model is clearly limited (only one-joint muscles are permitted, elastic forces are disregarded, etc.). However, when only joint torques, not muscle forces, are known, this is the simplest existing model that allows estimating MEE for motion in an uncontroversial way.

The MEE for movement of one body segment isolated from the entire body is

$$\left. \text{MEE} \right|_{t_1}^{t_2} = \int_{t_1}^{t_2} \left\{ \left| \mathbf{F}_p \dot{\mathbf{r}}_p \right| + \left| -\mathbf{F}_d \dot{\mathbf{r}}_d \right| + \left| T_p \dot{\theta} \right| + \left| -T_d \dot{\theta} \right| \right\} dt \tag{6.46}$$

which is an integral of the sum of the absolute values of the power of the joint forces and moments acting on the link (compare with equation 6.25). Consider an imaginary case where all body links are separated from each other but move in the same way as in walking. For this case, the sum of expressions for all the body links equals the total MEE. In walking, this sum is three to five times larger than the values obtained from equation 6.43. This difference is due to the transfer of energy among the body segments, which is not possible when the links are isolated.

6.3.3.2 Energy Expenditure to Move Body Links (Fraction Approach)

Representative papers: Fenn (1929, 1930a, 1930b); Pierrynowski, Winter, and Norman (1980); Williams and Cavanagh (1983)

The fraction approach is based on determining the changes in the mechanical energy of individual body segments (section **6.2.3**), from which the change in the mechanical energy of the entire body is computed. The method determines the work done *on* the body, that is, the work of the resultant forces and couples acting at the CoMs of individual body segments. Imagine strings like those used to manipulate puppets attached to the segments' CoMs. The work of the forces exerted by the strings on the body segments would be equal to the changes in the body's potential and kinetic translational energy. Precisely this work, plus the work done to rotate the body segments, is measured with the fraction approach. In reality, the sources of energy in the models analyzed in this chapter are joint torques. As explained previously in **6.1.2**, when some sources do positive work while other sources simultaneously do negative work, the work of the resultant forces and couples does not represent the mechanical energy expended for motion.

According to equation 6.31, when the work of external forces is absent, the work done *on* the entire body by the joint torques is

$$W\bigg|_{t_1}^{t_2} = \int_{t_1}^{t_2} \sum_{i=1}^{N-1} P_{i,i+1}\, dt \qquad (6.47)$$

When this equation is compared with equation 6.40*a*, it is apparent that the MEE of the sources equals the work done on the human body only if

- all the sources are intercompensated and recuperative or
- all the joint power values have the same sign, either positive or negative (there is no "tug-of-war" situation).

The intercompensation means that the positive power in some joints is compensated by the negative power in other joints (expended energy is compensated by the absorbed energy). For instance, for an antisymmetric movement, this means that the energy for raising one arm is supplied from the lowering of the other arm. Obviously, this assumption is unrealistic. When a subject does negative work in the knee joint and an equal amount of positive work in the elbow joint, it does not mean that he or she does no work at all and does not expend mechanical energy. Another assumption, that the muscles are ideal springs that recoil 100% of the stored elastic energy, is equally unrealistic. In this case, the arm raising would be performed without energy expenditure, at the expense of the recuperated energy of elastic deformation.

The second requirement is much more realistic. There are many movements in which all joint torques produce either only positive power (e.g., lifting loads, rising from a chair, walking upstairs) or only negative power (landing). In some movements (speed skating, cycling), the negative work is small and can be disregarded. In these motor tasks, the MEE equals the work done on the body (which in turn equals the change in the body's total mechanical energy) plus the work done on the environment.

A significant limitation of the fraction approach is the need to make additional assumptions about the conditions of the energy transfer between body segments. Suppose that the energy of one body segment increases while the energy of a second segment simultaneously decreases by the same amount. If these segments are adjacent, it is possible that the total energy is conserved and is simply transferred between the neighboring segments (see figure 6.10 again). As explained in **6.2.2.2** and table 6.1, to decide on whether such an energy transfer can take place, the joint torque must be known. However, in the fraction method, the joint torques are not known. In some publications on the topic, the energy transfer is assumed rather than computed. Hence, the effects rather than the causes are postulated. The cause–effect relationships on which the assumed energy transfer is based are explained in table 6.3. Some of the assumptions are not realistic but are included in the table for completeness of the discussion. For instance, the third assumption contradicts human anatomy. It implies that antisymmetric motion does not require energy. This would be possible only if all the joints were served by one conjoint muscle.

The fraction approach provides a tool for determining the work done *on* the human body. However, with the limitation that all joint torques produce either only positive or only negative power, it is not an adequate way to determine the MEE for motion.

Table 6.3   Potential Causes of Assumed Energy Transfers

Energy transfer	Potential causes	Comments
Not permitted	All body segments are disconnected.	The model provides an estimation of what would happen if the body segments were disconnected (see equation 6.46).
Permitted between the segments of the same extremity	All the segments of the same extremity are served by one conjoint muscle.	Compared with the other conditions, this is the most realistic model, though it is still too simple to be valid.
Permitted between all the body segments	All the body parts are served by one conjoint muscle.	Hardly a realistic assumption.

6.3.3.3 The CoM Model

Representative papers: Fenn (1930a, 1930b); Cavagna (1975)

The CoM method, which is similar to the fraction approach, is based on König's theorem. In the fraction approach, the kinetic energy of an individual

■ ■ ■ *From the Literature* ■ ■ ■

Different Methods of Estimating Total Power Yield Similar Results in Some Movements but Not in Others

Manual Lifting: Source and Fraction Techniques Yield Similar Results

Source: Looze, M.P. de, J.B.J. Bussmann, I. Kingma, and H.M. Toussaint. 1992. Different methods to estimate total power and its components during lifting. *J. Biomech.* 25 (9): 1089–95.

During lifting, joint torques produce only positive power. Four subjects were filmed while they lifted an 18.8-kg load, and ground reaction forces were measured. The total generated power was calculated (1) by summation of joint powers (source approach) and (2) as the rate of change of the summed energy of body segments (fraction approach). The results were compared. The resulting instantaneous power curves showed a high level of agreement.

Running: Difference Among Variants of the Fraction Approach

Source: Williams, K.R., and P.R. Cavanagh. 1983. A model for the calculation of mechanical power during distance running. *J. Biomech.* 16 (2): 115–28.

In running, the joint torques produce both the positive and negative power. In this study, mechanical power calculations were performed via a segmental energy analysis (fraction approach) using three-dimensional cinematic data from 31 well-trained subjects running over ground at 3.57 m/ s. The magnitude of power calculated was very dependent on assumptions involving the amount of permitted energy transfer between the segments and some other details. Mechanical power values ranged from 273 to 1,775 W, depending on the particular assumptions made and computational methods employed. The difference between methods was 6.5-fold.

segment is partitioned into the translational kinetic energy of the segment's CoM and the kinetic energy associated with the rotation of the segment around its CoM. In the CoM approach, the kinetic energy of the human body is

■ ■ ■ *FROM THE LITERATURE* ■ ■ ■

Energy of the CoM
in Pole Vaulting and Triple Jumping

Sources: Dillman, C.J., and R.C. Nelson. 1968. The mechanical energy transformations of pole vaulting with a fiberglass pole. *J. Biomech.* 1: 175–83.

Amadio, A.C., and W. Baumann. 1995. Energetic aspects of the triple jump. In *XVth congress of the International Society of Biomechanics, book of abstracts,* ed. K. Häkkinen, K.L. Keskinen, P.V. Komi, and A. Mero, 48–49. Jyväskylä, Finland: Jyväskylä University.

In these studies, the energy associated with the movement of the CoM was recorded. During vaults with a fiberglass pole, a decrease in kinetic energy and an increase in potential energy occurs (figure 6.29). The fiberglass pole provides an effective means of transforming kinetic energy to potential energy through the vault.

In the triple jump, 57 jumps resulting in an average performance of 15.57 m (14.38–17.33 m) were filmed. The energy decreased during the performance (figure 6.30). On average, the following energy loss was observed: hop 5%, step 15%, and jump 18%. Hence, the energy loss increased from one support phase to another.

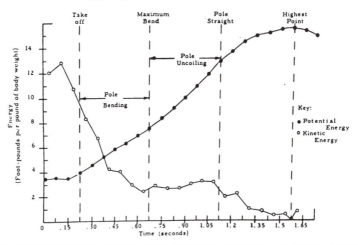

Figure 6.29 Transformation of kinetic and potential energy during a vault. The cleared height was 15 ft (4.572 m).

Reprinted from *Journal of Biomechanics*, 1, C.J. Dillman and R.C. Nelson, The mechanical energy transformation of pole vaulting with a fiberglass pole, 175–183, 1968, with permission from Elsevier Science.

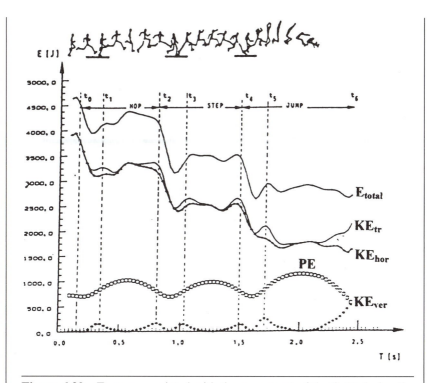

Figure 6.30   Energy associated with the movement of the CoM during the 17.33-m triple jump. A drop in the horizontal kinetic energy (KE_{hor}) is observed during the first halves of the support periods.

Adapted, by permission, from A.C. Amadio and W. Baumann, 1995, Energetic aspects of the triple jump. In *XVth Congress of the International Society of Biomechanics*, Book of Abstracts, edited by K. Häkkinen et al. (Juväskylä, Finland), 48–49.

represented as the sum of the energy associated with motion of the CoM of the body and the energy associated with motion of individual segments relative to the CoM of the body. In the literature, the energy associated with motion of the body's CoM is often called the *external energy,* and the energy associated with the movement of the body parts around the CoM is called the *internal energy.* Note that external forces can change both the external and internal energy of the body. The external energy changes when the external forces induce acceleration of the CoM. The internal energy changes when the external forces cause a rotation of the body or relative displacement of the body segments.

Occasionally, the analysis is limited to the movement of the body's CoM (the *single-mass model*).

6.3.3.3.1 Energy of the CoM

Representative papers: Cavagna (1978); Cavagna, Thys, and Zamboni (1976)

The energy associated with the position and motion of the CoM (*energy of the CoM*) can be further partitioned into potential energy and two fractions of kinetic energy, horizontal and vertical. The kinetic energy is a function of the CoM velocity, and the potential energy depends on the vertical location of the CoM. Changes in the potential energy are due to the lifting and lowering of the body, and changes in the kinetic energy are due to the accelerating and decelerating of the body. In walking and running, the CoM velocity in the vertical and mediolateral directions is relatively small and often ignored. After this simplification, the potential energy and the kinetic energy associated with CoM movement in only the anteroposterior direction are analyzed.

Consider the changes in the kinetic and potential energy of the CoM during one step in walking (figure 6.31) and running (figure 6.32). In one walking cycle, both the speed of the CoM and its height over the ground change. The maximal speed is reached immediately before the heel contacts the ground; the CoM assumes its lowest position close to this moment. The highest position of the CoM is at midstance, when the hip of the stance leg passes over the ankle. The CoM velocity is minimal near this instant. As a result, the kinetic and potential energy of the CoM change out of phase, and the fluctuations in the total mechanical energy associated with the CoM movement are small. The energy transformation is analogous to the movement of a ball on a wavelike surface. During this motion, energy is conserved due to the transformation of the potential gravitational energy into kinetic energy and vice versa.

In running, unlike walking, the CoM is lowest at midstance. The maximal velocity and maximal position of the CoM coincide in time. They are both observed immediately after the takeoff. The kinetic and potential energy of the CoM change in synchrony, and the fluctuations in the total mechanical energy during a cycle are large. To maintain a constant running speed, the mechanical energy loss must be offset by an equal energy supply. The energy comes both from the elastic energy stored previously (recuperated energy) and from metabolic processes.

To put it briefly, in walking the upper part of the body vaults over a supporting leg like an inverted pendulum, while in running the legs work to some extent as pogo sticks (figure 6.33).

The savings due to the transformation of energy from kinetic into potential form and vice versa can be estimated with the energy conservation coefficient. The coefficient is similar to the coefficient introduced previously for a single body segment (equation 6.28) and is defined as

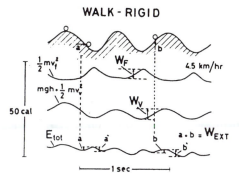

WALK-RIGID

Figure 6.31 Mechanical energy changes of the CoM of the body during walking at 4.5 km/h. The pattern is similar to the motion of a ball rolling up and down a hill (top). The kinetic energy of forward motion and the potential energy change in opposite phase. The kinetic energy of the vertical motion is negligible and is included in the potential energy. The positive change in the total mechanical energy, $\Sigma\Delta\uparrow(E_K + E_p)$, is less than the sum of the changes in the kinetic and potential energy, $\Sigma\Delta\uparrow E_K + \Sigma\Delta\uparrow E_p$.

Adapted, by permission, from G.A. Cavagna, 1978, Aspects of efficiency and inefficiency of terrestrial locomotion. In *Biomechanics VI-A*, edited by E. Asmussen and K. Jørgensen (Baltimore: University Park Press), 3–26.

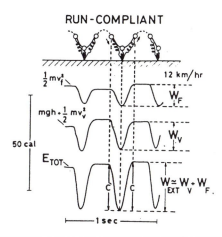

RUN-COMPLIANT

Figure 6.32 Mechanical energy changes of the CoM of the body during running at 12 km/h. Contrary to walking, the kinetic energy of forward motion and the sum of the potential and kinetic energy of vertical motion change in phase. The energy changes are similar to those in a bouncing body (top).

Adapted, by permission, from G.A. Cavagna, 1978, Aspects of efficiency and inefficiency of terrestrial locomotion. In *Biomechanics VI-A*, edited by E. Asmussen and K. Jørgensen (Baltimore: University Park Press), 3–26.

■ ■ ■ **FROM THE LITERATURE** ■ ■ ■

Carrying Loads on the Head Is Economical Due to Conservation of Energy

Source: Heglund, N.C., P.A. Willems, M. Penta, and G.A. Cavagna. 1995. Energy-saving gait mechanics with head-supported loads. *Nature* 375 (6526): 52–54.

In many areas of the world that lack a transportation infrastructure, people routinely carry extraordinary loads supported by their heads, for example the Sherpa of the Himalayas and the women of East Africa. It had been shown that African women from the Kikuyu and Luo tribes can carry loads substantially more cheaply than army recruits; however, the mechanism for their economy remained unknown. The authors investigated, using a force platform, the mechanics of carrying head-supported loads by Kikuyu and Luo women. The weight-specific mechanical work required to maintain the motion of the common CoM of the body and load decreases as load increases. The decrease in work is a result of a greater conservation of mechanical energy resulting from an improved pendulum-like transformation of the energy of the CoM during each step, back and forth between gravitational potential energy and kinetic energy.

$$CC_{CoM} = \frac{\left(\sum \Delta \uparrow E_K + \sum \Delta \uparrow E_P\right) - \sum \Delta \uparrow (E_K + E_P)}{\sum \Delta \uparrow E_K + \sum \Delta \uparrow E_P} \times 100\% \qquad (6.48)$$

where the vertical arrows $\uparrow$ stand for the increase in the kinetic energy E_K and potential E_P energy of the CoM. During walking, the coefficient depends on speed; it changes in a systematic manner and attains its maximal value (around 65%) at a speed around 2 m/s (figure 6.34). At this speed, the potential and kinetic energy of the CoM change out of phase during a stride (an increase in E_P corresponds to a decrease in E_K, and vice versa). In addition, the magnitude of the work against the weight of the body is close to the work necessary for maintaining the horizontal velocity of the CoM. In running, the values of the coefficient remain low.

The conservation coefficient should be interpreted and used with caution. It neglects the energy expenditure for negative work and characterizes only a necessary but not sufficient condition for the energy transformation (see a similar discussion in **6.2.3.2**). It also neglects elastic energy. The coefficient reflects possible energy saving in the work of an imaginary (effective) force acting at the CoM. It does not characterize the energy saving at the level of joint power and work.

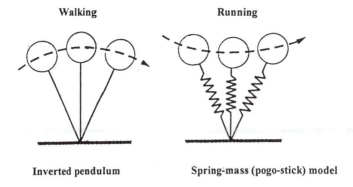

Figure 6.33 Single-mass models of walking and running. These conceptual models should not be taken literally. They are useful for grasping the main distinction between walking and running, but they oversimplify the real situation. In particular, the pogo-stick model of running (also known as the spring-mass model) does not explain the mechanism of the forward propulsion. Specifically, it completely neglects the work done for the leg rotation. In addition, during ordinary walking, the stance leg is not completely extended (in race walking, the rules require full extension of the supporting leg at the vertical position).

■ ■ ■ *FROM THE LITERATURE* ■ ■ ■

Storage of Elastic Energy in an Above-Knee Prosthesis

Source: Farber, B.S., and J.S. Jacobson. 1995. An above-knee prosthesis with a system of energy recovery: A technical note. *J. Rehabil. Res. Dev.* 32 (4): 337–48.

A new prosthesis that allows some knee flexion during the early stance phase of walking was developed. Potential energy is stored in the spring shock absorber of the knee unit. The coefficient of energy recovery increased by 30% in comparison with a conventional above-knee prosthesis. Energy costs to patients decreased an average of 35% during gait with the new prosthesis.

The energy of the CoM is only one fraction of the total energy of the body. Usually, excessive lifting of the body and excessive deceleration (braking) in each step increase the work performed and the metabolic energy expenditure. However, there are situations when the energy associated with the CoM does not represent the total energy expenditure.

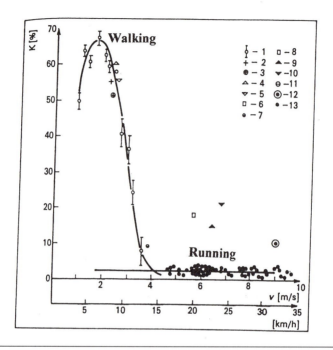

Figure 6.34 Conservation coefficient in walking and running. The numbers correspond to different subjects and methods, including optical methods and computation from the GRF.

Reprinted from V.M. Zatsiorsky and N.A. Yakunin, 1980, "Mechanical work and energy in human locomotion," *Human Physiology* 6(4):579–696.

When fluctuations of the energy of the CoM decrease or even disappear, it does not mean that the total energy expenditure decreases. Imagine a walk during which the CoM moves horizontally at a constant velocity. Such a gait pattern is called a comfortable walk and is of a special interest for designing bipedal walking machines. Future pilots and passengers of such walking devices will not experience unpleasant accelerations during each step if the machine uses a comfortable walk. Since horizontal acceleration of the CoM is zero and the CoM is moving at a constant height, the mechanical power and work of the effective force applied at the CoM is equal to zero in this case. However, a comfortable walk requires several times more energy than an ordinary walk. The reason is that to make the walk comfortable, the walker (person or walking device) should bend the knees during the support period and thus increase the magnitude of the joint torques. Ambulating with the knees bent offers lower vertical oscillation of the CoM, but only at a high cost of increased energy expenditure. In the biomechanics literature, running on flexed legs is called *Groucho running* (after comedian and movie star Groucho Marx, 1890–1977).

■ ■ ■ *FROM THE LITERATURE* ■ ■ ■

Why Do Endurance Runners With Equal Aerobic Potential Have Different Performance Times?

Source: Miyashita, M., M. Miura, Y. Murase, and K. Yamaji. 1978. Running performance from the viewpoint of aerobic power. In *Environmental stress*, ed. L.J. Follnsbee, 183–93. New York: Academic Press.

The authors measured maximal oxygen consumption, VO_2max, in 45 trained middle- and long-distance runners and compared the individual values of VO_2max with the best performance time for a 5,000-m race. There was a clear trend for greater VO_2max values to correspond with better performance results (figure 6.35a). Then, five subjects with a similar VO_2max of about 68 ml $\cdot$ kg^{-1} $\cdot$ min^{-1} were selected for biomechanical analysis of their running technique. The subjects were divided into two groups according to their running performance. Three of the subjects had the best records for the 5,000-m race, ranging from 14 min 48 s to 14 min 56 s. The other two had poorer records of 16 min 10 s and 16 min 52 s. There was no difference in height, weight, or VO_2max between the two groups. The subjects were asked to run 5,000 m at their maximum effort. The subjects were filmed when they were near the 3,000-m point, running at a nearly constant speed. The mean speed for the good runners was 5.48 m/s, and that for the poor runners was 5.06 m/s. There were evident differences in the running techniques of the good and poor runners. The average vertical displacement of the CoM in one step cycle was 10 cm for the poor runners but only 6 cm for the good runners. Hence, the poor runners lifted their body 4 cm higher in each step (i.e., used a "bouncier" run) than the good runners did. An average step length for the good runners was 1.77 m and 1.60 m for the poor runners. The good runners ran 5,000 m in 2,825 steps, while the poor runners required 3,125 steps. The vertical work expended in lifting the body was 9,407 kg $\cdot$ m for a good runner and 17,868 kg $\cdot$ m for a poor runner, a difference of 8,461 kg $\cdot$ m. The unnecessary vertical work performed by the poor runners was huge. It equaled the work of lifting a 57.5-kg runner's body 149 m high, to approximately the height of a 50-story building.

6.3.3.3.2 Work of the Effective Force

Representative papers: van Ingen Schenau (1998); Zatsiorsky (1998)

When the single-mass model is employed, all the external forces acting on the human body are replaced by an effective (resultant) force $\mathbf{F}_{eff}$ acting at the

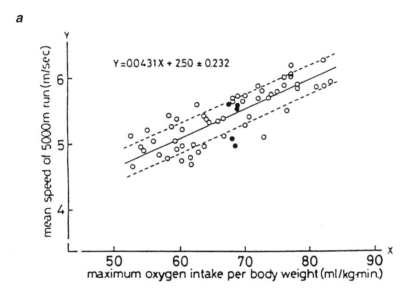

Figure 6.35 (*a*) The correlation between running performance and V̇O₂max among trained adult runners. Black dots (•) designate the subjects who participated in the biomechanics research (see text for details). (*b*) The CoM trajectories in the sagittal plane for two runners with similar V̇O₂max values and different performance results.

Adapted, by permission, from M. Miyashita et al., 1978. Running performance from the viewpoint of aerobic power. In *Environmental Stress. Individual Human Adaptations*, edited by L.J. Folinsbee et al. (New York: Academic Press), 183–193.

CoM. The $\mathbf{F}_{eff}$ is a resultant force, that is, an imaginary force that produces the same effect on the CoM (but not on the entire human body) as all the external forces combined. The effect of the external forces on body rotation and relative displacement of the body parts (the internal energy) is completely ignored with this approach. It is like attaching a massless rod at the CoM and then determining the power or work expended to push and pull it.

The work of the effective force acting at the CoM during a finite displacement is

$$W_{eff} = \int \left(F_x dx + F_y dy + F_z dz \right) \tag{6.49}$$

where the components of the effective force F_x, F_y, and F_z are computed as $F_x = mx_{CoM}$, $F_y = my_{CoM}$, and $F_z = mz_{CoM}$. When equation 6.49 is used, the magnitude of the imaginary (effective) force is multiplied by the elementary displacement of the imaginary particle (the CoM). Clearly, this procedure departs from the classical definition of work as a product of a real force and displacement of a real point of force action. In the physics literature, the integral 6.49 is sometimes called *pseudowork* (Sherwood 1983).

In locomotion, when the air resistance and friction forces are disregarded, only two external forces exist: ground reaction force and gravity force. Hence, the effective force acting at the CoM can easily be computed from the GRF. The horizontal components of the $\mathbf{F}_{eff}$ in this case equal the corresponding horizontal components of the GRF. The vertical component is

$$F_z \text{ (at the CoM)} = F_z \text{ (ground reaction)} - mg \tag{6.50}$$

In experiments, the acceleration of the CoM is calculated by dividing the GRF (minus body weight, if appropriate) by the subject's mass. The instantaneous velocity values are determined by integration of the acceleration. The integration constants are obtained from the boundary conditions of the system, in particular, from average horizontal velocity measured independently. Then the products of the effective force acting at the CoM of the body ($\mathbf{F}_{eff}$) and the CoM velocity ($\mathbf{v}_{CoM}$) are calculated and used as a measure of the rate of the mechanical energy associated with the movement of the CoM:

$$\mathbf{F}_{eff}\mathbf{v}_{CoM} = \frac{d}{dt}E_{CoM} \tag{6.51}$$

Note that the equation includes $\mathbf{F}_{eff}$ but not $\mathbf{F}_{GR}$ (for the horizontal forces, $F_{eff} = F_{GR}$). The GRF is not applied at the CoM, and hence, the product $\mathbf{F}_{GR}\mathbf{v}_{CoM}$ does not make much sense. As discussed in **6.1.1.1**, in the absence of skidding, the GRF does no work.

Equation 6.51 does *not* represent the total power produced at a takeoff; it neglects the kinetic energy of the body segments in their movement relative to the CoM. For instance, during a standing vertical jump, the lower legs do not move before the heel-off. Hence, the upper part of the body moves upward faster than the body's CoM and has a certain kinetic energy with respect to it. The situation at hand can be modeled as a two-mass system (figure 6.36). For simplicity, suppose that a vertical GRF is constant and the masses of the two bodies are equal. The change in the kinetic energy of the CoM can be computed from the elementary relations: $F = ma$, $v = at$, and $K = mv^2/2$, where F is the force above the body weight, t is the time of the force action, K is the kinetic energy, and m is the mass of the entire system. The change in the kinetic energy of the entire system can be deduced from $F = (m/2)a'$, $v' = 2v$, and $K' = [(0.5m)(2v)^2]/2 = mv^2$. Hence, the work done on the entire system in increasing its kinetic energy is twice the work done on the CoM. The difference is due to the movement of the upper mass with respect to the CoM of the entire system. During human movement, the difference is smaller because the mass of the lower legs is only about 11% of the total body mass (see chapter 4).

An integral of equation 6.51 over an appropriate time interval serves as a measure of work done *on* the CoM, W_{CoM}. The work can also be computed as an algebraic sum of the gains and losses of energy ΔE_{CoM} over a given time period, $W_{CoM} = \Sigma \Delta E_{CoM}$. If the CoM moves in such a way that the energy associated with it either increases or decreases monotonically, the work represents the energy expenditure for that motion. However, as discussed previously, for a motion that starts and ends at rest at the same level, the work done *on* the CoM is zero. This information does not tell anything about the amount of mechanical energy expended for motion.

The problem is in the recuperated energy. In contemporary biomechanics research, the magnitude of the recuperated energy usually remains unknown. If the sources of energy were completely compensated over time, the motion would be performed without energy expenditure (the work performed by the nonconservative forces would be zero). An assumption that the sources of energy are not compensated during time results in the summation of the absolute values of ΔE_{CoM}:

$$\text{MEE}_{CoM} = \sum \left| \Delta E_{CoM} \right| = \sum \Delta \uparrow E_{CoM} + \sum \left| \Delta \downarrow E_{CoM} \right| \qquad (6.52)$$

where $\uparrow$ and $\downarrow$ designate the energy gains and losses, respectively. Summation of only positive changes, $\text{MEE}_{CoM} = \Sigma \Delta \uparrow E_{CoM}$, corresponds to the assumption that the performers do not expend energy for negative work. None of these approaches is completely valid.

6.3.3.3.3 Rate of Energy of the CoM and Rate of Total Energy of the Body
Representative paper: Aleshinsky (1986c)

Examination of the mechanical energy of the CoM and its fractions can provide useful information about the studied motion (see examples in figures 6.31,

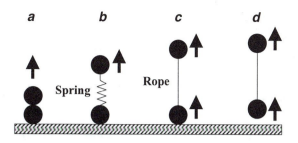

Figure 6.36 A two-mass system illustrating a sequence of events during a standing vertical jump. The masses are equal. The vertical ground reaction force is assumed constant. (*a*) The masses are close to each other. The upper mass starts moving upward. (*b*) The upper mass continues to move under the influence of a spring. The kinetic energy of the upper mass increases. (*c*) The upper mass reaches the maximal distance from the lower mass and pulls it up. The kinetic energy of the upper mass is transferred in part to the lower mass. (*d*) The upper and lower masses move up. The kinetic energy of the system is shared between the two masses.

6.32 and 6.34). This analysis can be complemented by the examination of internal energy of the body: the kinetic energy of the links associated with their motion relative to the CoM (figure 6.37).

Consider the system's total energy balance equation. For the particular case of single support without slip or air resistance, no external moments applied, the equation is

$$\frac{d}{dt}K_{tot} = \frac{d}{dt}K_{CoM} + \frac{d}{dt}K_{rel} = \left(\mathbf{F}_{eff} \cdot \mathbf{v}_{CoM}\right) + \left[\sum P_j - \left(\mathbf{F}_{eff} \cdot \mathbf{v}_{CoM}\right)\right] \quad (6.53)$$

where K_{tot} is the total kinetic energy of the human body, K_{CoM} is the energy associated with the movement of the CoM, K_{rel} is the kinetic energy of the links associated with their motion relative to the CoM, and P_j is joint power at a joint j. The equation relates the time rate of the fractions of the kinetic energy with the total joint power and the power of the effective force, $\mathbf{F}_{eff} \cdot \mathbf{v}_{CoM}$. An integral of equation 6.53 over time represents the sum of the work done on the CoM and the work done on the links to change their kinetic energy with respect to the CoM. It is subject to the zero-work paradox. Similar to other time integrals of the real values of joint power, the integral represents the MEE only when all the joints perform either exclusively positive or exclusively negative work (or the sources are completely intercompensated and compensated during time).

In attempts to determine the "work" expended in changing the energy of the CoM and the energy of the links in their motion relative to the CoM, some

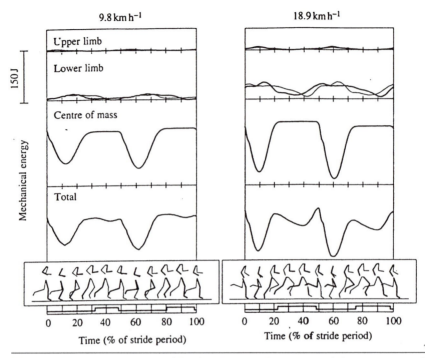

Figure 6.37 Kinetic energy of the CoM (external energy) and the energy associated with the movement of the body parts around the CoM (internal energy) during one stride of running at 9.8 and 18.9 km/h. The stick figures show the link position every 10% of the stride period. Thick lines indicate the position and E_K of the segments closest to the camera. The aerial phases are indicated by the upward shift in the bottom trace. Note the difference between the energy pattern of the CoM and the entire body at 18.9 km/h. At this speed, the energy of the CoM and the lower limbs change out of phase.

Adapted, by permission, from P.A. Willems, G.A. Cavagna, and N.C. Heglund, 1995, "External, internal and total work in human locomotion," *The Journal of Experimental Biology* 198:379–393. By permission of Company of Biologists Ltd.

authors integrate the absolute values of the expressions in the brackets in equation 6.53 and call them "external work" and "internal work."

$$\text{MEE} = \underbrace{\int_{t_1}^{t_2} \left|\mathbf{F}_{\text{eff}} \cdot \mathbf{v}_{\text{CoM}}\right| dt}_{\substack{\text{MEE for movement of} \\ \text{the CoM ("external work")}}} + \underbrace{\int_{t_1}^{t_2} \left|\sum P_j - \left(\mathbf{F}_{\text{eff}} \cdot \mathbf{v}_{\text{CoM}}\right)\right| dt}_{\substack{\text{MEE for movement of the body} \\ \text{parts relative to the CoM} \\ \text{("internal work")}}} \tag{6.54}$$

In the equation, the *real* values of joint power P_j are summed algebraically. Hence, complete intercompensation of the sources is (implicitly) assumed. At the same time, the absolute values are included as the integrands. The absolute values of power, rather than the real values, are integrated when the sources are not compensated during time (see equation 6.44). Hence, equation 6.54 represents the MEE for a system that is served by intercompensated and nonrecuperated sources of energy.

Equation 6.54 has been criticized in the literature. Because the absolute values of the power are integrated, the method prevents canceling out the products $\mathbf{F}_{\text{eff}} \cdot \mathbf{v}_{\text{CoM}}$ that appear in the two integrands with different signs (compare equations 6.53 and 6.54). Consequently, the products are computed twice, which may lead to wrong results. For instance if a subject "freezes" all of his or her joints (power in all joints is zero and consequently $\Sigma P_j = 0$) and moves on a support as an inverted pendulum, equation 6.54 yields nonzero power values. This result does not make much sense.

In summary, partitioning the total kinetic energy into the two fractions associated with the movement of the CoM and the movement of the links relative to the CoM is a straightforward operation, but the computation of the mechanical energy expended for the change in the fractions cannot be done without additional information about the sources of energy: their intercompensation and compensation during time.

6.3.3.4 A General Overview of the Models

The existing situation with respect to determining mechanical work and energy expended for human motion is disappointing. The results of calculations are unambiguous only in one case: when all sources of energy do either only positive or only negative work. In other situations, the work of some forces is

■ ■ ■ *FROM THE LITERATURE* ■ ■ ■

Work to Move the CoM and the Work to Move the Body Parts

Source: Cavagna, G.A., and M. Kaneko. 1977. Mechanical work and efficiency in level walking and running. *J. Physiol.* 268 (2): 647–81.

In this paper, the expressions "external work" and "internal work" are used with the meaning given in equation 6.54.

The mechanical power spent to accelerate the limbs relative to the trunk in level walking and running at various speeds, $\dot{W}_{\text{int}}$, was computed from the increments of the kinetic energy curves. $\dot{W}_{\text{int}}$ increased approximately as the square of the speed of walking and running (figure 6.38a). The total work was computed as the sum $W_{\text{tot}} = |W_{\text{ext}}| + |W_{\text{int}}|$, where W_{ext} is the work done to accelerate the CoM, and W_{int} is the work to move the body parts with respect to the trunk (figure 6.38b). Computation of the kinetic energy of the body links with respect to the trunk, rather than with respect to the CoM as required by König's theorem, is a first approximation of the internal energy. This technique does not take into account the displacement of the CoM with respect of the trunk due to the movement of the body limbs.

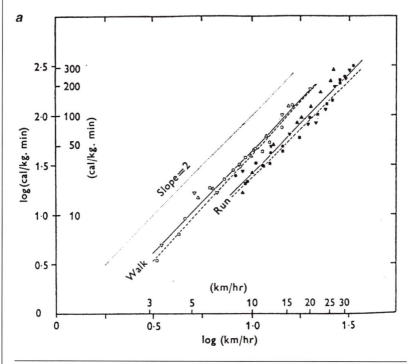

Figure 6.38 Estimates of the power and work expended for accelerating the CoM and the limbs relative to the trunk in level walking and running at various speeds (3–33 km/h). (a) The mechanical power $\dot{W}_{int}$ spent to accelerate the limbs relative to the trunk. The abscissa is the logarithm of speed; the ordinate is the logarithm of $\dot{W}_{int}$. The slope of the curves equals 2; hence, the relationships are quadratic. The continuous lines indicate the power calculated, assuming no transfer of energy between the limbs; the dotted lines fit the data obtained by assuming a complete transfer of kinetic energy between the upper and lower segments of each limb. (b) The internal work W_{int}, external work W_{ext}, and total work W_{tot} done per unit of distance during level walking and running, as a function of speed.

Adapted, by permission, from G.A. Cavagna and M. Kaneko, 1977, "Mechanical work and efficiency in human locomotion," *Journal of Physiology* (London) 268:467–481.

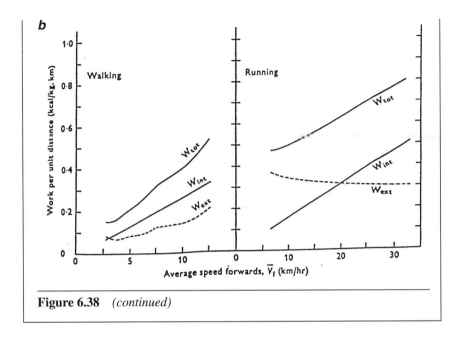

Figure 6.38   *(continued)*

disregarded. The ignored forces depend on the model employed, as summarized in table 6.4.

In the future, interest will most probably shift to determining the work and power of muscle forces. To perform such calculations, the muscle forces and the changes in muscle length during human movement must be known.

6.4 MINIMIZING THE ENERGY EXPENDITURE FOR MOTION

Representative paper: Dickinson et al. (2000)

Those animals that expend less energy for ambulating than their competitors have a better chance for survival in the process of evolution. Therefore, it is natural to expect that, at least in some situations, animals and people try to minimize the energy expended for motion. Research efforts in biomechanics on this subject cluster in three main areas:

1. Quantification of the energy expenditure for motion

2. Ways of minimizing the energy expenditure

3. The role of energy minimization in movement control

We limit ourselves to brief comments and several examples.

Table 6.4   Work Computed and Work Ignored in Different Techniques

Level of analysis	Forces whose work is	
	Computed	Disregarded
Muscles	Muscle forces	None
Joint power and work	Resultant moment produced by the agonistic and antagonistic muscles under the assumptions that the joints are served only by one-joint muscles and antagonists are not activated	Antagonistic muscles; specific action of two-joint muscles (energy transfer between the joints)
Body segments	Resultant forces and couples reduced to CoMs	Joint forces and moments that act opposite to the link displacement and do negative work
General CoM of the body	The resultant (effective) force at the CoM	All the forces that act in the direction opposite to the CoM displacement

6.4.1 Energy Expended for Motion

The energy expended for motion can be estimated as metabolic energy, mechanical energy, or both. The computation of metabolic energy is a part of physiology and is not covered in this book. The difficulties in determining the mechanical energy expended for human movement were discussed previously.

6.4.1.1 Worklike Measures and Movement Economy

Representative paper: Kram (2000)

Because mechanical work done on the body in such activities as walking and running does not provide useful information about the mechanical energy expended for motion, worklike measures, such as MEE, are sometimes used. It is convenient to compare the MEE for ambulating a distance L with the amount of work for lifting the body the same distance. A dimensionless MEE is computed as MEE$' = $ MEE$/BL$, where B is the body weight of and L is distance traveled. For instance, for a person with a body weight of 750 N walking at 1 m/s, the MEE per 1 m of distance is around 150 J. The work to lift the body 1 m is 750 N $\times$ 1 m = 750 J. Hence, the dimensionless MEE$'$ is around 20%.

■ ■ ■ *FROM THE LITERATURE* ■ ■ ■

MEE of the Muscle Forces

Source: Prilutsky, B.I., and V.M. Zatsiorsky. 1994. Tendon action of two-joint muscles: Transfer of mechanical energy between joints during jumping, landing, and running. *J. Biomech.* 27 (1): 25–34.

In the model used in this study, muscle forces rather than joint moments were considered the sources of mechanical energy. The expenditure of mechanical energy was calculated as

$$\text{MEE} = \int_{t_1}^{t_2} \left[\underbrace{\sum_{1,(j,j+1)} \left| P(t)_{j,j+1}^{m_k^{(2)}} \right|}_{\substack{\text{MEE of} \\ \text{two-joint muscles}}} + \underbrace{\sum_{1,j} \left| P(t)_j^{m_i^{(1)}} \right|}_{\substack{\text{MEE of} \\ \text{one-joint muscles}}} \right] dt \tag{6.55}$$

where

$$P(t)_{j,j+1}^{m_k^{(2)}}$$

is the power developed in joints j and $j + 1$ by the kth two-joint muscle $m_k^{(2)}$, and

$$P(t)_j^{m_i^{(1)}}$$

is the power generated in joint j by the lth one-joint muscle $m_l^{(1)}$.

The summation is done across all the muscles. To calculate the total MEE, several assumptions were made (one-joint antagonists are not active, the muscle forces are proportional to the cross-sectional areas of the muscles, the muscles generate the moments in only one plane, etc.). Some of the results of the modeling were reported previously (see From the Literature, p. 483).

In addition to the MEE, other methods have also been recommended. Because muscles spend energy not only when they act concentrically and do work but also when they act isometrically and eccentrically, the integral of the force over time may serve as a measure of the expended energy. It is known from engineering that for an electromechanical torque actuator that is not compensated during time, the integral

$$A_j = \int_{t_1}^{t_2} \left[T_j^2 + \left(-T_j \right)^2 \right] dt \tag{6.56}$$

is proportional to the energy expended by the joint torque T_j at joint j during the period from t_1 to t_2. It is not clear yet whether this approach is valid in biomechanics; further experimental and theoretical research is necessary. Another suggestion is to consider positive and negative work as two separate entities that cannot be added (in this case, two distinct parameters should be analyzed).

While it is not expected that these worklike measures are identical to work and can replace it in the energy balance equations, it is expected that these measures can be used to compare (1) different activities, (2) the same activity performed with different intensities (e.g., speed, rate of pedaling), and (3) different performers (e.g., patients with healthy subjects or good versus poor athletes). An example of such a measure is *economy*, the amount of energy expended per unit of distance. For a locomotion at a constant speed, economy can be determined from the following equation:

$$E = eD \qquad (6.57)$$

where E is the total amount of expended metabolic energy, e is the economy measured in joules per meter, and D stands for the distance. Hence,

$$\frac{dE}{dt} = e\frac{dD}{dt} = ev \qquad (6.58)$$

where v is speed. In experiments, subjects are asked to move at a constant speed, and the metabolic energy expenditure, or the oxygen uptake, is determined. In technically complex sports, the difference in economy between elite athletes and athletes with 1 to 2 years of training is large: in swimming at a speed of 1.03 m/s, the difference was 42.9%; in speed skating at a speed of 8.5 m/s, the difference was 17.9% (Zatsiorsky, Aleshinsky, and Yakunin 1982). Hence, elite athletes expend almost one half of the energy that average athletes do when swimming at the same speed. Note that the amount of mechanical work performed by athletes in these activities remains unknown. From the economy coefficients, we do not know whether the good athletes perform less work than less experienced athletes (e.g., by not displacing the body unnecessarily in the vertical direction) or whether they do the same amount of work but do it more efficiently.

6.4.1.2 Work of Muscle Forces

Representative papers: Komi (1990); Gregor and Abelew (1994); Herzog, Zatsiorsky, Prilutsky, and Leonard (1994)

By measuring both the length and force of a muscle in a living subject, it is possible to compute the amount of work done by the muscle. Direct measurement of muscle forces requires implanting sensors on the tendons. In vivo force measurements have been successfully performed on moving human subjects and animals. The measurements were complemented by the measurements of the muscle length. In a moving subject, the muscle forces are not constant within a cycle of movement. The force–length relationships have the peculiar form of *work loops*. Work loops render information about the mechanical energy absorbed and released by the muscles (figure 6.39).

Experiments have revealed that the kinematics of muscle-tendon complexes may differ from the kinematics of muscle fibers. In some animals (running turkeys, hopping wallabies), the tendons are very elastic. When the muscle-tendon complex stretches, the muscle fibers are nearly isometric or even shortening. The deformation of the entire complex in this case is caused by the tendon extension. Under these conditions, the tendons rather than muscles store and release the potential elastic energy (figure 6.40).

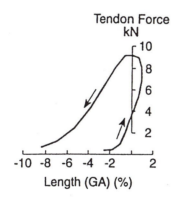

Figure 6.39 Force–length profile (work loop) of the human triceps surae muscles during ball-of-the-foot contact running at 5.78 m/s. The force was measured from a transducer implanted in the Achilles tendon. The muscle length was estimated from filming; it represents the combined length of the muscle belly and tendon of the gastrocnemius (GA) muscle. The arrows designate the time course of the stretch and shortening of the muscle. Immediately after the foot contact, the muscle is forcibly stretched while the force sharply increases (the arrow is directed upward). During this period, the muscle-tendon complex does negative work, and elastic energy can be stored in the muscle and tendon. During the second half of the stance period, the muscle shortens and does positive work. If the muscle-tendon complex behaves as an ideal spring (returns 100% of the stored elastic energy), the positive work done at the expense of metabolic energy equals the area of the work loop.

Adapted from *Journal of Biomechanics*, 23 (Suppl. 1), P.V. Komi, Relevance of in vivo force measurements to human biomechanics, 23–34, 1990, with permission from Elsevier Science.

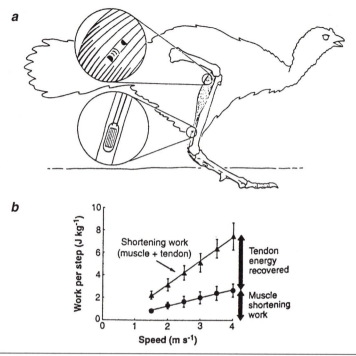

Figure 6.40 Direct measurement of the muscle and tendon work during stance. (*a*) Location of force transducers (lower inset) and length transducers (upper inset) implanted in the lateral gastrocnemius muscle of wild turkeys. (*b*) The work performed during shortening of the lateral gastrocnemius muscle-tendon unit (triangles) and active muscle fibers (circles) as a function of speed. Elastic energy recovery from the tendon, which equals the difference between the two lines, accounted for more than 60% of the work done by the muscle-tendon complex during shortening. Only the positive work is shown; lengthening (negative) work was ignored. Values are the mean and standard deviations for five animals, 10 strides per animal.

Adapted with permission from T.J. Roberts et al., 1997, "Direct measurement of the muscle and tendon work," *Science* 275:1113–1115. Copyright © 1997 American Association for the Advancement of Science.

6.4.2 Energy-Saving Mechanisms

Representative papers: Alexander (1991a, 1991b)

Terrestrial human locomotion is remarkably economical. There are many ways to decrease the energy expenditure for motion. Some of them have already been mentioned, including decreasing the air resistance (e.g., through proper posture in cycling or speed skating), energy transformation between the potential and kinetic form (pendulum-like motions), transfer of energy among the body segments, and energy recuperation through the use of the elastic energy. Other means include selection of the optimal speed (see figure 6.34) and optimal frequency of movement (e.g., on a bicycle ergometer).

Experienced endurance athletes perform less unnecessary work. For instance, as mentioned previously (see figure 6.35), good endurance runners exhibit smaller fluctuations of the CoM in the vertical direction and smaller fluctuations of CoM velocity in the horizontal direction (figure 6.41). In each step in running, the CoM velocity decreases during the first half of the support period and increases during the second half. The work expended to change the velocity of the CoM in a cycle from v_{min} to v_{max} is

$$W_{hor} = \frac{m}{2}\left(v_{max}^2 - v_{min}^2\right) = \frac{mv_{av}}{2}\left(v_{max} - v_{min}\right) \qquad (6.59)$$

Hence, the work done to maintain the CoM speed at the same average level is proportional to $\Delta v = v_{max} - v_{min}$.

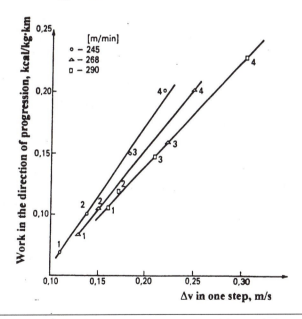

Figure 6.41 Relationship between the horizontal positive work done on the CoM and forward velocity change during one step. The subjects were four young distance runners. In the experiment, the subjects run at three different constant speeds: 245, 268, and 290 m/min. The numbers in the figure identify the subjects. Note the linear relationship between Δv and the horizontal positive work.

Reprinted from M. Kaneko et al., 1981, *Biomechanics VII-B*, edited by A. Morecki et al. (Warszawa: Polish Scientific Publishers. Baltimore: University Park Press), 234–240.

Since the positive work done on the CoM neglects the storage and recoil of the elastic potential energy, it is not a valid measure of work done at the expense of metabolic energy. However, when the object of interest is not an absolute amount of work done but a comparison of different subjects, positive work is a useful measure for ranking the subjects by skill level. For instance, elite distance runners perform less positive work than collegiate athletes when running at the same and faster speeds (figure 6.42).

6.4.3 Does the Central Nervous System Minimize the Energy Expenditure for Motion?

The answer to this question is not easy to obtain. Even if the central controller tries to minimize energy expenditure, most probably the expenditure is minimized in some motor tasks (e.g., endurance running) but not in others (e.g., sprinting or throwing for distance). We may expect that the criteria for various tasks are different. The human body has no specific receptors for energy expenditure. This information might be deduced from *proprioceptors* located within the skeletal muscles, tendons, and joints or *visceroceptors* located within the visceral organs and the circulatory system. Visceroceptors contribute to the sensation of intensity of physical effort and fatigue, but they do not provide immediate information about the cause of the increased energy spending. This information might be deduced from the proprioceptors.

Because muscles spend energy for static efforts, which are even more fatiguing than dynamic ones, it is highly improbable that the CNS tries to minimize the amount of work performed while completely neglecting the energy necessary for force production. Several criteria that the CNS could use to optimize movement have been suggested in the literature. The first criterion, *muscular effort,* was proposed by Nubar and Contini (1961). For one muscle, the muscular effort G_i equals

$$G_i = a_i T_i^2 t \qquad (6.60)$$

where T_i is the joint torque generated by muscle i, a_i is a constant, and t is the period during which the muscle is active. The assumption is that the CNS tries to minimize the total muscular effort, the sum of the efforts of the individual muscles. Whether this is true or not remains unknown.

6.5 SUMMARY

Although the methods described in this chapter are highly sophisticated, they do not satisfactorily resolve the problem of computing mechanical work expended for human motion.

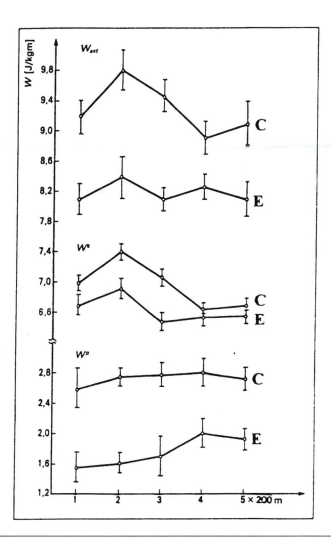

Figure 6.42 The positive work in running expended for changes in the potential and kinetic energy of the CoM by members of the Soviet Olympic team (E, $n = 5$) and college athletes (C, $n = 4$). The subjects ran 1,000 m (five indoor laps). The performance results of elite athletes were in the range of 2 min 37 s to 2 min 42 s, and the performance results of the collegiate athletes were from 2 min 51 s to 2 min 58 s. The potential and kinetic energy changes during 1 m of distance in a stance period were recorded. W^k and W^p are the fractions of the positive work expended in changing the kinetic and potential energy of the CoM, respectively. W_{ext} is the positive work done to change the total mechanical energy of the CoM. The elite athletes performed less work while running at a faster speed.

Adapted from Zatsiorsky, V.M., S.Y. Aleshinsky, and N.A. Yakunin. 1982. *Biomechanical basis of endurance* (in Russian). Moscow: FiS.

The human body is a multilink system powered by force actuators (muscles). The forces generated by the individual muscles are usually unknown. The muscles spend energy not only for work production but also for force generation. The muscles also spend energy when they resist forcible stretching. In addition, the muscle-tendon complexes and ligaments can store elastic potential energy, which can later be recoiled. The proportion of the stored and recoiled energy is generally unknown. All these phenomena make determining the mechanical energy expenditure in human movement a perplexing task.

In classical mechanics, the work of a force $\mathbf{F}$ on an infinitesimal displacement $d\mathbf{r}$ is defined as the scalar product of the vectors $\mathbf{F}$ and $d\mathbf{r}$. Power is the time rate of doing work. The expression "mechanical work" is ambiguous until the force under consideration is specified. No work is done when (1) the force is perpendicular to the displacement or (2) the force is applied at a motionless point. When the point of force application moves but the body itself does not, the work is zero. Work is invariant; it does not depend on the system of coordinates.

Conservative forces depend on the position of their point of application. Gravity force (weight) and elastic forces are conservative forces. Friction and hydrodynamic resistance are not conservative forces. A system where work is done only by conservative forces is called a conservative system. In a conservative system, the sum of the potential and kinetic energy of the system is constant.

The total work done by several forces on a rigid body equals the work of the resultant force and couple. It is also equal to the change in the body's kinetic energy (the work–energy principle for a rigid body). The total kinetic energy of a system of material particles equals the sum of the kinetic energy of the CoM and the kinetic energy of the particles in their motion relative to the CoM (König's theorem). In particular, the kinetic energy of a rigid body equals the sum of the energy associated with the movement of the CoM (translational kinetic energy) and the energy associated with the body's rotation around the CoM. The work of the resultant (the work done on the body) can sharply differ from the work of individual forces exerted on the body.

In human movement, the work or power done on external objects is called external work or power, and the work or power expended to move the body links (to change their total mechanical energy) is named internal work or power. The chapter deals mainly with determining internal work.

Mechanical power from individual muscles (muscle power) is the primary source of human movements. Unfortunately, at this time it cannot be determined except for certain muscles and under unique conditions. Instead, the joint power is computed. This approach neglects the activity of antagonistic muscles as well as energy losses for internal friction, displacement of the muscles with respect to the bones, and so on. Still, the basic level at which mechanical work is customarily analyzed is the level of the joint torques. The work of other forces, for instance, the work of resultant forces done on body segments, is compared with this benchmark.

The joint power P is computed as the scalar product of the vectors of the joint torques ($\mathbf{T}$) and the relative angular velocity ($\dot{\alpha}$) at the joints, $P = \mathbf{T} \cdot \dot{\alpha}$. For joints served only by one-joint muscles, the relationship between joint power and muscle power is plain. If the joint power is positive, the muscles produce positive work and the mechanical energy of the system increases. When the joint power is negative, the total mechanical energy of the adjacent segments decreases; the energy is used to forcibly stretch the muscles and tendons. In joints crossed only by one-joint muscles, the energy can be transferred between the adjacent body segments. In joints crossed by two-joint muscles, the energy can be transferred between the nonadjacent segments. The joint forces redistribute the mechanical energy among the body segments without changing the total mechanical energy of the whole human body.

The work done *by* individual joint forces and torques should be distinguished from the work done *on* the body segments (the total work done by all the forces and couples acting on the segment). The work done on the segments can be seen as the work of the resultant force and couple reduced to the CoM of the segment.

Forces and moments of force (joint torques) acting on the system are called the sources of mechanical energy. The kinetic energy of a link associated with the horizontal, vertical, and rotational motion and the potential energy of the link are called the fractions of the mechanical energy.

In human movement, the mechanical energy can transform from one fraction to another. Energy transformation from potential energy into kinetic energy and back is especially common. This does not mean, however, that the body segments behave as ideal pendulums. For any ideal pendulum, the total mechanical energy (the sum of the kinetic and potential energy) is constant over an oscillation cycle, and the peak values of the kinetic and potential energy are equal. These conditions are seldom satisfied in human motion. The energy conservation coefficient is intended to characterize the proportion of the mechanical energy conserved in a given movement for the work of the resultant force and couple reduced to the CoM of the segment.

A negative change in the body's mechanical energy is usually larger than the negative work of joint moments. The ratio of the negative work of joint moments to the reduction in total mechanical energy is called the index of softness of landing.

When a motor task involves several body segments, two main problems arise: (1) what is the total amount of work or power done *on* the entire system, and (2) what is the total amount of mechanical energy generated *by* the sources? If the total energy of the human body does not change, the work done on the body equals zero. For instance, in the absence of energy losses to the environment, the mechanical work done *on* the body in walking and running for an arbitrary number of steps at constant average speed is zero (the zero-work paradox). However, to execute such an activity, mechanical energy must be expended. Although the work done on the human body can always be computed, the computed values of work and power are often useless even if correct. The main topic of interest is not the total work done *on* the body but the total amount of mechanical energy expended by the sources. To compute the mechanical energy expended by the sources, summation of the absolute values is suggested. Presently, the total mechanical energy expended for moving a multilink system such as a human body can be determined unambiguously only for some movements, when all the joint torques produce either only positive or only negative power. In all other situations, additional information about the sources of energy is required.

Two characteristics of the sources of energy are defined: intercompensation of sources and compensation of sources during time (recuperation). When the sources are not intercompensated, the total power they produce equals the sum of the absolute values of the power that each develops. The joint torques generated by one-joint muscles are not intercompensated sources of mechanical energy. For a source of energy compensated during time, the mechanical energy expenditure (MEE) equals the absolute value of the integral over time of the power developed by the source. Energy compensation over time occurs when the sources of energy are conservative.

Three main methods are employed to measure the mechanical energy expended for human body motion. In the first method, known as the source approach, the joint power or work is determined. The second technique, the fraction approach, determines the change of mechanical energy of individual body segments. In the third approach, the CoM model, the energy associated with the movement of the CoM of the body is determined.

6.6 QUESTIONS FOR REVIEW

1. Give a definition of the work of a force.
2. Discuss workless forces in human movement.
3. Prove the invariance of the work of a force.
4. Give examples of conservative and nonconservative forces.
5. Discuss the work–energy principle for a rigid body.
6. Explain why the work done on a rigid body can be much less than the work of a force exerted on the body.
7. Discuss joint power.
8. How can joint power be determined?

9. Discuss the physical meaning of negative joint power when the joint is crossed only by (a) one-joint muscles or (b) two-joint muscles.

10. Explain the mechanisms of energy transfer between adjacent and nonadjacent body segments.

11. Discuss the work of joint forces.

12. Define sources and fractions of mechanical energy in human movement.

13. Discuss the transformation and conservation of energy during the motion of one link.

14. Why is the negative work of joint torques usually smaller than the negative work done on the body segments in human movement?

15. How can the total work done on the entire human body be computed?

16. Discuss the zero-work paradox.

17. Define intercompensation of sources of mechanical energy.

18. Discuss compensation of sources during time.

19. In a scientific paper, the total power was computed as the sum $P_{tot} = \Sigma P_j$, where P_j is the joint power at an individual joint. What is the rate of change of the total mechanical energy of the system? Under which condition does P_{tot} represent the total power generated by the joint torques? Discuss the case when $P_{tot} = 0$.

20. In a scientific paper, the absolute values of power were used in the equations $P_{tot} = \Sigma |P_s|$ and $MEE = \int |P_s| dt$. Explain the physical meaning of these equations.

21. Compare the source and fraction approaches for determining the amount of mechanical energy expended for motion.

22. In a class research project, a student used the fraction approach and assumed that energy is not transferred between the body segments. What is the physical meaning of this assumption?

23. Compare the changes in the fractions of energy associated with the movement of the CoM in walking and running.

24. Discuss the energy conservation coefficient.

25. Discuss the Froude number.

6.7 BIBLIOGRAPHY

Abbott, B.C., B. Bigland, and J.M. Ritchie. 1952. The physiologic cost of negative work. *J. Physiol. (Lond.)* 117: 380–90.

Ae, M., K. Miyashita, T. Yokoi, and Y. Hashihara. 1987. Mechanical power and work done by the muscles of the lower limb during running at different speeds. In *Biomechanics X-B,* ed. B. Johnson, 895–99. Champaign, IL: Human Kinetics.

Aissaoui, R., P. Allard, A. Junqua, L. Frossard, and M. Duhaime. 1996. Internal work estimation in three-dimensional gait analysis. *Med. Biol. Eng. Comput.* 34 (6): 467–71.

Aleshinsky, S.Y. 1986a. An energy "sources" and "fractions" approach to the mechanical energy expenditure problem: I. Basic concepts, description of the model, analysis of a one-link system movement. *J. Biomech.* 19 (4): 287–93.

Aleshinsky, S.Y. 1986b. An energy "sources" and "fractions" approach to the mechanical energy expenditure problem: II. Movement of the multi-link chain model. *J. Biomech.* 19 (4): 295–300.

Aleshinsky, S.Y. 1986c. An energy "sources" and "fractions" approach to the mechanical energy expenditure problem: III. Mechanical energy expenditure reduction during one link motion. *J. Biomech.* 19 (4): 301–6.

Aleshinsky, S.Y. 1986d. An energy "sources" and "fractions" approach to the mechanical energy expenditure problem: IV. Criticism of the concept of "energy transfers within and between links." *J. Biomech.* 19 (4): 307–9.

Aleshinsky, S.Y. 1986e. An energy "sources" and "fractions" approach to the mechanical energy expenditure problem: V. The mechanical energy expenditure reduction during motion of the multi-link system. *J. Biomech.* 19 (4): 311–15.

Alexander, R.M. 1984. Elastic energy stores in running vertebrates. *Am. Zool.* 24: 85–94.

Alexander, R.M. 1991a. Elastic mechanisms in primate locomotion. *Z. Morphol. Anthropol.* 78 (3): 315–20.

Alexander, R.M. 1991b. Energy-saving mechanisms in walking and running. *J. Exp. Biol.* 160: 55–69.

Alexander, R.M. 1997. A minimum energy cost hypothesis for human arm trajectories. *Biol. Cybern.* 76 (2): 97–105.

Alexander, R.M. 2000. Energy-minimizing choices of muscles and patterns of movement. *Motor Control* 4 (1): 45–47.

Alexander, R.M., and H.C. Bennet-Clark. 1977. Storage of elastic strain energy in muscle and other tissues. *Nature* 265 (5590): 114–17.

Alexander, R.M., and G.E. Goldspink. 1977. *Mechanics and energetics of animal locomotion.* London: Chapman and Hall.

Amadio, A.C., and W. Baumann. 1995. Energetic aspects of the triple jump. In *XVth congress of the International Society of Biomechanics, book of abstracts,* ed. K. Häkkinen, K.L. Keskinen, P.V. Komi, and A. Mero, 48–49. Jyväskylä, Finland: Jyväskylä University.

Anderson, F.C., and M.G. Pandy. 1993. Storage and utilization of elastic energy during jumping. *J. Biomech.* 26: 1413–27.

Antonutto, G., C. Capelli, M. Girardis, P. Zamparo, and P.E. di Prampero. 1999. Effects of microgravity on maximal power of lower limbs during very short efforts in humans. *J. Appl. Physiol.* 86 (1): 85–92.

Araki, K., T. Nakatani, K. Toda, Y. Taenaka, E. Tatsumi, T. Masuzawa, Y. Baba, A. Yagura, Y. Wakisaka, K. Eya, et al. 1995. Power of the fatigue resistant in situ latissimus dorsi muscle. *ASAIO J.* 41 (3): M768–M771.

Arampatzis, A., and G.P. Brüggemann. 1998. A mathematical high bar–human body model for analysing and interpreting mechanical-energetic processes on the high bar. *J. Biomech.* 31 (12): 1083–92.

Arampatzis, A., and G.P. Brüggemann. 1999. Mechanical energetic processes during the giant swing exercise before dismounts and flight elements on the high bar and the uneven parallel bars. *J. Biomech.* 32 (8): 811–20.

Arampatzis, A., A. Knicker, V. Metzler, and G.P. Brüggemann. 2000. Mechanical power in running: A comparison of different approaches. *J. Biomech.* 33 (4): 457–63.

Ardigo, L.P., C. Lafortuna, A.E. Minetti, P. Mognoni, and F. Saibene. 1995. Metabolic and mechanical aspects of foot landing type, forefoot and rearfoot strike, in human running. *Acta Physiol. Scand.* 155 (1): 17–22.

Asmussen, E., and F. Bonde-Petersen. 1974a. Apparent efficiency and storage of elastic energy in human muscles during exercise. *Acta Physiol. Scand.* 92: 537–45.

Asmussen, E., and F. Bonde-Petersen. 1974b. Storage of elastic energy in skeletal muscles. *Acta Physiol. Scand.* 91: 385–92.

Avis, F.J., A. Hoving, and H.M. Toussaint. 1985. A dynamometer for the measurement of force, velocity, work and power during an explosive leg extension. *Eur. J. Appl. Physiol.* 54 (2): 210–15.

Baron, R., N. Bachl, R. Petschnig, H. Tschan, G. Smekal, and R. Pokan. 1999. Measurement of maximal power output in isokinetic and non-isokinetic cycling: A comparison of two methods. *Int. J. Sports Med.* 20 (8): 532–37.

Barr, A.E., K.L. Siegel, J.V. Danoff, C.L. McGarvey, III, A. Tomasko, I. Sable, and S.J. Stanhope. 1992. Biomechanical comparison of the energy-storing capabilities of SACH and Carbon Copy II prosthetic feet during the stance phase of gait in a person with below-knee amputation. *Phys. Ther.* 72 (5): 344–54.

Bartonietz, K. 2000. Javelin throwing: An approach to performance development. In *The IOC encyclopaedia of sports medicine.* Vol. 9, *Biomechanics in sport: Performance enhancement and injury prevention,* ed. V.M. Zatsiorsky, 401–34. Oxford, UK: Blackwell Science.

Bassett, D.R., Jr., C.R. Kyle, L. Passfield, J.P. Broker, and E.R. Burke. 1999. Comparing cycling world hour records, 1967–1996: Modeling with empirical data. *Med. Sci. Sports Exerc.* 31 (11): 1665–76.

Baudinette, R.V. 1991. The energetics and cardiorespiratory correlates of mammalian terrestrial locomotion. *J. Exp. Biol.* 160: 209–31.

Becket, R., and K. Chang. 1968. An evaluation of the kinematics of gait by minimum energy. *J. Biomech.* 1: 147–59.

Beelen, A., and A.J. Sargeant. 1991. Effect of fatigue on maximal power output at different contraction velocities in humans. *J. Appl. Physiol.* 71 (6): 2332–37.

Behnke, H. 1997. Optimization models for the force and energy in competitive running. *J. Math. Biol.* 35 (4): 375–90.

Bejor, M., R. Lombardi, L. Panella, and L. Ricciardi. 1993. Power output during knee extension/flexion movement performed in controlled mechanical conditions. *G. Ital. Med. Lav.* 15 (5–6): 105–8.

Belli, A., J. Avela, and P.V. Komi. 1993. Mechanical energy assessment with different methods during running. *Int. J. Sports Med.* 14 (5): 252–56.

Bellizzi, M.J., K.A.D. King, S.K. Cushman, and P.G. Weyand. 1998. Does the application of ground force set the energetic cost of cross-country skiing? *J. Appl. Physiol.* 85 (5): 1736–43.

Berg, C., and J. Rayner. 1995. The moment of inertia of bird wings and the inertial power requirement for flapping flight. *J. Exp. Biol.* 198 (Pt. 8): 1655–64.

Bernardi, M., I. Canale, V. Castellano, L. Di Filippo, F. Felici, and M. Marchetti. 1995. The efficiency of walking of paraplegic patients using a reciprocating gait orthosis. *Paraplegia* 33 (7): 409–15.

Bertram, J.E., A. Ruina, C.E. Cannon, Y.H. Chang, and M.J. Coleman. 1999. A point-mass model of gibbon locomotion. *J. Exp. Biol.* 202 (Pt. 19): 2609–17.

Bianchi, L., D. Angelini, and F. Lacquaniti. 1998. Individual characteristics of human walking mechanics. *Pflugers Arch.* 436 (3): 343–56.

Bianchi, L., D. Angelini, G.P. Orani, and F. Lacquaniti. 1998. Kinematic coordination in human gait: Relation to mechanical energy cost. *J. Neurophysiol.* 79 (4): 2155–70.

Biewener, A.A. 1998. Muscle-tendon stresses and elastic energy storage during locomotion in the horse. *Comp. Biochem. Physiol. B Biochem. Mol. Biol.* 120 (1): 73–87.

Biewener, A.A., and T.J. Roberts. 2000. Muscle and tendon contributions to force, work, and elastic energy savings: A comparative perspective. *Exerc. Sport Sci. Rev.* 28 (3): 99–107.

Billat, L.V., J.P. Koralsztein, and R.H. Morton. 1999. Time in human endurance models: From empirical models to physiological models. *Sports Med.* 27 (6): 359–79.

Blickhan, R. 1989. The spring-mass model for running and hopping. *J. Biomech.* 22: 1217–27.

Blickhan, R., and R.J. Full. 1987. Locomotion energetics of the ghost crab: II. Mechanics of the centre of mass during walking and running. *J. Exp. Biol.* 130: 155–74.

Blickhan, R., and R.J. Full. 1992. Mechanical work in terrestrial locomotion. In *Biomechanics: Structures and systems,* ed. A.A. Biewener, 75–96. New York: Oxford University Press.

Bobbert, M.F., K.G. Gerritsen, M.C. Litjens, and A.J. Van Soest. 1996. Why is counter-movement jump height greater than squat jump height? *Med. Sci. Sports Exerc.* 28 (11): 1402–12.

Bobbert, M.F., P.A. Huijing, and G.J. van Ingen Schenau. 1986. An estimation of power output and work done by the human triceps surae muscle-tendon complex in jumping. *J. Biomech.* 19 (11): 899–906.

Bosco, C., A. Belli, M. Astrua, J. Tihanyi, R. Pozzo, S. Kellis, O. Tsarpela, C. Foti, R. Manno, and C. Tranquilli. 1995. A dynamometer for evaluation of dynamic muscle work. *Eur. J. Appl. Physiol.* 70 (5): 379–86.

Bosco, C., P.V. Komi, and K. Sinkkonen. 1980. Mechanical power, net efficiency and muscle structure in male and female middle-distance runners. *Scand. J. Sport Sci.* 2: 47–51.

Bosco, C., P. Luhtanen, and P.V. Komi. 1983. A simple method for measurement of mechanical power in jumping. *Eur. J. Appl. Physiol.* 50 (2): 273–82.

Brandt, R.A., and M.A. Pichowsky. 1995. Conservation of energy in competitive swimming. *J. Biomech.* 28 (8): 925–33.

Brisswalter, J., and P. Legros. 1995. Use of energy cost and variability in stride length to assess an optimal running adaptation. *Percept. Mot. Skills* 80 (1): 99–104.

Broker, J.P., and R.J. Gregor. 1994. Mechanical energy management in cycling: Source relations and energy expenditure. *Med. Sci. Sports Exerc.* 26 (1): 64–74.

Broker, J.P., C.R. Kyle, and E.R. Burke. 1999. Racing cyclist power requirements in the 4000-m individual and team pursuits. *Med. Sci. Sports Exerc.* 31 (11): 1677–85.

Brueckner, J.C., G. Atchou, C. Capelli, A. Duvallet, D. Barrault, E. Jousselin, M. Rieu, and P.E. di Prampero. 1991. The energy cost of running increases with the distance covered. *Eur. J. Appl. Physiol.* 62 (6): 385–89.

Buckley, J.G., W.D. Spence, and S.E. Solomonidis. 1997. Energy cost of walking: Comparison of "intelligent prosthesis" with conventional mechanism. *Arch. Phys. Med. Rehabil.* 78 (3): 330–33.

Buczek, F.L., T.M. Kepple, K.L. Siegel, and S.J. Stanhope. 1994. Translational and rotational joint power terms in a six degree-of-freedom model of the normal ankle complex. *J. Biomech.* 27 (12): 1447–57.

Bunc, V., S. Sprynarova, J. Parizkova, and J. Leso. 1984. Effects of adaptation on the mechanical efficiency and energy cost of physical work. *Hum. Nutr. Clin. Nutr.* 38 (4): 317–19.

Burdett, R.G., G.S. Skrinar, and S.R. Simon. 1983. Comparison of mechanical work and metabolic energy consumption during normal gait. *J. Orthop. Res.* 1 (1): 63–72.

Burstrom, L., and S.H. Bylund. 2000. Relationship between vibration dose and the absorption of mechanical power in the hand. *Scand. J. Work Environ. Health* 26 (1): 32–36.

Caldwell, G.E., and L.W. Forrester. 1992. Estimates of mechanical work and energy transfers: Demonstration of a rigid body power model of the recovery leg in gait. *Med. Sci. Sports Exerc.* 24 (12): 1396–412.

Candau, R., A. Belli, G.Y. Millet, D. Georges, B. Barbier, and J.D. Rouillon. 1998. Energy cost and running mechanics during a treadmill run to voluntary exhaustion in humans. *Eur. J. Appl. Physiol.* 77 (6): 479–85.

Capelli, C., C. Donatelli, C. Moia, C. Valier, G. Rosa, and P.E. di Prampero. 1990. Energy cost and efficiency of sculling a Venetian gondola. *Eur. J. Appl. Physiol.* 60 (3): 175–78.

Capelli, C., P. Zamparo, A. Cigalotto, M.P. Francescato, R.G. Soule, B. Termin, D.R. Pendergast, and P.E. di Prampero. 1995. Bioenergetics and biomechanics of front crawl swimming. *J. Appl. Physiol.* 78 (2): 674–79.

Capmal, S., and H. Vandewalle. 1997. Torque–velocity relationship during cycle ergometer sprints with and without toe clips. *Eur. J. Appl. Physiol.* 76 (4): 375–79.

Cappozzo, A., F. Figura, T. Leo, and M. Marchetti. 1978. Movements and mechanical energy changes of the upper part of the human body during walking. *Biomechanics VI-A*, ed. E. Asmussen and K. Jørgensen, 272–79. Baltimore: University Park Press.

Cappozzo, A., F. Figura, and M. Marchetti. 1976. The interplay of muscular and external forces in human ambulation. *J. Biomech.* 9 (1): 35–43.

Cavagna, G.A. 1970. Elastic bounce of the body. *J. Appl. Physiol.* 29 (3): 279–82.

Cavagna, G.A. 1975. Force platforms as ergometers. *J. Appl. Physiol.* 39: 174–91.

Cavagna, G.A. 1978. Aspects of efficiency and inefficiency of terrestrial locomotion. In *Biomechanics VI-A,* ed. E. Asmussen and K. Jørgensen, 3–26. Baltimore: University Park Press.

Cavagna, G.A., B. Dusman, and R. Margaria. 1968. Positive work done by previously stretched muscle. *J. Appl. Physiol.* 24 (1): 21–32.

Cavagna, G.A., and P. Franzetti. 1981. Mechanics of competition walking. *J. Physiol. (Lond.)* 315: 243–51.

Cavagna, G.A., P. Franzetti, and T. Fuchimoto. 1983. The mechanics of walking in children. *J. Physiol. (Lond.)* 343: 323–39.

Cavagna, G.A., P. Franzetti, N.C. Heglund, and P. Willems. 1988. The determinants of step frequency in running, trotting, and hopping in man and other vertebrates. *J. Physiol. (Lond.)* 399: 81–92.

Cavagna, G.A., N.C. Heglund, and C.R. Taylor. 1977. Mechanical work in terrestrial locomotion: Two basic mechanisms for minimizing energy expenditure. *Am. J. Physiol.* 233 (5): R243–R261.

Cavagna, G.A., and M. Kaneko. 1977. Mechanical work and efficiency in level walking and running. *J. Physiol. (Lond.)* 268 (2): 647–81.

Cavagna, G.A., L. Komarek, G. Citterio, and R. Margaria. 1971. Power output of the previously stretched muscle. In *Biomechanics II,* ed. J. Vredenbregt and J. Wartenweiler, 159–67. Basel: Karger.

Cavagna, G.A., L. Komarek, and S. Mazzoleni. 1971. The mechanics of sprint running. *J. Physiol. (Lond.)* 217 (3): 709–21.

Cavagna, G.A., M. Mantovani, P.A. Willems, and G. Musch. 1997. The resonant step frequency in human running. *Pflugers Arch.* 434 (6): 678–84.

Cavagna, G.A., and R. Margaria. 1966. Mechanics of walking. *J. Appl. Physiol.* 21 (1): 271–78.

Cavagna, G.A., F.P. Saibene, and R. Margaria. 1963. External work in walking. *J. Appl. Physiol.* 18: 1–9.

Cavagna, G.A., F.P. Saibene, and R. Margaria. 1964. Mechanical work in running. *J. Appl. Physiol.* 19: 249–56.

Cavagna, G.A., L. Tesio, T. Fuchimoto, and N.C. Heglund. 1983. Ergometric evaluation of pathological gait. *J. Appl. Physiol.* 55 (2): 607–13.

Cavagna, G.A., H. Thys, and A. Zamboni. 1976. The sources of external work in level walking and running. *J. Physiol. (Lond.)* 262 (3): 639–57.

Cavagna, G.A., P.A. Willems, P. Franzetti, and C. Detrembleur. 1991. The two power limits conditioning step frequency in human running. *J. Physiol. (Lond.)* 437: 95–108.

Cavagna, G.A., P.A. Willems, and N.C. Heglund. 1998. Walking on Mars [letter]. *Nature* 393 (6686): 636.

Cavagna, G.A., A. Zamboni, T. Farragiana, and R. Margaria. 1972. Jumping on the moon: Power output at different gravity values. *Aerospace Med.* 43 (4): 408–14.

Cavanagh, P.R., and R. Kram. 1985a. The efficiency of human movement: A statement of the problem. *Med. Sci. Sports Exerc.* 17 (3): 304–8.

Cavanagh, P.R., and R. Kram. 1985b. Mechanical and muscular factors affecting the efficiency of human movement. *Med. Sci. Sports Exerc.* 17 (3): 326–31.

Celentano, F., G. Cortili, P.E. di Prampero, and P. Cerretelli. 1974. Mechanical aspects of rowing. *J. Appl. Physiol.* 36 (6): 642–47.

Chadwick, E.K., A.C. Nicol, J.V. Lane, and T.G. Gray. 1999. Biomechanics of knife stab attacks. *Forensic Sci. Int.* 105 (1): 35–44.

Chang, Y.H., and R. Kram. 1999. Metabolic cost of generating horizontal forces during human running. *J. Appl. Physiol.* 86 (5): 1657–62.

Chapman, A.E., and G.E. Caldwell. 1983a. Factors determining changes in lower limb energy during swing in treadmill running. *J. Biomech.* 16 (1): 69–77.

Chapman, A.E., and G.E. Caldwell. 1983b. Kinetic limitations of maximal sprinting speed. *J. Biomech.* 16 (1): 79–83.

Chapman, A., G. Caldwell, R. Herring, R. Lonergan, and S. Selbie. 1987. Mechanical energy and the preferred style of running. In *Biomechanics X-B,* ed. B. Jonsson, 875–79. Champaign, IL: Human Kinetics.

Chapman, A.E., G.E. Caldwell, and W.S. Selbie. 1985. Mechanical output following muscle stretch in forearm supination against inertial loads. *J. Appl. Physiol.* 59 (1): 78–86.

Charteris, J. 1999. Effects of velocity on upper to lower extremity muscular work and power output ratios of intercollegiate athletes. *Br. J. Sports Med.* 33 (4): 250–54.

Chou, L.-S., L.F. Draganich, and S.-M. Song. 1997. Minimal energy trajectories of the swing ankle when stepping over obstacles of different heights. *J. Biomech.* 30 (2): 115–20.

Clayton, H.M., J.L. Lanovaz, H.C. Schamhardt, M.A. Willemen, and G.R. Colborne. 1998. Net joint moments and powers in the equine forelimb during the stance phase of the trot. *Equine Vet. J.* 30 (5): 384–89.

Clayton, H.M., H.C. Schamhardt, M.A. Willemen, J.L. Lanovaz, and G.R. Colborne. 2000. Net joint moments and joint powers in horses with superficial digital flexor tendinitis. *Am. J. Vet. Res.* 61 (2): 197–201.

Cleland, J. 1867. On the actions of muscles passing over more than one joint. *J. Anat. Physiol.* 1: 85–93.

Colborne, G.R., J.L. Lanovaz, E.J. Sprigings, H.C. Schamhardt, and H.M. Clayton. 1997a. Joint moments and power in equine gait: A preliminary study. *Equine Vet. J.* 23 (Suppl.): 33–36.

Colborne, G.R., J.L. Lanovaz, E.J. Sprigings, H.C. Schamhardt, and H.M. Clayton. 1997b. Power flow in the equine forelimb. *Equine Vet. J.* 23 (Suppl.): 37–40.

Colborne, G.R., J.L. Lanovaz, E.J. Sprigings, H.C. Schamhardt, and H.M. Clayton. 1998. Forelimb joint moments and power during the walking stance phase of horses. *Am. J. Vet. Res.* 59 (5): 609–14.

Colborne, G.R., S. Naumann, P.E. Longmuir, and D. Berbrayer. 1992. Analysis of mechanical and metabolic factors in the gait of congenital below knee amputees: A comparison of the SACH and Seattle feet. *Am. J. Phys. Med. Rehabil.* 71 (5): 272–78.

Cooper, P.G., G.J. Wilson, D.T. Hardman, O. Kawaguchi, Y.F. Huang, A. Martinez-Coll, R. Carrington, E. Puchert, R. Crameri, C. Horam, and S.N. Hunyor. 1999. In situ measurements of skeletal muscle power output using new capacitive strain gauge. *Med. Biol. Eng. Comput.* 37 (4): 451–55.

Corwin, C., V. Zatsiorsky, and V. Fortney. 1998. Work and power production in the vertical jump under various loads fixed to the upper body or ankles. In *Proceedings of NACOB '98. The third North American Congress on Biomechanics, August 14–18, 1998,* 417–18. Waterloo, ON: University of Waterloo.

Cotes, K.E., and F. Meade. 1960. The energy expenditure and mechanical work demand in walking. *Ergonomics* 3: 97–119.

Crompton, R.H., L. Yu, W. Weijie, M. Gunther, and R. Savage. 1998. The mechanical effectiveness of erect and "bent-hip, bent-knee" bipedal walking in *Australopithecus afarensis. J. Hum. Evol.* 35 (1): 55–74.

Czerniecki, J.M., and A. Gitter. 1992. Insights into amputee running: A muscle work analysis. *Am. J. Phys. Med. Rehabil.* 71 (4): 209–18.

Czerniecki, J.M., A.J. Gitter, and J.C. Beck. 1996. Energy transfer mechanisms as a compensatory strategy in below knee amputee runners. *J. Biomech.* 29 (6): 717–22.

Czerniecki, J.M., A. Gitter, and C. Munro. 1991. Joint moment and muscle power output characteristics of below knee amputees during running: The influence of energy storing prosthetic feet. *J. Biomech.* 24 (1): 63–75.

Dalleau, G., A. Belli, M. Bourdin, and J.R. Lacour. 1998. The spring-mass model and the energy cost of treadmill running. *Eur. J. Appl. Physiol.* 77 (3): 257–63.

Danoff, J.V. 1978. Power produced by maximal velocity elbow flexion. *J. Biomech.* 11: 481–86.

Davids, J.R., A.M. Bagley, and M. Bryan. 1998. Kinematic and kinetic analysis of running in children with cerebral palsy. *Dev. Med. Child Neurol.* 40 (8): 528–35.

Davies, C.T. 1980a. Effect of air resistance on the metabolic cost and performance of cycling. *Eur. J. Appl. Physiol.* 45 (2–3): 245–54.

Davies, C.T. 1980b. Effects of wind assistance and resistance on the forward motion of a runner. *J. Appl. Physiol.* 48 (4): 702–9.

Davies, C.T., and K. Young. 1985. Mechanical power output in children aged 11 and 14 years. *Acta Paediatr. Scand.* 74 (5): 760–64.

Dean, G.A. 1965. An analysis of the energy expenditure in level and grade walking. *Ergonomics* 8: 31–47.

de Boer, R.W., J. Cabri, W. Vaes, J.P. Clarijs, A.P. Hollander, G. de Groot, and G.J. van Ingen Schenau. 1987. Moments of force, power, and muscle coordination in speed-skating. *Int. J. Sports Med.* 8 (6): 371–78.

Demes, B., W.L. Jungers, T.S. Gross, and J.G. Fleagle. 1995. Kinetics of leaping primates: Influence of substrate orientation and compliance. *Am. J. Phys. Anthropol.* 96 (4): 419–29.

Derrick, T.R., J. Hamill, and G.E. Caldwell. 1998. Energy absorption of impacts during running at various stride lengths. *Med. Sci. Sports Exerc.* 30 (1): 128–35.

Devita, P., and W.A. Skelly. 1992. Effect of landing stiffness on joint kinetics and energetics in the lower extremity. *Med. Sci. Sports Exerc.* 24 (1): 108–15.

Devita, P., and J. Stribling. 1991. Lower extremity joint kinetics and energetics during backward running. *Med. Sci. Sports Exerc.* 23 (5): 602–10.

Dickinson, M.H., C.T. Farley, R.J. Full, M.A. Koehl, R. Kram, and S. Lehman. 2000. How animals move: An integrative view. *Science* 288 (5463): 100–106.

Dillman, C.J., and R.C. Nelson. 1968. The mechanical energy transformations of pole vaulting with a fiberglass pole. *J. Biomech.* 1: 175–83.

di Prampero, P.E. 2000. Cycling on earth, in space, on the moon. *Eur. J. Appl. Physiol.* 82 (5–6): 345–60.

di Prampero, P.E., G. Cortili, P. Mognoni, and F. Saibene. 1976. Energy cost of speed skating and efficiency of work against air resistance. *J. Appl. Physiol.* 40 (45): 584–91.

Duff-Raffaele, M., D.C. Kerrigan, P.J. Corcoran, and M. Saini. 1996. The proportional work of lifting the center of mass during walking. *Am. J. Phys. Med. Rehabil.* 75 (5): 375–79.

Ebbeling, C.J., J. Hamill, and J.A. Crussemeyer. 1994. Lower extremity mechanics and energy cost of walking in high-heeled shoes. *J. Orthop. Sports Phys. Ther.* 19 (4): 190–96.

Ehara, Y., M. Beppu, S. Nomura, Y. Kunimi, and S. Takahashi. 1993. Energy storing property of so-called energy-storing prosthetic feet. *Arch. Phys. Med. Rehabil.* 74 (1): 68–72.

Ekevad, M., and B. Lundberg. 1997. Influence of pole length and stiffness on the energy conversion in pole vaulting. *J. Biomech.* 30 (3): 259–64.

Elftman, H. 1939. Forces and energy changes in the leg during walking. *Am. J. Physiol.* 125: 339–56.

Elftman, H. 1940. The work done by muscles in running. *Am. J. Physiol.* 129: 672–84.

Epstein, Y., J. Rosenblum, R. Burstein, and M.N. Sawka. 1988. External load can alter the energy cost of prolonged exercise. *Eur. J. Appl. Physiol.* 57 (2): 243–47.

Ericson, M.O. 1988. Mechanical muscular power output and work during ergometer cycling at different loads and speeds. *Eur. J. Appl. Physiol.* 57: 382–87.

Ericson, M.O., A. Bratt, R. Nisell, U.P. Arborelius, and J. Ekholm. 1986. Power output and work in different muscle groups during ergometer cycling. *Eur. J. Appl. Physiol.* 55 (3): 229–35.

Falls, H.B., and L.D. Humphrey. 1976. Energy cost of running and walking in young women. *Med. Sci. Sports* 8 (1): 9–13.

Farber, B.S., and J.S. Jacobson. 1995. An above-knee prosthesis with a system of energy recovery: A technical note. *J. Rehabil. Res. Dev.* 32 (4): 337–48.

Farley, C.T. 1997. Maximum speed and mechanical power output in lizards. *J. Exp. Biol.* 200 (Pt. 16): 2189–95.

Farley, C.T., and D.P. Ferris. 1998. Biomechanics of walking and running: Center of mass movements to muscle action. *Exerc. Sport Sci. Rev.* 26: 253–85.

Farley, C.T., and T.A. McMahon. 1992. Energetics of walking and running: Insights from simulated reduced-gravity experiments. *J. Appl. Physiol.* 73 (6): 2709–12.

Fedak, M.A., N.C. Heglund, and C.R. Taylor. 1982. Energetics and mechanics of terrestrial locomotion: II. Kinetic energy changes of the limbs and body as a function of speed and body size in birds and mammals. *J. Exp. Biol.* 97: 23–40.

Fenn, W.O. 1924. The relation between the work performed and the energy liberated in muscular contraction. *J. Physiol. (Lond.)* 58: 373–95.

Fenn, W.O. 1929. Mechanical energy expenditure in sprint running as measured by moving pictures. *Am. J. Physiol.* 90: 343.

Fenn, W.O. 1930a. Frictional and kinetic factors in the work of sprint running. *Am. J. Physiol.* 92: 583–611.

Fenn, W.O. 1930b. Work against gravity and work due to velocity changes in running. *Am. J. Physiol.* 93: 433–62.

Fenn, W.O. 1932. Zur Mechanik des Radfahrens in Vergleich zu des Laufens. *Pflugers Arch.* 229: 354–66.

Ferretti, G., M. Gussoni, P.E. di Prampero, and P. Cerretelli. 1987. Effects of exercise on maximal instantaneous muscular power of humans. *J. Appl. Physiol.* 62 (6): 2288–94.

Flynn, T.W., and R.W. Soutas-Little. 1993. Mechanical power and muscle action during forward and backward running. *J. Orthop. Sports Phys. Ther.* 17 (2): 108–12.

Foerster, S.A., A.M. Bagley, C.D.J. Mote, and H.B. Skinner. 1995. The prediction of metabolic energy expenditure during gait from mechanical energy of the limb: A preliminary study. *J. Rehabil. Res. Dev.* 32 (2): 128–34.

Francescato, M.P., M. Girardis, and P.E. di Prampero. 1995. Oxygen cost of internal work during cycling. *Eur. J. Appl. Physiol.* 72 (1–2): 51–57.

Frederick, E.C. 1985. Synthesis, experimentation, and the biomechanics of economical movement. *Med. Sci. Sports Exerc.* 17 (1): 44–47.

Fregly, B.J., and F.E. Zajac. 1996. A state-space analysis of mechanical energy generation, absorption, and transfer during pedaling. *J. Biomech.* 29 (1): 81–90.

Fukashiro, S. 1988. Moment of force and mechanical power in joints during leg extension. In *Biomechanics XI-B,* ed. G. de Groot, A.P. Hollander, P.A. Huijing, and G.J. van Ingen Schenau, 938–42. Amsterdam: Free University Press.

Fukashiro, S., and P.V. Komi. 1987. Joint moment and mechanical power flow of the lower limb during vertical jump. *Int. J. Sports Med.* 8: 15–21.

Fukunaga, T., and A. Matsuo. 1980. Effect of running velocity on external mechanical power output. *Ergonomics* 23 (2): 123–36.

Fukunaga, T., A. Matsuo, and M. Ichikawa. 1981. Mechanical energy output and joint movements in sprint running. *Ergonomics* 24 (10): 765–72.

Fukunaga, T., A. Matsuo, K. Yamamoto, and T. Asami. 1986. Mechanical efficiency in rowing. *Eur. J. Appl. Physiol.* 55 (5): 471–75.

Fukunaga, T., A. Matsuo, K. Yamamoto, and T. Asami. 1987. Mechanical power in rowing. In *Biomechanics X-B,* ed. B. Johnson, 725–28. Champaign, IL: Human Kinetics.

Fukunaga, T., A. Matsuo, K. Yuasa, H. Fujimatsu, and K. Asahina 1980. Effect of running velocity on external mechanical power output. *Ergonomics* 23 (2): 123–36.

Full, R.J. 1989. Mechanics and energetics of terrestrial locomotion: From bipeds to polypeds. In *Energy transformations in cells and organisms,* ed. W. Wieser and E. Gnaiger, 175–82. Stuttgart, NY: Thieme.

Full, R.J. 1991. The concepts of efficiency and economy in land locomotion. In *Efficiency and economy in animal physiology,* ed. R.W. Blake, 97–131. New York: Cambridge University Press.

Full, R.J., R. Kram, and B. Wong. 1993. Instantaneous power at the leg joints of running cockroaches. *Am. Zool.* 33: 140A.

Full, R.J., D.R. Stokes, A.N. Ahn, and R.K. Josephson. 1998. Energy absorption during running by leg muscles in a cockroach. *J. Exp. Biol.* 201: 997–1012.

Full, R.J., D.A. Zuccarello, and A. Tullis. 1990. Effect of variation in form on the cost of terrestrial locomotion. *J. Exp. Biol.* 150: 233–46.

Funato, K., A. Matsuo, and T. Fukunaga. 2000. Measurement of specific movement power application: evaluation of weight lifters. *Ergonomics* 43 (1): 40–54.

Furusawa, K., A.V. Hill, and J.L. Parkinson. 1927. The energy used in "sprint" running. *Proc. R. Soc. Lond.* B102: 43–50.

Garhammer, J.J. 1980a. Evaluation of human power capacity through Olympic weightlifting analyses. Doctoral diss., University of California, Los Angeles.

Garhammer, J.J. 1980b. Power production by Olympic weightlifters. *Med. Sci. Sports Exerc.* 12: 50–60.

Garhammer, J. 1982. Energy flow during Olympic weight lifting. *Med. Sci. Sports Exerc.* 14 (5): 353–60.

Garhammer, J. 1993. A review on power output studies of Olympic and powerlifting: Methodology, performance, prediction, and evaluation tests. *J. Strength Conditioning Res.* 7 (2): 76–89.

Geil, M.D., M. Parnianpour, and N. Berme. 1999. Significance of nonsagittal power terms in analysis of a dynamic elastic response prosthetic foot. *J. Biomech. Eng.* 121 (5): 521–24.

Gersten, J.W., W. Orr, A.V. Sexton, and D. Okin. 1969. External work in level walking. *J. Appl. Physiol.* 26: 286–89.

Gielen, C.C.A.M., and G.J. van Ingen Schenau. 1992. The constrained control of force and position in multi-link manipulators. *IEEE Trans. Syst. Man Cybern.* 22: 1214–19.

Gitter, A., J. Czerniecki, and M. Meinders. 1997. Effect of prosthetic mass on swing phase work during above-knee amputee ambulation. *Am. J. Phys. Med. Rehabil.* 76 (2): 114–21.

Gitter, A., J. Czerniecki, and K. Weaver. 1995. A reassessment of center-of-mass dynamics as a determinate of the metabolic inefficiency of above-knee amputee ambulation. *Am. J. Phys. Med. Rehabil.* 74 (5): 332–38.

Glasheen, J.W., and T.A. McMahon. 1995. Arms are different from legs: Mechanics and energetics of human hand-running. *J. Appl. Physiol.* 78 (4): 1280–87.

Gregersen, C.S., N.A. Silverton, and D.R. Carrier. 1998. External work and potential for elastic storage at the limb joints of running dogs. *J. Exp. Biol.* 201 (Pt. 23): 3197–210.

Gregoire, L., H.E. Veeger, P.A. Huijing, and G.J. van Ingen Schenau. 1984. Role of mono- and biarticular muscles in explosive movements. *Int. J. Sports Med.* 5 (6): 301–5.

Gregor, R.J., and T.A. Abelew. 1994. Tendon force measurements and movement control: A review. *Med. Sci. Sports Exerc.* 26: 1359–72.

Gregor, R.J., and D. Kirkendall. 1978. Performance efficiency of world class female marathon runners. In *Biomechanics VI-B*, ed. E. Asmussen and K. Jorgensen. Baltimore: University Park Press.

Grenier, S.G., and D.G.E. Robertson. 1998. Comparison of mechanical and physiological energy costs of walking. In *Proceedings of NACOB '98: The third North American Congress on Biomechanics, August 14–18, 1998,* 129–30. Waterloo, ON: University of Waterloo.

Grieve, D.W., and J. van der Linden. 1986. Force, speed and power output of the human upper limb during horizontal pulls. *Eur. J. Appl. Physiol.* 55: 425–30.

Griffin, T.M., N.A. Tolani, and R. Kram. 1999. Walking in simulated reduced gravity: Mechanical energy fluctuations and exchange. *J. Appl. Physiol.* 86 (1): 383–90.

Groot, G. de, E. Welbergen, L. Clijsen, J. Clarijs, J. Cabri, and J. Antonis. 1994. Power, muscular work, and external forces in cycling. *Ergonomics* 37 (1): 31–42.

Gussoni, M., V. Margonato, R. Ventura, and A. Veicsteinas. 1990. Energy cost of walking with hip joint impairment. *Phys. Ther.* 70 (5): 295–301.

Hartmann, U., A. Mader, K. Wasser, and I. Klauer. 1993. Peak force, velocity, and power during five and ten maximal rowing ergometer strokes by world class female and male rowers. *Int. J. Sports Med.* 14 (Suppl. 1): S42–S45.

Hatze, H. 1998. Validity and reliability of methods for testing vertical jumping performance. *J. Appl. Biomech.* 14 (2): 127–40.

Hatze, H., and J.D. Buys. 1977. Energy-optimal controls in the mammalian neuromuscular system. *Biol. Cybern.* 27 (1): 9–20.

Hautier, C.A., M.T. Linossier, A. Belli, J.R. Lacour, and L.M. Arsac. 1996. Optimal velocity for maximal power production in non-isokinetic cycling is related to muscle fibre type composition. *Eur. J. Appl. Physiol.* 74 (1–2): 114–18.

Hawkins, D., and P. Mole. 1997. Modeling energy expenditure associated with isometric, concentric, and eccentric muscle action at the knee. *Ann. Biomed. Eng.* 25 (5): 822–30.

He, J.P., R. Kram, and T.A. McMahon. 1991. Mechanics of running under simulated low gravity. *J. Appl. Physiol.* 71 (3): 863–70.

Heglund, N.C., and G.A. Cavagna. 1985. Efficiency of vertebrate locomotory muscles. *J. Exp. Biol.* 115:283–92.

Heglund, N.C., G.A. Cavagna, and C.R. Taylor. 1982. Energetics and mechanics of terrestrial locomotion: III. Energy changes of the centre of mass as a function of speed and body size in birds and mammals. *J. Exp. Biol.* 97: 41–56.

Heglund, N.C., M.A. Fedak, C.R. Taylor, and G.A. Cavagna. 1982. Energetics and mechanics of terrestrial locomotion: IV. Total mechanical energy changes as a function of speed and body size in birds and mammals. *J. Exp. Biol.* 97: 57–66.

Heglund, N.C., and C.R. Taylor. 1988. Speed, stride frequency and energy cost per stride: How do they change with body size and gait? *J. Exp. Biol.* 138: 301–18.

Heglund, N.C., P.A. Willems, M. Penta, and G.A. Cavagna. 1995. Energy-saving gait mechanics with head-supported loads. *Nature* 375 (6526): 52–54.

Heise, G.D., and P.E. Martin. 1998. "Leg spring" characteristics and the aerobic demand of running. *Med. Sci. Sports Exerc.* 30 (5): 750–54.

Herzog, W., V. Zatsiorsky, B.I. Prilutsky, and T.R. Leonard. 1994. Variations in force-time histories of cat gastrocnemius, soleus and plantaris muscles for consecutive walking steps. *J. Exp. Biol.* 191: 19–36.

Heus, R., A.H. Wertheim, and G. Havenith. 1998. Human energy expenditure when walking on a moving platform. *Eur. J. Appl. Physiol.* 77 (4): 388–94.

Hof, A.L. 1998. In vivo measurement of the series elasticity release curve of human triceps surae muscle. *J. Biomech.* 31 (9): 793–800.

Hof, A.L., B.A. Geelen, and J. Van den Berg. 1983. Calf muscle moment, work and efficiency in level walking: Role of series elasticity. *J. Biomech.* 16 (7): 523–37.

Hof, A.L., E.R. van der Knaap, M.A.A. Schallig, and D.P. Struwe. 1993. Calf muscle work and energy changes in human treadmill walking. *J. Electromyogr. Kinesiol.* 4: 203–16.

Hoffman, J.R., D. Liebermann, and A. Gusis. 1997. Relationship of leg strength and power to ground reaction forces in both experienced and novice jump trained personnel. *Aviat. Space Environ. Med.* 68 (8): 710–14.

Holt, K.G., J. Hamill, and R.O. Andres. 1991. Predicting the minimal energy costs of human walking. *Med. Sci. Sports Exerc.* 23 (4): 491–98.

Houdijk, H., J.J. de Koning, G. de Groot, M.F. Bobbert, and G.J. van Ingen Schenau. 2000. Push-off mechanics in speed skating with conventional skates and klapskates. *Med. Sci. Sports Exerc.* 32 (3): 635–41.

Hubley, C.L., and R.P. Wells. 1983. A work-energy approach to determine individual joint contributions to vertical jump performance. *Eur. J. Appl. Physiol.* 50 (2): 247–54.

Hull, M.L., and D.A. Hawkins. 1990. Analysis of muscular work in multisegmental movements: Application to cycling. In *Multiple muscle systems,* ed. J. Winters and S.-Y. Woo, 621–38. New York: Springer.

Hull, M.L., S. Kautz, and A. Beard. 1991. An angular velocity profile in cycling derived from mechanical energy analysis. *J. Biomech.* 24 (7): 577–86.

Inman, V.T. 1967. Conservation of energy in ambulation. *Arch. Phys. Med. Rehabil.* 48: 484–88.

Inman, V.T., H.J. Ralston, and F. Todd. 1981. *Human walking.* Baltimore: Williams and Wilkins.

Iossifidou, A.N., and V. Baltzopoulos. 2000. Peak power assessment in isokinetic dynamometry. *Eur. J. Appl. Physiol.* 82 (1–2): 158–60.

Ito, A., P.V. Komi, B. Sjodin, C. Bosco, and J. Karlsson. 1983. Mechanical efficiency of positive work in running at different speeds. *Med. Sci. Sports Exerc.* 15 (4): 299–308.

James, C., P. Sacco, and D.A. Jones. 1995. Loss of power during fatigue of human leg muscles. *J. Physiol. (Lond.)* 484 (Pt. 1): 237–46.

James, R.S., I.S. Young, V.M. Cox, D.F. Goldspink, and J.D. Altringham. 1996. Isometric and isotonic muscle properties as determinants of work loop power output. *Pflugers Arch.* 432 (5): 767–74.

Jaskolska, A., P. Goossens, B. Veenstra, A. Jaskolski, and J.S. Skinner. 1999. Comparison of treadmill and cycle ergometer measurements of force-velocity relationships and power output. *Int. J. Sports Med.* 20 (3): 192–97.

Jensen, R.L., P.S. Freedson, and J. Hamill. 1996. The prediction of power and efficiency during near-maximal rowing. *Eur. J. Appl. Physiol.* 73 (1–2): 98–104.

Kaneko, M. 1990. Mechanics and energetics in running with special reference to efficiency. *J. Biomech.* 23 (Suppl. 1): 57–63.

Kaneko, M., A. Ito, T. Fuchimoto, and J. Toyooka. 1981. Mechanical work and efficiency of young distance runners during level running. In *Biomechanics VII-B*, ed. A. Morecki, K. Fidelus, K. Kedzior and A. Wit, 234–40. Warszawa: Polish Scientific Publishers. Baltimore: University Park Press.

Kaneko, M., A. Ito, T. Fuchimoto, and J. Toyooka. 1983. Effects of running speed on the mechanical power and efficiency of sprint- and distance-runners. *Nippon Seirigaku Zasshi* 45 (12): 711–13.

Kaneko, M., T. Yamazaki, and J. Toyooka. 1979. Direct determination of the internal mechanical work and the efficiency in bicycle pedalling (in Japanese). *Nippon Seirigaku Zasshi* 41 (3): 68–69.

Kaplan, M.L., and J.H. Heegaard. 2000. Energy-conserving impact algorithm for the heel-strike phase of gait. *J. Biomech.* 33 (6): 771–75.

Katch, V.L., J.F. Villanacci, and S.P. Sady. 1981. Energy cost of rebound-running. *Res. Q. Exerc. Sport* 52 (2): 269–72.

Katz, J.S., and L. Katz. 1999. Power laws and athletic performance. *J. Sports Sci.* 17 (6): 467–76.

Kautz, S.A., M.L. Hull, and R.R. Neptune. 1994. A comparison of muscular mechanical energy expenditure and internal work in cycling. *J. Biomech.* 27 (12): 1459–67.

Khosravi-Sichani, B., H. Hemami, and S. Yurkovich. 1992. Energy transformations in human movement by contact. *J. Biomech.* 25 (8): 881–89.

Klinger, A., and M. Adrian. 1987. Power output as a function of fencing technique. In *Biomechanics X-B*, ed. B. Johnson, 791–95. Champaign, IL: Human Kinetics.

Komi, P.V. 1990. Relevance of in vivo force measurements to human biomechanics. *J. Biomech.* 23 (Suppl. 1): 23–34.

Komi, P.V., and C. Bosco. 1978. Utilization of stored elastic energy in leg extensor muscles by men and women. *Med. Sci. Sports Exerc.* 10: 261–65.

Komi, P.V., and M. Virmavirta. 2000. Determinants of successful ski-jumping performance. In *The IOC encyclopaedia of sports medicine.* Vol. 9, *Biomechanics in sport: Performance enhancement and injury prevention*, ed. V.M. Zatsiorsky, 349–62. Oxford, UK: Blackwell Science.

Koning, J.J. de, and G.J. van Ingen Schenau. 1994. On the estimation of mechanical power in endurance sports. *Sport Sci. Rev.* 3 (2): 34–54.

Kram, R. 2000. Muscular force or work: What determines the metabolic energy cost of running? *Exerc. Sport Sci. Rev.* 28 (3): 138–43.

Kram, R., and T.J. Dawson. 1998. Energetics and biomechanics of locomotion by red kangaroos (*Macropus rufus*). *Comp. Biochem. Physiol. B Biochem. Mol. Biol.* 120 (1): 41–49.

Kram, R., and C.R. Taylor. 1990. Energetics of running: A new perspective. *Nature* 346 (6281): 265–67.

Kram, R., B. Wong, and R.J. Full. 1997. Three-dimensional kinematics and limb kinetic energy of running cockroaches. *J. Exp. Biol.* 200 (Pt. 13): 1919–29.

Kramer, P.A. 1999. Modelling the locomotor energetics of extinct hominids. *J. Exp. Biol.* 202 (Pt. 20): 2807–18.

Kramer, P.A., and G.G. Eck. 2000. Locomotor energetics and leg length in hominid bipedality. *J. Hum. Evol.* 38 (5): 651–66.

Kubo, K., H. Kanehisa, Y. Kawakami, and T. Fukunaga. 2000. Elastic properties of muscle-tendon complex in long-distance runners. *Eur. J. Appl. Physiol.* 81 (3): 181–87.

Kyle, C.R. 1979. Reduction of wind resistance and power output of racing cyclists and runners travelling in groups. *Ergonomics* 22 (4): 387–97.

Kyrolainen, H., P.V. Komi, and A. Belli. 1995. Mechanical efficiency in athletes during running. *Scand. J. Med. Sci. Sports* 5 (4): 200–208.

Landesberg, A., and S. Sideman. 2000. Force–velocity relationship and biochemical-to-mechanical energy conversion by the sarcomere. *Am. J. Physiol. Heart Circ. Physiol.* 278 (4): H1274–H1284.

Lanshammar, H. 1982. Variation of mechanical energy levels for normal and prosthetic gait. *Prosth. Orth. Int.* 6: 97–103.

Laursen, B., D. Ekner, E.B. Simonsen, M. Voigt, and G. Sjogaard. 2000. Kinetics and energetics during uphill and downhill carrying of different weights. *Appl. Ergon.* 31 (2): 159–66.

Le Chevalier, J.M., H. Vandewalle, C. Thepaut-Mathieu, J.F. Stein, and L. Caplan. 2000. Local critical power is an index of local endurance. *Eur. J. Appl. Physiol.* 81 (1–2): 120–27.

Lejeune, T.M., P.A. Willems, and N.C. Heglund. 1998. Mechanics and energetics of human locomotion on sand. *J. Exp. Biol.* 201 (Pt. 13): 2071–80.

Lieber, R.L. 1990. Hypothesis: Biarticular muscles transfer moments between joints. *Dev. Med. Child Neurol.* 32 (5): 456–58.

Lloyd, B.B., and R.M. Zacks. 1972. The mechanical efficiency of treadmill running against a horizontal impeding force. *J. Physiol. (Lond.)* 223 (2): 355–63.

Looze, M.P. de, J.B.J. Bussmann, I. Kingma, and H.M. Toussaint. 1992. Different methods to estimate total power and its components during lifting. *J. Biomech.* 25 (9): 1089–95.

Luhtanen, P., and P.V. Komi. 1978. Mechanical energy states during running. *Eur. J. Appl. Physiol.* 38 (1): 41–48.

Luhtanen, P., and P.V. Komi. 1979. Mechanical power and segmental contribution to force impulses in long jump take-off. *Eur. J. Appl. Physiol.* 41 (4): 267–74.

Luhtanen, P., and P.V. Komi. 1980. Force-, power-, and elasticity-velocity relationships in walking, running, and jumping. *Eur. J. Appl. Physiol.* 44 (3): 279–89.

Luhtanen, P., P. Rahkila, H. Rusko, and J.T. Viiasalo. 1990. Mechanical work and efficiency in treadmill running at aerobic and anaerobic thresholds. *Acta Physiol. Scand.* 139 (1): 153–59.

Lundström, R., P. Kolmlund, and L. Lindberg. 1998. Absorption of energy during vertical whole-body vibration exposure. *J. Biomech.* 31 (4): 317–26.

MacIntosh, B.R., R.R. Neptune, and J.F. Horton. 2000. Cadence, power, and muscle activation in cycle ergometry. *Med. Sci. Sports Exerc.* 32 (7): 1281–87.

Magnusson, S.P., P. Aagaard, B. Larsson, and M. Kjaer. 2000. Passive energy absorption by human muscle-tendon unit is unaffected by increase in intramuscular temperature. *J. Appl. Physiol.* 88 (4): 1215–20.

Mansour, J.M., M.D. Lesh, M.D. Nowak, and S.R. Simon. 1982. A three dimensional multi-segmental analysis of the energetics of normal and pathological human gait. *J. Biomech.* 15 (1): 51–59.

Margaria, R. 1976. *Biomechanics and energetics of muscular exercise.* Oxford: Clarendon Press.

Margaria, R. 1968. Capacity and power of the energy processes in muscle activity: Their practical relevance in athletics. *Int. Z. Angew. Physiol.* 25 (4): 352–60.

Martin, J.C., D.L. Milliken, J.E. Cobb, K.L. McFadden, and A.R. Coggan. 1998. Validation of a mathematical model for road cycling power. *J. Appl. Biomech.* 14 (3): 276–91.

Martin, P.E., and P.R. Cavanagh. 1990. Segment interactions within the swing leg during unloaded and loaded running. *J. Biomech.* 23 (6): 529–36.

Martin, P.E., G.D. Heise, and D.W. Morgan. 1993. Interrelationships between mechanical power, energy transfers, and walking and running economy. *Med. Sci. Sports Exerc.* 25 (4): 508–15.

Martindale, W.O., and D.G. Robertson. 1984. Mechanical energy in sculling and in rowing an ergometer. *Can. J. Appl. Sport Sci.* 9 (3): 153–63.

Matsuo, A., T. Fukunaga, and T. Asami. 1985. Relation between external work and running performance in athletes. In *Biomechanics IX-B,* ed. D. Winter, R.W. Norman, R.P. Wells, K.C. Hayes, and A.E. Patla, 319–24. Champaign, IL: Human Kinetics.

Matsuo, A., T. Fukunaga, and T. Hirata. 1987. The effect of the additional weight on external work and energy consumption in running. In *Biomechanics X-B,* ed. B. Johnson, 849–53. Champaign, IL: Human Kinetics.

McClay, I., and K. Manal. 1999. Three-dimensional kinetic analysis of running: Significance of secondary planes of motion. *Med. Sci. Sports Exerc.* 31 (11): 1629–37.

McFadyen, B.J. 1994. A geometric analysis of muscle mechanical power with applications to human gait. *J. Biomech.* 27 (9): 1189–93.

McFadyen, B.J., S. Lavoie, and T. Drew. 1999. Kinetic and energetic patterns for hindlimb obstacle avoidance during cat locomotion. *Exp. Brain Res.* 125 (4): 502–10.

McGibbon, C.A., and D.E. Krebs. 1998. The influence of segment endpoint kinematics on segmental power calculations. *Gait Posture* 7 (3): 237–42.

McGill, S.M., and D.A. Dainty. 1984. Computer analysis of energy transfers in children walking with crutches. *Arch. Phys. Med. Rehabil.* 65 (3): 115–20.

McMahon, T.A. 1985. The role of compliance in mammalian running gaits. *J. Exp. Biol.* 115: 263–82.

McMahon, T.A., G. Valiant, and E.C. Frederick. 1987. Groucho running. *J. Appl. Physiol.* 62 (6): 2326–37.

McMillan, D.W., G. Garbutt, S. Oliver, and M.A. Adams. 1995. The effect of sustained loading on a lumbar spine motion segment: Finite element simulation and experimental validation. In *XVth congress of the International Society of Biomechanics, book of abstracts,* ed. K. Häkkinen, K.L. Keskinen, P.V. Komi, and A. Mero, 614–15. Jyväskylä, Finland: Jyväskylä University.

Menard, M.R., M.E. McBride, D.J. Sanderson, and D.D. Murray. 1992. Comparative biomechanical analysis of energy-storing prosthetic feet. *Arch. Phys. Med. Rehabil.* 73 (5): 451–58.

Miller, C.A., and M.C. Verstraete. 1996. Determination of the step duration of gait initiation using a mechanical energy analysis. *J. Biomech.* 29 (9): 1195–99.

Miller, C.A., and M.C. Verstraete. 1999. A mechanical energy analysis of gait initiation. *Gait Posture* 9 (3): 158–66.

Millet, G.P., G.Y. Millet, M.D. Hofmann, and R.B. Candau. 2000. Alterations in running economy and mechanics after maximal cycling in triathletes: Influence of performance level. *Int. J. Sports Med.* 21 (2): 127–32.

Minetti, A.E. 1995. Optimum gradient of mountain paths. *J. Appl. Physiol.* 79 (5): 1698–703.

Minetti, A.E. 1998a. The biomechanics of skipping gaits: A third locomotion paradigm? *Proc. R. Soc. Lond. B Biol. Sci.* 265 (1402): 1227–35.

Minetti, A.E. 1998b. A model equation for the prediction of mechanical internal work of terrestrial locomotion. *J. Biomech.* 31 (5): 463–68.

Minetti, A.E., and R.M. Alexander. 1997. A theory of metabolic costs for bipedal gaits. *J. Theor. Biol.* 186 (4): 467–76.

Minetti, A.E., L.P. Ardigo, E. Reinach, and F. Saibene. 1999. The relationship between mechanical work and energy expenditure of locomotion in horses. *J. Exp. Biol.* 202 (Pt. 17): 2329–38.

Minetti, A.E., L.P. Ardigo, and F. Saibene. 1994a. Mechanical determinants of the minimum energy cost of gradient running in humans. *J. Exp. Biol.* 195: 211–25.

Minetti: A.E., L.P. Ardigo, and F. Saibene. 1994b. The transition between walking and running in humans: Metabolic and mechanical aspects at different gradients. *Acta Physiol. Scand.* 150 (3): 315–23.

Minetti, A.E., L.P. Ardigo, F. Saibene, S. Ferrero, and A. Sartorio. 2000. Mechanical and metabolic profile of locomotion in adults with childhood-onset GH deficiency. *Eur. J. Endocrinol.* 142 (1): 35–41.

Minetti, A.E., C. Capelli, P. Zamparo, P.E. di Prampero, and F. Saibene. 1995. Effects of stride frequency on mechanical power and energy expenditure of walking. *Med. Sci. Sports Exerc.* 27 (8): 1194–202.

Minetti, A.E., and F. Saibene. 1992. Mechanical work rate minimization and freely chosen stride frequency of human walking: a mathematical model. *J. Exp. Biol.* 170: 19–34.

Minetti, A.E., F. Saibene, L.P. Ardigo, G. Atchou, F. Schena, and G. Ferretti. 1994. Pygmy locomotion. *Eur. J. Appl. Physiol.* 68 (4): 285–90.

Miyashita, M., M. Miura, Y. Murase, and K. Yamaji. 1978. Running performance from the viewpoint of aerobic power. In *Environmental stress,* ed. L.J. Folinsbee, 183–93. New York: Academic Press.

Mochon, S., and T.A. McMahon. 1980. Ballistic walking. *J. Biomech.* 13 (1): 49–57.

Moffroid, M.T., and E.T. Kusiak. 1975. The power struggle: Definition and evaluation of power of muscular performance. *Phys. Ther.* 55 (10): 1098–104.

Morgan, D.L., U. Proske, and D. Warren. 1978. Measurement of muscle stiffness and the mechanism of elastic storage of energy in hopping kangaroos. *J. Physiol. (Lond.)* 282: 253–61.

Morgan, D.W., P.E. Martin, and G.S. Krahenbuhl. 1989. Factors affecting running economy. *Sports Med.* 7 (5): 310–30.

Morgan, D.W., P.E. Martin, G.S. Krahenbuhl, and F.D. Baldini. 1991. Variability in running economy and mechanics among trained male runners. *Med. Sci. Sports Exerc.* 23 (3): 378–83.

Morrison, J.B. 1970. The mechanics of muscle function in locomotion. *J. Biomech.* 3 (4): 431–51.

Morton, R.H. 1985. Comment on "A model for the calculation of mechanical power during distance running." *J. Biomech.* 18: 161–62.

Muir, G.D., J.M. Gosline, and J.D. Steeves. 1996. Ontogeny of bipedal locomotion: Walking and running in the chick. *J. Physiol. (Lond.)* 493 (Pt. 2): 589–601.

Myers, M.J., and K. Steudel. 1985. Effect of limb mass and its distribution on the energetic cost of running. *J. Exp. Biol.* 116: 363–73.

Nadeau, S., D. Gravel, and A.B. Arsenault. 1997. Relationships between torque, velocity and power output during plantarflexion in healthy subjects. *Scand. J. Rehabil. Med.* 29 (1): 49–55.

Nagano, A., Y. Ishige, and S. Fukashiro. 1998. Comparison of new approaches to estimate mechanical output of individual joints in vertical jumps. *J. Biomech.* 31 (10): 951–55.

Nagle, F.J., P. Webb, and D.M. Wanta. 1990. Energy exchange in downhill and uphill walking: A calorimetric study. *Med. Sci. Sports Exerc.* 22 (4): 540–44.

Navarro, E., J. Campos, P. Vera, and E. Chillaron. 1995. A kinetic energy model of human body applied to 3-D javelin throwing. In *XVth congress of the International Society of Biomechanics, book of abstracts,* ed. K. Häkkinen, K.L. Keskinen, P.V. Komi, and A. Mero, 668–69. Jyväskylä, Finland: Jyväskylä University.

Neptune, R.R., and A.J. van den Bogert. 1998. Standard mechanical energy analyses do not correlate with muscle work in cycling. *J. Biomech.* 31 (3): 239–45.

Neptune, R.R., and W. Herzog. 1999. The association between negative muscle work and pedaling rate. *J. Biomech.* 32 (10): 1021–26.

Neptune, R.R., S.A. Kautz, and F.E. Zajac. 2000. Muscle contributions to specific biomechanical functions do not change in forward versus backward pedaling. *J. Biomech.* 33 (2): 155–64.

Nigg, B.M., and M. Anton. 1995. Energy aspects for elastic and viscous shoe soles and playing surfaces. *Med. Sci. Sports Exerc.* 27 (1): 92–97.

Nishii, J. 2000. Legged insects select the optimal locomotor pattern based on the energetic cost. *Biol. Cybern.* 83 (5): 435–42.

Nonweiler, T., ed. 1956. *Air resistance in cycling, Technical report no. 106.* Cranfield, England: The College of Aeronautics.

Norman, R.W., M.T. Sharratt, J.C. Pezzack, and E.G. Noble. 1975. Reexamination of the mechanical efficiency of horizontal treadmill running. In *Biomechanics V-B,* ed. P.V. Komi, 87–93. Baltimore: University Park Press.

Nubar, Y., and R. Contini. 1961. A minimal principle in biomechanics. *Bulletin Math. Biophys.* 23: 379–90.

Olney, S.J., P.A. Costigan, and D.M. Hedden. 1987. Mechanical energy patterns in gait of cerebral palsied children with hemiplegia. *Phys. Ther.* 67 (9): 1348–54.

Olney, S.J., H.E. MacPhail, D.M. Hedden, and W.F. Boyce. 1990. Work and power in hemiplegic cerebral palsy gait. *Phys. Ther.* 70 (7): 431–38.

Olney, S.J., T.N. Monga, and P.A. Costigan. 1986. Mechanical energy of walking of stroke patients. *Arch. Phys. Med. Rehabil.* 67 (2): 92–98.

Osternig, L.R., J. Hamill, J.A. Sawhill, and B.T. Bates. 1983. Influence of torque and limb speed on power production in isokinetic exercise. *Am. J. Phys. Med.* 62 (4): 163–71.

Pandy, M.G., and F.E. Zajac. 1991. Optimal muscular coordination strategies for jumping. *J. Biomech.* 24 (1): 1–10. See also comment in *J. Biomech.* (1992) 25 (2): 207–9.

Pennycuick, C.J., A. Hedenstrom, and M. Rosen. 2000. Horizontal flight of a swallow (*Hirundo rustica*) observed in a wind tunnel, with a new method for directly measuring mechanical power. *J. Exp. Biol.* 203 (Pt. 11): 1755–65.

Pierrynowski, M.R., R.W. Norman, and D.A. Winter. 1981. Mechanical energy analyses of the human during local carriage on a treadmill. *Ergonomics* 24 (1): 1–14.

Pierrynowski, M.R., D.A. Winter, and R.W. Norman. 1980. Transfers of mechanical energy within the total body and mechanical efficiency during treadmill walking. *Ergonomics* 23 (2): 147–56.

Prilutsky, B.I., W. Herzog, and T. Leonard. 1996. Transfer of mechanical energy between ankle and knee joints by gastrocnemius and plantaris muscles during cat locomotion. *J. Biomech.* 29 (4): 391–403.

Prilutsky, B.I., W. Herzog, T.R. Leonard, and T.L. Allinger. 1996. Role of the muscle belly and tendon of soleus, gastrocnemius and plantaris in mechanical energy absorption and generation during cat locomotion. *J. Biomech.* 29 (4): 417–34.

Prilutsky, B.I., L.N. Petrova, and L.M. Raitsin. 1996. Comparison of mechanical energy expenditure of joint moments and muscle forces during human locomotion. *J. Biomech.* 29 (4): 405–16.

Prilutsky, B.I., and V.M. Zatsiorsky. 1991. A quantitative assessment of the "tendon" action of two-joint muscles (in Russian). *Biofizika* 36 (1): 154–56.

Prilutsky, B.I., and V.M. Zatsiorsky. 1992. Mechanical energy expenditure and efficiency of walking and running. *Hum. Physiol.* 18 (3): 118–27.

Prilutsky, B.I., and V.M. Zatsiorsky. 1994. Tendon action of two-joint muscles: Transfer of mechanical energy between joints during jumping, landing, and running. *J. Biomech.* 27 (1): 25–34.

Prilutsky (Prilutskii), B.I., V.M. Zatsiorsky, and L.N. Petrova. 1991. Mechanical energy expenditure on the movement of man and the anthropomorphic mechanism. *Biophysics* 37 (6): 1001–5.

Prilutsky, B.I., V.M. Zatsiorsky, and L.N. Petrova. 1992. Mechanical energy expenditure on human movement and the anthropomorphic mechanism (in Russian). *Biofizika* 37 (6): 1101–5.

Prilutsky, B.I., V.M. Zatsiorsky, and L.N. Petrova. 1993. "Tendon" action of two-joint muscles during human locomotion: Transfer of mechanical energy between links in shock-absorbing and push-off phases (in Russian). *Biofizika* 38 (4): 719–25.

Prince, F., D.A. Winter, G. Sjonnensen, C. Powell, and R.K. Wheeldon. 1998. Mechanical efficiency during gait of adults with transtibial amputation: A pilot study comparing the SACH, Seattle, and Golden-Ankle prosthetic feet. *J. Rehabil. Res. Dev.* 35 (2): 177–85.

Pugh, L.G. 1971. The influence of wind resistance in running and walking and the mechanical efficiency of work against horizontal or vertical forces. *J. Physiol.* 213 (2): 255–76.

Quesada, P.M., L.J. Mengelkoch, R.C. Hale, and S.R. Simon. 2000. Biomechanical and metabolic effects of varying backpack loading on simulated marching. *Ergonomics* 43 (3): 293–309.

Ralston, H.J., and L. Lukin. 1969. Energy levels of human body segments during level walking. *Ergonomics* 12: 39–46.

Rayner, J.M. 1999. Estimating power curves of flying vertebrates. *J. Exp. Biol.* 202 (Pt. 23): 3449–61.

Roberts, T.J., M.S. Chen, and C.R. Taylor. 1998. Energetics of bipedal running: II. Limb design and running mechanics. *J. Exp. Biol.* 201 (Pt. 19): 2753–62.

Roberts, T.J., R. Kram, P.G. Weyand, and C.R. Taylor. 1998. Energetics of bipedal running: I. Metabolic cost of generating force. *J. Exp. Biol.* 201 (Pt. 19): 2745–51.

Roberts, T.J., R.L. Marsh, P.G. Weyand, C.R. Taylor. 1997. Muscular force in running turkeys: The economy of minimizing work. *Science* 275: 1113–15.

Robertson, D.G.E., and D.A. Winter. 1980. Mechanical energy generation, absorption and transfer amongst segments during walking. *J. Biomech.* 13 (10): 845–54.

Robertson, S., H. Frost, H. Doll, and J.J. O'Connor. 1998. Leg extensor power and quadriceps strength: An assessment of repeatability in patients with osteoarthritic knees. *Clin. Rehabil.* 12 (2): 120–26.

Rodman, P.S., and H.M. McHenry. 1980. Bioenergetics and the origin of hominid bipedalism. *Am. J. Phys. Anthropol.* 52 (1): 103–6.

Rovick, J.S., and D.S. Childress. 1988. Pendular model of paraplegic swing-through crutch ambulation. *J. Rehabil. Res. Dev.* 25 (4): 1–16.

Saibene, F. 1990. The mechanisms for minimizing energy expenditure in human locomotion. *Eur. J. Clin. Nutr.* 44 (Suppl 1): 65–71.

Saini, M., D.C. Kerrigan, M.A. Thirunarayan, and M. Duff-Raffaele. 1998. The vertical displacement of the center of mass during walking: A comparison of four measurement methods. *J. Biomech. Eng.* 120 (1): 133–39.

Sakurai, S., and M. Miyashita. 1985. Mechanical energy changes during treadmill running. *Med. Sci. Sports Exerc.* 17 (1): 148–52.

Sargeant, A.J. 1994. Human power output and muscle fatigue. *Int. J. Sports Med.* 15 (3): 116–21.

Sargeant, A.J., E. Hoinville, and A. Young. 1981. Maximum leg force and power output during short-term dynamic exercise. *J. Appl. Physiol.* 51 (5): 1175–82.

Savolainen, S., and R. Visuri. 1994. A review of athletic energy expenditure, using skiing as a practical example. *J. Appl. Biomech.* 10: 253–69.

Schepens, B., P.A. Willems, and G.A. Cavagna. 1998. The mechanics of running in children. *J. Physiol. (Lond.)* 509 (Pt. 3): 927–40.

Seck, D., H. Vandewalle, N. Decrops, and H. Monod. 1995. Maximal power and torque–velocity relationship on a cycle ergometer during the acceleration phase of a single all-out exercise. *Eur. J. Appl. Physiol.* 70 (2): 161–68.

Seroussi, R.E., A. Gitter, J.M. Czerniecki, and K. Weaver. 1996. Mechanical work adaptations of above-knee amputee ambulation. *Arch. Phys. Med. Rehabil.* 77 (11): 1209–14.

Shadwick, R.E. 1990. Elastic energy storage in tendons: Mechanical differences related to function and age. *J. Appl. Physiol.* 68 (3): 1033–40.

Sharp, A. 1898. *Bicycles and tricycles.* London: Longmans, Green & Co.

Sherwood, B.A. 1983. Pseudowork and real work. *Am. J. Phys.* 51 (7): 597–602.

Shorten, M.R. 1993. The energetics of running and running shoes. *J. Biomech.* 26 (Suppl. 1): 41–51.

Siegel, K.L., T.M. Kepple, and G.E. Caldwell. 1996. Improved agreement of foot segmental power and rate of energy change during gait: Inclusion of distal power terms and use of three-dimensional models. *J. Biomech.* 29 (6): 823–37.

Simon, C. 1998. Effectiveness and economy in running. In *ISBS '98: XVI International Symposium on Biomechanics in Sports, University of Konstanz, proceedings 1,* ed. H.J. Riehle and M.M. Vieten, 247–50. Konstanz, Germany: UVK-Universitätverlag Konstanz.

Stefanyshyn, D.J., and B.M. Nigg. 1997. Mechanical energy contribution of the metatarsophalangeal joint to running and sprinting. *J. Biomech.* 30 (11–12): 1081–85.

Stefanyshyn, D.J., and B.M. Nigg. 1998. Contribution of the lower extremity joints to mechanical energy in running vertical jumps and running long jumps. *J. Sports Sci.* 16 (2): 177–86.

Stefanyshyn, D.J., and B.M. Nigg. 2000. Influence of midsole bending stiffness on joint energy and jump height performance. *Med. Sci. Sports Exerc.* 32 (2): 471–76.

Steudel, K. 1990. The work and energetic cost of locomotion: II. Partitioning the cost of internal and external work within a species. *J. Exp. Biol.* 154: 287–303.

Steudel, K. 1996. Limb morphology, bipedal gait, and the energetics of hominid locomotion. *Am. J. Phys. Anthropol.* 99 (2): 345–55.

Stevens, E.D. 1993. Relation between work and power calculated from force–velocity curves to that done during oscillatory work. *J. Muscle Res. Cell Motil.* 14 (5): 518–26.

Sun, M., and J.O. Hill. 1993. A method for measuring mechanical work and work efficiency during human activities. *J. Biomech.* 26 (3): 229–41.

Swain, D.P. 1994. The influence of body mass in endurance bicycling. *Med. Sci. Sports Exerc.* 26 (1): 58–63.

Swaine, I.L. 2000. Arm and leg power output in swimmers during simulated swimming. *Med. Sci. Sports Exerc.* 32 (7): 1288–92.

Takarada, Y., Y. Hirano, Y. Ishige, and N. Ishii. 1997. Stretch-induced enhancement of mechanical power output in human multijoint exercise with countermovement. *J. Appl. Physiol.* 83 (5): 1749–55.

Taylor, C.R. 1985. Force development during sustained locomotion: A determinant of gait, speed and metabolic power. *J. Exp. Biol.* 115: 253–62.

Tesio, L., P. Civaschi, and L. Tessari. 1985. Motion of the center of gravity of the body in clinical evaluation of gait. *Am. J. Phys. Med.* 64 (2): 57–70.

Thomas, M., M.A. Fiatarone, and R.A. Fielding. 1996. Leg power in young women: Relationship to body composition, strength, and function. *Med. Sci. Sports Exerc.* 28 (10): 1321–26.

Thompson, C.J., and M.G. Bemben. 1999. Reliability and comparability of the accelerometer as a measure of muscular power. *Med. Sci. Sports Exerc.* 31 (6): 897–902.

Thys, H., P.A. Willems, and P. Saels. 1996. Energy cost, mechanical work and muscular efficiency in swing-through gait with elbow crutches. *J. Biomech.* 29 (11): 1473–82.

Tibarewala, D.N., A.K. Ghosh, and S. Ganguli. 1980. An integrated biomechanical-bioenergetic technique for evaluation of human locomotion. *J. Med. Eng. Technol.* 4 (5): 241–46.

Too, D., and G.E. Landwer. 2000. The effect of pedal crank arm length on joint angle and power production in upright cycle ergometry. *J. Sports Sci.* 18 (3): 153–61.

Toussaint, H.M., W. Knops, G. de Groot, and A.P. Hollander. 1990. The mechanical efficiency of front crawl swimming. *Med. Sci. Sports Exerc.* 22 (3): 402–8.

Unnithan, V.B., J.J. Dowling, G. Frost, and O. Bar-Or. 1999. Role of mechanical power estimates in the O_2 cost of walking in children with cerebral palsy. *Med. Sci. Sports Exerc.* 31 (12): 1703–8.

van der Linde, R.Q. 1999a. Passive bipedal walking with phasic muscle contraction. *Biol. Cybern.* 81 (3): 227–37.

van der Linde, R.Q. 1999b. Towards applicable ballistic walking. *Technol. Health Care* 7 (6): 449–53.

Vandewalle, H., G. Peres, J. Heller, J. Panel, and H. Monod. 1987. Force–velocity relationship and maximal power on a cycle ergometer: Correlation with the height of a vertical jump. *Eur. J. Appl. Physiol.* 56 (6): 650–56.

van Ingen Schenau, G.J. 1984. An alternative view on the concept of utilization of elastic energy in human movement. *Hum. Mov. Sci.* 3: 301–6.

van Ingen Schenau, G.J. 1998. Positive work and its efficiency are at their dead-end: Comments on a recent discussion. *J. Biomech.* 31 (2): 195–97.

van Ingen Schenau, G.J., M.F. Bobbert, and A. de Haan. 1997. Does elastic energy enhance work and efficiency in the stretch-shortening cycle? *J. Appl. Biomech.* 13 (4): 389–415.

van Ingen Schenau, G.J., and P.R. Cavanagh. 1990. Power equations in endurance sports. *J. Biomech.* 23 (9): 865–81.

van Ingen Schenau, G.J., R. Jacobs, and J.J. de Koning. 1991. Can cycle power predict sprint running performance? *Eur. J. Appl. Physiol.* 63 (3–4): 255–60.

van Ingen Schenau, G.J., W.W. van Woensel, P.J. Boots, R.W. Snackers, and G. de Groot. 1990. Determination and interpretation of mechanical power in human movement: Application to ergometer cycling. *Eur. J. Appl. Physiol.* 61 (1–2): 11–19.

Vanlandewijck, Y.C., A.J. Spaepen, and R.J. Lysens. 1994. Wheelchair propulsion efficiency: Movement pattern adaptations to speed changes. *Med. Sci. Sports Exerc.* 26 (11): 1373–81.

Vaughan, C.L. 1984. Biomechanics of running gait. *Crit. Rev. Biomed. Eng.* 12 (1): 1–48.

Veeger, H.E., L.H. van der Woude, and R.H. Rozendal. 1992. Effect of handrim velocity on mechanical efficiency in wheelchair propulsion. *Med. Sci. Sports Exerc.* 24 (1): 100–107.

Viitasalo, J.T., L. Osterback, M. Alen, P. Rahkila, and E. Havas. 1987. Mechanical jumping power in young athletes. *Acta Physiol. Scand.* 131 (1): 139–45.

Voronov, A.V., E.K. Lavrovsky, and V.M. Zatsiorsky. 1995. Modelling of rational variants of the speed-skating technique. *J. Sports Sci.* 13 (2): 153–70.

Walsh, E.G., and G.W. Wright. 1987. Inertia, resonant frequency, stiffness and kinetic energy of the human forearm. *Q. J. Exp. Physiol.* 72: 161–70.

Wang, W., R.H. Crompton, and M.M. Gunther. 1996. Comparison of the powers at the lower limb joints during walking at different velocities and their significance for a possible optimal walking velocity. In *Engineering in sport,* ed. S. Haake, 71–76. Rotterdam: Balkema.

Ward-Smith, A.J. 1985. A mathematical theory of running, based on the first law of thermodynamics, and its application to the performance of world-class athletes. *J. Biomech.* 18 (5): 337–49.

Ward-Smith, A.J. 1997. A mathematical analysis of the bioenergetics of hurdling. *J. Sports Sci.* 15 (5): 517–26.

Ward-Smith, A.J. 1999a. The bioenergetics of optimal performances in middle-distance and long-distance track running. *J. Biomech.* 32 (5): 461–65.

Ward-Smith, A.J. 1999b. New insights into the effect of wind assistance on sprinting performance. *J. Sports Sci.* 17 (4): 325–34.

Webb, P., W.H. Saris, P.F. Schoffelen, G.J. van Ingen Schenau, and F. Ten Hoor. 1988. The work of walking: A calorimetric study. *Med. Sci. Sports Exerc.* 20 (4): 331–37.

Weber, W., and E. Weber. 1992. *Mechanics of the human walking apparatus.* Berlin: Springer-Verlag. Originally published in German, *Mechanik der menschlichen gehwerkzeuge,* 1836, Berlin.

Wells, R.P. 1988. Mechanical energy costs of human movement: An approach to evaluating the transfer possibilities of two-joint muscles. *J. Biomech.* 21 (11): 955–64.

White, A.P. 1994. Factors affecting speed in human-powered vehicles. *J. Sports Sci.* 12 (5): 419–21.

Whiting, W.C., R.J. Gregor, R.R. Roy, and V.R. Edgerton. 1984. A technique for estimating mechanical work of individual muscles in the cat during treadmill locomotion. *J. Biomech.* 17 (9): 685–94.

Whitt, F.R., and D.G. Wilson. 1974. *Bicycling science.* Cambridge, MA: MIT Press.

Whittlesey, S.N., R.E. van Emmerik, and J. Hamill. 2000. The swing phase of human walking is not a passive movement. *Motor Control* 4 (3): 273–92.

Widrick, J.J., P.S. Freedson, and J. Hamill. 1992. Effect of internal work on the calculation of optimal pedaling rates. *Med. Sci. Sports Exerc.* 24 (3): 376–82.

Wilkie, D.R. 1960. Man as a source of mechanical power. *Ergonomics* 3 (1): 1–8.

Willems, P.A., G.A. Cavagna, and N.C. Heglund. 1995. External, internal and total work in human locomotion. *J. Exp. Biol.* 198 (Pt. 2): 379–93.

Williams, K.R. 1985. The relationship between mechanical and physiological energy estimates. *Med. Sci. Sports Exerc.* 17 (3): 317–25.

Williams, K.R. 2000. The dynamics of running. In *Biomechanics in sport,* ed. V.M. Zatsiorsky, 161–83. Oxford, UK: Blackwell Science.

Williams, K.R., and P.R. Cavanagh. 1983. A model for the calculation of mechanical power during distance running. *J. Biomech.* 16 (2): 115–28.

Williams, K.R., and P.R. Cavanagh. 1987. Relationship between distance running mechanics, running economy, and performance. *J. Appl. Physiol.* 63 (3): 1236–45.

Winter, D.A. 1978. Calculation and interpretation of mechanical energy of movement. *Exerc. Sport Sci. Rev.* 6: 183–201.

Winter, D.A. 1979. A new definition of mechanical work done in human movement. *J. Appl. Physiol.* 46 (1): 79–83.

Winter, D.A. 1980. Energetics of the mechanics of human locomotion. *J. Am. Osteopath. Assoc.* 80 (4): 295–97.

Winter, D.A. 1983a. Energy generation and absorption at the ankle and knee during fast, natural, and slow cadences. *Clin. Orthop.* 175: 147–54.

Winter, D.A. 1983b. Moments of force and mechanical power in jogging. *J. Biomech.* 16 (1): 91–97.

Winter, D.A. 1987. Mechanical power in human movement: Generation, absorption and transfer. *Med. Sci. Sports Exerc.* 25: 34–45.

Winter, D.A., A.O. Quanbury, and G.D. Reimer. 1975. Instantaneous energy and power flow in normal human gait. In *Biomechanics V-A,* ed. P.V. Komi, 334–40. Baltimore: University Park Press.

Winter, D.A., A.O. Quanbury, and G.D. Reimer. 1976. Analysis of instantaneous energy of normal gait. *J. Biomech.* 9 (4): 253–57.

Winter, D.A., and D.G.E. Robertson. 1978. Joint torque and energy pattern in normal gait. *Biol. Cybern.* 29: 137–42.

Witte, H., S. Recknagel, J.G. Rao, M. Wüthrich, and C. Lesch. 1997. Is elastic energy storage of quantitative relevance for the functional morphology of the human locomotor apparatus? *Acta Anat.* 158: 106–11.

Woodson, C., W.D. Bandy, D. Curis, and D. Baldwin. 1995. Relationship of isokinetic peak torque with work and power for ankle plantar flexion and dorsiflexion. *J. Orthop. Sports Phys. Ther.* 22 (3): 113–15.

Wretenberg, P., and U.P. Arborelius. 1994. Power and work produced in different leg muscle groups when rising from a chair. *Eur. J. Appl. Physiol.* 68 (5): 413–17.

Yack, H.J. 1984. Techniques for clinical assessment of human movement. *Phys. Ther.* 64 (12): 1821–30.

Yoshihuku, Y., Y. Ikegami, and S. Sakurai. 1987. Energy flow from the trunk to the upper limb in Tsuki motion of top-class players of the martial arts, Shorinji Kempo. In *Biomechanics X-B,* ed. B. Johnson, 733–37. Champaign, IL: Human Kinetics.

Young, R.P., and R.G. Marteniuk. 1995. Changes in inter-joint relationships of muscle moments and powers accompanying the acquisition of a multi-articular kicking task. *J. Biomech.* 28 (8): 701–14.

Yu, B. 1999. Horizontal-to-vertical velocity conversion in the triple jump. *J. Sports Sci.* 17 (3): 221–29.

Zamparo, P., G. Antonutto, C. Capelli, and P.E. di Prampero. 2000. Effects of different after-loads and knee angles on maximal explosive power of the lower limbs in humans. *Eur. J. Appl. Physiol.* 82 (5–6): 381–90.

Zamparo, P., G. Antonutto, C. Capelli, M. Girardis, L. Sepulcri, and P.E. di Prampero. 1997. Effects of elastic recoil on maximal explosive power of the lower limbs. *Eur. J. Appl. Physiol.* 75 (4): 289–97.

Zamparo, P., C. Capelli, B. Termin, D.R. Pendergast, and P.E. di Prampero. 1996. Effect of the underwater torque on the energy cost, drag and efficiency of front crawl swimming. *Eur. J. Appl. Physiol.* 73 (3–4): 195–201.

Zamparo, P., R. Perini, C. Orizio, M. Sacher, and G. Ferretti. 1992. The energy cost of walking or running on sand. *Eur. J. Appl. Physiol.* 65 (2): 183–87.

Zarrugh, M.Y. 1981a. Kinematic prediction of intersegment loads and power at the joints of the leg in walking. *J. Biomech.* 14: 713–25.

Zarrugh, M.Y. 1981b. Power requirements and mechanical efficiency of treadmill walking. *J. Biomech.* 14: 157–165.

Zarrugh, M.Y., F.N. Todd, and H.J. Ralston. 1974. Optimization of energy expenditure during level walking. *Eur. J. Appl. Physiol.* 33 (4): 293–306.

Zatsiorsky, V.M. 1998. Can total work be computed as a sum of the "external" and "internal" work? *J. Biomech.* 31 (2): 191–92.

Zatsiorsky, V.M. 1986. Mechanical work and energy expenditure in human motion (in Russian). In *Contemporary problems of biomechanics: 3. Optimization of biomechanical movement,* ed. I.V. Knets, 14–32. Riga: Zinatne.

Zatsiorsky (Zatsiorskii), V.M., S.Y. Aleshinskii, N.G. Mikhailov, V.V. Tyupa, and N.A. Yakunin. 1980. Expenditure of mechanical energy during walking. *Hum. Physiol.* 6 (4): 233–38.

Zatsiorsky, V.M., S.Y. Aleshinsky, and N.A. Yakunin. 1982. *Biomechanical basis of endurance* (in Russian). Moscow: FiS. Also published in German translation as *Biomechanische Grundlagen der Ausdauer.* Berlin: Sportverlag, 1987.

Zatsiorsky, V.M., and R.J. Gregor. 2000. Mechanical power and work in human movement. In *Energetics of human activity,* ed. W.A. Sparrow, 195–227. Champaign, IL: Human Kinetics.

Zatsiorsky, V.M., and N.A. Yakunin. 1980. Mechanical work and energy during locomotion (in Russian). *Fiziol. Cheloveka (Hum. Physiol.)* 6 (4): 579–696.

Zatsiorsky, V.M., N.A. Yakunin, and I.I. Kotelevskaia. 1987. Energy expenditure during human walking and running (a meta-analysis) (in Russian). *Fiziol. Cheloveka (Hum. Physiol.)* 13 (1): 133–38.

Zatsiorsky, V.M., and B.I. Prilutsky. 1987. Soft and stiff landing. In *Biomechanics X-B,* ed. B. Johnson, 739–44. Champaign, IL: Human Kinetics.

Zatsiorsky, V.M., S. Werner, and M.A. Kaimin. 1994. Basic kinematics of walking: Step length and step frequency, a review. *J. Sport Med. Phys. Fitness* 34 (2): 109–34.

Zernicke, R.F., and K. Schneider. 1993. Biomechanics and developmental neuromuscular control. *Child Dev.* 64: 982–1004.

Zoladz, J.A., A.C. Rademaker, and A.J. Sargeant. 2000. Human muscle power generating capability during cycling at different pedalling rates. *Exp. Physiol.* 85 (1): 117–24.

APPENDICES

Appendices **1** and **2** contain data on body-segment parameters and regression equations for subject-specific estimation. Interested readers should also consult figures 4.18 and 4.21 and table 4.4 in chapter **4** for additional information. Appendix **3** gives equations used for the geometric modeling of human body segments.

The inertial properties of body segments are characterized by five main variables: segment mass, CoM location on the long axis of the segment, and three moments of inertia (or three radii of gyration) about three orthogonal axes. The moments of inertia and radii of gyration are given with respect to the cardinal anatomic axes.

1
APPENDIX

INERTIAL PROPERTIES OF CADAVERS

The studies in table A1.1 were performed on the cadavers of middle-aged and old men. Some of the men were chronically ill before death. The heights of the subjects were close to the average for the corresponding population group (in the 55th to 57th percentile). However, body weights were much below the average (in the 14th to 23rd percentile).

All the references mentioned in this appendix are included in the bibliography in chapter 4.

Table A1.1 Major Cadaveric Studies, Adult Males

Study	Subjects, n	Age, years	Height, cm	Weight, kg	Moments of inertia, plane
Dempster 1955[a]	8	68.5 (52–83)	169.4 (155.3–186.6)	59.8 (49–72)	Transverse
Clauser et al. 1969[a]	13	49.3 ± 13.7 (28–74)	172.7 ± 5.94 (162.5–184.9)	66.5 ± 8.70 (54.0–87.9)	None
Chandler et al. 1975	6	54.3 ± 7.4 (45–65)	172.1 ± 5.7 (164.5–181.7)	65.17 ± 13.2 (50.6–89.2)	X, Y, Z principal axes[b]

[a]The incomplete data are at the Canadian Society for Biomechanics Web site, **http://www.health.uottawa.ca/biomech/csb/ARCHIVES/**. As the endpoints of the segments Dempster (1955) used the joint centers of rotation, while Clauser et al. (1969) and Chandler et al. (1975) used bony landmarks.

[b]Inertia tensors (three principal moments of inertia, orientation of the principal axes) were determined for all main body segments. The technique involved supporting each cadaver segment rigidly in a lightweight frame and measuring the moment of inertia about six axes. After that, the principal moments of inertia were computed. Unfortunately, something went wrong with the calculation in this otherwise excellent study. The requirement that the sum of the two moments of inertia be larger than the third moment of inertia (equation 4.11) was not satisfied for some body segments. For instance, the following average data were presented for the torso (trunk plus neck): $I_{xx} = 16{,}193 \times 10^3$ gm · cm², $I_{yy} = 10{,}876 \times 10^3$ gm · cm², and $I_{zz} = 3{,}785 \times 10^3$ gm · cm². The required inequality $I_{yy} + I_{zz} > I_{xx}$ is evidently not satisfied. This lapse was observed in five subjects out of six. For the entire body, the data are incorrect for two subjects out of three. The reader is advised to use the data on the principal moments of inertia from this study with caution. According to the authors (p. 97): "The study design did not attempt to provide a statistically valid sampling for establishing population estimates of these parameters [moments of inertia], and no attempt should be made to use the results reported as such." Linear regression equations based on these data (Hinrichs 1985) yield negative values of the moments of inertia for some subjects (Veeger et al. 1991). Clearly, this makes no sense.

Table A1.2 Relative Mass of Body Segments in Cadavers and Living People As Percentage of Body Mass

Segment	Studies on cadavers						Studies on living people		
	Harless 1860 (2)	Braune and Fischer 1889 (3)	Fischer 1906 (1)	Dempster 1955 (8)[a]	Clauser et al. 1969 (13)	Bernstein et al. 1931 (76)	Plagenhoef 1971 (35)	Zatsiorsky and Seluyanov 1979, 1983 (100)
Head and neck	7.6	7.0	8.8	7.9	7.3	—	—	6.94 ± 0.70
Trunk	44.2	46.1	45.2	56.5[b] 46.9[c]	50.7	52.95	55.4	43.46 ± 1.57[d]
Upper arm	3.2	3.3	2.8	2.7 r 2.61	2.6	2.65	3.3	2.70 ± 0.24
Forearm	1.7	2.1	2.6[e]	1.6 r 1.51	1.6	1.82	1.9	1.63 ± 0.14
Hand	0.9	0.8		0.6	0.7	0.7	0.65	0.61 ± 0.08

(continued)

Table A1.2 (*continued*)

Segment	Studies on cadavers					Studies on living people		
	Harless 1860 (2)	Braune and Fischer 1889 (3)	Fischer 1906 (1)	Dempster 1955 (8)[a]	Clauser et al. 1969 (13)	Bernstein et al. 1934 (76)	Plagenhoef 1971 (35)	Zatsiorsky and Seluyanov 1979, 1983 (100)
Thigh	11.9	10.7	11.0	9.6 r 9.7 1	10.3	12.2	10.5	14.16 ± 0.10[d]
Calf	4.6	4.8	4.5	4.5	4.3	4.65	4.5	4.33 ± 0.31
Foot	2.0	1.8	2.1	1.4	1.5	1.46	1.45	1.37 ± 0.16

The number of subjects is given in parentheses.

[a]The sum is not 100% due to "unavoidable errors [that] resulted from dismemberment" (Dempster 1955:185). Because the cadavers were stored for 10 to 24 days before the segmentation, significant emaciation and weight loss may have occurred. The corrected values can be found in Miller and Nelson (1973), who redistributed the mass proportionally throughout the body parts to 100%.

[b]With head and neck, without limbs.

[c]Without shoulders.

[d]The difference from the other studies is mainly due to the different segmentation of the trunk and thigh.

[e]With hand.

Table A1.3   Relative Distance Between Center of Gravity and Joint Axis or Other Landmark

Segment and reference landmarks	n	Distance from center of gravity to reference landmark as % of segment length
Hand (rest position), wrist axis to knuckle III	16	50.6% to wrist axis 49.4% to knuckle III
Forearm, elbow axis to wrist axis	16	43.0% to elbow axis 57.0% to wrist axis
Upper arm, glenohumeral axis to elbow axis	16	43.6% to glenohumeral axis 56.4% to elbow axis
Forearm plus hand, elbow axis to ulnar styloid	16	67.7% to elbow axis 32.3% to ulnar styloid
Whole upper limb, glenohumeral axis to ulnar styloid	16	51.2% to glenohumeral axis 48.8% to ulnar styloid
Shoulder mass, sternal end of clavicle to glenohumeral axis	14	84.0% to sternal end of clavicle (oblique) 71.2% to glenohumeral axis (oblique)
Foot, heel to toe II	16	24.9% to ankle axis (oblique)[a] 43.8% to heel (oblique)[a] 59.45% to toe II (oblique)[a]
Lower leg, knee axis to ankle axis	16	43.4% to knee axis 56.7% to ankle axis
Thigh, hip axis to knee axis	16	43.7% to hip axis 56.7% to knee axis
Leg plus foot, knee axis to medial malleolus	16	43.4% to knee axis 56.6% to medial malleolus
Whole lower limb, hip axis to medial malleolus	16	43.4% to hip axis 56.6% to medial malleolus
Head and trunk minus limbs, vertex to transverse line through hip axes	7	60.4% to vertex 39.6% to hip axes
Head and trunk minus limbs and shoulders, vertex to line through hip axes	7	64.3% to vertex 35.7% to hip axes
Head and neck, vertex to C7 centrum	6	43.3% to vertex 56.7% to C7 centrum

(continued)

Table A1.3 *(continued)*

Segment and reference landmarks	*n*	Distance from center of gravity to reference landmark as % of segment length
Thorax, T1 centrum to T12 centrum	6	62.75% to T1 centrum 37.3% to T12 centrum
Abdominopelvic mass, L1 centrum to hip axes	5	59.9% to L1 centrum 40.1% to hip axes

After Dempster (1955); see also figure 4.18*a* in chapter **4**.

[a]Alternatively, a ratio of 42.9 to 57.1 along the heel-to-toe distance establishes a point above which the center of gravity lies; the center of gravity lies on a line between ankle axis and the ball of the foot.

NONLINEAR REGRESSION EQUATIONS FOR SEGMENTAL MOMENTS OF INERTIA

The equations are from Yeadon and Morlock (1989), based on data obtained by Chandler et al. (1975).

The equations for the nontorso segments are based on the length h and a mean perimeter p. For segments with three perimeters, the mean perimeter p is calculated as $p = (p_1 + 2p_2 + p_3)/4$; for the hand, $p = (p_1 + p_2)/2$; for the head, $p = p_1$. For the torso, the mean width is computed as $w = (w_1 + 2w_2 + w_3)/4$. The principal axes of inertia x, y, and z are aligned with the anteroposterior, mediolateral, and longitudinal anatomic axes, correspondingly. Because the moments of inertia about the anteroposterior and mediolateral axes are close for the nontorso segments, they are replaced by the transverse moment of inertia, $I_t = I_x = I_y$. For these segments, the following regression equations were developed:

$$I_z = k_1 p^4 h$$
$$I_t = 1/2 I_z + k_2 p^2 h^3 \tag{A.1}$$

where linear measurements are in meters and moments of inertia are in $kg \cdot m^2$. Anatomic dimensions used in the equations are described in table A1.4 and values of k_1 and k_2 are given in table A1.5.

Table A1.4   Anatomic Dimensions Used in the Equations

Segment	Variable	Definition
Head	h	Length: chin to vertex
	p_1	Perimeter: above ear
Torso	h	Length: trochanter to acromion
	p_1, w_1	Perimeter, width: nipple
	p_2, w_2	Perimeter, width: umbilicus
	p_3, w_3	Perimeter, width: hip
Upper arm	h	Length: shoulder center to elbow center
	p_1	Perimeter: below axilla
	p_2	Perimeter: maximum
	p_3	Perimeter: elbow
Forearm	h	Length: elbow center to wrist center
	p_1	Perimeter: elbow
	p_2	Perimeter: maximum
	p_3	Perimeter: wrist
Hand (in a flexed, relaxed position)	h	Length: wrist center to tip of finger III
	p_1	Perimeter: wrist
	p_2	Perimeter: metacarpophalangeal joints
Thigh	h	Length: hip center to knee center
	p_1	Perimeter: below gluteal furrow
	p_2	Perimeter: midthigh
	p_3	Perimeter: knee
Shank	h	Length: knee center to ankle center
	p_1	Perimeter: knee
	p_2	Perimeter: maximum
	p_3	Perimeter: minimum near ankle
Foot	h	Length: heel to toe
	p_1	Perimeter: minimum near ankle
	p_2	Perimeter: arch
	p_3	Perimeter: ball

Table A1.5   Values of Coefficients k_1 and k_2 for Body Segments Other Than the Torso

Segment	k_1	k_2
Head	0.701	2.33
Upper arm	0.979	6.11
Forearm	0.810	4.98
Hand	1.309	7.68
Thigh	1.593	8.12
Calf	0.853	5.73
Foot	1.001	3.72

From Yeadon and Morlock (1989).

For the trunk, the segmental moments of inertia are given by

$$I_x = dwh(c_2 w^2 + c_3 h^2)$$
$$I_y = dwh(c_1 d^2 + c_3 h^2) \qquad (A.2)$$
$$I_z = dwh(c_1 d^2 + c_2 w^2)$$

where d is the mean depth computed as $d = (p - 2w)/(\pi - 2)$. The values of the coefficients are $c_1 = 49.4$, $c_2 = 55.0$ and $c_3 = 68.8$.

2
APPENDIX

INERTIAL PROPERTIES MEASURED
IN LIVING SUBJECTS

At the time of this writing, only one comprehensive study has been published in which the inertial properties of all main body segments were investigated in living subjects (Zatsiorsky and Seluyanov 1979, 1983, 1985; Zatsiorsky, Aruin, and Seluyanov 1981; Zatsiorsky, Raitsin, Seluyanov, Aruin, and Prilutsky 1983; Tischler, Zatsiorsky and Seluyanov 1981; Zatsiorsky, Seluyanov, and Chugunova 1990a, 1990b).

AN UNPROFESSIONAL DIGRESSION
ABOUT THIS STUDY

Readers who are not interested in the history of this study can skip this section. This study was performed in 1972 through 1976. It was first published in Russian in a small number of copies in 1978, in an anthropology journal in 1979, and as a book chapter in 1981. It has never been published in full in English.

The main reason that we did not publish our research in English is that this was very difficult to do in the Soviet Union. Every totalitarian society needs enemies and secrecy. The general philosophy of any totalitarian society is that enemies and spies are everywhere and they are trying to steal secrets. Because of this, a special government agency called the Government Agency for the Protection of Secrets in Media existed in the former Soviet Union. The main goal of this agency was censorship, predominantly political censorship. Without obtaining clearance from the agency, publication abroad was strictly prohibited. Such unauthorized publication was considered treason, and the authors were punished accordingly. To make the situation worse, individual researchers were not allowed to approach the agency. A request for clearance had to be sent by a ministry and signed by a vice minister. The vice minister did not sign a request until a special committee in the ministry decided that the paper did not contain secret material and could be published abroad. In turn, the ministerial committee would not accept the manuscript for consideration without a special letter signed by a president or vice president of a university, and the vice president of the university could not sign an application letter without clearance from a special committee in the university. Thus, to publish a paper abroad, an author first had to obtain signatures from seven university professors who served on the committee. None of these professors even looked at the paper; they simply signed the clearance (they were usually normal, reasonable people). The problem was that they were busy people with different schedules; hence, it could take several days to locate them all. Then, the author had to make an appointment with a vice president of the university, obtain his signature, and send the manuscript to the ministry. In the ministry, the committee that issued clearances held meetings once a month. The author thus waited. The clearance from the committee was directed to a vice minister, who either signed it or invited the author for a discussion. Only after that could the manuscript be directed to the Government Agency for the Protection of Secrets in Media.

When we prepared our paper on body-segment parameters and obtained all the signatures and clearances, I was invited to speak with a vice minister. This man was a typical party boss and understood nothing about science. But he firmly knew that if he did not sign a clearance, nothing would happen to him, so he was reluctant to sign. When we met, he asked, "Is it really a good study?" How can one answer this question about one's own study? Sure, it is a good study. After all, it is *my* study. Of course it is good. His reaction was immediate and typical: "If this is such a good study, we must keep it *secret.*" After that, I backpedaled: "No, no, you did not understand me. The study is not so good that we need to make it secret. However, it is not extremely bad either." I really did not know how to defend our product. Finally, he signed the letter, and we submitted the manuscript for publication. In several months, we received a response from the journal. The review was positive, but we were asked to make several corrections. We made the corrections in one day. But after that, I was told that according to the rules, I had to repeat all the steps again: obtain the signatures of the seven committee members, and so on and so forth. I decided that I did not want to go through all that trouble again, and the study was never published in full in English.

During the perestroika period, the restrictions were eased, and we published this material in several conference proceedings, 12 to 15 years after we finished the project. During that time, new technology became available. When the project was being performed, neither personal computers nor floppy disks existed. We recorded information either on punch tapes (gamma-scanner data) or on punch cards and magnetic tape (anthropometric data). We used a Wang 2200 computer with 32 kilobytes of RAM and a tape recorder instead of a hard drive. (When I tell this to my students, they laugh.) After a few years, this technology became obsolete, and we were not able to retrieve some information.

GAMMA-SCANNER METHOD

Basic Principle and Experimental Technique

When a gamma-radiation beam passes through a substance, it becomes weaker. If the intensity of the gamma-radiation beam is measured before and after passage through a body, the mass of the underlying tissues can be evaluated by the intensity of absorption (figure A2.1). During the experiment, the subjects were scanned. The data on *surface density* (mass per unit of surface, g/cm^2) and segment boundaries were put in a computer and analyzed. The radiation dose during the experiment did not exceed 10 millirads (50 times lower than the maximum permissible dose; 20 times smaller than the dose obtained by a patient during a single diagnostic X-ray of the thorax).

During scanning, the subject was in a supine position. The borderlines between segments were marked on the skin, and when the gamma beam was above a mark, the researcher sent a signal to a computer. Because the distance between some anatomic points is different in a supine position from the distance in an upright posture, some anthropometric measurements were made in

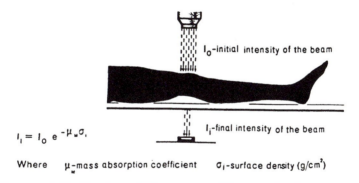

I_0-initial intensity of the beam

I_i-final intensity of the beam

$$I_i = I_0 \, e^{-\mu_\bullet \sigma_i}$$

Where μ-mass absorption coefficient σ_i-surface density (g/cm^2)

Figure A2.1   The gamma-scanner method. The mass absorption coefficient (μ) depends on the energy of gamma quanta and on the chemical elements that make up the body. With the energy of gamma quanta in the range 0.5 to 1.3 megaelectronvolts (MeV), the mass absorption coefficients for the main chemical elements of the human body are practically identical. For instance, at the energy of 1 MeV, the values of μ for oxygen, carbon, and nitrogen equal 0.0606. The value for calcium is 0.0608. The only exception is hydrogen, for which $\mu = 0.122$. However, it is known that hydrogen is distributed in the body evenly. This allows computation of the surface density of the body (in g/cm^2) for any location of the radiation beam.

Adapted, by permission, from V. Zatsiorsky and V. Seluyanov, 1983, The mass and inertia characteristics of the main segments of the human body. In *Biomechanics VIII-B*, edited by H. Matsui and K. Kobayashi (Champaign, IL: Human Kinetics), 1153.

the supine position. The feet and the arm were scanned separately from the rest of the body. During scanning, the arm was abducted 90° to clearly distinguish the arm mass from the trunk mass.

Sample

The subjects were physically fit men and women (table A2.1). The mean height, leg length, and chest circumference of the male subjects were similar to the mean values observed in college students (Zatsiorsky and Seluyanov 1979). However, the mean body weight was 3.5 kg larger. The estimated muscle mass was approximately 4 kg larger than the population average. All the female subjects were national athletes (nine swimmers and six fencers). Due to the large height and relatively low body weight of the female subjects, they are *not* representative of the general female population of the same age.

Table A2.1   Basic Anthropometric Data (mean ± standard deviation)

Parameter	Male ($n = 100$)	Female ($n = 15$)
Age, years	23.8 ± 6.2	19.0 ± 4.0
Mass, kg	73.0 ± 9.1	61.9 ± 7.3
Height, cm	174.1 ± 6.2	173.5 ± 3.3
Chest circumference, cm	91.9 ± 5.3	88.8 ± 3.8
Body fat, %	11.4 ± 3.4	17.6 ± 4.0

DATA PRESENTATION

The data are for a human in an anatomic posture: standing upright on a horizontal surface with arms hanging straight down at the sides of the body, palms turned forward, and head erect. However, the moments of inertia of the foot are given for an imaginary position with the foot rotated outward; that is, an axis that is parallel to the flexion axis at the ankle joint is considered an antero-posterior axis. The axis from the heel to the tip of the second toe is the longitudinal axis.

Table A2.2 Anatomic Landmarks Used in the Project

Acromion	The most lateral point on the lateral edge of the acromial process of the scapula
Acropodion	The tip of the longest toe (the first or second)
Bispinous breadth	The distance between the two iliospinales
Cervicale	The superior tip of the spine of the seventh cervical vertebra
Dactylion	The tip of the middle finger
Gluteal furrow landmark	The most inferior point of the gluteal furrow formed by the protrusion of the buttock beyond the back of the leg
Gonion	The most lateral point on the inferior posterior tip of the gonial angle formed by the intersection of the vertical and horizontal portions of the jawbone
Iliospinale (or anterior iliospinale)	The inferior point on the anterior superior iliac spine
Inguinal ligament	The ligament that extends from the anterior superior iliac spine to the pubic tubercle and forms the groin crease
Malleoli	Lateral and medial bony protrusions of the ankle
Metacarpale	The juncture of a metacarpal with the first phalanx of a finger
Third metacarpale	The distal palpable point on the metacarpal of the third digit on the dorsal surface of the hand
Midgonion, midhip, and midshoulder	The points midway between the gonions, hip joint centers, and shoulder joint centers, respectively
Omphalion	The central point of the navel (umbilicus)
Pternion	The rearmost point on the heel
Radiale	The lateral tip on the proximal head of the radius
Sphyrion	The distal tip of the tibia
Sphyrion fibulare	The distal tip of the fibula
Stylion	The distal tip of the styloid process of the radius
Suprasternale	The most caudal point on the jugular notch of the breastbone (sternum)
Tibiale mediale (tibiale)	The most proximal point on the medial superior border of the head of the tibia
Tibiale laterale	The most proximal point on the lateral superior border of the head of the tibia

(continued)

Table A2.2 *(continued)*

Trochanterion	The superior point on the greater trochanter of the femur
Vertex	The uppermost point of the head, when the head is held in the Frankfort plane (looking directly forward with the gazing line parallel to the floor)
Xiphion (or substernale)	The lowermost end of the sternum

Table A2.3   Main Anthropometric Measurements

Segments	Length (distance between anthropometric landmarks)		Circumference,[a] level of measurement
	From	To	
Foot	Acropodion	Pternion	Metatarsus
Shank	Tibiale	Sphyrion	Maximum calf
Thigh	Anterior iliospinale	Tibiale	Below the gluteal fold
Hand	Stylion	Dactylion	Metacarpus
Forearm	Radiale	Stylion	Maximum circumference
Upper arm	Acromion (arm abducted 90°)	Radiale	Maximum circumference
Head and neck	Vertex	Cervicale	Maximum circumference
Thorax	Cervicale	Xiphion	Immediately below the breast nipples (in men) or immediately below the breasts (in women)
Abdomen	Xiphion	Omphalion	Omphalion
Pelvis	Omphalion	Anterior iliospinale	Trochanterion

Note: The length measurements from this table are used mainly for segmentation purposes. They characterize the length of the segments whose inertial properties are determined. Specific measurements necessary for the best predictive regression equations are described in table A2.8. Specific measurements used in the nonlinear regression equations are listed in table A2.10.

[a]Measured perpendicularly to the longitudinal axis of a segment.

Table A2.4   Segmentation Pattern

| Segment | Borders of the segments | | Comments |
	Proximal or cranial	Distal or caudal	
Head (head and neck)	Vertex	Cervicale	The segmentation is performed in the transverse plane.
Upper trunk (thorax)	Cervicale	Xiphion	
Middle trunk (abdomen)	Xiphion	Omphalion	
Lower trunk (pelvis)	Omphalion	The planes through the iliospinales at an angle of 37° to the midsagittal plane	The point of intersection of the dissecting planes defines the quasi length (l) of the pelvis, $l = b/2 \tan 37°$, where b is the bispinous breadth. The quasi length was introduced to avoid palpating the subjects' groin area.
Thigh	A plane through the iliospinale at an angle of 37° to the midsagittal plane	Tibiale	
Shank	Tibiale	Sphyrion	
Foot	Sphyrion	—	The foot length is measured from pternion to acropodion.
Upper arm	Acromion	Radiale	The dissection is performed through the acromion by the vertical (sagittal) planes when the arm is abducted 90°.
Forearm	Radiale	Stylion	
Hand	Stylion	—	

Table A2.5 Mass-Inertial Characteristics of an Average Female Athlete

Segment	m, kg	CoM, %	m, %	I_{ap}, kg · cm²	I_{ml}, kg · cm²	I_{lg}, kg · cm²	R_{ap}, %	R_{ml}, %	R_{lg}, %
Foot	0.88	59.86	1.29	41.1	35.6	8.9	29.9	27.9	13.9
Shank	3.00	40.30	4.81	409.9	399.7	48.6	29.7	29.3	10.2
Thigh	9.16	46.08	14.78	1,690.1	1,647.3	324.2	27.4	27.0	12.0
Hand	0.35	64.98	0.56	6.0	4.4	2.4	24.1	20.6	15.2
Forearm	0.86	57.42	1.38	40.9	39.7	5.3	27.9	27.5	10.1
Upper arm	1.58	55.99	2.55	92.3	80.7	26.2	32.4	30.3	17.3
Head and neck	4.20	48.41	6.68	183.6	216.5	172.7	27.1	29.5	26.1
Upper part of trunk (thorax)	9.58	50.50	15.45	1,082.8	490.3	1,005.0	46.6	31.4	44.9
Middle part of trunk (abdomen)	9.08	45.12	14.65	717.3	480.2	658.7	43.3	35.4	41.5
Lower part of trunk (pelvis)	7.73	34.78	12.47	477.3	411.0	503.0	30.6	28.4	31.4

Symbols: m (kg) is the mass of the segment; CoM (%) is the location of the CoM of the segment along its longitudinal axis as a percentage of the segment length; m (%) is the ratio of the mass of the segment to the total body mass; I_{ap}, I_{ml}, and I_{lg} are the moments of inertia about the anteroposterior, mediolateral, and longitudinal axes of the segments, correspondingly; R_{ap}, R_{ml}, and R_{lg} are the radii of gyration as percentages of the segment length about the anteroposterior, mediolateral, and longitudinal axes of the segment, respectively. (The data for males are in table 4.4, chapter **4**.)

Table A2.6   Coefficients of Multiple Regression Equations for Estimating the Inertial Properties of Male Body Segments From Known Body Mass and Height

The equations are in the form $Y = B_0 + B_1X_1 + B_2X_2$, where X_1 is body weight in kilograms, X_2 is body height in centimeters, and Y is a mass-inertia characteristic. For example, for a 70-kg subject who is 173 cm tall, the estimated mass of the shank equals $Y = -1.592 + 0.0362 \times 70 + 0.0121 \times 173 = 3.03$ kg. $n = 100$. R is a multiple coefficient of correlation; σ is the standard error. During the height measurements, the subjects stood erect, heels together; the head was in the Frankfort plane. Note that the coefficients of correlation for some body segment parameters, notably for the CoM location, are low. When the coefficients are below 0.5, instead of regression equations the average data presented in figure 4.18 and table A1.3 may be used.

	Mass of segments, kg				
Segment	B_0	B_1	B_2	R	σ
Foot	−0.829	0.0077	0.0073	0.702	0.101
Shank	−1.592	0.0362	0.0121	0.872	0.219
Thigh	−2.649	0.1463	0.0137	0.891	0.721
Hand	−0.1165	0.0036	0.00175	0.516	0.063
Forearm	0.3185	0.01445	−0.00114	0.786	0.101
Upper arm	0.250	0.03012	−0.0027	0.837	0.178
Head and neck	1.296	0.0171	0.0143	0.591	0.322
Upper part of trunk	8.2144	0.1862	−0.0584	0.798	1.142
Middle part of trunk	7.181	0.2234	−0.0663	0.828	1.238
Lower part of trunk	−7.498	0.0976	0.04896	0.743	1.020

(continued)

Table A2.6   *(continued)*

Position of the CoM along the longitudinal axis, cm

The CoM position is measured from the following anthropometric points: foot—the tip of the second toe; shank—tibiale; thigh—iliospinale; hand—third dactylion; forearm—stylion; upper arm—radiale; head—vertex; upper part of trunk—cervicale; middle part of trunk—xiphion; lower part of trunk—omphalion.

Segment	B_0	B_1	B_2	R	σ
Foot	3.767	0.065	0.033	0.530	1.1
Shank	–6.05	–0.039	0.142	0.510	1.25
Thigh	–2.42	0.038	0.135	0.600	1.31
Hand	4.11	0.026	0.033	0.383	1.12
Forearm	0.192	–0.028	0.093	0.371	1.14
Upper arm	1.67	0.03	0.054	0.368	1.4
Head and neck	8.357	–0.0025	0.023	0.288	0.69
Upper part of trunk	3.32	0.0076	0.047	0.258	1.19
Middle part of trunk	1.398	0.0058	0.045	0.437	1.18
Lower part of trunk	1.182	0.0018	0.0434	0.320	1.0

Moment of inertia about the anteroposterior axis, $kg \cdot cm^2$

Segment	B_0	B_1	B_2	R	σ
Foot	–100	0.480	0.626	0.75	6.8
Shank	–1,105	4.59	6.63	0.85	48.6
Thigh	–3,557	31.7	18.61	0.84	248
Hand	–19.5	0.17	0.116	0.50	3.7
Forearm	–64	0.95	0.34	0.71	10.2
Upper arm	–250.7	1.56	1.512	0.62	27.6
Head and neck	–78	1.171	1.519	0.40	42.5
Upper part of trunk	81.2	36.73	–5.97	0.73	297
Middle part of trunk	618.5	39.8	–12.87	0.81	237
Lower part of trunk	–1,568	12	7.741	0.69	156

Moment of inertia about the mediolateral axis, kg · cm²

Segment	B_0	B_1	B_2	R	σ
Foot	−97.09	0.414	0.614	0.77	5.77
Shank	−1,152	4.594	6.815	0.85	49
Thigh	−3,690	32.02	19.24	0.85	244
Hand	−13.68	0.088	0.092	0.43	2.7
Forearm	−67.9	0.855	0.376	0.71	9.6
Upper arm	−232	1.525	1.343	0.62	26.6
Head and neck	−112	1.43	1.73	0.49	40
Upper part of trunk	367	18.3	−5.73	0.66	171
Middle part of trunk	263	26.7	−8.0	0.78	175
Lower part of trunk	−934	11.8	3.44	0.73	117

Moment of inertia about the longitudinal axis, kg · cm²

Segment	B_0	B_1	B_2	R	σ
Foot	−15.48	0.144	0.088	0.55	2.7
Shank	−70.5	1.134	0.3	0.47	2.2
Thigh	−13.5	11.3	−2.28	0.89	49
Hand	−6.26	0.0762	0.0347	0.43	1.8
Forearm	5.66	0.306	−0.088	0.66	2.9
Upper arm	−16.9	0.662	0.0435	0.44	12.5
Head and neck	61.6	1.72	0.0814	0.42	35.6
Upper part of trunk	561	36.03	−9.98	0.81	212
Middle part of trunk	1,501	43.14	−19.8	0.87	188
Lower part of trunk	−775	14.7	1.685	0.78	116

Table A2.7   Coefficients of Multiple Regression Equations for Estimating the Inertial Properties of Female Body Segments From Known Body Mass and Height

The equations are in the form $Y = B_0 + B_1X_1 + B_2X_2$, where X_1 is body weight in kilograms, X_2 is body height in centimeters, and Y is a mass-inertia characteristic. $n = 15$. R is a multiple coefficient of correlation, σ is the standard error.

	Mass of segments, kg				
Segment	B_0	B_1	B_2	R	σ
Foot	−1.207	−0.0175	0.0057	0.71	0.11
Shank	−0.436	−0.011	0.0238	0.42	0.36
Thigh	5.185	0.183	−0.042	0.73	0.81
Hand	−0.116	0.0017	0.002	0.48	0.03
Forearm	0.295	0.009	0.0003	0.38	0.11
Upper arm	0.206	0.0053	0.0066	0.27	0.21
Head and neck	2.388	−0.001	0.015	0.24	0.49
Upper part of trunk	−16.593	0.14	0.0995	0.64	1.47
Middle part of trunk	−2.741	0.031	0.056	0.45	1.09
Lower part of trunk	−4.908	0.124	0.0272	0.61	0.9

Position of the CoM along the longitudinal axis, %

The CoM position is measured from the following anthropometric points: foot—the tip of the second toe; shank—tibiale; thigh—iliospinale; hand—third dactylion; forearm—stylion; upper arm—radiale; head—vertex; upper part of trunk—cervicale; middle part of trunk—xiphion; lower part of trunk—omphalion.

Segment	B_0	B_1	B_2	R	σ
Foot	30.25	−0.103	0.200	0.34	3.9
Shank	41.94	−0.102	0.025	0.42	1.0
Thigh	50.90	−0.090	0.0072	0.35	1.2
Hand	41.74	−0.120	0.172	0.53	1.9
Forearm	61.40	0.096	−0.062	0.36	1.4
Upper arm	44.96	0.034	0.051	0.36	1.2
Head and neck	21.50	0.181	−0.085	0.38	3.2
Upper part of trunk	34.50	0.012	0.084	0.24	2.7
Middle part of trunk	36.68	0.025	0.037	0.17	2.0
Lower part of trunk	26.10	−0.020	0.056	0.23	1.7

(continued)

Table A2.7 *(continued)*

Moment of inertia about the anteroposterior axis, kg · cm²				

Segment	B_0	B_1	B_2	R	σ
Foot	−92.24	0.486	0.558	0.60	7.6
Shank	−963.1	−3.57	9.04	0.71	63.8
Thigh	−4,033.4	44.99	17.08	0.83	205.9
Hand	−5.71	0.122	0.035	0.44	1.47
Forearm	−132.1	0.620	0.825	0.73	7.58
Upper arm	−151.4	0.107	1.554	0.25	45.3
Head and neck	217.8	−0.032	0.059	0.01	53.5
Upper part of trunk	−4,038.5	28.6	20.0	0.71	249.5
Middle part of trunk	−368.7	−6.22	8.86	0.24	245.8
Lower part of trunk	−987.6	14.90	3.76	0.58	129.5

Moment of inertia about the mediolateral axis, kg · cm²				

Segment	B_0	B_1	B_2	R	σ
Foot	−61.4	0.348	0.406	0.72	4.0
Shank	−943.3	−2.51	8.47	0.70	62.2
Thigh	−2,659.4	50.35	6.96	0.75	252.0
Hand	−5.79	0.087	0.034	0.47	1.1
Forearm	−138.5	0.533	0.887	0.72	7.9
Upper arm	−330.4	−0.461	2.67	0.61	25.6
Head and neck	66.4	−0.447	1.29	0.18	49.1
Upper part of trunk	−2,075.0	15.6	9.4	0.66	147.2
Middle part of trunk	−546.0	2.87	5.1	0.36	125.0
Lower part of trunk	−633.3	10.8	2.26	0.64	78.8

Moment of inertia about the longitudinal axis, $kg \cdot cm^2$

Segment	B_0	B_1	B_2	R	σ
Foot	23.9	0.337	−0.059	0.56	2.3
Shank	−53.2	0.284	0.489	0.33	12.7
Thigh	1,339.8	6.30	−8.28	0.53	94.4
Hand	−2.138	0.053	0.0073	0.39	0.7
Forearm	7.4	0.21	−0.08	0.61	1.3
Upper arm	−118.6	1.19	0.44	0.83	5.3
Head and neck	−35.48	2.43	0.237	0.41	29.9
Upper part of trunk	−2,823.2	25.8	12.8	0.77	163.2
Middle part of trunk	−672.9	1.47	7.53	0.28	211.6
Lower part of trunk	−715.9	23.5	−1.106	0.64	140.0

Table A2.8 Best Predictive Regression Equations for Estimating Inertial Properties of Body Segments in Males

Coefficients of linear multiple regression equations $Y = B_0 + B_1X_1 + \ldots + B_nX_n$

Inertial properties of the foot

Parameter	B_0	B_1	B_2	B_3	R	σ
Mass, kg	−0.6286	0.066	−0.0136	0.0048	0.723	0.099
CoM, cm	−1.267	0.519	0.176	0.061	0.711	0.92
I_s, kg · cm²	−91.17	5.25	−0.335	0.386	0.810	6.0
I_f, kg · cm²	−89.1	4.788	0.477	0.271	0.835	5.0
I_p, kg · cm²	−11.9	0.771	0.047	0.243	0.518	2.8

CoM position is determined from the end of the second toe. The longitudinal axis is the axis from the heel to the tip of the second toe.

X_1—length of the foot, cm. The length is measured as a projected distance between the rearmost point on the heel and the tip of the most prominent toe (in an overwhelming majority of the subjects, it was the second toe).

X_2—maximal width of the foot, cm. The width is measured as the distance between the heads of the first and the fifth metatarsals.

X_3—body fat, kg. Estimated from skinfold measurements.

Parameter	B_0	B_1	B_2	B_3	R	σ
Mass, kg	−6.017	0.0675	0.0145	0.205	0.963	0.121
CoM, cm	0.0937	0.396	0.064	−0.041	0.645	1.1
I_s, kg · cm²	−1,437	28.64	3.202	21.6	0.964	24.3
I_f, kg · cm²	−1,489	28.97	6.48	21.5	0.968	23.1
I_t, kg · cm²	−194.8	0.214	−3.64	8.9	0.583	20.5

CoM position is determined from the tibiale.

X_1—length of the shank, cm.

X_2—lower diameter of the shank (projected transverse distance between the tips of the malleoli), cm.

X_3—mean circumference, cm. $X_3 = \dfrac{C_1 + C_2 + C_3}{3}$, where C_1 is the proximal circumference of the shank, C_2 is the distal circumference of the shank, and C_3 is the maximal circumference of the shank.

Inertial properties of the thigh

Parameter	B_0	B_1	B_2	B_3	R	σ
Mass, kg	−17.819	0.153	0.23	0.367	0.933	0.572
CoM, cm	−3.655	0.478	−0.07	0.088	0.800	0.99
I_s, kg · cm²	−6,729	87.8	50.3	75.3	0.893	206
I_f, kg · cm²	−6,774	88.4	38.6	78.0	0.896	205
I_t, kg · cm²	−1,173	4.06	6.0	26.8	0.878	52

The CoM position is determined from the iliospinale.

X_1—projected length of the thigh from the iliospinale to tibiale in the supine position, cm.

X_2—lower diameter of the thigh (maximal horizontal distance between the tips of the inner and outer tuberosities of the femur, i.e., the femoral *epicondyles*), cm.

X_3—mean circumference, cm. $X_3 = \dfrac{C_1 + C_2 + C_3}{3}$, where C_1 is the distal circumference of the thigh measured 10 cm above the center of the patella, C_2 is the proximal circumference of the thigh measured approximately 2 cm below the gluteal furrow landmark (the gluteal fold), and C_3 is the median circumference of the thigh measured halfway between the C_1 and C_2 measurements. *(continued)*

	Inertial properties of the hand					
Parameter	B_0	B_1	B_2	B_3	R	σ
Mass, kg	−0.594	0.041	−0.035	0.029	0.744	0.054
CoM, cm	−3.005	0.596	0.264	0.091	0.687	0.887
I_s, kg · cm²	−41.05	2.29	−1.62	1.27	0.743	2.9
I_f, kg · cm²	−26.6	1.818	−1.083	0.527	0.782	1.85
I_p, kg · cm²	−14.9	0.596	−0.814	0.818	0.540	1.72

The CoM position is determined from the end of the third digit with the fingers extended.
X_1—length of the hand, cm.
X_2—width of the hand (at the level of metacarpus) with fingers adducted.
X_3—mean circumference, cm. $X_3 = \dfrac{C_1 + C_2}{2}$, where C_1 is the circumference of the hand and C_2 is the smallest circumference of the distal forearm.

	Inertial properties of the forearm					
Parameter	B_0	B_1	B_2	B_3	R	σ
Mass, kg	−2.04	0.05	−0.0049	0.087	0.874	0.08
CoM, cm	0.732	0.588	−0.0857	−0.0187	0.691	0.89
I_s, kg · cm²	−229	7.12	−0.049	5.066	0.910	6.0
I_f, kg · cm²	−220	7.06	−0.082	4.544	0.930	5.1
I_p, kg · cm²	−39.2	0.560	−0.972	1.996	0.676	2.7

The CoM position is determined from the stylion.
X_1—length of the forearm, cm.
X_2—width of the hand (same as for the hand), cm.
X_3—mean circumference of the forearm, cm. $X_3 = \dfrac{C_1 + C_2 + C_3}{3}$, where C_1 is the smallest circumference of the distal forearm, C_2 is the maximal circumference of the forearm, and C_3 is the median circumference measured halfway between the C_1 and C_2 measurements.

Inertial properties of the upper arm

Parameter	B_0	B_1	B_2	B_3	R	σ
Mass, kg	−2.58	0.0471	0.104	0.0651	0.889	0.144
CoM, cm	−2.004	0.566	0.056	−0.016	0.924	0.618
I_s, kg · cm^2	−359	10.2	6.4	8.5	0.913	14.4
I_f, kg · cm^2	−331	10.3	5.5	5.6	0.927	13.6
I_l, kg · cm^2	−106	0.4	3.8	4.2	0.648	11.2

The CoM position is determined from the radiale.

X_1—biomechanical length of the upper arm (the projected distance between the acromion and the radiale measured with the arm abducted 90°), cm. It can be estimated from the anatomic length of the forearm X (the distance from the acromion to the radiale measured in the anatomic posture) using the regression equation $Y = 4.64 + 0.586X$ ($r = 0.413$; $\sigma = 2.3$ cm).

X_2—circumference of the relaxed upper arm with the arm hanging straight down at the side of the body, measured at the level where the maximal circumference is observed during muscle contraction.

X_3—mean diameter, cm. $X_3 = \dfrac{D_1 + D_2}{2}$, where D_1 is the lower diameter of the upper arm and D_2 is the lower diameter of the forearm. D_1 is the distance between the outward edges of the internal and external condyles of the humerus (the epicondyle and epitrochlea). D_2 is the distance between the styloid processes (bony protuberances) of the radius and ulna at the distal end of the forearm.

(continued)

Table A2.8 *(continued)*

Inertial properties of the head and neck

Parameter	B_0	B_1	B_2	B_3	B_4	R	σ
Mass, kg	−7.385	0.146	0.071	0.0356	0.199	0.791	0.245
CoM, cm	0.201	0.503	0.0027	0.043	−0.158	0.752	0.53
I_s, kg · cm²	−987	23.74	3.97	3.46	18.58	0.754	31.1
I_f, kg · cm²	−983	19.9	8.43	3.22	10.2	0.748	30.9
I_t, kg · cm²	−721	7.36	6.14	2.28	18.25	0.603	31.6

The CoM position is determined from the vertex.

X_1—projected length from the vertex to the cervicale in the supine position, cm.

X_2—maximal horizontal circumference of the head, measured with the head in the Frankfort plane, cm.

X_3—mean circumference, cm. $X_3 = \dfrac{C_1 + C_2}{2}$, where C_1 is the circumference of the head (same as X_2) and C_2 is the circumference of the neck measured beneath the thyroid cartilage (Adam's apple).

X_4—head diameter measured in the mediolateral direction, cm.

Inertial properties of the thorax (the upper part of the trunk)

Parameter	B_0	B_1	B_2	B_3	B_4	R	σ
Mass, kg	−18.91	0.421	0.199	0.078	0.065	0.928	0.715
CoM, cm	−2.854	0.567	0.0067	0.032	0.0152	0.927	0.51
I_s, kg · cm²	−5,175	105.4	45.8	4.01	8.65	0.888	201
I_f, kg · cm²	−2,650	65.6	17.12	5.83	9.8	0.908	96
I_t, kg · cm²	−4,149	54.8	43.7	8.88	9.63	0.867	184

The CoM position is determined from the cervicale.

X_1—projected length from the cervicale to the xiphoid process, measured in the supine position, cm.

X_2—circumference of the chest during a breathing pause, cm. The measuring tape goes immediately below the breast nipples (in men) and below the inferior angles of the scapulas. During the measurement the tape is slightly inclined with respect to the transverse plane of the body.

X_3—mesosternal diameter of the chest, cm. The diameter is measured in the anteroposterior direction at the middle of the sternum.

X_4—body fat, kg. Estimated from skinfold measurements.

(*continued*)

Table A2.8 *(continued)*

Inertial properties of the abdomen (the middle part of the trunk)

Parameter	B_0	B_1	B_2	B_3	B_4	R	σ
Mass, kg	−13.62	0.444	0.195	−0.017	0.0887	0.931	0.694
CoM, cm	−0.742	0.485	0.0007	−0.002	0.001	0.931	0.44
I_s, kg · cm²	−3,271	76.7	30.3	10.2	18.3	0.914	141
I_f, kg · cm²	−2,354	65.3	21.5	−2.3	10.57	0.945	82
I_l, kg · cm²	−2,657	43.0	33.3	1.6	20.6	0.887	145

The CoM position is determined from the xiphoid process.

X_1—length of the middle part of the trunk, between the xiphoid and the omphalion, in the supine position, cm.

X_2—circumference of the abdomen at the level of the navel, cm.

X_3—diameter coxae, the distance between the anterior borders of the right and left ossa coxae, cm.

X_4—body fat, kg. Estimated from skinfold measurements.

Inertial properties of the pelvis (the lower part of the trunk)

Parameter	B_0	B_1	B_2	B_3	R	σ
Mass, kg	−15.18	0.182	0.243	0.0216	0.791	0.938
CoM, cm	0.205	0.064	0.134	−0.08	0.388	0.970
I_s, kg · cm²	−2,354	22.6	34.37	4.41	0.745	144
I_f, kg · cm²	−1,816	18.0	23.6	7.29	0.766	111
I_l, kg · cm²	−2,009	20.1	24.9	11.2	0.824	105

The CoM position is measured from the omphalion (the center of the navel).

X_1—maximal circumference of the buttocks, cm. Excessive pressing that may deform the buttocks should be avoided during the measurement.

X_2—bispinous breadth, the distance between the two iliospinales, cm.

X_3—body fat, kg. Estimated from skinfold measurements.

Table A2.9   Nonlinear Regression Equations: Coefficients for Calculating Inertial Properties of Body Segments From Anthropometric Data

Segment	K^b	Males ($n = 100$)				Females ($n = 15$)			
		K_m $\times$ 10^{-5}	K_s $\times$ 10^{-2}	K_f $\times$ 10^{-2}	K_l $\times$ 10^{-2}	K_m $\times$ 10^{-5}	K_s $\times$ 10^{-2}	K_f $\times$ 10^{-2}	K_l $\times$ 10^{-2}
1 Foot	1.000	6.14	7.86	7.14	1.60	6.35	8.98	7.79	2.09
2 Shank	1.000	5.85	8.77	8.44	1.44	6.59	8.80	8.58	1.42
3 Thigh	1.083	6.64	7.18	7.18	1.33	6.48	7.49	7.30	1.24
4 Hand	1.000	5.54	6.65	4.86	2.29	4.56	5.85	4.32	1.58
5 Forearm	1.000	6.26	7.55	7.03	1.51	6.43	7.81	7.59	1.14
6 Upper arm	0.730	9.67	10.81	9.71	2.06	9.49	10.50	9.18	2.34
7 Head	0.760	6.37	8.68	9.38	1.25	5.39	7.36	8.68	1.33
8 Thorax	1.456	5.72	21.83	9.35	1.35	5.33	21.71	9.83	1.33
9 Abdomen	1.035	8.49	20.65	12.60	1.43	8.55	18.73	12.54	1.40
10 Pelvis	2.305	3.60	10.90	8.92	0.76	3.43	9.37	8.07	0.74
11 Trunk	1.465	5.64	6.23	5.27	1.18	—	—	—	—
12 Abdomen and pelvis together	1.470	5.59	7.80	6.73	1.08	—	—	—	—

Example: For a thigh (segment 3) of length (L) 50 cm and circumference (C) 60 cm, the inertia parameters can be computed as follows:

$m_3 = K_{m3} \times L \times C^2 \times 10^{-5} = 6.64 \times 50 \times 60^2 \times 10^{-5} = 11.95$ kg

$I_{s3} = K_{s3} \times m_3 \times L^2 \times 10^{-2} = 7.18 \times 11.95 \times 50^2 \times 10^{-2} = 2{,}145$ kg · cm^2

$I_{f3} = I_{s3} = 2{,}145$ kg · cm^2

$I_{l3} = K_{l3} \times m_3 \times C^2 \times 10^{-2} = 1.33 \times 11.95 \times 60^2 \times 10^{-2} = 572$ kg · cm^2

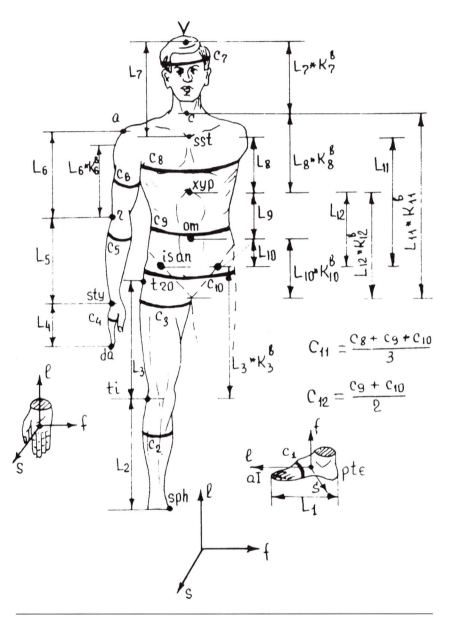

Figure A2.2 The anthropometric measurements used for estimating subject-specific values in the nonlinear regression equations. The biomechanical lengths of segments are also shown. See explanation in table A2.10.

Adapted, by permission, from V. Zatsiorsky, V. Seluyanov, and L. Chugunova, 1990, In vivo body segment inertial parameters determination using a gamma-scanner method. In *Biomechanics of Human Motion: Applications in Rehabilitation, Sports and Ergonomics*, edited by N. Berme and A. Cappozzo (Worthington, OH: Betec Corporation), 186–202.

Table A2.10 Anthropometric Measurements Used With the Nonlinear Regression Equations

Segment	Parameter	Symbol	Measurement or computation
Foot	Length, an	L_1	From acropodion to pternion
	Length, bi	—	Same
	Circumference	C_1	At the level of metatarsus
Shank	Length, an	L_2	From sphyrion to tibiale (The sphyrion is located proximally about 2 cm relative to the sphyrion fibulare. Tibiale laterale and tibiale mediale are at about the same height.)
	Length, bi	—	Same
	Circumference	C_2	Maximal circumference
Thigh	Length, an	L_3	From tibiale to trochanterion
	Length, bi	L_{bi}	$L_{bi} = L_3 \times K_3^b = L_3 \times 1.083$
	Circumference	C_3	Beneath the gluteal furrow landmark
Hand	Length, an	L_4	From third dactylion to stylion
	Length, bi	—	Same
Forearm	Length, an	L_5	From stylion to radiale
	Length, bi	—	Same
Upper arm	Length, an	L_6	From radiale to acromion (during upright standing)
	Length, bi	L_{bi}	$L_{bi} = L_6 \times K_6^b = L_6 \times 0.730$ (Alternatively, it can be measured as the distance from acromion to tibiale with the arm abducted 90°.)
	Circumference	C_6	Maximal circumference
Head and neck	Length, an	L_7	From vertex to suprasternale
	Length, bi	L_{bi}	$L_{bi} = L_7 \times K_7^b = L_7 \times 0.760$ (Alternatively, it can be measured as the distance from vertex to cervicale.)
	Circumference	C_7	Maximal circumference, measured above the brow ridges and parallel to the Frankfort plane

(continued)

Table A2.10 *(continued)*

Segment	Parameter	Symbol	Measurement or computation
Thorax	Length, an Length, bi	L_8 L_{bi}	From trochanterion to xiphion $L_{bi} = L_8 \times K_8{}^b = L_8 \times 1.456$ (Alternatively, it can be measured as the distance from xiphion to cervicale.)
	Circumference	C_8	Immediately below the breast nipples
Abdomen	Length, an Length, bi	L_9 L_{bi}	From omphalion to xiphion $L_{bi} = L_9 \times K_9{}^b = L_9 \times 1.035$ (This small correction accounts for the difference in length in upright standing and in supine position.)
	Circumference	C_9	At the level of omphalion
Pelvis	Length, an Length, bi	L_{10} L_{bi}	From omphalion to iliospinale $L_{bi} = L_{10} \times K_{10}{}^b = L_{10} \times 2.305$ (The quasi length of the pelvis, i.e., the projected distance from the omphalion to the intersection of the hip segmentation planes, is estimated.)
	Circumference	C_{10}	Maximal circumference of the buttocks
Trunk	Length, an Length, bi	L_{11} L_{bi}	From suprasternale to iliospinale $L_{bi} = L_{11} \times K_{11}{}^b = L_{11} \times 1.465$ (The length from cervicale to the intersection of the hip segmentation planes is estimated.)
	Circumference	C_{11}	$C_{11} = \dfrac{C_8 + C_9 + C_{10}}{3}$
Abdomen and pelvis together	Length, an Length, bi	L_{12} L_{bi}	From xiphion to iliospinale $L_{bi} = L_{12} \times K_{12}{}^b = L_{12} \times 1.470$ (The length from xiphion to the intersection of the hip segmentation planes is estimated.)
	Circumference	C_{12}	$C_{12} = \dfrac{C_9 + C_{10}}{2}$

The commonly accepted anthropometric measurements are used. All lengths are measured as projected distances. an—anatomic length; bi—biomechanical length.

Table A2.11 Adjusted Values of the Body Segment Parameters

Adjusted parameters for females (F; body mass = 61.9 kg, height = 1.735 m) and males (M; 73.0 kg, 1.741 m). Values reported in tables 4.4 and A2.5 are adjusted in order to reference them to the joint centers or other commonly used landmarks rather than the original anatomic landmarks (see also figure A2.3). Segment CoM positions are referenced either to proximal or cranial endpoints (origin). Both segment CoM positions and radii of gyration (r) are relative to the respective segment lengths. A set of easy-to-use endpoints is used in the first part of the table; for some segments, values using alternative endpoints are given in the second part. Relative segment masses were not adjusted (see tables 4.4 and A2.5).

| Segment | Endpoints | | Length, mm | | Longitudinal CoM position, mm | | Sagittal r, % | | Transverse r, % | | Longitudinal r, % | |
	Origin	Other	F	M	F	M	F	M	F	M	F	M
Head	VERT*	MIDG*	200.2	203.3	58.94	59.76	33.0	36.2	35.9	37.6	31.8	31.2
Trunk	SUPR*	MIDH*	529.3	531.9	41.51	44.86	35.7	37.2	33.9	34.7	17.1	19.1
Thorax	SUPR*	XIPH*	142.5	170.7	20.77	29.99	74.6	71.6	50.2	45.4	71.8	65.9
Abdomen	XIPH*	OMPH*	205.3	215.5	45.12	45.02	43.3	48.2	35.4	38.3	41.5	46.8
Pelvis	OMPH*	MIDH*	181.5	145.7	49.20	61.15	43.3	61.5	40.2	55.1	44.4	58.7
Upper arm	SJC†	EJC†	275.1	281.7	57.54	57.72	27.8	28.5	26.0	26.9	14.8	15.8
Forearm	EJC†	WJC†	264.3	268.9	45.59	45.74	26.1	27.6	25.7	26.5	9.4	12.1
Hand	WJC†	MET3†	78.0	86.2	74.74	79.00	53.1	62.8	45.4	51.3	33.5	40.1
Thigh	HJC†	KJC†	368.5	422.2	36.12	40.95	36.9	32.9	36.4	32.9	16.2	14.9
Shank	KJC†	LMAL*	423.3	434.0	44.16	44.59	27.1	25.5	26.7	24.9	9.3	10.3
Foot	HEEL*	TTIP*	228.3	258.1	40.14	44.15	29.9	25.7	27.9	24.5	13.9	12.4

(continued)

609

Table A2.11 (continued)

			Using alternative endpoints									
Head	VERT*	CERV*	243.7	242.9	48.41	50.02	27.1	30.3	29.5	31.5	26.1	26.1
Trunk	CERV*	MIDH*	614.8	603.3	49.64	51.38	30.7	32.8	29.2	30.6	14.7	16.9
Trunk	MIDS†	MIDH†	497.9	515.5	37.82	43.10	37.9	38.4	36.1	35.8	18.2	19.7
Thorax	CERV*	XIPH*	228.0	242.1	50.50	50.66	46.6	50.5	31.4	32.0	44.9	46.5
Forearm	EJC†	STYL*	262.4	266.9	45.92	46.08	26.3	27.8	25.9	26.7	9.5	12.2
Hand	WJC†	DAC3†	170.1	187.9	34.27	36.24	24.4	28.8	20.8	23.5	15.4	18.4
Hand	STYL*	DAC3†	172.0	189.9	35.02	36.91	24.1	28.5	20.6	23.3	15.2	18.2
Hand	STYL*	MET3*	79.9	88.2	75.34	79.48	51.9	61.4	44.3	50.2	32.7	39.2
Shank	KJC†	AJC†	438.6	440.3	43.52	43.95	26.7	25.1	26.3	24.6	9.2	10.2
Shank	KJC†	SPHY*	426.0	427.7	44.81	45.24	27.5	25.8	27.1	25.3	9.4	10.5

*Normal projection on the segment longitudinal axis.

†Assumed to lie on the segment longitudinal axis.

After P. de Leva. 1996. Adjustments to Zatsiorsky-Seluyanov's segment inertia parameters. *J. Biomech.* 29 (9): 1223–30.

Abbreviations in Table A2.11

AJC, EJC, HJC, KJC, SJC, WJC—the joint centers of ankle, elbow, hip, knee, shoulder, and wrist, respectively (after de Leva 1996b; see *Kinematics of Human Motion*, section 5.1); CERV—cervicale; DAC3—third dactylion; HEEL—pternion; LMAL—the most lateral point on the lateral malleolus; MET3—third metacarpale; MIDG, MIDH, MIDS—midgonion, midhip, and midshoulder, respectively; OMPH—omphalion; SPHY—sphyrion; STYL—stylion; SUPR—suprasternale; TTIP—the tip of the longest toe (acropodion); VERT—vertex; XIPH—xiphion

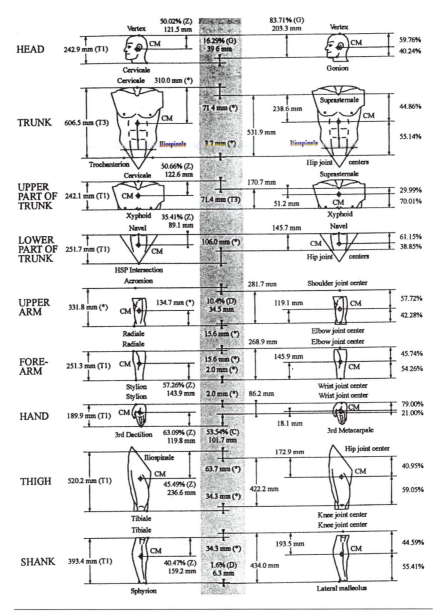

Figure A2.3 A graphic description of the main adjustments to the relative CoM positions for males. The original anatomical reference points (Zatsiorsky and Seluyanov 1983) are shown at the left; the suggested reference points are on the right. The adjusted distances are shown on the right of the corresponding segments. All percentage values are relative to the segment lengths indicated on their left.

Adapted from *Journal of Biomechanics*, 29 (9), de Leva, Adjustments to Zatsiorsky-Seluyanov's segment inertia parameters, 1223–1230, 1996, with permission from Elsevier Science.

REFERENCES

de Leva, P. 1996. Adjustments to Zatsiorsky-Seluyanov's segment inertia parameters. *J. Biomech.* 29 (9): 1223–30.

Tischler, W.A., V.M. Zatsiorsky, and V.N. Seluyanov. 1981. Study of the mass-inertial characteristics of human body by gamma-scanning during 6-month hypokinesia (in Russian). *Space Biology and Aerospace Medicine [Kosmicheskaja biologija i aviakosmicheskaja medicina]* 15 (1): 36–42.

Zatsiorsky, V.M., A.S. Aruin, and V.N. Seluyanov. 1981. *Biomechanics of musculoskeletal system* (in Russian). Moscow: FiS Publishers. (Also available in German: Saziorski, W.M., A.S. Aruin, and W.N. Selujanov. 1984. *Biomechanik des menschlichen Bewegungsapparates.* Berlin: Sportverlag.)

Zatsiorsky, V.M., L.M. Raitsin, V.N. Seluyanov, A.S. Aruin, and B.I. Prilutsky. 1983. Biomechanical characteristics of the human body. In W. Baumann, ed., *Biomechanics and performance in sport* (Cologne, Germany: Bundesinstitut für Sportwissenschaft), 71–83.

Zatsiorsky, V.M., and V.N. Seluyanov. 1985. Estimation of the mass and inertia characteristics of the human body by means of the best predictive regression equations. In Winter, D.A., R.W. Norman, R.P. Wells, K.C. Hayes, and A.E. Patla, eds, *Biomechanics IX-B* (Champaign, IL: Human Kinetics), 233–39.

Zatsiorsky, V.M., and V.N. Seluyanov. 1983. The mass and inertia characteristics of the main segments of the human body. In Matsui, H., and K. Kabayashi, eds, *Biomechanics VIII-B* (Champaign, IL: Human Kinetics), 1152–59.

Zatsiorsky, V.M., and V.N. Seluyanov. 1979. Mass-inertia characteristics of human body segments and their relationship with anthropometric parameters (in Russian). *Problems in Anthropology [Voprosy Anthropologii]* 62: 91–103.

Zatsiorsky, V.M., V.N. Seluyanov, and L.G. Chugunova. 1990a. In vivo body segment inertial parameters determination using a gamma-scanner method. In Berme, N., and A. Cappozzo, eds, *Biomechanics of human movement: Application in rehabilitation, sports, and ergonomics* (Worthington, OH: Bertec Corporation), 187–202.

Zatsiorsky, V.M., V.N. Seluyanov, and L.G. Chugunova. 1990b. Methods of determining mass-inertial characteristics of human body segments. In Chernyi, G.G., and S.A. Regirer, eds, *Contemporary problems of biomechanics* (Moscow/Boston: Mir Publishers/CRC Press), 272–91.

GEOMETRIC MODELING OF HUMAN BODY SEGMENTS

Table A3.1   Inertial Parameters of the Simple Homogeneous Solids Used for Modeling Human Body Segments

Body	Mass	Center of mass, h_c	Principal moments of inertia
Circular cylinder	$\rho \cdot \pi r^2 h$	$h/2$	$I_x = I_y = m\dfrac{1}{12}\left(3r^2 + h^2\right)$ $I_z = \dfrac{1}{2}mr^2$
Elliptical cylinder	$\rho \cdot \pi a b h$	$h/2$	$I_x = \dfrac{1}{4}ma^2 + \dfrac{1}{12}mh^2$ $I_y = \dfrac{1}{4}mb^2 + \dfrac{1}{12}mh^2$ $I_z = \dfrac{1}{4}m\left(a^2 + b^2\right)$
Semi-ellipsoid	$\rho \cdot \dfrac{2\pi}{3}abc$	$\dfrac{3}{8}c$	$I_x = \dfrac{1}{5}m\left(b^2 + c^2\right) - mh_c^2$ $I_y = \dfrac{1}{5}m\left(a^2 + c^2\right) - mh_c^2$ $I_z = \dfrac{1}{5}m\left(a^2 + b^2\right)$

(continued)

Table A3.1 *(continued)*

Body	Mass	Center of mass, h_c	Principal moments of inertia
Truncated elliptical cone	$\rho \cdot \pi h G_{20}(a,b)$	$\dfrac{G_{21(a,b)}}{G_{20(a,b)}} h$	$I_x = \rho \cdot \pi h \left(\dfrac{G_{40}(a,b,b,b)}{4} + h^2 G_{22}(a,b) \right) - mh_c^2$
			$I_y = \rho \cdot \pi h \left(\dfrac{G_{40}(a,a,a,b)}{4} + h^2 G_{22}(a,b) \right) - mh_c^2$
			$I_z = \rho \cdot \dfrac{\pi}{4} h \left[G_{40}(a,a,a,b) + G_{40}(a,b,b,b) \right]$
Truncated circular cone	$\rho \cdot \dfrac{1}{3}\pi h \cdot \left(R^2 + r^2 + Rr \right)$	$\dfrac{1}{3}\left(\dfrac{2r+R}{r+R} \right) h$	Same as for the truncated elliptical cone with $A = B$ and $a = b$.

Notes. ρ is the density of the segment. To find the mass of a solid, its volume is multiplied by the density ρ. Geometric dimensions are shown in figure 4.27. The CoM location is indicated from the "proximal" end, that is, the end with larger area. Formulas for computing the inertial parameters of a stadium solid can be found in the studies by Yeadon (1990) and Kwon (1999).

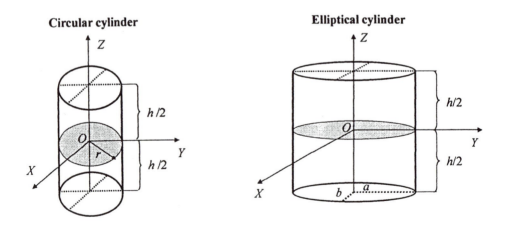

Figure A3.1 Geometric forms most commonly used for modeling human body segments.

Semiellipsoid

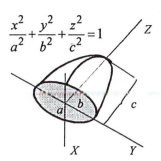

$$\frac{x^2}{a^2}+\frac{y^2}{b^2}+\frac{z^2}{c^2}=1$$

Truncated elliptical cone

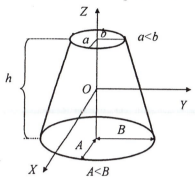

Stadium solid

Figure A3.1 *(continued)*

These are auxiliary parameters for computing the internal properties of a truncated elliptical cone (Kwon 1999).

$$G_{20}(a,b)=\frac{F_{22}(a,b)}{3}+\frac{F_{21}(a,b)}{2}+F_{20}(a,b)$$

$$G_{21}(a,b)=\frac{F_{22}(a,b)}{4}+\frac{F_{21}(a,b)}{3}+\frac{F_{20}(a,b)}{2}$$

$$G_{22}(a,b)=\frac{F_{22}(a,b)}{5}+\frac{F_{21}(a,b)}{4}+\frac{F_{20}(a,b)}{3}$$

$$G_{40}(a,b,c,d)=\frac{F_{44}(a,b,c,d)}{5}+\frac{F_{43}(a,b,c,d)}{4}+\frac{F_{42}(a,b,c,d)}{3}+\frac{F_{41}(a,b,c,d)}{2}+F_{40}(a,b,c,d)$$

where

$$F_{22}(a, b) = (a - A)(b - B)$$
$$F_{21}(a, b) = A(b - B) + B(a - A)$$
$$F_{20}(a, b) = AB$$
$$F_{44}(a, b, c, d) = (a - A)(b - B)(c - C)(d - D)$$
$$F_{43}(a, b, c, d) = A(b - B)(c - C)(d - D) + (a - A)B(c - C)(d - D) +$$
$$(a - A)(b - B)C(d - D) + (a - A)(b - B)(c - C)D$$
$$F_{42}(a, b, c, d) = AB(c - C)(d - D) + A(b - B)C(d - D) + A(b - B)(c - C)D$$
$$F_{41}(a, b, c, d) = (a - A)BCD + A(b - B)CD + AB(c - C) + ABC(d - D)$$
$$F_{40}(a, b, c, d) = ABCD$$

Reprinted, by permission, from Kwon, Y.-H. 1999. Kwon3D Motion Analysis Web site. **http://kwon3d.com/**

REFERENCES

Kwon, Y.-H. 1999. Kwon3D Motion Analysis Web site. **http://kwon3d.com/**

Yeadon, M.R. 1990. The simulation of aerial movement: II. A mathematical inertia model of the human body. *J. Biomech.* 23 (1): 67–74.

GLOSSARY

abduction—The movement of a body part away from the midsagittal plane.

acceleration—The rate of change of velocity, a vector quantity.

>*angular acceleration*—The rate of change of angular velocity.

>*centripetal acceleration*—Acceleration pointing toward the center of curvature; the rate of change of the direction of velocity.

>*Coriolis acceleration*—The acceleration that is acting when a body moves with regard to a rotating reference frame.

>*normal acceleration,* see *centripetal acceleration*

>*tangential acceleration*—Acceleration pointing along the path; the rate of change of the magnitude of the velocity.

Achilles tendon—The tendon connecting the gastrocnemius muscle to the calcaneus.

acromion—The most lateral point on the lateral edge of the acromial process of the scapula.

acropodion—The tip of the longest toe (the first or second).

active elements (of the musculoskeletal system)—Muscles and joints that can change their length (or angle) without external forces.

actuator—A device that generates forces or converts one form of energy into another.

>*force actuator*—A device that is able to generate a force.

>*linear actuator,* see *force actuator*

>*rotary actuator,* see *torque actuator*

>*torque actuator*—A real or imaginary generator that produces a moment of force with respect to a joint axis.

adduction—The movement of a body part toward the midsagittal plane.

aerodynamic drag—The resistive force impressed on a body along an airflow.

afferent—Conveying or transmitting toward the central nervous system.

affine scaling—A geometric transformation, such as shift, rotation, and stretching. Lines that are parallel before affine transformation remain parallel after the transformation.

agonist—A muscle that produces a moment of force in the direction of the desired joint motion.

allometry—Systematic change of body proportions with body size.

Amonton's law—The force of friction is directly proportional to the load and does not depend on the contact area or the skidding velocity.

anatomic position—An erect body stance with the arms at the sides and the palms of the hands facing forward.

angle

pointing angle—The angle between the pointing axis and a reference axis.

polar angle—The angle between the radial axis and a reference (polar) axis.

angular momentum (of a rigid body)—The product of a body's moment of inertia and angular velocity.

conservation of angular momentum—The angular momentum of a system is constant unless external torque is exerted on the system.

antagonist—A muscle that generates a joint moment in an opposite direction to the desired joint motion.

anterior iliospinale, see *iliospinale*

antisymmetric movement—Synchronous movement of two extremities in a vertical plane in opposite directions.

apparent mass—A measure of the chain inertia referred to a point of force application. Apparent mass can be seen as a coefficient m in the equation $F = ma$, where F is force and a is the acceleration of the point of force application. Apparent mass can be a scalar or a matrix.

arm movement

reaching movement—Movement of the hand in the radial direction.

whipping movement— Movement of the arm in the polar direction.

axes

pointing axis—The axis along a distal body segment (usually the forearm).

polar axis—A reference line in a polar coordinate system.

radial axis—An axis from the most proximal joint of a chain to the endpoint.

axes of the human body

anteroposterior axis—Intersection of the sagittal and transverse planes.

lateromedial (frontal) axis—Intersection of the frontal and transverse planes.

longitudinal axis—Intersection of the sagittal and frontal planes.

Bernstein's problem—The mastering of the redundant (abundant) degrees of freedom as a basis for the control of human and animal movements.

biarticular muscles, see *two-joint muscles*

bispinous breadth—The distance between the two iliospinales.

body

heterogeneous body—A nonhomogeneous body.

homogeneous body—A body in which the masses contained in any two equal volumes are equal.

rigid body—A body in which the distance between any two points within the body remains constant.

body link—A representation of a body segment in which a central straight line extends longitudinally through the body segment and terminates at both ends in axes about which the adjacent members rotate.

body segment—A part of the body considered separately from the adjacent parts.

body segment parameters—Inertial characteristics of a human body segment, such as mass, tensors of inertia, and location of the center of mass.

cardinal axes (of the human body), see *principal axes*

caudal—Referring to a position more toward the tail.

center of buoyancy—The centroid of the volume of liquid displaced by a body. For a body completely submerged in a liquid, the center of buoyancy coincides with the center of volume of the body.

center of gravity—The point of application of the resultant gravity force. For the human body, the center of gravity coincides with the center of mass.

center of mass (CoM)—A point with respect to which the mass moment of the first order in any plane equals zero.

center of mass model (of total body energy)—A model in which the kinetic energy of a body is represented as the sum of the kinetic energy of the center of mass and the kinetic energy of the body links in their motion relative to the center of mass. See also *König's theorem.*

center of pressure (CoP)—The point where the resultant of vertical force components intersects a support surface.

central nervous system (CNS)—The brain and the spinal cord.

centrifugal—Directed away from the center.

centripetal—Directed toward the center.

centripetal coupling coefficients—The coefficients in the dynamic equations of kinematic chains representing the effects of the angular velocity at one joint on the torques at other joints.

centroid—Center of volume.

centroidal axis—A line through the center of mass.

cervical—Pertaining to the neck.

cervicale—The superior tip of the spine of the seventh cervical vertebra.

chain inertia matrix—A square matrix of which the elements represent the inertial resistance felt at one joint to the angular acceleration at another joint.

characteristic equation of matrix $[A]$—The equation $[A]\mathbf{V} - \lambda\mathbf{V} = ([A] - \lambda[I])\mathbf{V} = \mathbf{0}$, where $\mathbf{V}$ is the eigenvector and λ is the eigenvalue of matrix $[A]$.

characteristic value—A root of the characteristic equation; see also *eigenvalue.*

characteristic vector, see *eigenvector*

closed-loop problem—Indeterminacy in solving an inverse problem of dynamics when a kinematic chain is closed via the environment. For instance, when two legs are on the ground, the ground reaction forces exerted by each leg cannot be determined. Only the total force exerted by the two legs can be computed exactly.

closure—A grasp that immobilizes the held object completely. Closure imposes six constraints on the object.

force closure—A grasp that can be broken under external forces without changing the finger position.

form closure—A grasp that cannot be broken without changing the finger configuration.

CNS, see *central nervous system*

coccyx—The caudal termination of the vertebral column.

coefficient

coefficient of dynamic friction—Ratio of shear force to normal force as registered during sliding.

coefficient of rolling resistance—The ratio of the horizontal force necessary to induce wheel rotation to the weight (vertical force) impinged on the wheel.

coefficient of static friction—The ratio of shear friction force to normal force that corresponds to the transition from rest to motion.

kinetic coefficient of friction—A coefficient of friction that depends on the sliding velocity, normal force, and the time of force application.

CoM model, see *center of mass model*

commutative—Independent of the order of the terms ($A + B = B + A$; $AB = BA$).

compliance—The amount of deformation per unit of force. Compliance is the inverse of stiffness.

compliance matrix—The inverse of the stiffness matrix.

joint compliance—The amount of joint angular deflection per unit of torque increment.

components (of a matrix of the apparent endpoint stiffness)

antisymmetric component, see *rotational component* (of a postural force field)

symmetric component, see *conservative component* (of a postural force field)

components (of a postural force field)

conservative component—The component of a field (restoring force) along the direction of deflection.

curl component, see *rotational component*

rotational component—The component of the field (restoring force) that is at a right angle to the displacement.

components (of a vector)—Elements into which a vector quantity can be resolved.

conservation of energy (in human movement)—Transformation of potential energy into kinetic energy and back.

conservative system—A system in which work is done only by conservative forces and therefore the sum of the kinetic and potential energies remains constant amid all changes.

contact—A collection of adjacent points where two surfaces touch each other over a contiguous area. An individual contact point is described by its position, orientation, and curvature.

coordinates—A set of numbers that defines the state of an object or a system.

Cartesian coordinates—Three fixed planes that intersect one another at right angles. The position of any point P in space is uniquely defined by the three perpendiculars from P on these planes.

generalized coordinates—Any set of quantities that specify the state of a system.

line coordinates, see *Plücker's coordinates*

Plücker's coordinates (in a plane)—An ordered triple of real numbers $\{L, M; R\}$, where L and M have dimensions of length, while R has dimension of area.

polar coordinates (in a plane)—A pair of numbers that locate a point: (1) the distance of the point from the origin and (2) the angle made by the radius to the point and the polar axis (the fixed line through the origin).

screw coordinates—The coordinates used for wrenches and twists. In screw coordinates, the unit wrench is defined by a unit force (a unit vector pointing in the direction of the screw axis), the position vector of any point on the screw axis, and the pitch.

Coriolis coupling coefficients—Coefficients in the dynamic equations of kinematic chains that characterize the inertial resistance to Coriolis force.

Coulomb's law, see *Amonton's law*

couple—Two equal, opposite, and parallel forces acting concurrently at a distance d apart. The couple makes the body rotate.

couple of the rolling friction—Friction experienced during body rotation about an axis lying in the tangent plane.

couple of the twisting friction—Friction experienced during body rotation about the axis perpendicular to the two contacting surfaces.

couple vector—A free vector that represents a couple. The couple vector is normal to the plane of the couple; the sense of the couple vector is determined by the right-hand rule.

coupling inertia—Off-diagonal elements of the inertia matrix of a kinematic chain, which represent the inertial effects of the accelerated motion at one joint on the torque at another joint.

coupling (joint) forces and torques—Joint forces and torques determined from Newton's third law. The adjective "coupling" in this context refers to the method of obtaining these quantities.

cranial—Pertaining to the cranium, the bones of the skull.

cross-coupling terms (of a joint stiffness matrix)—The terms that relate changes in the torque at one joint to the angular displacements at another joint.

cross product (of vectors **P** and **Q**)—The vector of magnitude $P \times Q \sin\alpha$ (where α is the angle formed by **P** and **Q**), with direction perpendicular to the plane containing **P** and **Q** according to the right-hand rule.

curl (of a vector field)—A measure of circulation (the tendency to rotate) at a given point.

dactylion—The tip of the middle finger.

d'Alembert's principle—The sum of external forces and the reversed effective force is zero.

decubitus ulcers—Pressure sores.

density—The amount of mass per unit of volume.

surface density—The amount of mass per unit of surface.

determinant (of a square matrix)—A certain scalar associated with a matrix.

direction cosines—Cosines of the angles that a vector makes with each of the coordinate axes of a reference system.

direct terms (of a joint stiffness matrix)—The terms of the matrix that relate changes in the torque at a given joint to the angular deflection at this joint.

distal—Away from the center of the human body or origin; the opposite of proximal.

distributive (law)—This law is expressed mathematically as $(A + B)C = AC + BC$.

dorsal—Pertaining to the posterior portion of a body part; the opposite of ventral.

dot product, see *scalar product*

drafting (in cycling)—Riding behind another athlete.

drag, see *aerodynamic drag*

dual—Vector whose first three elements and last three elements have different units of measurement. Wrenches and twists are duals.

action dual, see *wrench*

motion dual, see *twist*

duality of statics and kinematics—Similarity between these two branches of mechanics.

dynamic coefficients—Coefficients representing the effects of inertia and gravity in the dynamic equations.

dynamics—The study of the relationships between forces and the resulting motion, a branch of kinetics.

economy—The amount of energy expended per unit of distance.

effective force—The resultant of all the forces acting on a rigid body applied at the center of mass, which equals the vector *ma*, where *m* is the mass of the body and a is the acceleration of the center of mass.

> *reversed effective force*—The vector *–ma*.

effective mass—The ratio of the time integral of the force to the velocity of the impactor immediately before a collision.

effective moment—The resultant moment of all the forces and couples acting on a rigid body. Effective moment is a free moment.

efferent—Conveying away from the central nervous system.

> *efferent neuron*, see *motoneuron*

eigenvector and **eigenvalue** of a square matrix [*A*]—A nonzero vector **V** and a scalar λ that satisfy the equation $[A]\mathbf{V} = \lambda\mathbf{V}$.

elasticity—Ability of a body or material to resist deformation and to restore the original shape and size after being deformed.

electromyogram (EMG)—A recording of the electrical activity of a muscle.

ellipsoid of inertia—A geometric representation of the moments of inertia of a body around any axis through the center of mass.

EMG, see *electromyogram*

end effector—The last link of an open kinematic chain.

endpoint inertia—Inertial resistance to a force impressed at the endpoint of a kinematic chain.

energy (in physics)—The capacity for doing work.

> *energy fractions*, see *fractions of mechanical energy*

> *energy of the CoM*—Energy associated with the position and motion of the center of mass.

> *external energy* (in the CoM model)—Kinetic energy associated with the movement of the center of mass.

> *internal energy* (in the CoM model)—Kinetic energy associated with the movement of the body parts around the center of mass.

> *kinetic energy*—Energy due to motion.

>> *rotational kinetic energy* (of a rigid body)—Kinetic energy associated with rotation of a body around its center of mass; see also *König's theorem*.

>> *translational kinetic energy* (of a rigid body)—Kinetic energy associated with the movement of the center of mass of a body; see also *König's theorem*.

> *mechanical energy* (of a rigid body)—The sum of the kinetic and potential energy of the body.

potential energy—Energy stored in the system in latent form.

 elastic potential energy—Potential energy due to deformation.

 gravitational potential energy—Potential energy due to the location of the body in a gravity field.

total mechanical energy (of a rigid body), see *mechanical energy*

energy balance equation (for a single link)—Equation expressing the relationship between the power of the sources and the time rate of the fractions of mechanical energy.

energy compensation during time—Transformation of kinetic energy into potential energy and its subsequent use.

energy conservation coefficient—Proportion of the mechanical energy conserved in a given movement.

energy recuperation, see *energy compensation during time*

energy transfer—Redistribution of kinetic energy among body links. One link loses energy and the second gains energy in the same amount.

epicondyle—A projection of a bone above a condyle.

equations

 closed-form dynamic equations—Dynamic equations that contain all the variables in explicit input-output form.

 dynamic equations—Equations relating the forces and moments acting on a body with the changes in the body motion.

 equations of motion, see *dynamic equations*

 state-space equations—Dynamic equations in matrix-vector form.

equilibrium (of a material particle)—A particle is said to be at equilibrium when it either is at rest or moves along a straight line with constant velocity.

equilibrium-point hypothesis—The central nervous system controls movement by changing the equilibrium states of the effector-plus-load system.

equilibrium position (of a mechanical system)—The position of a system at rest under a given set of forces and couples.

 intended equilibrium (of a human body), see *naturally unstable equilibrium*

 naturally stable equilibrium (of a human body)—An equilibrium position to which the body is returned by external forces, for instance, by force of gravity.

 naturally unstable equilibrium (of a human body)— Equilibrium maintained by the restoring joint torques.

 neutral equilibrium—An equilibrium position from which a small displacement does not cause forces acting either away from or toward the original position.

 stable equilibrium—An equilibrium position from which a small displacement generates forces that tend to return the system back to the original position.

unintended equilibrium (of a human body), see *naturally stable equilibrium*

unstable equilibrium—An equilibrium position from which a small displacement generates forces that tend to move the system away from the original position.

equipollent systems (of vectors)—Two or more systems of vectors with equal resultants and resultant moments.

equivalent systems (of forces)—Two or more systems of forces that produce the same effect on a rigid body.

Euclidean norm, see *norm*

Euler dynamic equations—The dynamic equations expressed in a frame attached to the body and having its origin at the center of mass.

field—A function that assigns a scalar or a vector to each point in some region of space.

conservative field—Field for which line integrals depend only on the initial and final coordinates and not on the particular path between them. The field of gravity is conservative.

force field—A function that assigns a force vector to each point in a region.

scalar field—A function that assigns a number to each point in a region.

vector field—A function that assigns a vector to each point in a region.

first law of thermodynamics, see *law of conservation of energy*

force—A measure of the action of one body on another. Force is a vector quantity that is defined by its magnitude, direction, and point of application (or the line of force action).

braking forces—External forces acting opposite to the direction of progression.

centrifugal force—The force of reaction to the centripetal force. The centripetal and centrifugal force are applied to different bodies.

centripetal force—The force that induces centripetal acceleration.

concurrent forces—Forces meeting at the same point.

conservative forces—Forces that depend only upon the position of their point of application.

Coriolis force—The force associated with Coriolis acceleration.

coupling forces—The forces of action and reaction acting according to Newton's third law between adjacent segments in a kinematic chain.

driving forces, see *propelling forces*

effective force—An imaginary force that produces the same effect on the center of mass (but not on the entire body) as all the external forces combined.

external force—Force acting between the human body and environment.

friction force—The tangential component of a contact force between two bodies.

generalized contact force—Three orthogonal force components and three orthogonal moment components acting at a contact and considered together.

internal force—Force acting between body parts.

joint resultant force, see *joint force*

propelling forces—External forces that act in the direction of progression.

resultant force—Force that produces the same effect on a rigid body as two or more forces. Resultant force is obtained by summation of the forces according to the parallelogram rule.

slip force—The minimal grip force required to prevent slippage of a handheld object.

force couple, see *couple*

force ellipsoid—The graphical representation of a force that can be generated by an end effector in all directions.

force-moment transformation matrix—A 6×6 matrix that represents transformation of the forces and moments from one frame of reference to another.

force of buoyancy—The upward force acting on a body submerged in liquid.

force platform—A device for measuring the ground reaction forces.

fraction approach (of computing the mechanical energy expenditure)—A method of computing mechanical energy expenditure from the changes in the fractions of the mechanical energy of a body.

fraction effective force—The ratio of tangential force to total force exerted on the hand rim in manual wheelchair propulsion.

fractions of mechanical energy—The kinetic energy of a link associated with the horizontal, vertical, and rotational motion and potential energy of the link.

frame, see *reference system*

free-body diagram—A schematic representation of an isolated link showing all the forces and couples acting on the link.

free moment, see *moment of a couple*

friction, see *friction force*

dynamic friction—The friction between bodies when they slide with respect to each other.

friction angle—A geometric representation of a coefficient of friction in a plane.

friction cone—A geometric representation of a coefficient of friction in space.

kinetic friction—Friction that depends on movement velocity.

static friction—The friction that must be overcome to set a body in motion.

Froude number (Fr)—The speed adjusted to the body size; $Fr = v(gh)^{-0.5}$, where v is speed, h is a linear dimension of the body (e.g., leg length), and g is acceleration due to gravity.

fulcrum—The axis about which a lever turns.

gamma-scanner method—A method of determining body-segment parameters by measuring the intensity of a gamma-radiation beam before and after it passes through a human body.

gluteal furrow landmark—The most inferior point of the gluteal furrow formed by the protrusion of the buttock beyond the back of the leg.

gonion—The most lateral point on the inferior posterior tip of the gonial angle formed by the intersection of the vertical and horizontal portions of the jawbone.

gradient [of a function $z = f(x, y)$]—The largest directional derivative of the function at a point.

gravity line—The vertical line through the center of gravity.

grip—A system of unit wrenches and wrench intensities exerted on a grasped object.

grip matrix—A $6 \times n$ matrix whose columns are the unit wrenches of the grip.

Groucho running—Running on flexed legs.

ground reaction force (GRF)—The reaction force exerted on the performer by the supporting surface.

holonomic constraints—Constraints that restrict movement in certain directions; geometric constraints.

Hooke's law—The linear relationship between an applied force and the amount of deformation.

hydrodensitometric method, see *underwater weighing*

iliospinale—The inferior point on the anterior superior iliac spine.

impact—Collision of two bodies during a very short interval of time.

impactor—The body part that is in immediate contact with the environment during an impact (e.g., the foot during landing, the hand during boxing blows).

impulse

    linear impulse—The definite integral of force over time.

index of softness of landing—The ratio of the negative work of joint moments to the reduction of total mechanical energy during landing.

inertial force—Reversed effective force.

inertial moment—Reversed effective moment.

inertia matrix (of a rigid body)—A 3×3 matrix of which the diagonal terms are the moments of inertia and the off-diagonal elements are the products of inertia.

inguinal ligament—The ligament that extends from the anterior superior iliac spine to the pubic tubercle and forms the groin crease.

inner product, see *scalar product*

insertion (of a muscle)—The more movable, usually more distal, attachment of a muscle.

interactive inertia, see *coupling inertia*

invariant—Independent of the chosen reference frame.

ischemia—A lack of blood supply in an organ or tissue.

ischemic pallor—Blanching (loss of color) due to an absence of blood in a tissue.

isometry—Geometric similarity.

isotropic transformation—The situation in which a unit joint torque vector produces an endpoint force of equal magnitude in all directions.

iterated integral—An integral of integrals, for example, a double or triple integral.

Jacobian—A matrix of partial derivatives.

>*chain Jacobian*—Jacobian relating the differential endpoint displacement with the differential angular displacements at the joints.

>*muscle Jacobian*—Jacobian relating the differential changes in muscle length with the differential angular displacements at the joints.

>*orientation Jacobian*—The matrix whose elements are the partial derivatives of the end-effector orientation with respect to the joint angles.

>*positional Jacobian*—The matrix whose elements are the partial derivatives of the end-effector position with respect to the joint angles.

joint

>*ideal joint*—Joint without energy dissipation due to friction or deformation.

>*revolute joint*—Joint allowing only rotation of the adjacent segments.

joint force—The force at a joint that represents the effects of gravity and inertia. Because the force does not account for the muscle forces, it does not represent the real forces acting on the articular surfaces.

joint strength curve—A plot of maximal voluntary force versus joint angle.

joint torque, see *torque*

joule (J)—The SI unit of work.

kinematic chain—Two or several rigid bodies connected by joints.

kinetic diagram—Free-body diagram of a body with the reversed effective force and couple added.

kinetics—The study of the relationships between forces and their effects on bodies at rest (statics) and bodies in motion (dynamics).

König's theorem—The total kinetic energy of a system of material particles equals the sum of the kinetic energy of the center of mass (assuming the entire mass is concentrated at the center) and the kinetic energy of the particles in their motion relative to the center of mass.

Lagrangian—The difference between the kinetic energy and potential energy of the system.

Lagrangian formulation—The equations of motion that are based on the Lagrangian.

lateral—Pertaining to the side; farther from the midsagittal plane.

law of conservation of energy—Energy can be neither created nor destroyed, only transformed from one kind to another.

length (of a body member)

> *anatomic length* (of a body segment)—The distance between predetermined anatomic landmarks.

> *biomechanical length* (of a body segment or link)—The distance between joint axes.

lever—A rigid body revolving about a fulcrum and subjected to three forces: a resistance force, an effort force, and a force exerted on the fulcrum.

> *first-class lever*—Lever in which the resistance and effort forces are applied on opposite sides of the fulcrum.

> *second-class lever*—Lever in which the resistance force is applied between the fulcrum and the effort force.

> *third-class lever*—Lever in which the effort force is applied between the fulcrum and resistance. Such an arrangement is typical for muscles in the human body.

ligament—A fibrous band of connective tissue that binds bone to bone.

ligamentous action (of muscles), see *tendon action*

line integral—The integral of the tangential component of a vector field along a curve.

malleoli—Lateral and medial bony protrusions of the ankle.

mass—Property of matter to resist a change in velocity. Mass also manifests itself in a gravitational attraction between bodies.

mass moment of the first order (of a body)—The integral $\int mx\,dV$, where m is mass, x is the distance from the yz plane, and V is the volume of the body.

mass moment of the second order, see *moment of inertia*

matrix (in anatomy)—The intercellular substance of a tissue.

matrix (in mathematics)—A rectangular array of numbers.

> *diagonal matrix*—A square matrix in which all the off-diagonal elements are zero.

> *rotation matrix*—A matrix whose elements are direction cosines used to represent the rotation of one reference frame with respect to another.

symmetric matrix—A square matrix in which all the off-diagonal elements are symmetric with respect to the diagonal, $a_{ij} = a_{ji}$ for $i, j = 1, 2, \ldots, n$.

matrix of inertia, see *inertia matrix*

matrix of joint stiffness

cross-coupling terms (of the matrix of joint stiffness)—Off-diagonal terms of the matrix, which relate changes in the torque at one joint to the angular displacements at another joint.

direct terms (of the matrix of joint stiffness)—Diagonal elements of the matrix, which relate changes in the torque at a given joint to the angular deflection at this joint.

maximum advantage of using a component of exertion (MACE)—A measure of the effectiveness of static force efforts in a given direction.

mechanical analogy—A model in which a biological object is replaced by a simple mechanical object for the purpose of study.

mechanical energy expenditure (MEE)—The total amount of mechanical energy expended by the sources for a given motion.

mechanical model—A model in which a complex mechanical object is replaced by a simple one for the purpose of study.

medial—Toward or nearer the midsagittal plane.

metacarpale—The juncture of a metacarpal with the first phalanx of the finger.

third metacarpale—The distal palpable point on the metacarpal of the third digit on the dorsal surface of the hand.

metacarpus—The region of the hand between the wrist and the phalanges.

metatarsus—The region of the foot between the ankle and the phalanges.

midgonion (midhip, midshoulder)—The points midway between the gonions, hip joint centers, and shoulder joint centers, respectively.

midsagittal plane—The plane dividing the human body into right and left halves.

mixed triple product—A scalar product of a vector **P** and a vector $\mathbf{V} = \mathbf{Q} \times \mathbf{R}$.

mobility—The inverse of mass.

mobility ellipse (of a kinematic chain)—An ellipse representing the endpoint mobility of a chain.

mobility tensor (of a kinematic chain)—The inverse of the inertia matrix of a chain.

model

basic model (of a human body)—Modeling a human body as a kinematic chain consisting of rigid links with ideal revolute joints.

CoM model—The kinetic energy of the body is represented as a sum of the energy associated with motion of the total CoM of the body and the energy associated with motion of individual segments relative to the total CoM.

stick-figure model—Visual representation of the basic model of the human body.

moment arm (of a couple)—The shortest distance between the lines of action of the two parallel forces that form a couple.

moment arm (of a force)—A coefficient relating a force and the moment produced by this force about a certain center or axis.

moment arm in two dimensions—The perpendicular distance from a center of rotation to the line of force action.

moment arm in three dimensions—The product of the perpendicular distance from an axis to the line of force action and the sine of the angle between these two lines.

moment of a couple—A measure of the turning effect of a couple.

moment of force—A measure of the turning effect of a force.

joint moment, see *joint torque*

moment of force about an axis (in three dimensions)—A component of the moment along the axis.

moment of force **F** *about a point O*—A cross product of vectors **r** and **F**, where **r** is the position vector from O to the point of force application, $\mathbf{M}_O = \mathbf{r} \times \mathbf{F}$.

moment of inertia—A measure of the resistance of a body to a change in angular velocity.

area moment of inertia—Geometric property of a homogeneous body with density 1 computed as a moment of the second order.

axial moment of inertia—Moment of inertia with respect to an axis.

polar moment of inertia—Moment of inertia with respect to a point.

moment of the first order (of a material particle)—The product of mass concentrated at a point and the distance of the point from a given plane.

momentum (of a rigid body)

angular momentum (of a rigid body)—The product of a body's moment of inertia and angular velocity.

linear momentum (of a rigid body)—The product of a body's mass and the velocity of the center of mass.

motoneuron, or **motor neuron**—An efferent nerve cell that conducts impulses to muscles.

motor abundance, see *motor redundancy*

motor redundancy—The phenomenon that the same movement goal can be achieved in many different ways. Redundant motor systems are described mathematically by underdetermined sets of equations.

motor unit—A single motoneuron and the muscle fibers it innervates.

muscle action—Production of muscle force.

> *concentric muscle action*—The muscle produces force while shortening.

> *eccentric muscle action*—The muscle produces force while lengthening.

> *isometric muscle action*—The muscle produces force without change in length.

muscle spindles—Sensory organs within skeletal muscles that are sensitive to muscle stretch.

muscular effort—$G_i = a_i T_i^2 t$, where T_i is the joint torque generated by muscle i, a_i is a constant, and t is the period during which the muscle is active.

newton (N)—The SI unit of force.

Newton-Euler equations—Dynamic equations relating the sums of all the external forces and couples acting on a rigid body with the acceleration of the center of mass and the time rate of the angular momentum of the body (equations 5.1). The equations affirm that any system of external forces acting on a rigid body is equivalent to the effective force and couple.

norm (of a vector)—The magnitude of a vector.

normal—Perpendicular.

null space (of transformation $[M]$ from a vector space R^n into a vector space R^m)—The subspace of R^n containing all vectors **X** such that $[M]\mathbf{X} = \mathbf{0}$, where **X** is any vector in R^n.

obese—Excessively fat.

omphalion—The central point of the navel (umbilicus).

origin (of a muscle)—The more stationary, usually more proximal, attachment of a muscle.

orthogonal—Perpendicular.

overdetermined systems (or overconstrained systems)—The systems that are described by *overdetermined sets of equations* that have more equations than unknowns.

parallel axis theorem—The moment of inertia about any axis equals the moment with respect to the parallel centroidal axis plus the mass times the square of distance between the axes.

parallel manipulator—A manipulator with two or more end effectors, for example, the fingers grasping a rigid object. The links in a parallel manipulator form closed kinematic chains.

parallelogram rule—The rule of addition and subtraction of concurrent forces.

pascal (Pa)—A unit of pressure, 1 N/m^2.

passive elements (of the musculoskeletal system)—Ligaments, tendons, inactive muscles, and inactive joints. Passive elements cannot change their length (or angle) without external forces.

pendular motion—Motion of an ideal pendulum, in which the total mechanical energy is constant over an oscillation cycle.

pitch (in kinematics)—The magnitude of translation along the helical axis per unit of rotation around this axis.

pitch (in statics)—The ratio of the magnitude of the moment about a wrench axis (wrench moment) to the magnitude of the force along this axis.

planes, anatomic

frontal plane—A plane that divides the body into anterior and posterior sections.

sagittal plane—The plane dividing a human body into left and right sections.

transverse plane—For a subject in an upright posture, the horizontal plane.

point of wrench application (PWA)—A point where a wrench vector intersects the area of contact between a body and a support surface.

postmortem—After death.

postural force field—A vector field of the restoring forces at an endpoint.

potential, see *potential function*

potential function—A scalar function that assigns a certain amount of potential energy to any point in the field.

power—The rate of doing work.

joint power—Power of joint torques. The fraction of the muscle power generated in joints and expended to overcome external resistance and to change the mechanical energy of body segments.

muscle power—Power of muscle forces.

net power—The time rate of the mechanical energy of the system upon which the sources of energy act.

source power—Power produced by the sources.

power equations—Equations expressing the relationship between the power of the sources and the time rate of the mechanical energy of the system.

pressure—The amount of force per unit of area.

principal axes (in anatomy)—The axes at the intersection of the sagittal, transverse, and frontal planes of the body.

principal axes of inertia—Axes with respect to which the products of inertia are zero.

principal moments of inertia—The moments of inertia about the principal axes.

principal transformation (of the endpoint stiffness matrix)—Expressing the stiffness matrix in the principal directions.

principle of linear impulse and momentum—Linear impulse equals the increment in linear momentum.

problem of dynamics

> *direct problem of dynamics*—Determining the changes in motion caused by the given forces.

> *inverse problem of dynamics*—Determining the forces that produced a given motion.

product of inertia—A measure of the resistance of a body to centripetal acceleration.

proprioceptors—Receptors located within the skeletal muscles, tendons, and joints.

proximal—Closer to the center of the human body or to the origin of a link; the opposite of distal.

pseudowork—The scalar product of the effective force and the displacement of the center of mass. When this product is computed, an *imaginary* force is multiplied by the displacement of an *imaginary* point.

pternion—The rearmost point on the heel.

quadratic form, see *quadric*

quadric—A function of two, three, or more variables of the second degree; a quadratic form.

quasi-static solution (of an inverse problem of dynamics)—Estimation of the joint torques and forces during movement using the methods of statics. The inertial forces are neglected.

radiale—The lateral tip on the proximal head of the radius.

radius of gyration—Distance from the axis at which a particle of the same mass as the mass of a body has the same moment of inertia as the original body.

rambling (in the rambling-trembling hypothesis)—Migration of the reference point with respect to which the balance during standing is instantly maintained.

range space (of transformation [M] from a vector space R^n into a vector space R^m)—The image of R^n in R^m under the linear transformation [M].

reciprocal product (of screws)—The dot product of dual vectors in one of which the first and the last three elements are interchanged.

recursive—Involving recurrent calculations, one link at a time.

reference system (reference frame; space)—System of coordinates.

reflex—A rapid, involuntary action resulting from a response of the central nervous system to a stimulus.

> *heterogenic reflex*—Reflex in which stretching one muscle changes the activity of other muscles.

> *nonautogenic reflex*, see *heterogenic reflex*

> *stretch reflex*—Reflex induced by stretching spindle receptors within a muscle.

monosynaptic stretch reflex—Reflex whose pathway involves only one neuron and one synapse.

short-latency stretch reflex, see *monosynaptic stretch reflex*

tonic stretch reflex—Reflex induced by a slow stretching that causes the stretched muscle to maintain sustained contraction.

relative mass—The mass of a body segment expressed as a percentage of the total body mass.

repeated integral, see *iterated integral*

right-hand (thumb) rule—The convention used to determine the direction of a vector, in particular, the vector of moment of force. When the fingers of the right hand curl in the direction of the induced rotation, the vector points in the direction of the thumb.

rigidity, see *stiffness*

safety margin—The difference between the produced force and the minimal force necessary to prevent slippage of a handheld object.

scalar product (of two vectors **P** and **Q**)—A scalar that equals the magnitude of **P** times the component of **Q** in the direction of **P**; $\mathbf{P} \cdot \mathbf{Q} = PQ \cos\theta$.

scaling (in allometric analysis)

ontogenetic scaling—The data on children of different ages (and hence different sizes) are analyzed.

static scaling—The data on adults of different sizes are analyzed.

screw—A line with a pitch, such as a wrench or a twist (helical motion).

secondary moment—A resultant moment that is unnecessary for the task.

secondary moment hypothesis—The hypothesis that the central nervous system tries to minimize secondary moment in order to avoid needless muscle activity.

second moment, see *moment of inertia*

set of equations

determined set of equations—A set of equations that has as many equations as unknowns. The determinant of the coefficient matrix is not zero, and the system has a unique solution.

overdetermined set of equations—A set of equations that has more equations than unknowns.

underdetermined set of equations—A set of equations that has more unknowns than equations.

single-mass model—Analysis of human motion limited to the movement of the body's center of mass.

singularity—A specific position for which a given mathematical operation is not defined.

source approach (of determining mechanical energy expenditure)—A method of computing mechanical energy expenditure as the time integral of joint power.

sources of mechanical energy—In biomechanics, forces and moments of force (joint torques) acting on the system.

intercompensated sources—Sources that compensate for each other. The energy absorbed by one source compensates for the energy expended by the second. For instance, a two-joint muscle can absorb energy in one joint and generate energy in the second.

nonintercompensated sources—The energy expended by one source is not compensated by the energy absorbed by another source. For instance, joint torques in joints served by single-joint muscles are not intercompensated.

recuperative sources, see *sources compensated during time*

sources compensated during time—Sources that can absorb mechanical energy for subsequent use.

space, see *reference system*

space cross—Two nonintersecting forces.

sphyrion—The distal tip of the tibia.

sphyrion fibulare—The distal tip of the fibula.

square-cube law—Volume increases as the cube of linear dimensions, but strength increases only as their square.

standard position (of an ellipse)—The position in which the principal axes of an ellipse are along the coordinate axes.

static analysis

direct static analysis—Computing the end-effector force from known joint torques.

forward static analysis, see *direct static analysis*

inverse static analysis—Computing the joint torques from known external forces.

statics—A branch of kinetics dealing with bodies at rest.

stiffness—The amount of force per unit of deformation. Stiffness is the inverse of compliance.

apparent stiffness—The stiffness-like measurements obtained from active objects. The word "apparent" stresses that these measurements differ from the analogous parameter of passive objects.

endpoint stiffness—The relationship between the force differential increment and the linear differential displacement of the end effector. Endpoint stiffness is usually described by a matrix.

incremental stiffness (of an active object)—The amount of force or torque increment per unit of deflection or deformation.

intrinsic stiffness—Resistance to a perturbation that is due to purely peripheral factors.

joint stiffness—The amount of torque increment per unit of joint angular deflection.

preflex stiffness, see *intrinsic stiffness*

reflex stiffness—Reflex contribution to the apparent joint or muscle stiffness.

stiffness matrix—A matrix relating the differential force or torque increments with the differential deflection or deformation.

stiffness ellipse (at the endpoint)—The ellipse obtained by connecting the tips of the restoring forces generated in response to a unit deflection in all directions.

ellipse orientation—The angle between the major axis of an ellipse and the X axis of the fixed reference system.

ellipse shape—The ratio of the major to minor axes (the ratio of the maximal to minimal stiffness) of an ellipse.

ellipse size—The ellipse area.

strain—A relative elongation $\Delta l/l$, where l is the initial length of the object and Δl is its elongation.

stress—The amount of force per unit of area.

striking mass, see *effective mass*

stylion—The distal tip of the styloid process of the radius.

substernale, see *xiphion*

suprasternale—The most caudal point on the jugular notch of the breastbone (sternum).

systems of equations, see *set of equations*

tangential—Acting along a line or a surface.

tendon—A band of connective tissue that attaches muscle to bone.

tendon action (of muscles)—Muscles' behavior as nonextensible struts.

tendon action of two-joint muscles—Isometric muscle action that does not do any work. The amount of positive power generated by the muscle in one joint equals the amount of negative power absorbed by the same muscle in the second joint.

tensors—Mathematical objects that undergo certain types of transformations under changes of the coordinate system.

tensor of inertia, see *inertia matrix*

terms of the dynamic equations

acceleration-related terms—Inertia times the joint angular acceleration. In the literature, some authors call this term the *net torque*.

centrifugal terms—Inertia times angular velocity squared.

Coriolis terms—Inertia times the product of the joint angular velocities.

velocity-related terms—The centrifugal and Coriolis terms combined.

theorem of moments, see *Varignon's theorem*

theorem of the six constants of a body—The theorem that defines the moment of inertia of the body with respect to any axis through the center of mass.

thrust, see *propelling forces*

tibiale laterale—The most proximal point on the lateral superior border of the head of the tibia.

tibiale mediale (tibiale)—The most proximal point on the medial superior border of the head of the tibia.

torque, see *moment of a couple*

active joint torques—Joint torques generated by joint actuators.

equilibrating joint torques (of an endpoint force **F**)—The joint torques that counterbalance the force and couple exerted externally on an end effector.

equivalent joint torques—Joint torques calculated for joints served by two-joint muscles.

interactive torques (forces)—The joint torques and forces induced by motion in other joints.

joint torque—Two moments of force about the joint rotation axis acting on the adjacent segments. The moments are equal in magnitude and opposite in direction.

muscle-tendon torques—The active joint torques produced by muscle-tendon complexes.

net joint torque—The total torque produced at a joint; the sum of the active and passive torques. In the literature, some authors use this term with another meaning.

passive joint torques—Joint torques resisted by passive tissues, in particular, by the skeleton and ligaments.

total center of mass—The center of mass of the human body (as opposed to the centers of mass of individual body segments).

transformation analysis—Establishing a relationship between forces and couples acting on an end effector with the forces and torques acting at the joints. See also *static analysis.*

transmissibility

principle of transmissibility—The resultant effects of a force acting on a rigid body do not alter if the force is applied along the line of force action at any point.

trembling (in the rambling-trembling hypothesis)—Oscillation of the body around the rambling trajectory while standing.

tremor—Small oscillation of body parts around an equilibrium position.

tribology—The field of science that studies friction.

triple scalar product, see *mixed triple product*

trochanter—A prominent process on a bone, specifically on the femur.

trochanterion—The superior point on the greater trochanter of the femur.

twist—A combination of angular velocity and the linear velocity parallel to the axis of rotation

two-joint muscles—Muscles that cross and serve two joints.

underwater weighing—A method to measure the percentage of body fat.

unit vectors—Vectors of unit magnitude.

Varignon's theorem—The sum of moments equals the moment of the resultant.

vector—A quantity possessing magnitude and direction.

> *dual vectors*, see *dual*
>
> *free vectors*—Vectors that can be freely translated in space while maintaining a constant orientation, for example, couple vectors.

vector product, see *cross product*

velocity—The time rate of change of position. Velocity is a vector quantity.

vertex —The uppermost point of the head, when the head is held in the Frankfort plane (looking directly forward with the gazing line parallel to the floor).

virtual coefficient—The power transmitted by a unit wrench exerted on a body with unit twist (1 N of force acting on the body rotating at 1 rad/s).

virtual displacement—A hypothetical, small displacement of a body or a system from an equilibrium position.

virtual position (of a limb)—In the equilibrium-point hypothesis, the position toward which the muscular activity generated by the tonic stretch reflex drives the limb.

virtual trajectory (of a limb)—In the equilibrium-point hypothesis, the trajectory toward which the muscular activity generated by the tonic stretch reflex drives the limb.

virtual work—The work done by a force over a virtual displacement.

visceroceptors—Receptors located within the visceral organs and the circulatory system.

watt (W)—The SI unit of power.

whiplike movement—Acceleration of the endpoint of a moving kinematic chain induced by the deceleration of the proximal segments.

work (of a force)—Scalar product of the vectors of force ($\mathbf{F}$) and the infinitesimal displacement of the point of force application ($d\mathbf{r}$), $W = \mathbf{F}d\mathbf{r}$.

elementary work, see *work*

external work (in biomechanics)—Work done by a performer on external objects (environment).

internal work (in biomechanics)—Work done by a performer on the body links.

joint work—Work of the joint torques.

negative work (of a force)—Work done by a force directed opposite to the displacement.

negative work (of a muscle)—Work done by a muscle to resist its elongation.

positive work—work done by a force directed along the displacement.

quasi-mechanical work—The sum of the gains in kinetic energy and gravitational potential energy of a body.

work done on a body—The total work done by all the forces impressed on a body; see also *work–energy principle for a rigid body*.

work–energy principle for a rigid body—The total work done by all the forces acting on a rigid body is equal to the change in the body's kinetic energy. It also equals the work of the resultant force and couple.

working envelope—The set of boundary points that can be reached by the end effector.

working range (volume)—The set of all points that can be reached by the end effector.

work loop—Force–length relationship of a muscle during a working cycle.

wrench—A force and a couple of which the vectors are along the same line.

xiphion—The lowermost end of the sternum.

Young's modulus—The stress required to elongate the body to double its original length.

zero-work paradox—For movements beginning and ending at rest at the same vertical location, energy is expended, but the work is zero.

INDEX

Figures and tables are indicated by an italic *f* and *t*, respectively.

ABOUT THE AUTHOR

Vladimir M. Zatsiorsky, PhD, is a world-renowned expert in the biomechanics of human motion. He has been a professor in the department of kinesiology at Pennsylvania State University since 1991 and is director of the university's biomechanics laboratory.

Before coming to North America in 1990, Dr. Zatsiorsky served for 18 years as professor and chair of the department of biomechanics at the Central Institute of Physical Culture in Moscow. He has received several awards for his achievements, including the Geoffrey Dyson Award from the International Society of Biomechanics in Sports (the society's highest honor) and the USSR's National Gold Medal for the Best Scientific Research in Sport in 1976 and 1982.

Dr. Zatsiorsky has been awarded an honorary *honoris causa* doctoral degree for The Academy of Physical Education (Poland, 1999). He is an honorary member of the International Association of Sports Kinetics (since 1999) and an active fellow of the American Academy of Kinesiology and Physical Education.

Dr. Zatsiorsky has authored or coauthored more than 250 scientific papers. He has also authored or coauthored 14 books on various aspects of biomechanics that have been published in English, Russian, German, Italian, Spanish, Portuguese, Chinese, Japanese, Polish, Romanian, Czech, Serbo-Croatian, and Bulgarian. His latest books are *Science and Practice of Strength Training, Kinematics of Human Motion, Biomechanics in Sports: Performance Enhancement and Injury Prevention,* and *Classics in Movement Science* (coeditor).